# PURE MATHEMATICS

A UNIVERSITY AND
COLLEGE COURSE

BY

F. GERRISH, M.A.

VOLUME I
CALCULUS

CAMBRIDGE
AT THE UNIVERSITY PRESS
1970

Published by the Syndics of the Cambridge University Press
Bentley House, 200 Euston Road, London N.W. 1
American Branch: 32 East 57th Street, New York, N.Y. 10022

Standard Book Number: 521 05069 3


First printed 1960
Reprinted 1970

Printed in Great Britain
at the University Printing House, Cambridge
(Brooke Crutchley, University Printer)

# CONTENTS

# GENERAL PREFACE

This two-volume text-book on Pure Mathematics has been designed to cover completely the requirements of the revised regulations for the B.Sc. General Degree (Part I) of the University of London. It presents a serious treatment of the subject, written to fill a gap which has long been evident at this level. The author believes that there is no other book addressed primarily to the General Degree student which covers the ground with the same self-contained completeness and thoroughness, while also indicating the way to further progress. On the principle that 'the correct approach to any examination is from above', the book has been constructed so that those students who do not intend to take the subject Mathematics in Part II of their degree course will find included some useful matter a little beyond the prescribed syllabus (which throughout has been interpreted as an examination schedule rather than a teaching programme); while those who continue with Mathematics will have had sound preparation. As it is the author's experience that many students who begin a degree course have received hasty and inadequate training, a complete knowledge of previous work has *not* been assumed.

Although written for the purpose just mentioned, this book will meet the needs of those taking any course of first-year degree work in which Pure Mathematics is studied, whether at University or Technical College. For example, most of the Pure Mathematics required for a one-year ancillary subject to the London Special Degrees in Physics, Chemistry, etc. is included, and also that for the first of the two years' work ancillary to Special Statistics. The relevant matter for Part I (and some of Part II) of the B.Sc. Engineering Degree is covered. The book provides an introduction to the first year of an Honours Degree in Mathematics at most British universities, and would serve as a basis for the work of the mathematical specialist in the Grammar School. Much of the material is suitable for pupils preparing for scholarships in Natural Sciences.

By a natural division the subject-matter falls conveniently into two volumes which, despite occasional cross-references, can be used independently as separate text-books on Calculus (Vol. I) and on Algebra, Trigonometry and Coordinate Geometry (Vol. II). According to the plan of study chosen, the contents may be dealt with in turn, or else split up into two or even three courses of reading in Calculus, Algebra-Trigonometry and Geometry taken concurrently. Throughout it has been borne in mind that many students necessarily work without much direct supervision, and it is hoped that those of even moderate ability will be able to use this book alone.

A representative selection of worked examples, with explanatory remarks, has been included as an essential part of the text, together with many sets of 'exercises for the reader' spread throughout each chapter and *carefully graded* from easy applications of the bookwork to 'starred' problems (often with hints for solution) slightly above the ultimate standard required. In a normal use of the book there will not be time or need to work through every 'ordinary' problem in each set; but some teachers welcome a wide selection. To each chapter is appended a Miscellaneous Exercise, both backward- and forward-looking in scope, for revision purposes. Answers are provided at the end of each volume. It should be clear that, although practice in solving problems is an important part of the student's training, in no sense is this a cram-book giving drill in examination tricks. However, those who are pressed for time (as so many part-time and evening students in the Technical Colleges unfortunately are) may have to postpone the sections in small print and all 'starred' matter for a later reading.

Most of the problems of 'examination type' have been taken from Final Degree papers set by the University of London, and I am grateful to the Senate for permission to use these questions. Others have been collected over a number of years from a variety of unrecorded (and hence unacknowledged) sources, while a few are home-made.

It is too optimistic to expect that a book of this size will be completely free from typographical errors, or the Answers from mathematical ones, despite numerous proof readings. I shall be grateful if readers will bring to my notice any such corrections or other suggestions for possible improvements.

Finally, I thank the staff of the Cambridge University Press for the way in which they have met my requirements, and for the excellence of their printing work.

F. GERRISH

DEPARTMENT OF MATHEMATICS
THE POLYTECHNIC
KINGSTON-UPON-THAMES

# PREFACE TO VOLUME I

This volume deals with Calculus and some of its applications to topics like areas and arc-lengths, centroids and moments of inertia, and the geometry of plane curves.

The discursive introductory Chapter 1, which assembles ideas of use in the sequel, should help the reader to decide what he is expected to know from previous work. In the past he has probably regarded Mathematics as a collection of techniques for solving 'problems'; now he has to be persuaded that there is a deeper aspect of the subject —a system of thought as well as a process of action. Although appreciation of the need for rigour comes only gradually, yet the ideas presented in Chapter 2 are fundamental to a genuine understanding of Calculus. The third chapter employs these ideas in a re-examination of the process (here called *derivation*) of finding the derivative of a function, and many familiar results are systematically proved from the definitions without appeal to graphical appearances.

The remaining chapters in this volume need not be read in numerical order. For example, the early part of Chapter 9 on partial derivatives may well follow Chapter 3; the rest of it can be read whenever required. Further, only Part (A) of the long chapter on integration is necessary in order to start differential equations (Chapter 5), and Part (B) can be taken later as revision.

In treating linear differential equations with constant coefficients, a direct method for finding the complementary function has been given as an alternative to the usual 'trial exponentials'; complex numbers are easily avoided until the formal section on the symbolic use of $D$ for calculating a particular solution. However, the customary methods can be employed without inconvenience by teachers who prefer them. It may be felt, particularly by those who favour use of the now fashionable Laplace transform (which is not considered in this book), that too much has been said about symbolic $D$; but the author's teaching experience does not confirm this.

The early parts of Chapters 6 and 7 will undoubtedly be found difficult, but they contain important matters which will repay careful study. Chapter 6 leads up to Taylor's theorem, a result so often merely stated with the remark that a proof is beyond the reader's range; the present treatment may dispel this illusion. (The corresponding *infinite series* finds its natural place in Volume II.) Chapter 7 opens with a descriptive introduction to Riemann's theory of the definite integral. No rigorous approach can be made in a book of this kind, but it is essential for the student to understand definite integration as a limiting summation of contributions from elements, and be able to use it thus.

Chapter 8 continues the geometrical applications, and concludes with a discussion of 'curvature' and 'envelopes' more comprehensive than is usual at this level.

# REFERENCES AND ABBREVIATIONS

In a decimal reference such as 12.73 (2),

| | | |
|---|---|---|
| 12 | denotes | chapter (Ch. 12), |
| 12.7 | denotes | section, |
| 12.73 | denotes | sub-section, |
| 12.73 (2) | denotes | part. |

| | | |
|---|---|---|
| (ii) | refers to | equation (ii) in the *same* section. |
| ex. (ii) | refers to | worked example (ii) in the *same* section. |
| 4.64, ex. (ii) | refers to | worked example (ii) in sub-section 4.64. |
| Ex. 12 (*b*), no. 6 | refers to | problem number 6 in Exercise 12 (*b*). |
| wo | means | with respect to. |

**In the text,** matter in small type (*other than 'ordinary' worked examples*) and in 'starred' worked examples is subsidiary, and may be omitted at a first reading if time is short.

**In an exercise**

| | | |
|---|---|---|
| no. 6 | refers to | problem number 6 in the *same* Exercise. |
| a 'starred' problem | | *either* depends on matter in small type in the text, or on ideas in a later chapter; |
| | | *or* is above the general standard of difficulty. |
| matter in [...] | is | a hint for the solution of a problem. |
| matter in (...) | is | explanatory comment. |

# 1

## REVIEW OF SOME FACTS, DEFINITIONS AND METHODS

### 1.1 Numbers, variables and functions

### 1.11 Numbers

When we speak of a 'number', our meaning depends on the stage which we have reached in the study of mathematics. In early arithmetic we are concerned with the 'natural numbers' $1, 2, 3, \ldots$, together with the number 0; later we deal with fractions or 'rational numbers', and learn how to express a given fraction as a decimal (either terminating or recurring) and conversely. When the need has arisen in algebra, we meet 'signed numbers' like $+2$, $-5$, $-\frac{2}{3}$.

However, we soon find that these types of number are not adequate for all mathematical purposes. For example, the theorem of Pythagoras shows that a right-angled isosceles triangle whose equal sides are of unit length has hypotenuse of length $x$ units, where $x^2 = 2$. It is easy to prove (see below) that $x$ cannot be a rational number; so that, in particular, it cannot be expressed as a decimal which terminates or recurs. The length of the hypotenuse therefore corresponds to a new kind of number, which is denoted by $\sqrt{2}$ and called an 'irrational number'.

Suppose that the number $x$ satisfying $x^2 = 2$ were rational; then it could be written in the form $x = p/q$ where $p$, $q$ are natural numbers. Without loss of generality we may assume that the fraction $p/q$ is already in its lowest terms, i.e. that $p$ and $q$ have no common factor. Then $p^2/q^2 = 2$, so that $p^2 = 2q^2$, and hence $p^2$ is even (i.e. divisible by 2). Therefore $p$ must be even, say $p = 2r$. Thus $4r^2 = 2q^2$, $q^2 = 2r^2$, and by the same argument, $q$ must be even, say $q = 2s$. This shows that $p$, $q$ have the common factor 2, contradicting the hypothesis. Hence $x$ cannot be expressed in the form $p/q$, i.e. it is not a rational number.

It is helpful to represent numbers geometrically. Take a line (for convenience drawn 'horizontally' across the page) and a point $O$ on it. Starting from $O$, there are two directions in which we could proceed along the line; let us agree (as in all graphical work) to take the right-hand one as positive. Choose a point $I$ on this part, and let $OI$ be taken as the unit of length. Then all rational numbers can be represented uniquely by points of the line. Our remarks above about $\sqrt{2}$ can now be expressed as follows: if we construct a right-angled triangle with sides of length $OI$, and lay off its hypotenuse along the line in fig. 1, with one end at $O$, then the other end will not fall on any point of the line which has already been labelled with a rational number. In other words, although every rational number can be represented by a point of the line, not every point on the line corresponds to a rational number. To complete the correspondence between points and numbers we have to admit irrational numbers (i.e. those that are not rational).

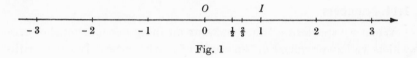

Fig. 1

The term 'irrational number' includes all numbers like $\sqrt{2}, \sqrt[3]{5}, \ldots$ (called *surds*) which arise from the need to solve equations like $x^2 = 2$, $x^3 = 5, \ldots$ whose solutions cannot be expressed rationally. However, it includes more than these: the number $\pi$, met at an early stage as the length of the circumference of a circle with unit diameter, and accepted on trust, is an example of an irrational number which is not a surd (this can be proved, but not easily); others will be met in this book. In practice the existence of irrational numbers causes no difficulty if we are able to obtain approximations as near to them as we please by means of rational numbers. For example, the square root process can be used to express $\sqrt{2}$ approximately as a decimal to as many places as required; experiments with circular objects show that $\pi$ lies between 3·14 and 3·15, and later theoretical work enables us (12.74) to obtain a decimal approximation as accurate as we please. It is easy to see that between any two rational numbers there lies another rational number, and therefore infinitely many rational numbers; and it can be shown (but not here) that between any two rational numbers lies also an irrational one. Thus the 'rational points' of the line are packed indefinitely closely, yet 'between' any two of these lies an 'irrational point'.

All the sorts of number mentioned above are included under the title *real number*;† so that by 'number' we may mean

(i) *integers* (the numbers 0, ± 1, ± 2, ...);

(ii) *rational numbers* (those which can be expressed in the form $p/q$ where $p$, $q$ are integers and $q \neq 0$; the integers are included, since any integer $p$ can be written $p/1$);

† The reason for this curious name will appear in 13.12.

(iii) *irrational numbers* (these consist of all the real numbers which are not rational, e.g. $\sqrt{2}, \sqrt[3]{5}, 3 - \sqrt{6}, \pi$).

We shall not attempt to discuss further the concept of 'real number', a matter for a book on the foundations of mathematics. Here we shall be concerned with developing the subject from approximately the stage which the reader has attained prior to beginning General Degree work. So we continue to use numbers with the confidence which we have shown in the past, noting the types of number mentioned in (i)–(iii) above (especially the need for type (iii)), to which we shall refer in the sequel.

## 1.12 Constants and variables

In algebra, letters are used to denote unspecified numbers. When using them we learn to think of some (called *constants*) as representing the same number throughout the work, while others (called *variables*) are regarded as successively representing many numbers (possibly in some limited range). In some contexts the variables may be restricted to take integral values only, or rational values only; in others they may range over the real numbers.

All the values of $x$ for which $a \leqslant x \leqslant b$ form what is called a *closed interval*. It would be represented in fig. 1 by the segment between $a$ and $b$, end-points included. When the end-points are excluded, we obtain the *open interval* $a < x < b$.

## 1.13 Functions

Throughout mathematical work we meet the situation of one variable being dependent on another. For example, in the kinematics of straight line motion, the distance moved may depend on the time; in a graph of $y$ against $x$, the ordinate $y$ depends on the abscissa $x$; the volume of a gas depends on the pressure to which it is subjected. In all cases we understand that, when a definite value for one variable is assigned, then one or more definite values for the other are determined. We do not imply that *every* assignment of the first variable must give rise to a value of the other; thus there is no pressure which will produce a negative volume. The choice may therefore be restricted to those values for which the variable to be calculated has a meaning.

The variable whose value we choose to select is called the *independent variable*, and the one whose value or values are determined thereby is called the *dependent variable*. The relation is expressed by saying that the dependent variable is a *function* of the independent variable. If $x$ is the independent and $y$ the dependent variable, the reader will know that we write this general relationship as $y = f(x)$. Thus $f(x)$ denotes some (unspecified) variable whose value depends on that of $x$, much in the way that $x$ denotes some (unspecified) number. When we need to consider more than one function of $x$ in the same piece of work, we naturally use different functional symbols such as $g(x)$, $F(x)$, $\phi(x)$, etc.

Although elementary work is concerned with functions of one independent variable, yet many examples arise in which several independent variables are present. For instance, the volume of a right circular cone depends on both the radius and the height. In general, if $x, y, z, \ldots$ are independent variables and $u$ depends on them, we write $u = f(x, y, z, \ldots)$ to express this functional relationship.

In elementary work the relationship between dependent and independent variables is almost always expressed by a mathematical formula (valid perhaps over a limited range). However, the general concept of 'function' is wider than that of 'formula'; all that is necessary is a rule to relate the two variables. Thus one could define $y$ as a function of $x$ as follows:

*if $x$ is prime, then $y = 0$; if $x$ is not prime, then $y = 1$.*

Further, a function may need more than one formula to specify it; e.g. the function whose graph is shown in fig. 2 would have to be defined as

$$y = 0 \quad \text{if} \quad x > 1 \quad \text{or if} \quad x < -1,$$

$$y = 1 + x \quad \text{if} \quad -1 \leqslant x \leqslant 0,$$

$$y = 1 - x \quad \text{if} \quad 0 \leqslant x \leqslant 1.$$

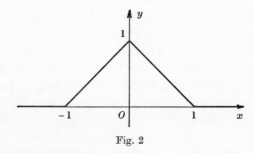

Fig. 2

In this book we shall be concerned only with functions expressible by one or more mathematical formulae.

## 1.14 The function $|x|$

The symbol $|x|$, called the *modulus*, *absolute value*, or *numerical value* of $x$, denotes the value of $x$ regardless of sign. Thus

$$|x| = \begin{cases} x & \text{if } x \geqslant 0, \\ -x & \text{if } x \leqslant 0. \end{cases}$$

Its graph is shown in fig. 3. The properties

$$|xy| = |x| \cdot |y|$$

and

$$\left|\frac{x}{y}\right| = \frac{|x|}{|y|} \quad (y \neq 0)$$

are easily verified.

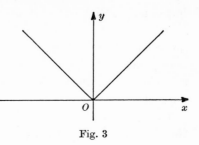

Fig. 3

The important result

$$|x+y| \leqslant |x| + |y| \tag{i}$$

(sometimes called the *triangle inequality*, for a reason which will be clear in 13.33) can also be verified from the above definition.† For if both $x$ and $y$ are positive, each side reduces to $x+y$; if both are negative, each side is $-x-y$; if $x$, $y$ have opposite signs, say $x < 0$ and $y > 0$, then $|x+y| < |x| + |y|$.

By writing $x-y$ instead of $x$, it follows from (i) that

$$|(x-y)+y| \leqslant |x-y| + |y|,$$

so

$$|x-y| \geqslant |x| - |y|.$$

By interchanging $x$ and $y$, and noting that $|y-x| = |x-y|$, we get

$$|x-y| \geqslant |y| - |x|.$$

These two results can be combined to give

$$|x-y| \geqslant \big||x| - |y|\big|. \tag{ii}$$

Replacing $y$ by $-y$ in (ii) and noting that $|-y| = |y|$ gives

$$|x+y| \geqslant \big||x| - |y|\big|.$$

Finally, by applying (i) twice we have

$$|x+y+z| \leqslant |x| + |y| + |z|;$$

and there are similar extensions for any number of variables.

† A neater proof is given in 1.21, ex. (iv).

## 1.2 Simple inequalities

### 1.21 Fundamental results

In elementary mathematics much prominence is given to equations, but in more advanced work *inequalities* (statements that one number is greater or less than another) become of increasing importance. Some have just been given in 1.14. We state here some principles for manipulating inequalities; many of these are analogous to those for equations, but there are important distinctions which should be noticed.

The relation $a > b$ ($a$ is greater than $b$) is equivalent to the statement that $a - b$ is positive. We can interpret $a < b$ ($a$ is less than $b$) to mean either that $b > a$ or that $a - b$ is negative. The following results are given for $>$; similar ones can be formulated for $<$.

I. *If $a > b$, then $a + x > b + x$ for any number $x$.* ('We can add or subtract the same number on both sides of an inequality.') For

$$(a + x) - (b + x) = a - b = \text{positive number because } a > b.$$

COROLLARY. *A term can be transferred from one side of an inequality to the other provided that its sign is changed.*

For example, if $a + b > c + d$, then subtraction of $b$ from both sides gives $a > c + d - b$.

II. *If $a > b$, then $ax \gtrless bx$ according as $x \gtrless 0$.*

For $ax - bx = (a - b)x$ which is positive if $x$ is positive, and is negative if $x$ is negative.

COROLLARY II (*a*). *If $a > b > 0$, then $1/a < 1/b$.*

Take $x = 1/ab$ in II.

COROLLARY II (*b*). *If $a_r > b_r > 0$ for $r = 1, 2, ..., n$, then*

$$a_1 a_2 ... a_n > b_1 b_2 ... b_n.$$

('Inequalities of the same type between *positive* numbers can be multiplied.') For, successive applications of II give

$$a_1 a_2 a_3 ... a_n > b_1 a_2 a_3 ... a_n > b_1 b_2 a_3 ... a_n > ... > b_1 b_2 ... b_n.$$

COROLLARY II (*c*). *If $a > b > 0$, then $a^n \gtrless b^n$ according as $n \gtrless 0$* (where $n$ is rational, and $a^{p/q}$ denotes the *positive* $q$th root of $a^p$ in the case when $n$ is the fraction $p/q$ with $q$ even).

*Proof.* If $n$ is a positive integer, the result follows from Corollary II (*b*) by putting $a_r = a, b_r = b$ for each $r$.

If $n$ is a positive rational number $p/q$,† then we have $a^{1/q} > b^{1/q}$; for $a^{1/q} \leqslant b^{1/q}$ would imply $a \leqslant b$, by applying to this the case just considered, with $n = q$. Hence $(a^{1/q})^p > (b^{1/q})^p$.

If $n$ is a negative rational number $-p/q$, then since $1/b > 1/a$ by Corollary II $(a)$, we can apply the above to this inequality with index $+p/q$ to give the result.

The above corollaries may be false if some or all of the numbers are negative. This is easily verified by numerical examples.

III. *If $a > b$ and $c > d$, then $a + c > b + d$.* ('Inequalities of the same type can be added.') For

$$(a + c) - (b + d) = (a - b) + (c - d) = \text{positive number.}$$

Observe that

   $(a)$ *inequalities cannot be subtracted*: $a > b$ and $c > d$ do not imply $a - c > b - d$; for $(a - c) - (b - d) = (a - b) - (c - d)$, which may be negative or zero;

   $(b)$ *inequalities cannot be divided*: $a > b$ and $c > d$ do not necessarily imply $a/c > b/d$; e.g. take $a = 4, b = 3, c = 2, d = 1$.

## Examples

   (i) *If $a < b + c$ and $a, b, c$ are positive, prove*

$$\frac{a}{1+a} < \frac{b}{1+b} + \frac{c}{1+c}.$$

We have $\qquad\qquad \dfrac{1}{a} > \dfrac{1}{b+c}, \quad \text{so} \quad 1 + \dfrac{1}{a} > 1 + \dfrac{1}{b+c},$

i.e. $\qquad\qquad\qquad \dfrac{1+a}{a} > \dfrac{1+b+c}{b+c}.$

$$\therefore \; \frac{a}{1+a} < \frac{b+c}{1+b+c} = \frac{b}{1+b+c} + \frac{c}{1+b+c} < \frac{b}{1+b} + \frac{c}{1+c}.$$

   (ii) *If $a_1, a_2, \ldots, a_n$ are positive numbers whose sum is $s$, then (if $n > 1$)*

$$(1 + a_1)(1 + a_2)\ldots(1 + a_n) > 1 + s.$$

For $\qquad (1 + a_1)(1 + a_2) = 1 + (a_1 + a_2) + a_1 a_2 > 1 + (a_1 + a_2);$

hence $\quad (1 + a_1)(1 + a_2)(1 + a_3) > \{1 + (a_1 + a_2)\}(1 + a_3) > 1 + (a_1 + a_2 + a_3)$

as in the previous step; and so on.

In particular, taking $a_1 = a_2 = \ldots = a_n = a$, we obtain

$$(1 + a)^n > 1 + na \quad (a > 0, n \text{ an integer}, n > 1),$$

a result sometimes called *Bernoulli's inequality.*

† Without loss of generality we may assume $p, q$ are both *positive* integers.

(iii) *With the notation of ex.* (ii), *and* $0 < a_r < 1$ *for* $r = 1, 2, ..., n$, *we have*

$$(1-a_1)(1-a_2)\ldots(1-a_n) < \frac{1}{1+s}.$$

For $$(1-a_r)(1+a_r) = 1-a_r^2 < 1,$$

so $1-a_r < 1/(1+a_r)$, and the result follows from ex. (ii).

(iv) *Prove that* $|x+y| \leqslant |x|+|y|$.

From the definition of $|x|$ (1.14) it follows immediately that

$$-|x| \leqslant x \leqslant |x|.$$

Similarly $$-|y| \leqslant y \leqslant |y|.$$

Adding these, $$-(|x|+|y|) \leqslant x+y \leqslant |x|+|y|.$$

Therefore $$|x+y| \leqslant |x|+|y|.$$

## 1.22 Arithmetic, geometric, and harmonic means

In this section all letters denote *positive* numbers.

**(1)** Given two positive numbers $a$, $b$, write

$$A = \tfrac{1}{2}(a+b), \quad G = \sqrt{(ab)}, \quad H = \frac{2ab}{a+b}.$$

Then $A$, $G$, $H$ are called the *arithmetic, geometric,* and *harmonic means* of $a$ and $b$. We shall prove that

$$A \geqslant G, \quad \text{where equality occurs only if } a = b.$$

For $(\sqrt{a}-\sqrt{b})^2 \geqslant 0$, with $=$ only when $a = b$; hence

$$a+b-2\sqrt{(ab)} \geqslant 0, \quad \text{i.e.} \quad A \geqslant G.$$

Since $1/H$ is the arithmetic mean of $1/a$ and $1/b$, the preceding result with $a$, $b$ replaced by $1/a$, $1/b$ shows that $G \geqslant H$.

## Examples

*If $a$, $b$, $c$ are not all equal, prove that*

(i) $a^2+b^2+c^2 > bc+ca+ab$;

(ii) $2(a^3+b^3+c^3) > bc(b+c)+ca(c+a)+ab(a+b)$.

Since $b^2+c^2 \geqslant 2bc$, etc., result (i) follows by adding. The relation is $>$, not $\geqslant$, because in at least one of the three separate inequalities the relation is certainly $>$. *Alternatively*,

$$a^2+b^2+c^2-bc-ca-ab = \tfrac{1}{2}\{(b-c)^2+(c-a)^2+(a-b)^2\} > 0.$$

Also $b^2+c^2-bc \geqslant bc$, so $b^3+c^3 \geqslant bc(b+c)$ on multiplying both sides by $b+c$. Adding the three such results gives (ii), with $>$ for the same reason as before.

(2) More generally, if $a_1, a_2, ..., a_n$ are all positive,

$$A = \frac{a_1 + a_2 + ... + a_n}{n} \text{ is their } \textit{arithmetic mean}$$

and
$$G = \sqrt[n]{(a_1 a_2 ... a_n)} \text{ is their } \textit{geometric mean.}$$

It is still true that $A \geqslant G$, *with equality occurring only when*

$$a_1 = a_2 = ... = a_n.$$

This is the 'theorem of the means'; the following proof was given by Cauchy.

By direct calculation,

$$a_1 a_2 = \left(\frac{a_1 + a_2}{2}\right)^2 - \left(\frac{a_1 - a_2}{2}\right)^2$$

$$< \left(\frac{a_1 + a_2}{2}\right)^2 \quad \text{if} \quad a_1 \neq a_2.$$

Applying this type of result twice,

$$a_1 a_2 a_3 a_4 \leqslant \left(\frac{a_1 + a_2}{2}\right)^2 \left(\frac{a_3 + a_4}{2}\right)^2 \leqslant \left(\frac{a_1 + a_2 + a_3 + a_4}{4}\right)^4,$$

with $=$ occurring in the first place if $a_1 = a_2$ and $a_3 = a_4$, and in the second place if $a_1 + a_2 = a_3 + a_4$. Hence, unless $a_1 = a_2 = a_3 = a_4$, we have

$$a_1 a_2 a_3 a_4 < \left(\frac{a_1 + a_2 + a_3 + a_4}{4}\right)^4.$$

Similarly, if $n$ is a power of 2, we can prove step by step that, unless all $a$'s are equal,

$$a_1 a_2 ... a_n < \left(\frac{a_1 + a_2 + ... + a_n}{n}\right)^n. \tag{i}$$

If $n$ is not a power of 2, then it lies between two consecutive powers of 2, say $2^{m-1} < n < 2^m$. Put $k = (a_1 + a_2 + ... + a_n)/n$, and apply (i) to the numbers $a_1, a_2, ..., a_n$ together with the $2^m - n$ numbers $k$:

$$a_1 a_2 ... a_n k^{2^m - n} < \left\{\frac{a_1 + a_2 + ... + a_n + (2^m - n) k}{2^m}\right\}^{2^m} = k^{2^m},$$

i.e.
$$a_1 a_2 ... a_n < k^n = \left(\frac{a_1 + a_2 + ... + a_n}{n}\right)^n,$$

giving $G < A$.

## Examples

(iii) *Prove $9a^2b^2c^2 < (bc+ca+ab)(a^4+b^4+c^4)$ if $a$, $b$, $c$ are not all equal.*
By the theorem of the means applied to $bc$, $ca$, $ab$,

$$\tfrac{1}{3}(bc+ca+ab) > \sqrt[3]{(bc.ca.ab)} = \sqrt[3]{(a^2b^2c^2)}.$$

By the same theorem applied to $a^4$, $b^4$, $c^4$,

$$\tfrac{1}{3}(a^4+b^4+c^4) > \sqrt[3]{(a^4b^4c^4)}.$$

Multiplying, the result follows.

(iv) *If $2x+5y = 3$, $x > 0$ and $y > 0$, find the greatest value of $x^3y^2$.*

As $x > 0$ and $y > 0$, we can apply the theorem to the five positive numbers
$\tfrac{2}{3}x, \tfrac{2}{3}x, \tfrac{2}{3}x, \tfrac{5}{2}y, \tfrac{5}{2}y$, getting

$$\sqrt[5]{\{(\tfrac{2}{3}x)^3 (\tfrac{5}{2}y)^2\}} < \tfrac{1}{5}(2x+5y) = \tfrac{3}{5}$$

unless $\tfrac{2}{3}x = \tfrac{5}{2}y$, in which case $<$ is replaced by $=$. Hence, provided that $x > 0$
and $y > 0$, the greatest value of $(\tfrac{2}{3}x)^3 (\tfrac{5}{2}y)^2$ is $(\tfrac{3}{5})^5$, and that of $x^3y^2$ is

$$(\tfrac{3}{5})^5 (\tfrac{3}{2})^3 (\tfrac{2}{5})^2 = 3^8/(5^7.2).$$

It occurs when $\tfrac{2}{3}x = \tfrac{5}{2}y$; this together with $2x+5y = 3$ gives $x = \tfrac{9}{10}, y = \tfrac{6}{25}$.

## Exercise 1(*a*)

**1** Find for what values of $x$

$$\text{(i)} \quad -2 < \frac{2x+1}{x-2} < 1; \quad \text{(ii)} \quad \frac{1}{2-x} < \frac{1}{x-3}.$$

**2** If $a$, $b$ are unequal positive numbers, prove the following:

(i) $a^3-a^2 > a^{-2}-a^{-3}$ $(a \neq 1)$. [Consider left-hand side minus right-hand side.]

(ii) $a^{m+n}+b^{m+n} > a^m b^n + a^n b^m$, $m$ and $n$ being positive integers.

**3** The two sets $a_1, a_2, ..., a_n$; $b_1, b_2, ..., b_n$ of positive numbers are such that
$m < \dfrac{a_r}{b_r} < M$ $(r = 1, 2, ..., n)$. Prove that

$$m < \frac{a_1+a_2+...+a_n}{b_1+b_2+...+b_n} < M, \quad m < \sqrt[n]{\left(\frac{a_1 a_2...a_n}{b_1 b_2...b_n}\right)} < M.$$

**4** If $1 \leqslant r \leqslant n$, prove $r(n+1-r) \geqslant n$, and deduce that $(n!)^2 > n^n$ for $n > 2$.

**5** If $0 < a < 1$ and $n$ is a positive integer, prove $(1-a)^n < 1/(1+na)$. [Use 1.21, ex. (iii).]

*In nos. 6, 7, $a_1, a_2, ..., a_n$ are positive numbers whose sum is $s$.*

**6** If $0 < a_r < 1$ $(r = 1, 2, ..., n)$, prove $(1-a_1)(1-a_2)...(1-a_n) > 1-s$ when $n > 1$. [Method of 1.21, ex. (ii).]

**7** If $s < 1$, prove $(1+a_1)(1+a_2)...(1+a_n) < 1/(1-s)$. [1.21, ex. (iii); use no. 6.]

**8** Prove 
$$a^4+b^4 \geqslant 2a^2b^2, \quad a^4+b^4+c^4+d^4 \geqslant 4abcd,$$

and 
$$(a^2+b^2)^2+(c^2+d^2)^2 \geqslant 2(ab+cd)^2.$$

Also prove 
$$a^4+b^4+c^4 \geqslant b^2c^2+c^2a^2+a^2b^2 \geqslant abc(a+b+c).$$

**9** Prove $\quad \dfrac{bc}{b+c} + \dfrac{ca}{c+a} + \dfrac{ab}{a+b} \leqslant \frac{1}{2}(a+b+c)$,

where $a, b, c$ are positive. $[(b+c)^2 \geqslant 4bc$, so $bc/(b+c) \leqslant \frac{1}{4}(b+c).]$

**10** (i) Prove $(y+z)(z+x)(x+y) \geqslant 8xyz$, where $x, y, z$ are positive.

(ii) If $a, b, c$ are the sides of a triangle, prove that
$$abc \geqslant (b+c-a)(c+a-b)(a+b-c).$$

**11** If $a > 0$ and $b > 0$, prove that the least value of $ax + b/x$ for $x > 0$ is $2\sqrt{(ab)}$.

**12** (i) If $x, y$ are positive, and $m, n$ are positive integers, prove
$$\frac{x^m y^n}{(x+y)^{m+n}} \leqslant \frac{m^m n^n}{(m+n)^{m+n}}.$$

(ii) If $p, q$ are positive integers, prove
$$\sin^{2p}\theta \cos^{2q}\theta \leqslant \frac{p^p q^q}{(p+q)^{p+q}}.$$

[For (i), apply $A \geqslant G$ to the $m$ numbers $x/m$ together with the $n$ numbers $y/n$.]

**13** If $w, x, y, z$ are all positive and $w+x+y+z = 1$, prove $wxyz^3 \leqslant 1/1728$ and find the values of $w, x, y, z$ for which equality is obtained.

**14** Find the greatest value of $x^2 y^3 z^4$ when $x, y, z$ are all positive and

(i) $x^2 + y^2 + z^2 = 1$;   (ii) $x^3 + 2y^3 + 3z^3 = 1$.

[(i) Consider $x^4 y^6 z^8$ and take
$$a_1 = a_2 = \tfrac{1}{2}x^2, \quad a_3 = a_4 = a_5 = \tfrac{1}{3}y^2, \quad a_6 = \dots = a_9 = \tfrac{1}{4}z^2.]$$

***15** Prove that the length of the shortest line which can be drawn to bisect the area of triangle $ABC$ is $\sqrt{(2bc)} . \sin\frac{1}{2}A$, where $A$ is the smallest angle.

## 1.3 Quadratic functions and quadratic inequalities

### 1.31 Sign of a quadratic function

We consider the quadratic function $y = ax^2 + bx + c$. The reader will be familiar with the general method of solving the corresponding quadratic equation
$$ax^2 + bx + c = 0$$

by 'completing the square', and will know how the sign of $b^2 - 4ac$ decides the nature of the roots. The expression $b^2 - 4ac$ is called the *discriminant* of the quadratic equation and also of the quadratic function.

(a) *If $b^2 > 4ac$*, the equation $y = 0$ has distinct roots, say $\alpha$ and $\beta$ where $\alpha < \beta$, and hence
$$y = a(x-\alpha)(x-\beta).$$

Thus $y$ will be positive for some values of $x$, zero for $x = \alpha$ and $x = \beta$, and negative for others. For example, if $a > 0$, then $y > 0$ for those values of $x$ which give the factors the same sign, viz. $x < \alpha$ or $x > \beta$; and $y < 0$ when the factors have opposite signs, i.e. when $\alpha < x < \beta$. Similar results can be stated when $a < 0$.

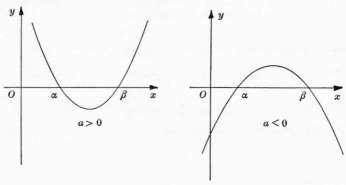

Fig. 4

(b) *If* $b^2 = 4ac$, *then*

$$y = a\left(x^2 + \frac{b}{a}x + \frac{c}{a}\right)$$

$$= a\left(x^2 + \frac{b}{a}x + \frac{b^2}{4a^2}\right)$$

since

$$\frac{c}{a} = \frac{b^2}{4a^2};$$

hence

$$y = a\left(x + \frac{b}{2a}\right)^2.$$

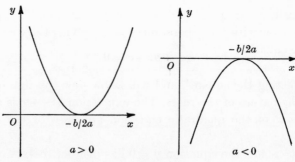

Fig. 5

Therefore $y$ is zero when $x = -b/2a$, and has the same sign as $a$ for all other values of $x$.

(c) *If $b^2 < 4ac$, then*

$$\frac{y}{a} = x^2 + \frac{b}{a}x + \frac{c}{a}$$

$$= \left(x + \frac{b}{2a}\right)^2 + \frac{c}{a} - \frac{b^2}{4a^2}$$

on completing the square, so

$$y = a\left\{\left(x + \frac{b}{2a}\right)^2 + \frac{4ac - b^2}{4a^2}\right\}. \tag{i}$$

Both terms in the outer bracket are positive, the first being zero when $x = -b/2a$; hence the contents of this bracket are always positive. Therefore *y has the same sign as a for all values of x.*

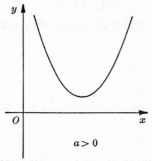

$$a > 0 \qquad\qquad a < 0$$

Fig. 6

If the quadratic expression $ax^2 + bx + c$ is positive for all values of $x$, it is said to be *positive definite*. The work under (c) shows that

*if $a > 0$ and $b^2 < 4ac$, then $ax^2 + bx + c$ is positive definite.*

The two conditions imply $c > 0$, since $a > 0$ and $4ac > b^2 \geqslant 0$. Similarly the two conditions $c > 0$, $b^2 < 4ac$ imply $a > 0$. Hence the two sets of conditions are equivalent. Conversely,

*if $ax^2 + bx + c$ is positive definite, then $a > 0$, $b^2 < 4ac$, and $c > 0$.*

For since $ax^2 + bx + c > 0$ for all values of $x$, the equation

$$ax^2 + bx + c = 0$$

has no roots; hence $b^2 < 4ac$. When $x = 0$, the hypothesis gives $c > 0$. These conditions imply $a > 0$, as above.

If the expression $ax^2 + bx + c$ is negative for all values of $x$, it is said

**2**

to be *negative definite*. The reader should show that *equivalent necessary and sufficient conditions for negative definiteness are*

$$a < 0, \ b^2 < 4ac \quad or \quad b^2 < 4ac, \ c < 0.$$

## 1.32 Cauchy's inequality

If we are given two sets $a_1, a_2, \dots, a_n$; $b_1, b_2, \dots, b_n$ each of $n$ numbers (not necessarily positive), we shall prove that

$$(a_1 b_1 + a_2 b_2 + \dots + a_n b_n)^2 < (a_1^2 + a_2^2 + \dots + a_n^2)(b_1^2 + b_2^2 + \dots + b_n^2) \qquad \text{(ii)}$$

*unless*
$$\frac{a_1}{b_1} = \frac{a_2}{b_2} = \dots = \frac{a_n}{b_n},$$

*in which case there is = instead of <.*

Consider the expression

$$y = (a_1 x + b_1)^2 + (a_2 x + b_2)^2 + \dots + (a_n x + b_n)^2. \qquad \text{(iii)}$$

For all $x$, $y \geqslant 0$; and we can have $y = 0$ only if there is an $x$ such that $a_r x + b_r = 0$ for each $r = 1, 2, \dots, n$, i.e. if

$$\frac{a_1}{b_1} = \frac{a_2}{b_2} = \dots = \frac{a_n}{b_n} = -\frac{1}{x}.$$

It is then easily verified that the two sides of (ii) are equal.

Excluding this case, we have (after expanding the brackets in (iii) and rearranging) that for all $x$,

$$(a_1^2 + a_2^2 + \dots + a_n^2) x^2 + 2(a_1 b_1 + \dots + a_n b_n) x + (b_1^2 + \dots + b_n^2) > 0;$$

i.e. this quadratic expression is positive definite. Hence by the converse result in 1.31,
$$4(a_1 b_1 + \dots + a_n b_n)^2 < 4(a_1^2 + \dots + a_n^2)(b_1^2 + \dots + b_n^2),$$

which gives the result stated.

## Exercise 1(*b*)

*Find for what values of x the following expressions are positive.*

1  $2x^2 - 7x + 3$.          2  $2 - x - 3x^2$.          3  $3x^2 - 2x + 5$.

4  $4x^2 + 4x + 1$.          5  $(x^2 + 2)(x^2 - 1)$.          6  $(x - 1)(x - 2)(x - 3)$.

7  $(x - 1)^2 (3 - x)$.          8  $x^3 - 3x^2 - x + 3$.          9  $\dfrac{(x - 1)(x - 2)}{(x + 1)(x - 4)}$.

10  Find for what values of $x$ the function $x^2 - 6x + 7$ lies between $\pm 1$.

11  Find the greatest value of $4 + 2x - 3x^2$, and the least value of $2x^2 - 3x + 1$.

12  Use equation (i) of 1.31 to prove that (i) if $a > 0$, then $y$ has a least value, viz. $(4ac - b^2)/4a$; (ii) if $a < 0$, then $y$ has a greatest value, viz. $(4ac - b^2)/4a$; (iii) in either case this value is attained when $x = -b/2a$.

13  Prove that the quadratic equation

$$(a^2 + b^2) x^2 + 2(a^2 + b^2 + c^2) x + (b^2 + c^2) = 0$$

always has roots.

**14** If $a > b > 0$, prove that $x(x-a) = \lambda(x-b)$ has roots for any $\lambda$. Can these roots ever be equal?

**15** Find the values of $\lambda$ for which the expression

$$5x^2 + 8x + 14 + \lambda(x^2 + 10x + 7)$$

is a perfect square. Hence find constants $A, B, C, D, p, q$ such that

$$5x^2 + 8x + 14 = A(x-p)^2 + B(x-q)^2 \quad \text{and} \quad x^2 + 10x + 7 = C(x-p)^2 + D(x-q)^2.$$

*16 Prove $(a^2 + b^2 + c^2)^2 < (a+b+c)(a^3+b^3+c^3)$ when $a, b, c$ are not all equal. [Apply Cauchy's inequality to the sets $a^{\frac{1}{2}}, b^{\frac{1}{2}}, c^{\frac{1}{2}}; a^{\frac{3}{2}}, b^{\frac{3}{2}}, c^{\frac{3}{2}}$.]

*17 If $l^2 + m^2 + n^2 = 1$ and $l'^2 + m'^2 + n'^2 = 1$, prove $-1 \leqslant ll' + mm' + nn' \leqslant 1$.

## 1.4 Graphs

To fix ideas in a problem it is frequently helpful to sketch a graph of the function concerned; thus in 1.31 we used sketch-graphs to illustrate the behaviour of a quadratic function. The reader will already be familiar with the general forms of graphs representing such simple functions as $x^2$, $x^3$, $1/x$ and the trigonometric functions. In this section we propose to illustrate some considerations which are useful in graph-sketching.

### 1.41 Examples

(i) *Sketch the graph of*

$$y = \frac{x}{x^2 + 1}.$$

There is one value of $y$ for each value of $x$. Since $y = 0$ when and only when $x = 0$, the graph cuts the axes only at the origin.

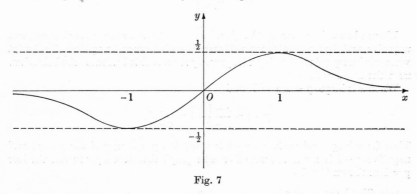

Fig. 7

If we change the signs of both $x$ and $y$, the equation is essentially unaltered; hence the graph is 'symmetrical about the origin $O$'.

When $x$ is large, $y \doteqdot x/x^2 = 1/x$, and this becomes small when $x$ increases. If $x$ is large and positive, $y$ is small and positive, so that the graph approaches $Ox$ from above. Similarly, when $x$ is large and negative, the graph approaches $Ox$ from below.

It can be shown that the value of $y$ is never numerically greater than $\frac{1}{2}$. For by considering $1/y$ and applying the theorem of the means, we have when $x > 0$ that

$$\frac{1}{y} = x + \frac{1}{x} \geqslant 2\sqrt{\left(x \times \frac{1}{x}\right)} = 2,$$

with $=$ only when $x = 1/x$, i.e. $x = 1$. Hence $y \leqslant \frac{1}{2}$, and $y = \frac{1}{2}$ when $x = 1$. When $x < 0$, the symmetry about $O$ shows that $y \geqslant -\frac{1}{2}$, with $y = -\frac{1}{2}$ only when $x = -1$. The graph therefore lies between the lines $y = \pm\frac{1}{2}$.

(ii) *Sketch the graph of*

$$y = \frac{x-1}{x-2}.$$

There is one value of $y$ for each value of $x$ except $x = 2$: there is no point on the graph corresponding to $x = 2$.

The graph cuts the $y$-axis where $x = 0$, and then $y = \frac{1}{2}$. It cuts the $x$-axis where $y = 0$, and then $x = 1$.

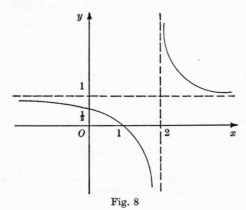

Fig. 8

When $x$ has values near to 2 but just less than 2, the denominator is small and negative, and the numerator is positive; hence $y$ is large and negative. Similarly, when $x$ is just greater than 2, $y$ is large and positive. This indicates the behaviour near the line $x = 2$.

When $x$ is large, $y \doteqdot x/x = 1$; and

$$y - 1 = \frac{x-1}{x-2} - 1 = \frac{1}{x-2}.$$

Thus for $x$ large and positive, $y - 1$ is small but positive; and for $x$ large and negative, $y - 1$ is small but negative. The graph therefore approaches the line $y = 1$ as shown (fig. 8).

(iii) *Sketch the graph of*

$$y = \frac{x}{x^2 - 1}.$$

There is one value of $y$ for each $x$ except $x = \pm 1$: the graph has no points corresponding to these values. Writing

$$y = \frac{x}{(x-1)(x+1)},$$

we see that when $x$ is just less than $+1$,

$$y = \frac{\text{number near} + 1}{(\text{small negative number})(\text{number near} + 2)}$$

$$= \text{large negative number.}$$

Similarly, when $x$ is just greater than $+1$, $y$ is large and positive. The neighbourhood of $x = -1$ can be discussed similarly.

If we now apply considerations illustrated in exs. (i), (ii), we obtain fig. 9.

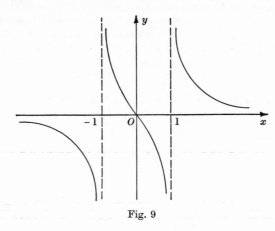

Fig. 9

*Remarks*

($\alpha$) In ex. (ii), the lines $x = 2$ and $y = 1$ are each approached indefinitely closely by the curve, yet are never crossed or even reached. They are called *asymptotes*† of the curve. In ex. (i) the $x$-axis is an asymptote; for although the curve does actually cross it at $O$, it also approaches $Ox$ indefinitely closely (without meeting it) as $x$ becomes large (positive or negative). Similarly, in ex. (iii) the lines $x = \pm 1$ and $Ox$ are asymptotes. The essential property of an asymptote is that, as we recede along it, the curve approaches it indefinitely closely yet never reaches it; whether the curve may cross the asymptote at a 'finite' point is immaterial.

($\beta$) The graph in ex. (i) illustrates that the function can take values only in the interval $-\frac{1}{2} \leqslant y \leqslant \frac{1}{2}$, whatever value $x$ may have. Those in exs. (ii), (iii) show that the function can take any value whatever, with the exception of $y = 1$ in ex. (ii). We may enquire whether these properties could have been discovered without first sketching the graphs.

(iv) *Find the range of possible values of*

$$\frac{3x^2 + 2}{2x^2 - 2x + 1},$$

*and use this information to assist in sketching the graph.*

If $k$ is a possible value, then the equation

$$k = \frac{3x^2 + 2}{2x^2 - 2x + 1},$$

† See 18.12(2) for a general definition.

i.e.                              $(2k-3)\,x^2 - 2kx + (k-2) = 0,$

must have roots (possibly coincident).† Hence, applying the condition '$b^2 \geqslant 4ac$',

$$4k^2 \geqslant 4(2k-3)\,(k-2),$$

i.e.                              $k^2 \geqslant 2k^2 - 7k + 6,$

i.e.                          $0 \geqslant k^2 - 7k + 6 = (k-1)\,(k-6).$

This product will be zero if $k = 1$ or $6$, and will be negative if and only if the factors $k-1$, $k-6$ have opposite signs, i.e. if $k > 1$ and $k < 6$. Hence we must have $1 \leqslant k \leqslant 6$, so *the expression can take all values between 1 and 6 inclusive.*

When $k = 1$ or $6$, the above quadratic equation satisfies the condition '$b^2 = 4ac$' for equal roots; the root is then '$x = -b/2a$', i.e.

$$x = -\frac{-2k}{2(2k-3)} = \frac{k}{2k-3}.$$

When $k = 1$, $x = -1$; and when $k = 6$, $x = \frac{2}{3}$. Hence the greatest value 6 is attained when $x = \frac{2}{3}$, and the least value 1 when $x = -1$.

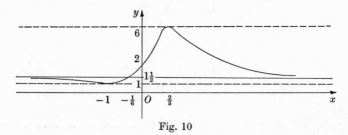

Fig. 10

To sketch the graph of

$$y = \frac{3x^2 + 2}{2x^2 - 2x + 1},$$

we first see from the above work that the graph lies entirely between the lines $y = 1$, $y = 6$, and touches these at the points $(-1, 1)$, $(\frac{2}{3}, 6)$ respectively; these are the turning-points on the curve.

The curve cuts $Oy$ where $x = 0$, and then $y = 2$. It does not cut $Ox$ since $3x^2 + 2 > 0$ for all $x$. Since

$$2x^2 - 2x + 1 = 2(x - \tfrac{1}{2})^2 + \tfrac{1}{2} > 0 \quad \text{for all } x,$$

the denominator can never be zero, and hence there is one value of $y$ for each value of $x$.

When $x$ is large, $y \doteqdot 3x^2/2x^2 = \frac{3}{2}$. A closer approximation is

$$y \doteqdot \frac{3x^2}{2x^2 - 2x} = \frac{3x}{2(x-1)},$$

which shows that when $x$ is large and positive, then $y > \frac{3}{2}$; and that when $x$ is large and negative, then $y < \frac{3}{2}$. Hence the graph approaches the horizontal asymptote $y = \frac{3}{2}$ from above when $x$ is large positive, and from below when $x$ is large negative. It cuts $y = \frac{3}{2}$ where $x = -\frac{1}{6}$; see footnote † below. We can now sketch the curve.

† If $k = \frac{3}{2}$ the equation is not quadratic, but becomes $-3x - \frac{1}{2} = 0$. Thus $\frac{3}{2}$ is the value of the function when $x = -\frac{1}{6}$.

The preceding examples illustrate the steps to be taken before sketching the graph of a function of the type $(ax^2+bx+c)/(Ax^2+Bx+C)$:

(a) Find where (if at all) the graph cuts $Ox$ and $Oy$.

(b) Find how the graph behaves when $x$ is large (positive and negative), and how it approaches the horizontal asymptote; also whether it cuts this asymptote.

(c) If the denominator has factors, there will be asymptotes parallel to $Oy$ through the points which make the denominator zero. Find how the graph behaves when $x$ approaches such points from both left and right.

(d) Find the possible range of values of $y$, and where the curve reaches the extreme positions (if any); these are the turning points.

## 1.42 Further examples

The functions in the preceding examples were 'algebraic fractions', and the steps (a)–(d) just indicated can be taken before sketching the graph of any such function, even when the denominator is not linear or quadratic. We now consider some simple functions involving root extractions.

(i) $y = \sqrt{x}$.

Since this equation can be written $x = y^2$, we can sketch the graph just as we would for $y = x^2$, but with the roles of $x$, $y$ interchanged; i.e. we regard $x$ as a function of $y$. See fig. 11.

(ii) $y = \sqrt{(1-x^2)}$.

If we write this free of the root sign as $y^2 = 1-x^2$, or better as $x^2+y^2 = 1$, we see that the equation is the condition for the point $(x,y)$ to lie at unit distance from the origin. The graph is therefore the circle with centre $O$ and radius 1.

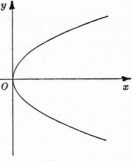

Fig. 11

In each of these examples we have assumed that both the positive and negative square roots are possible values of $y$. If we interpret $\sqrt{x}$ to mean 'the *positive* square root of $x$', then only the upper half of each graph would be required.

## Exercise 1(c)

*Sketch the graphs of the following functions.*

| | | | |
|---|---|---|---|
| 1  $x^2$. | 2  $x^3$. | 3  $1/x$. | 4  $1/x^2$. |

5  $x+\dfrac{1}{x}$.     6  $\dfrac{x+1}{x-2}$.     7  $\dfrac{1}{(x-1)(x-2)}$.     8  $\dfrac{x}{(x-1)(x-2)}$.

9  $\dfrac{x^2}{(x-1)(x-2)}$.     10  $\sqrt[3]{x}$.     11  $\dfrac{1}{\sqrt{x}}$.     12  $x^{\frac{3}{2}}$.

13  $x^{\frac{2}{3}}$.     14  $\dfrac{1}{\sqrt{(x^2-1)}}$.     15  $\dfrac{1}{\sqrt{(1-x^2)}}$.

16  Prove that $(x-2)/\{(x-1)(x-3)\}$ can take any value. Sketch the graph.

17  Show that $(x^2-2x+4)/(x^2+2x+4)$ lies between $\frac{1}{3}$ and 3. Sketch the graph.

18  Prove that $\{(x-5)(5x-13)\}/\{(2x-7)(4x-11)\}$ has no values between 1 and 4, and sketch the graph.

19  Find the possible range of values of $(x+1)/(x-1)^2$, and sketch the graph.

20  Prove that $(1-x^2)/(ax^2+bx+c)$ can take any value if $b^2 > (a+c)^2$.

21  Prove that $(x^2+2x+c)/(x^2+4x+3c)$ can take any value if $0 < c < 1$.

22  Show that $\{(x-a)(x-b)\}/(x-c)$ takes any value if $c$ lies between $a$ and $b$.

*23  Prove that the maximum and minimum values of

$$\frac{ax^2+bx+c}{Ax^2+Bx+C}$$

are the values of $k$ (if any) for which $ax^2+bx+c-k(Ax^2+Bx+C)$ is a perfect square. [This is the condition for the line $y = k$ to touch the curve.]

*24  If $a \neq c$, prove that $(ax^2+bx+c)/(cx^2+bx+a)$ can take any value if $b^2 > (a+c)^2$. [This implies $b^2 > 4ac$; use the conditions for positive definiteness.] Show also that there will be two values between which it cannot lie if $4ac < b^2 < (a+c)^2$, and two values between which it must lie if $b^2 < 4ac$. [This implies $(a+c)^2 > b^2$.]

## 1.5  Types of function

### 1.51  Classification by structure

(1) Functions can be classified according to the manner in which they are formed. If we start with a variable $x$ and write down its positive integral powers $x^0 = 1$, $x^1 = x, x^2, x^3, ..., x^n$, and then combine any constant multiples of these by addition or subtraction, we obtain a *polynomial function of $x$* (in short, a *polynomial in $x$*) of *degree n*. Thus $3x^4-2x^3+\frac{3}{8}x+5$, $2x^3+ax^2+bx+1$ are polynomials in $x$ of degrees 4, 3 respectively, the latter having $a$, $b$ as literal coefficients.

If we divide one polynomial in $x$ by another polynomial in $x$, we obtain a *rational function of $x$* (i.e. a 'ratio' of polynomials—an algebraic fraction). Thus

$$\frac{x}{x^2+1}, \quad \frac{1}{x}, \quad \frac{3x^3+5x+\sqrt{2}}{2x+7\pi}$$

are rational functions of $x$. Notice that the term 'rational' makes no reference to the *coefficients* of powers of $x$ in the function: these can be any sort of number. The rules of algebra show that rational functions of $x$ are generated by applying to $x$ and numbers the operations of addition, subtraction, multiplication and division in any finite combination.

Similar considerations apply for polynomial and rational functions of several independent variables. For example, if $x$, $y$ are independent variables, then $3x^4-2x^3y+5y^3$ is a polynomial in $x$ and $y$, and $(x^2+3y)/(6x^3-\sqrt{2}y^2)$ is a rational function of $x$ and $y$.

(2) Consider now the equation $3x^4-2x^3y+5y^3 = 0$. This can be regarded as a cubic equation in $y$ whose coefficients are functions of $x$.

It will determine $y$ as a function of $x$; for when a numerical value is assigned to $x$, we obtain an ordinary cubic for $y$ which determines at least one† numerical value of $y$. We say that the equation defines $y$ as an *implicit function* of $x$. If we were to solve the equation for $y$ in terms of $x$ by a mathematical formula (which in fact is possible, although not easy) we should have the same function $y$ expressed as an *explicit function* of $x$.

All the above types of function can be included under the heading *algebraic*; that is, they can all be defined (explicitly or implicitly) by polynomial equations in $y$ whose coefficients are polynomials in $x$. For example, the rational function $y = x/(x^2+1)$ can be defined (implicitly) by the polynomial equation

$$x^2 y - x + y = 0,$$

and is easily obtained explicitly by solving for $y$. Similarly, the equation $x^2 + y^2 = 1$ defines the (two-valued) function $y = \pm\sqrt{(1-x^2)}$ for $-1 \leqslant x \leqslant 1$; in 1.42, ex. (ii) we sketched its graph by actually using the defining equation instead of the explicit expression for $y$. On the other hand, the equation

$$y^5 - xy + 1 = 0$$

defines $y$ as a function of $x$ which cannot be obtained explicitly by any formula involving roots, powers, sums, differences, products or quotients (this fact can be proved, but we shall not do so in this book). This sort of example shows that consideration of implicit functions will be necessary.

We may wonder how information about the last function can be obtained, and in particular how its graph can be sketched. It happens in this case that the defining equation can easily be solved explicitly for $x$ in terms of $y$:

$$x = \frac{y^5+1}{y}.$$

Hence, if we choose values for $y$ and calculate the corresponding ones for $x$, we shall be able to plot a graph of $x$ *considered as a function of $y$*; see Ex. 1 (*d*), no. 10. A much simpler example of this method was given in 1.42, ex. (i).

In general, a polynomial equation in $x$ and $y$ can be regarded in two ways: (*a*) as a polynomial equation in $y$ whose coefficients are polynomials in $x$, which defines $y$ as an algebraic function $f(x)$ of $x$; or

† We shall prove later that every cubic equation has at least one root, and may have two or three roots.

(b) as a polynomial equation in $x$ whose coefficients are polynomials in $y$, which defines $x$ as an algebraic function $g(y)$ of $y$. The two functions $f(x)$, $g(x)$ so obtained† are called *inverses* of each other. Our graphical method above amounts to sketching the inverse function of $y$ when this is easily done. Consideration of inverse functions is thus seen to be useful.

(3) Any function which is not 'algebraic' in the above sense is called a *transcendental function*. It can be proved (but not in this book) that $\log_{10} x$, $\sin x$ and the other trigonometrical functions are of this type; and we shall meet others later. They too may be explicit (like the examples just given) or implicit (e.g. the function $y$ of $x$ defined by $xy = \sin y$); and as in the case of algebraic functions, they can be associated in inverse pairs: thus $10^x$ is the inverse of $\log_{10} x$, and $\sin x/x$ is that of the implicit function $y$ given by $xy = \sin y$.

Irrational numbers are classified similarly: those such as $\sqrt{2}$, $\sqrt[3]{5}$, $\sqrt{2+3\sqrt{5}}$, $\sqrt{(2+\sqrt{3})}$, ... that can be obtained as roots of polynomial equations in one variable $x$ with integer coefficients are called *algebraic numbers*; all others (and this can be proved to include $\pi$) are *transcendental numbers*.

## 1.52 Classification by properties

A different sort of classification (which may cut across the one just given) can be made by considering general properties which functions may possess.

(1) *Oddness, evenness.* If $f(x)$ is defined for pairs of equal and opposite values of $x$, and $f(-x) = f(x)$, then $f(x)$ is an *even function*: it is un-altered by changing the sign of $x$ throughout. The graph of $y = f(x)$ is therefore symmetrical about the $y$-axis (e.g. fig. 12 (a)).

Similarly, if $f(-x) = -f(x)$, then $f(x)$ is an *odd function*. Its graph is 'symmetrical about the origin' (e.g. fig. 12 (b)).

For example, $x^4 + 3x^2 - 2$, $\cos x$, $\tan^2 x$ are even; $x^3 + 5x$, $x^3/(x^4 + 1)$, $\sin x$, $\tan^3 x$ are odd. Functions like $x^3 + x^2 + 4x$, $3\sin x + 2\cos x$, in which some terms are even and others odd, are neither odd nor even; also see Ex. 1 (d), no. 3.

(2) *Periodicity.* If there is a positive number $p$ such that

$$f(x+p) = f(x)$$

† It is immaterial whether we write $g(x)$ or $g(y)$ when the *functions* are being discussed. The *relation* between them is that if $y = f(x)$, then $x = g(y)$, and conversely.

for all $x$, and if $p$ is the smallest such number, then $f(x)$ is *periodic*, with *period $p$*. Thus the period of $\sin x$, $\cos x$, is† $2\pi$; of $\tan x$, $\cot x$ is $\pi$; and of $\sin nx$ is $2\pi/n$. The graph of a periodic function consists of an

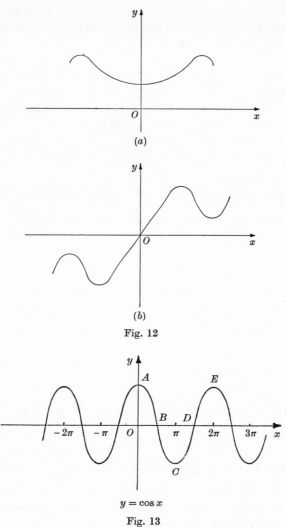

(a)

(b)

Fig. 12

$$y = \cos x$$

Fig. 13

arc of a curve repeated infinitely often in both directions of the $x$-axis. Thus the graph of $y = \cos x$ (fig. 13) consists of the curve $ABCDE$, which represents the function for $0 \leqslant x \leqslant 2\pi$, repeated over subsequent intervals of $2\pi$ in both directions of $Ox$.

† The proof that $2\pi$ is in fact the *smallest* number $p$ is difficult, and will not be given in this book.

It is known that $(a)$ no periodic function can be rational (unless, trivially, it is constant); $(b)$ no periodic function can be algebraic. Such a function must therefore be transcendental.

**(3)** *Many-valuedness.* If $n$ different possible values of $f(x)$ correspond to a general value of $x$, then $f(x)$ is an *$n$-valued function.*

### Examples

(i) $$y = \frac{2x-1}{3x+4}$$

is *single-valued* because there is just one value of $y$ corresponding to each value of $x$ except $x = -\frac{4}{3}$ (for which $y$ is not defined).

(ii) The function $y$ defined implicitly by $x^2 + y^2 = 9$ is *two-valued* because there are two values of $y$ for each $x$ when $-3 < x < 3$, even though there is only one when $x = \pm 3$ and $y$ is not defined when $x > 3$ or $x < -3$. (All this follows because we can find $y$ explicitly as $\pm\sqrt{(9-x^2)}$.)

(iii) The equation $y^3 - 6y^2 + 11y = x$ gives three values of $y$ for some values of $x$; e.g. if $x = 6$, the equation can be written $(y-1)(y-2)(y-3) = 0$, so that $y = 1$, 2 or 3 when $x = 6$. Hence $y$ is a *three-valued* function of $x$.

Graphically, a line parallel to $Oy$ will cut the curve (if at all) in (i) one, (ii) two, (iii) three points, in general.

### *The inverse circular functions.*

(iv) If $-1 \leqslant x \leqslant 1$, then the equation $\cos y = x$ defines $y$ (implicitly) as a function of $x$, written $y = \mathrm{Cos}^{-1}x$. Since $x = \cos y$, the graph (fig. 14) can be obtained from that of $y = \cos x$ by interchanging the axes of $x$ and $y$, and then reversing the sense of $Ox$ to restore right-handedness. A line parallel to $Oy$ cuts the graph (if at all) infinitely often. Thus $y$ is *infinitely many-valued.*

When $x$ is given, let $y = \alpha$ be the smallest positive angle for which $\cos y = x$; $\alpha$ is acute if $x > 0$ (fig. 15$(a)$), obtuse if $x < 0$ (fig. 15$(b)$). It is called the *principal value* of $\mathrm{Cos}^{-1}x$, and is written $\cos^{-1}x$. Thus (using radian measure) $0 \leqslant \cos^{-1}x \leqslant \pi$.

All other angles having their cosine equal to $x$ are bounded by $OX$ and one of the rays $OP$, $OP'$ (figs. 15$(a)$, $(b)$). Those bounded by $OP$ can be expressed as

$$2\pi + \alpha, \quad 4\pi + \alpha, \quad 6\pi + \alpha, \quad \ldots \quad \text{or} \quad \alpha - 2\pi, \quad \alpha - 4\pi, \quad \ldots,$$

according as we add complete positive or negative revolutions to $\alpha$; and those by $OP'$ as

$$2\pi - \alpha, \quad 4\pi - \alpha, \quad \ldots \quad \text{or} \quad -2\pi - \alpha, \quad -4\pi - \alpha, \quad \ldots.$$

All these are given by the expression $2n\pi \pm \alpha$, where $n$ is any integer (positive, negative, or zero). Hence

$$\mathrm{Cos}^{-1}\,x = 2n\pi \pm \cos^{-1}\,x.$$

Each value of $n$ determines a *branch* of the many-valued function $\mathrm{Cos}^{-1}x$. The *principal branch* is shown thickened in fig. 14.

(v) For $-1 \leqslant x \leqslant 1$, $\sin y = x$ similarly defines a many-valued function $y = \mathrm{Sin}^{-1}x$. The graph (fig. 16) is obtained from that of $y = \sin x$ as described in (iv).

Given $x$, the *principal value* $\sin^{-1}x$ is the smallest acute angle $y = \beta$ (positive or negative) for which $\sin y = x$; thus $-\frac{1}{2}\pi \leqslant \sin^{-1}x \leqslant \frac{1}{2}\pi$.

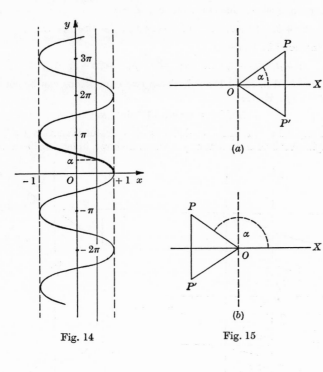

Fig. 14

Fig. 15

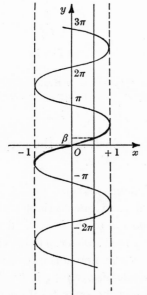

Fig. 16

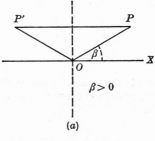

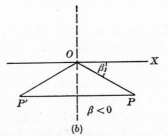

Fig. 17

Any other angle whose sine is $x$ is one of *either*

$$2\pi + \beta, \quad 4\pi + \beta, \quad \ldots \quad \text{or} \quad -2\pi + \beta, \quad -4\pi + \beta, \quad \ldots,$$

which are included in $2m\pi + \beta$,

*or*     $\pi - \beta, \quad 3\pi - \beta, \quad \ldots \quad \text{or} \quad -\pi - \beta, \quad -2\pi - \beta, \quad \ldots,$

which are included in $(2m+1)\pi - \beta$. These two expressions can be combined into $n\pi + (-1)^n \beta$. Hence

$$\text{Sin}^{-1} x = n\pi + (-1)^n \sin^{-1} x.$$

(vi)  For any value of $x$, $\tan y = x$ defines a many-valued function $y = \text{Tan}^{-1} x$ whose graph, obtained from that of $y = \tan x$, is shown in fig. 18. The *principal*

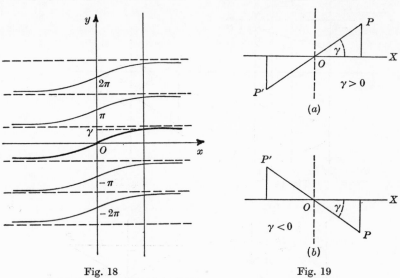

Fig. 18                               Fig. 19

*value* $\tan^{-1} x$ is the smallest acute angle $y = \gamma$ (positive or negative) for which $\tan y = x$; thus $-\frac{1}{2}\pi < \tan^{-1} x < \frac{1}{2}\pi$. All angles whose tangent is $x$ are given by $n\pi + \gamma$; hence

$$\text{Tan}^{-1} x = n\pi + \tan^{-1} x.$$

In Ch. 2 we shall consider functions which have the property of *continuity*, and in Ch. 3 those that are *derivable*.

**(4)** *Homogeneous polynomials and functions.*

Turning now to functions of more than one variable, we may enquire whether there is a useful extension of the idea of 'degree of a polynomial in $x$' for polynomials in two variables $x, y$ (which consist of the sum of a number of terms like $ax^p y^q$, where $a$ is a constant).

The expression $ax^p y^q$ is said to have *total degree* $p + q$ in $(x, y)$.

A polynomial each of whose terms has the same total degree $n$ is said to be *homogeneous of degree n*.

These definitions extend in the obvious way to cases of three or more variables. Thus the polynomials

$$x^2 + xy + 2y^2, \quad x - y + z, \quad 2x^4 + 3x^2y^2 - 5xy^2z$$

are each homogeneous in their variables, with degrees 2, 1, 4 respectively.

*If $f(x, y)$ is a homogeneous polynomial of degree $n$, then*

$$f(tx, ty) = t^n f(x, y)$$

*for all values of $x$, $y$, $t$.*

*Proof.* Each term of $f(x, y)$ is of the form $ax^p y^q$, where $p + q = n$ and $a$ is a constant. The corresponding term of $f(tx, ty)$ is therefore of the form

$$a(tx)^p (ty)^q = t^{p+q} ax^p y^q = t^n ax^p y^q.$$

Hence $t^n$ is a factor of $f(tx, ty)$, and the other factor is clearly $f(x, y)$.

The theorem generalises obviously for more than two variables. It can be used to extend the concept of homogeneity to functions other than polynomials.

*Definition.* If $f(tx, ty) = t^n f(x, y)$ for all values of $x$, $y$, $t$ for which the function is defined, then $f(x, y)$ is said to be *homogeneous* of degree $n$ in $(x, y)$.

For example, the functions

$$\frac{x^3 - y^3}{x^2 + xy - y^2}, \quad \frac{y^3}{x^5 + x^3y^2}, \quad \sqrt[3]{(x^2 - y^2 + 2z^2)}, \quad \tan\left(\frac{x}{y}\right)$$

are each homogeneous in their variables, with degrees $1$, $-2$, $\frac{2}{3}$, $0$, respectively.

In 10.22 we shall consider functions having the property of *symmetry* or of *skewness*.

## 1.53 Inadequacy of graphical representation

We may enquire whether a graphical representation of a given function is always possible.

(i) Consider first the function

$$y = \frac{x^2 - 9}{x - 3}.$$

Provided that $x \neq 3$, this can be simplified to give $y = x + 3$; but the latter is not the same as the given function because the first is not defined when $x = 3$ (it takes the meaningless form $0/0$), while the second is defined for all values of $x$, and in particular has the value 6 when $x = 3$. The graph of the given function would have no point corresponding to $x = 3$; but for all other values of $x$, however near to 3, it would be the same as that of $y = x + 3$. Thus the graph would be the line $y = x + 3$ *with the single point* (3, 6) *omitted*; and this situation cannot be represented adequately in a diagram.

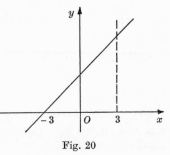

Fig. 20

(ii) A more complicated example is the function $y$ of $x$ defined by the rule:

$y = 0$   when   $x$ is rational,

$y = 1$   when   $x$ is irrational.

Its 'graph' would consist of an indefinitely closely packed row of points along the $x$-axis $y = 0$, and another such row along the line $y = 1$, neither row making up a complete 'continuous' line. No adequate diagram can be given, yet a formula can be obtained to give $y$ explicitly in terms of $x$: see Ex. 2 (c), no. 12.

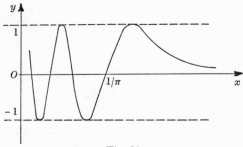

Fig. 21

(iii) Finally, consider the function $y = \sin(1/x)$, which is defined for all values of $x$ except $x = 0$. Since the sine of any angle always lies between $\pm 1$ inclusive, the graph lies between the lines $y = \pm 1$. It cuts $Ox$ at points for which $1/x = n\pi$, where $n$ is any integer (positive or negative), i.e. where $x = 1/n\pi$. Similarly, it meets the line $y = 1$ where $x = 1/(2n + \frac{1}{2})\pi$ and $y = -1$ where $x = 1/(2n - \frac{1}{2})\pi$. The curve oscillates between these lines, *and does so more and more rapidly as $x$ becomes closer to zero*. The curve does not cut $Ox$ for $x > 1/\pi$ or for $x < -1/\pi$; and when $x$ becomes large, $1/x$ and hence also $\sin(1/x)$ becomes small. Hence the $x$-axis is an asymptote. The deficiency in the graph (of which only the part for $x$ positive is shown in fig. 21— the rest is easily supplied since the function is odd) is that it cannot indicate clearly the behaviour of the function near $x = 0$.

These illustrations should convince the reader that, although graphical representation of a function is usually helpful, it has its limitations, and that any systematic study of the properties of functions cannot be based on graphical appearances only.

## Exercise 1($d$)

1 Classify the following functions as ($a$) odd, even, or neither; ($b$) periodic or not, and state the period if it exists.

(i) $\sin x$;      (ii) $\cos 2x$;      (iii) $x \sin x$;      (iv) $x + \sin x$;

(v) $x + \cos(1/x)$;    (vi) $\sin(x^2)$;      (vii) $\cos x + \tan^2 x$;    (viii) $|x|$;

(ix) $\sqrt{(1-x^2)}$;      (x) $x\sqrt{(1-x^2)}$;      (xi) $\sqrt{(1+x^3)}$;      (xii) $\tan^{-1} x$.

2 If an odd function $f(x)$ is defined at $x = 0$, show that it must be zero there.

3 If $f(x)$ is defined for all values of $x$, verify that $\phi(x) = f(x) + f(-x)$ is even and that $\psi(x) = f(x) - f(-x)$ is odd. Deduce that $f(x)$ can be expressed as the sum of an odd and an even function of $x$.

4 Verify that the product of two even or of two odd functions is even; but that of an even and an odd function is odd. State corresponding results for sums.

*5 Construct polynomial equations in $x$, $y$ which are satisfied by the following algebraic functions $y$.

(i) $x + \sqrt{x}$;    (ii) $\sqrt{(x + \sqrt{x})}$;    (iii) $\dfrac{\sqrt{(x+1)} - 1}{\sqrt{(x+1)} + 1}$;    (iv) $\dfrac{\sqrt{(x+1)} - \sqrt{x}}{\sqrt{(x+1)} + 2\sqrt{x}}$.

6 Pick out from the following functions those which are homogeneous in their variables, and state the degree in each such case.

(i) $x^5 - 3x^2y^3 + 5xy^4$;      (ii) $x^3 + 3x^2y + 3xy + y^3$;      (iii) $\sqrt{(x^2 + y^2 - z^2)}$;

(iv) $1/\sqrt{(3xyz)}$;      (v) $\dfrac{1}{x^2} + \dfrac{2}{y^2} + \dfrac{3}{z^2} + \dfrac{4}{t^2}$;      (vi) $\sin\left(\dfrac{y}{x}\right)$;

(vii) $\tan(xy)$.

7 (i) *Prove that every homogeneous function of degree $n$ in $(x, y)$ can be written in the form $x^n g(y/x)$.* [Take $t = 1/x$ in the definition in 1.52(4); write $g(y/x) = f(1, y/x)$.]

(ii) *Conversely, verify that every function of the form $x^n g(y/x)$ is homogeneous of degree $n$ in $(x, y)$.* [Replace $x$, $y$ by $tx$, $ty$ in the function.]

8 If $f(x, y)$, $g(x, y)$ are homogeneous of degree $m$, $n$ respectively, what can be said about: (i) $f(x, y) g(x, y)$; (ii) $f(x, y)/g(x, y)$; (iii) $f(x, y) \pm g(x, y)$?

*9 If $f(u, v)$ is homogeneous of degree $n$ in $(u, v)$, and each of $u$, $v$ is a homogeneous function of degree $m$ in $(x, y)$, prove that when $f(u, v)$ is expressed in terms of $x$ and $y$, it is homogeneous of degree $mn$ in $(x, y)$. [Let $f(u, v)$ become $\phi(x, y)$; then $\phi(tx, ty) = f(t^m u, t^m v) = (t^m)^n f(u, v) = t^{mn} \phi(x, y)$.]

10 Sketch the graph of the implicit function $y$ defined by $y^5 - xy + 1 = 0$ by first finding the inverse function $x$.

11 Consider the behaviour near $x = 0$ of the following functions, and explain why those in (ii), (iii) cannot be fully represented by a graph.

(i) $x \sin x$;      (ii) $x \sin(1/x)$;      (iii) $x^2 \sin(1/x)$.

## 1.6   Plane curves

### 1.61   Parametric equations

In 1.5 we have illustrated various properties which a function may possess by means of sketch-graphs. Given two graduated coordinate axes $Ox$, $Oy$, each such property of a function can be interpreted as a geometrical property of a curve in the plane $xOy$.

By *plane curve* we mean the set of points in the plane $xOy$ whose coordinates $(x, y)$ satisfy some equation $F(x, y) = 0$. This equation can be thought of as determining $y$ (implicitly) as a function $y = f(x)$ of $x$; or, if more convenient, $x$ as a function $x = g(y)$ of $y$.

It may be possible to discover functions $\phi(t)$, $\psi(t)$ of a third variable $t$ which are such that
$$x = \phi(t), \quad y = \psi(t) \tag{i}$$
will satisfy the equation $F(x, y) = 0$ for all values of $t$ (or, at any rate, for all $t$ in some range). Geometrically this means that the point whose coordinates are $\phi(t)$, $\psi(t)$ will lie on the curve for all relevant values of $t$. As $t$ varies, this point will trace out the curve (or part of the curve, since there may also be points of the curve not expressible in the form (i)). We therefore say that equations (i) are *parametric equations* of the curve $F(x, y) = 0$, and call $t$ the *parameter*. If each value of $t$ gives just one point on the curve, and if also *every* point of the curve can be obtained from (i) by giving $t$ just one suitable value, then equations (i) are called a *proper* parametric representation of the curve. If *rational* functions $\phi(t)$, $\psi(t)$ can be found, the curve is said to be *unicursal*.

The reader will already have used such a representation for some simple kinds of curve; e.g. $x = at^2$, $y = 2at$ are proper parametric equations of the parabola $y^2 = 4ax$ (see 16.12); and $x = a \cos t$, $y = a \sin t$ properly represent the circle $x^2 + y^2 = a^2$ if $0 \leqslant t < 2\pi$. Sometimes the parametric equations arise naturally.

### Example

*The cycloid*

The path of a point on the circumference of a circle which rolls without slipping along a fixed straight line is a plane curve called a *cycloid*.

Let the circle of centre $C$ and radius $a$ roll along $Ox$, and let the tracing point $P$ start from $O$, so that the circle begins by touching $Ox$ at $O$. Let the coordinates of $P$ be $(x, y)$ when the circle has turned through an angle $\theta$. Since no slipping occurs, arc $NP = ON$. After constructing $PK$ as shown, we have
$$x = OM = ON - PK = \text{arc } NP - PC \sin \theta = a\theta - a \sin \theta,$$
$$y = MP = NC - KC = a - a \cos \theta.$$

Parametric equations of the cycloid are therefore

$$x = a(\theta - \sin\theta), \quad y = a(1 - \cos\theta).$$

From the definition we see that the cycloid is composed of an infinity of identical arches all lying above $Ox$, having height $2a$ (the diameter of the

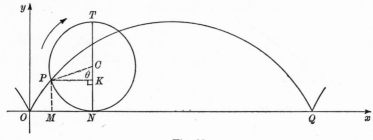

Fig. 22

generating circle) and base of length $2\pi a$ (the circumference of this circle). The parametric equations confirm all this. An $(x, y)$-equation for the cycloid could be obtained by eliminating $\theta$; but it is much less simple and less useful than the parametric representation.

Sometimes (e.g. in the dynamics of the cycloidal pendulum) we require the equations of the curve when placed base upwards (fig. 23). The reader may verify that they are

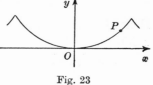

Fig. 23

$$x = a(\theta + \sin\theta), \quad y = a(1 - \cos\theta).$$

## 1.62 Polar coordinates

Although in elementary work the position of a point $P$ in a plane is specified by coordinates† $(x, y)$ referred to perpendicular axes $Ox$, $Oy$, this is not the only way available. If $O$ is a fixed point (the *pole*) and $OX$ is a fixed line (the *initial line*), we may join $OP$ and consider the angle $POX = \theta$ (measured positively in the counterclockwise sense from $OX$, as in trigonometry) and the distance $OP = r$ (fig. 24). Then to any given pair of numbers $(r, \theta)$ corresponds a unique point $P$ in the plane; but, unless we make some restrictions, a given point $P$ will be specified by many different pairs. For example, $(c, \alpha)$, $(c, \alpha + 2\pi)$, $(-c, \alpha + \pi)$ give the same point, where in the last case we locate the point by first turning the radius $OP$ through angle $\alpha + \pi$, and then measuring off a distance $c$ along it from $O$ *in the sense from $P$ towards $O$*.

If we restrict $r$ to be positive and $\theta$ to lie in the range $-\pi < \theta \leqslant \pi$,

† Called *cartesian coordinates* after Descartes who introduced their use in 1637.

then the numbers $r$, $\theta$ corresponding to a given point are unique.† In certain applications (e.g. to complex numbers, Ch. 13) it is desirable to have these limitations, but in geometry it is usually convenient to leave $r$ and $\theta$ unrestricted. In either case we call $(r, \theta)$ *polar coordinates* of $P$; $r$ is the *radius vector*, $\theta$ the *vectorial angle*.

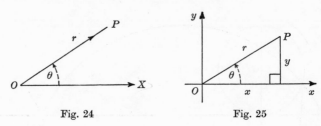

Fig. 24                    Fig. 25

*If the origin $O$ is taken for pole and $Ox$ for initial line* (and this is usually done, without comment), the relations between the cartesian and polar coordinates of the same point $P$ are seen (fig. 25) to be

$$x = r\cos\theta, \quad y = r\sin\theta. \tag{ii}$$

Hence if polar coordinates are given, these formulae determine $x$, $y$ uniquely.

Conversely, when $x$, $y$ are given, we have

$$r = \sqrt{(x^2+y^2)}, \quad \theta = \text{Tan}^{-1}\frac{y}{x}. \tag{iii}$$

This last formula will not determine $\theta$ uniquely, even if the restrictions are imposed, for it gives *two* distinct values of $\theta$ in the range

$$-\pi < \theta \leqslant \pi;$$

we should require the one for which

$$\cos\theta : \sin\theta : 1 = x : y : r. \tag{iv}$$

## 1.63 Polar equation of a curve

A relation $F(r, \theta) = 0$ between polar coordinates corresponds to some curve in the plane, and is called the *polar equation* of the curve. As in the case of cartesian equations, it can be thought of in the form $r = f(\theta)$ or $\theta = g(r)$ when convenient, or even parametrically as $r = \phi(t)$, $\theta = \psi(t)$.

† The range $0 \leqslant \theta < 2\pi$, or any range covering an interval of $2\pi$, would serve equally well for our purpose.

We now illustrate by a few examples how a curve can be sketched from its polar equation. First observe that

(a) if the equation $F(r, \theta) = 0$ is unaffected when $\theta$ is replaced by $-\theta$, the curve is symmetrical about $Ox$ (for if $(r_1, \theta_1)$ lies on the curve, so does $(r_1, -\theta_1)$);

(b) if $F(r, \theta) = 0$ is unaffected when $\theta$ is replaced by $\pi - \theta$, the curve is symmetrical about $Oy$;

(c) the locus $r = c$ is a circle with centre $O$ and radius $c$; $\theta = \alpha$ is a 'half-line' or *ray* through $O$ making angle $\alpha$ with $Ox$.

### Examples

(i) *The cardioid.* Given a fixed point $O$ on a fixed circle of diameter $a$, draw any chord $OP'$ and produce it to $P$ so that $P'P = a$. The locus of $P$ as $P'$ varies on the circle is called a *cardioid* (fig. 26). Since

$$r = OP = OP' + P'P = a\cos\theta + a,$$

the polar equation is            $r = a(1 + \cos\theta).$

The equation shows that
(a) the curve is symmetrical about $Ox$;
(b) as $\theta$ increases from 0 to $\frac{1}{2}\pi$, $r$ decreases from $2a$ to $a$;
(c) as $\theta$ increases from $\frac{1}{2}\pi$ to $\pi$, $r$ decreases from $a$ to 0.

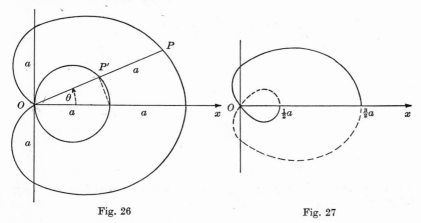

Fig. 26                    Fig. 27

More generally, if we take $P'P = c$ in the above definition, the locus of $P$ is a *limaçon*, whose polar equation is

$$r = a\cos\theta + c.$$

If $c > a$, then $r$ is always greater than 0. If $c < a$, $r$ can take negative values; we illustrate with the case $c = \frac{1}{2}a$.

(ii) *Sketch the limaçon* $r = \frac{1}{2}a(1 + 2\cos\theta).$
It is symmetrical about $Ox$. As $\theta$ increases from 0 through the values $\frac{1}{3}\pi$, $\frac{1}{2}\pi$, $\frac{2}{3}\pi$, to $\pi$, $r$ decreases from $\frac{3}{2}a$ through $a$, $\frac{1}{2}a$, 0 to $-\frac{1}{2}a$. To plot the last point, we

face along $Ox$ and rotate through angle $\pi$ counterclockwise, then mark off a distance $\frac{1}{2}a$ *in the opposite sense*, i.e. actually in the direction $Ox$. The curve is completed by using the symmetry about $Ox$ (fig. 27). It is seen to have a loop, with a 'double point' at $O$. The 'cusp' at $O$ of the cardioid (the case $c = a$) can be thought of as a loop which has shrunk to a point.

(iii) *The lemniscate* $(x^2 + y^2)^2 = a^2(x^2 - y^2)$.

The given equation is unchanged by writing $-x$ instead of $x$, or $-y$ instead of $y$. The curve is therefore symmetrical about $Ox$ and $Oy$. To sketch it, first transform to polar coordinates by putting $x = r\cos\theta$, $y = r\sin\theta$:

$$(r^2)^2 = a^2(r^2\cos^2\theta - r^2\sin^2\theta),$$

i.e.     $$r^2 = a^2\cos 2\theta.$$

The polar equation shows that

(*a*) as $\theta$ increases from 0 to $\frac{1}{4}\pi$, $r$ decreases from $a$ to 0;

(*b*) as $\theta$ increases from $\frac{1}{4}\pi$ to $\frac{3}{4}\pi$, $\cos 2\theta < 0$ and so no part of the curve lies in the region between the lines $\theta = \frac{1}{4}\pi$, $\theta = \frac{3}{4}\pi$;

(*c*) as $\theta$ increases from $\frac{3}{4}\pi$ to $\pi$, $r$ increases from 0 to $a$.

We use symmetry about $Ox$ to complete the curve (fig. 28).

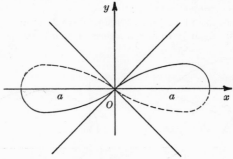

Fig. 28

## Exercise 1(*e*)

**1** When the graph of $y = f(x)$ is given, how can those of (i) $y = f(x) + c$; (ii) $y = cf(x)$; (iii) $y = f(x + c)$ be deduced?

**2** Verify that     $$x = a\frac{1 - t^2}{1 + t^2}, \quad y = \frac{2at}{1 + t^2}$$

are parametric equations of the circle $x^2 + y^2 = a^2$. Is *every* point of the circle given by these equations?

**3** Prove that the line $y = tx$ cuts the curve $x^3 + y^3 = axy$ at the origin and the point $\left(at/(1 + t^3), at^2/(1 + t^3)\right)$. Hence write down parametric equations for this curve.

**4** A circle of radius $a$ rolls along $Ox$ without slipping. Prove that parametric equations of the locus of the mid-point of a given radius are $x = a(\theta - \frac{1}{2}\sin\theta)$, $y = a(1 - \frac{1}{2}\cos\theta)$, and sketch the curve (a *trochoid*).

*Transform the following equations into polar coordinates.*

**5** $x^2 - y^2 = a^2$. **6** $xy = c^2$. **7** $(x^2 + y^2)^3 = a^2xy(x^2 - y^2)$.

**8** Transform the equation $r = a \sin 2\theta$ into cartesian coordinates.

*Sketch the curves given by the following polar equations. (Label the results and retain them for future use.) Negative values of r are allowed.*

**9** $r = a \cos \theta$. **10** $r = a\theta$. **11** $r\theta = a$. **12** $r^2 = a^2 \sin 2\theta$.

**13** $r = a \sin 2\theta$. **14** $r = a \cos 3\theta$. **15** $r \cos \theta = a \cos 2\theta$.

**16** $r = a \sec \theta + b$ $(a > 0, b > 0)$.

**17** Answer the questions in no. 1 for the polar graph $r = f(\theta)$.

**18** A circle of radius $a$ rolls without slipping on the outside of an equal fixed circle with centre $O$. Show that a point $P$ on the circumference of the rolling circle has coordinates $x = a(2 \cos \theta - \cos 2\theta)$, $y = a(2 \sin \theta - \sin 2\theta)$, the origin being $O$, $Ox$ the line through the original position $A$ on the fixed circle of the tracing point $P$, and the $\theta$ angle turned through by the the line of centres.

By taking $A$ for pole and $Ox$ as initial line, show that the locus has polar equation $r = 2a(1 - \cos \theta)$, and hence sketch the curve.

## Miscellaneous Exercise 1 ($f$)

**1** If $a_1 + a_2 + \ldots + a_n = s$, prove $1/a_1 + 1/a_2 + \ldots + 1/a_n \geqslant n^2/s$ (all $a$'s positive).

**2** If $s_n = 1 + \frac{1}{2} + \frac{1}{3} + \ldots + 1/n$, prove that for $n \geqslant 2$,

$$\frac{n}{n+1-s_{n+1}} < (n+1)^{1/n} < \frac{s_n}{n} + 1.$$

[Apply $A > G$ to $\frac{2}{1}, \frac{3}{2}, \frac{4}{3}, \ldots, (n+1)/n$ and to their reciprocals.]

**3** Prove

(i) $\frac{1}{2}(n+1) > \sqrt[n]{n!}$; (ii) $(n!)^3 < n^n \{\frac{1}{2}(n+1)\}^{2n}$;

(iii) $n! > (n+1)^{\frac{1}{2}(n-1)}$.

[Apply $A > G$ to

(i) $1, 2, \ldots, n$; (ii) $1^3, 2^3, \ldots, n^3$;

(iii) $1/(1.2), 1/(2.3), \ldots, 1/\{n(n+1)\}$.]

**4** If $x^3y = at^2 + bt^4$ and $t \geqslant x$, where $a, b, t, x, y$ are all positive, prove that $y \geqslant 2\sqrt{(ab)}$. [$x^3y/t^3 = a/t + bt \geqslant 2\sqrt{(ab)}$ by the theorem of the means.]

**5** Prove $ab \leqslant \{\frac{1}{2}(a+b)\}^2$ and deduce that if $a, b, c, d$ are all positive, then $abcd \leqslant \{\frac{1}{4}(a+b+c+d)\}^4$, with equality only when $a = b = c = d$. By giving $d$ a suitable value in terms of $a, b, c$, prove $abc \leqslant \{\frac{1}{3}(a+b+c)\}^3$.

**\*6** Prove

$$(a_1 b_1 c_1 + a_2 b_2 c_2 + \ldots + a_n b_n c_n)^2 < (a_1^2 + \ldots + a_n^2)(b_1^2 + \ldots + b_n^2)(c_1^2 + \ldots + c_n^2).$$

**7** Find necessary and sufficient conditions for $ax^2 + bx + c$ to be positive for all values of $x$. By expressing $x^4 - 8x^3 + 32x^2 - 64x + 48$ in the form

$$(x^2 + \lambda x + \mu)^2 - \rho,$$

show it is not negative for any value of $x$.

*Investigate the range of values and sketch the graph of*

8  $\dfrac{x^2 - 3x - 1}{x^2 - 6x + 8}$.

9  $\dfrac{7x^2 + 8x + 10}{x^2 + 8x - 2}$.

10  $\dfrac{3x^2 - 2x + 7}{x^2 - 2x + 3}$.

11  Prove that $(cx^2 + 3x - 4)/(c + 3x - 4x^2)$ can take all values if $1 \leqslant c \leqslant 7$.

12  Find the values of $x$ for which $(x^2 - x - 2)/(x^2 - 2) < 2$.

13  Find the ranges of $x$ for which $(4x^2 + 6x + 1)/(x^2 + 3x + 2) > 4$.

14  Writing $\qquad f(x) = \dfrac{x^2 + 2ax + b}{x^2 + 1} \qquad (a \neq 0),$

prove that there are two numbers $k$ such that $f(x) - k$ is of the form

$$\frac{(Ax + B)^2}{D(x^2 + 1)}.$$

If these numbers are $k_1$, $k_2$ $(k_1 < k_2)$, prove

$$f(x) - k_r = \frac{\{(1 - k_r) x + a\}^2}{(1 - k_r)(x^2 + 1)} \quad (r = 1, 2).$$

Prove also that $(1 - k_1)(1 - k_2) = -a^2$, and deduce that for all $x$, $k_1 \leqslant f(x) \leqslant k_2$.

Sketch the curve $y = f(x)$ when $a > 0$, indicating its position with respect to the lines $y = 1$, $y = k_1$, $y = k_2$.

15  Show that for all $k \neq 0$, the line $y = k$ meets the curve

$$y(x - 3)(x + 1) = ax + b$$

if $b$ lies between $-3a$ and $a$.

Sketch the curve when (i) $a = -3$, $b = 7$; (ii) $a = +3$, $b = 7$.

16  If $(ax^2 + bx + c)y + (a'x^2 + b'x + c') = 0$, find the condition for $x$ to be expressible as a rational function of $y$.

17  If $\qquad \dfrac{y - \beta}{\alpha - y} = K \dfrac{(x - b)^2}{(x - a)^2} \quad (\alpha < \beta),$

prove that (i) if $K > 0$, then $\alpha \leqslant y \leqslant \beta$; (ii) if $K < 0$, then $y \leqslant \alpha$ or $y \geqslant \beta$. Deduce that $y$ is a function of the form $(Ax^2 + Bx + C)/(px^2 + qx + r)$ which has turning-points $(a, \alpha)$, $(b, \beta)$. Taking $\alpha = 2$, $\beta = 5$, $a = -1$, $b = 2$, $K = 2$, construct the example in no. 10.

18  If $u = ax^2 + 2hxy + by^2 + 2gx + 2fy + c$, $a \neq 0$ and $ab - h^2 \neq 0$, show by two applications of 'completing the square' that

$$u = \frac{1}{a}(ax + hy + g)^2 + \frac{ab - h^2}{a}\left(y + \frac{af - gh}{ab - h^2}\right)^2 + \frac{\Delta}{ab - h^2},$$

where $\qquad \Delta = abc + 2fgh - af^2 - bg^2 - ch^2.$

Hence show that *sufficient* conditions for $u$ to be positive for all $x$, $y$ are $a > 0$, $ab - h^2 > 0$, $\Delta > 0$. (Given that $a$, $ab - h^2$ are non-zero, these conditions are also *necessary*.)

If $ab - h^2 = 0$, show that sufficient conditions are $a > 0$, $af - gh = 0$, $ac > g^2$.

19  If $a \neq 0$, $ab - h^2 \neq 0$, $\Delta = 0$ in no. 18, show that $u$ is of the form

$$\frac{1}{a}\{X^2 + (ab - h^2) Y^2\},$$

and deduce that $u$ can be resolved into linear factors if $\Delta = 0$ and $ab - h^2 < 0$.

Write down conditions sufficient for $u$ to be a perfect square.

**20** A circle of radius $a$ rolls without slipping on the outside of a fixed circle of radius $3a$ and centre $O$. Taking $O$ for origin and $Ox$ along the initial position of the line of centres, show that the coordinates of the point $P$ of the rolling circle which was initially on the fixed circle are

$$x = a(4\cos\theta - \cos 4\theta), \quad y = a(4\sin\theta - \sin 4\theta),$$

where $\theta$ is the angle turned through by the line of centres. Sketch the curve (an *epicycloid*).

**21** Answer the question in no. 20 when the small circle rolls *inside* the large one. (The curve generated is a *hypocycloid*.)

# 2

## LIMITS. CONTINUOUS FUNCTIONS

### 2.1 Limits: some examples from previous work

**2.11** Given a function $y = f(x)$, early work in Calculus is concerned with the following process.

(*a*) Let $\delta x$ denote any change (positive or negative) in the value of $x$, and let $\delta y$ be the corresponding change caused in $y$. This change $\delta y$ is calculated from the functional relationship $y = f(x)$, and will in general depend on both $\delta x$ and $x$.

(*b*) Write down and simplify the ratio $\delta y/\delta x$ (which in general also depends on $\delta x$ and $x$).

(*c*) Letting $\delta x$ approach zero, see whether the ratio $\delta y/\delta x$ approaches a definite value (in general a function of $x$). If this is the case, we call this value the *derivative* of $y$ with respect to $x$, and denote it by any one of the symbols

$$\frac{dy}{dx}, \quad Dy, \quad D_x y, \quad y', \quad f'(x), \quad \frac{d}{dx}f(x),$$

using whichever is most convenient for our purpose.

We express the property in (*c*) by saying that $\delta y/\delta x$ *tends to* $dy/dx$ when $\delta x$ tends to zero, and write

$$\frac{\delta y}{\delta x} \to \frac{dy}{dx} \quad \text{when} \quad \delta x \to 0. \tag{i}$$

We also call $dy/dx$ the *limit* of $\delta y/\delta x$ when $\delta x$ tends to zero, and write

$$\lim_{\delta x \to 0} \frac{\delta y}{\delta x} = \frac{dy}{dx}. \tag{ii}$$

The statements (i), (ii) are equivalent.

**2.12** A well-known limit is that of $\sin x/x$ when $x \to 0$ (angles here being in radian measure). If $\sin x$ and the other trigonometrical ratios are defined in the usual way by means of a right-angled triangle, and certain assumptions are made about the 'area' of a circular sector, it is proved in any book on elementary trigonometry or calculus that

$$\sin x < x < \tan x \quad \text{when} \quad 0 < x < \tfrac{1}{2}\pi. \tag{iii}$$

It follows that

$$1 < \frac{x}{\sin x} < \frac{1}{\cos x} \quad \text{when} \quad 0 < x < \tfrac{1}{2}\pi. \tag{iv}$$

When $x$ decreases towards zero (or, as we may say, approaches 0 through positive values—conveniently written $x \to 0+$), $\cos x$ increases towards 1, and so $1/\cos x$ decreases towards 1. Hence $x/\sin x$ also decreases towards 1, by (iv). This will also be true when $x$ approaches zero through negative values ($x \to 0-$), because $x/\sin x$ is an even function of $x$. We conclude that

$$\frac{x}{\sin x} \to 1 \quad \text{when} \quad x \to 0 \text{ in any manner;}$$

i.e. in the limit-notation,

$$\lim_{x \to 0} \frac{x}{\sin x} = 1. \tag{v}$$

Notice that in the above statements nothing is said about

(a) whether or not $x/\sin x$ actually takes the value 1 for some value of $x$; or

(b) whether or not $x/\sin x$ has a value when we put $x = 0$.

In fact for (b), $x/\sin x$ is *not defined* when $x = 0$, being of the form $0/0$. For (a), the inequality (iv) shows that $x/\sin x > 1$ however small and positive $x$ may be (and hence also for small negative $x$, since $x/\sin x$ is even): there is no value of $x$ for which $x/\sin x = 1$, and we say that the limit 1 is *unattained*. The statement (v) means that when $x$ becomes close to 0, $x/\sin x$ becomes close to 1, and that we can make $x/\sin x$ *as close as we please to* 1 *for all values of x sufficiently near to* 0.

**2.13** Another example is given by the function (1.53, (i))

$$y = \frac{x^2 - 9}{x - 3},$$

which is not defined when $x = 3$. For any other value of $x$, however close to 3, we have $y = x + 3$; and when $x \to 3$, we see that $y \to 6$. Yet there is no value of $y$ when $x = 3$, nor is there any value of $x$ which makes $y = 6$.

**2.14** On the other hand, there are many functions whose limit is also an actual value of the function. For example, if $y = 3x^2$, then when $x \to 2$ we see that $y \to 12$; and $y = 12$ when we put $x = 2$. Such functions (for which the limit is *attained*) form an important class which will be discussed in 2.6.

**2.15** A well-known geometrical example of a limit is the *tangent to a curve* at a point $P$, defined as the position to which a chord $PQ$ approaches when $Q$ approaches $P$ along the curve. If the curve has an 'angle' at $P$, as in fig. 29, then the chord $PQ$ approaches different positions according to whether $Q$ approaches $P$ from left or right, and in this case there is no 'tangent at $P$' in the ordinary sense.

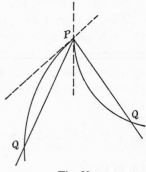

Fig. 29

## 2.2  The general idea of a limit

### 2.21  Informal definition

The idea pervading the preceding examples is expressible as follows.

If the values of $f(x)$ can be made as close as we please to the number $l$ by making $x$ sufficiently close to the number $a$, then we say that $f(x)$ *approaches* or *tends to* or *converges to* the limit $l$ when $x$ tends to $a$.

We write either
$$f(x) \to l \quad \text{when} \quad x \to a$$

or
$$\lim_{x \to a} f(x) = l,$$

but of course no sort of mixture of these two statements.

*Remarks*

($\alpha$) By the words '$x$ tends to $a$' we understand that $x$ must be allowed to approach $a$ from either side, i.e. through values less than $a$ ($x \to a-$) and also through values greater than $a$ ($x \to a+$). If $f(x)$ can be made arbitrarily close to $l$ only when $x \to a+$, or only when $x \to a-$, then $l$ would not be 'the limit' of $f(x)$ in the sense envisaged. (It was for this reason that we mentioned what happens when $x \to 0-$ in the case of $x/\sin x$ (2.12) before writing down the statement (v).)

($\beta$) We do not imply that $f(x)$ must approach steadily down to the limit $l$ (as does $x/\sin x$ towards 1) or steadily up to it (as for $\sin x/x$); when $x$ is close to $a$, $f(x)$ may take values some of which are greater and some less than $l$. For example, $x \sin(1/x) \to 0$ when $x \to 0$, although this function is positive for some values of $x$ near 0 and is negative for others (cf. Ex. 1 ($d$), no. 11 (ii)). The essential requirement is that the *difference* between $f(x)$ and $l$ can be made as (numerically) small as we please.

($\gamma$) Moreover, it is not enough that we can make this difference as small as we please for *some* (but not all) values of $x$ sufficiently near $a$. Thus from 1.53, (iii) the function $\sin(1/x)$ is zero for all values $x = 1/n\pi$ where $n$ is an integer (positive or negative, however large), and is as small as we please if we consider values of $x$ sufficiently close to these; but this function does not have 0 as limit when $x \to 0$ because it does not *remain* close to 0 for *all* small $x$. In fact, the function oscillates in value between $\pm 1$, and does not approach a limit when $x \to 0$.

($\delta$) As in 2.12, no mention is made of $f(a)$, because the function may not even be defined when $x = a$. Hence when we speak of 'all values of $x$ sufficiently close to $a$', we mean all $x$ in some small interval $a - \eta < x < a + \eta$ *excluding $a$ itself*; this is conveniently written

$$0 < |x - a| < \eta.$$

We call such values of $x$ a *neighbourhood* of $a$. Clearly a number $a$ has infinitely many neighbourhoods, one corresponding to each choice of $\eta$.

## 2.22 Formal definition

We can now express the requirements of our informal definition more precisely. Suppose we are given some positive number $\epsilon$ as our 'standard of closeness' of $f(x)$ to $l$; then we must be able to make

$$-\epsilon < f(x) - l < \epsilon$$

for *all* values of $x$ sufficiently close to $a$. That is, we must be able to give some positive number $\eta$ such that the above condition is satisfied for all $x$ for which $0 < |x - a| < \eta$. Our final definition of 'limit' is accordingly as follows.

$f(x)$ *tends to the limit $l$ when $x$ tends to $a$* if, given any positive number $\epsilon$ however small, we can find a corresponding positive number $\eta$ such that

$$|f(x) - l| < \epsilon \quad \text{for all $x$ for which} \quad 0 < |x - a| < \eta.$$

*Remark.* As a simple though useful deduction we have: if $f(x) = k$ for all $x$ in some neighbourhood of $a$, where $k$ is constant, then

$$\lim_{x \to a} f(x) = k.$$

## 2.3   Some general properties of limits

We now prove, directly from the definition just given, some properties of limits which would be expected on intuitive grounds. The results of 1.14 are required.

*If $f(x) \to \alpha$ and $g(x) \to \beta$ when $x \to a$, then*

(i) $kf(x) \to k\alpha$ *where $k$ is constant;*

(ii) $f(x) + g(x) \to \alpha + \beta$;

(iii) $f(x)\,g(x) \to \alpha\beta$;

(iv) $1/f(x) \to 1/\alpha$ *provided that $\alpha \neq 0$;*

(v) $f(x)/g(x) \to \alpha/\beta$ *provided that $\beta \neq 0$.*

*Proofs.* Our hypotheses are that

(*a*) to *any* number $\epsilon_1 > 0$, however small, corresponds a number $\eta_1$ such that

$$|f(x) - \alpha| < \epsilon_1 \quad \text{for all } x \text{ for which} \quad 0 < |x - a| < \eta_1;$$

(*b*) to *any* number $\epsilon_2 > 0$, however small, corresponds a number $\eta_2$ such that

$$|g(x) - \beta| < \epsilon_2 \quad \text{for all } x \text{ for which} \quad 0 < |x - a| < \eta_2.$$

In (i) we have to prove that, given any number $\epsilon > 0$, however small, we can find a number $\eta$ such that

$$\left|kf(x) - k\alpha\right| < \epsilon \quad \text{for all } x \text{ for which} \quad 0 < |x - a| < \eta.$$

Now
$$\left|kf(x) - k\alpha\right| = \left|k\big(f(x) - \alpha\big)\right| = |k| \cdot |f(x) - \alpha|$$
$$< |k|\,\epsilon_1 \quad \text{whenever} \quad 0 < |x - a| < \eta_1;$$

and we can choose $\epsilon_1$ in the first place so that $|k|\,\epsilon_1 < \epsilon$. Then

$$\left|kf(x) - k\alpha\right| < \epsilon \quad \text{whenever} \quad 0 < |x - a| < \eta_1.$$

Taking $\eta = \eta_1$, we have result (i).

For (ii), we have

$$\left|\big(f(x) + g(x)\big) - (\alpha + \beta)\right| = \left|\big(f(x) - \alpha\big) + \big(g(x) - \beta\big)\right|$$
$$\leqslant |f(x) - \alpha| + |g(x) - \beta|$$
$$< \epsilon_1 + \epsilon_2,$$

whenever $0 < |x - a| <$ the smaller of $\eta_1, \eta_2$. We can choose $\epsilon_1 = \epsilon_2 < \tfrac{1}{2}\epsilon$ and $\eta$ to be the smaller of $\eta_1, \eta_2$; then

$$|f(x) + g(x) - \alpha - \beta| < \epsilon \quad \text{whenever} \quad 0 < |x - a| < \eta,$$

and this proves (ii).

For (iii) we write

$$|f(x)\,g(x) - \alpha\beta| = \left|\big(f(x) - \alpha\big)\big(g(x) - \beta\big) + \alpha\big(g(x) - \beta\big) + \beta\big(f(x) - \alpha\big)\right|$$
$$\leqslant |f(x) - \alpha| \cdot |g(x) - \beta| + |\alpha| \cdot |g(x) - \beta| + |\beta| \cdot |f(x) - \alpha|$$
$$< \epsilon_1\epsilon_2 + |\alpha|\,\epsilon_2 + |\beta|\,\epsilon_1$$

whenever $0 < |x-a| < \eta$, where $\eta$ is the smaller of $\eta_1$, $\eta_2$. We can choose $\epsilon_1$ and $\epsilon_2$ so that $|\alpha|\,\epsilon_2 < \tfrac{1}{3}\epsilon$, $|\beta|\,\epsilon_1 < \tfrac{1}{3}\epsilon$ and also $\epsilon_1\epsilon_2 < \tfrac{1}{3}\epsilon$; then

$$|f(x)\,g(x) - \alpha\beta| < \epsilon \quad \text{whenever} \quad 0 < |x-a| < \eta,$$

and (iii) follows.

If $\alpha \neq 0$,

$$\left|\frac{1}{f(x)} - \frac{1}{\alpha}\right| = \left|\frac{\alpha - f(x)}{\alpha f(x)}\right| = \frac{|f(x) - \alpha|}{|\alpha|\,.\,|f(x)|}.$$

Now $\quad |f(x)| = |\alpha + (f(x) - \alpha)| \geqslant |\alpha| - |f(x) - \alpha| > |\alpha| - \epsilon_1$

when $0 < |x-a| < \eta_1$, and also $|f(x) - \alpha| < \epsilon_1$ under the same condition. Hence

$$\left|\frac{1}{f(x)} - \frac{1}{\alpha}\right| < \frac{\epsilon_1}{|\alpha|\,(|\alpha| - \epsilon_1)} \quad \text{when} \quad 0 < |x-a| < \eta_1.$$

We can choose $\epsilon_1 < \tfrac{1}{2}|\alpha|$, so that then $|\alpha| - \epsilon_1 > \tfrac{1}{2}|\alpha|$, and

$$\left|\frac{1}{f(x)} - \frac{1}{\alpha}\right| < \frac{\epsilon_1}{\tfrac{1}{2}|\alpha|^2}.$$

We can also choose $\epsilon_1 < \tfrac{1}{2}|\alpha|^2\epsilon$ if this is not already implied by the requirement $\epsilon_1 < \tfrac{1}{2}|\alpha|$ above; so that

$$\left|\frac{1}{f(x)} - \frac{1}{\alpha}\right| < \epsilon \quad \text{whenever} \quad 0 < |x-a| < \eta_1,$$

proving (iv).

Result (v) follows from (iii) and (iv) since $f(x) \to \alpha$ and $1/g(x) \to 1/\beta$ if $\beta \neq 0$.

## 2.4  Other ways in which a function can behave

(1) If $\alpha = 0$ in 2.3, (iv), then the hypothesis means that $f(x)$ is as small as we please for all $x$ sufficiently near $a$. Hence $1/f(x)$ will become numerically large when $x \to a$. For example, let $f(x) = x^2$; then when $x$ is small (positive or negative), $x^2$ is small and positive, and so $1/x^2$ is large and positive. In fact we can make $1/x^2$ as large as we please for all values of $x$ sufficiently close to $0$: given any positive number $K$ however big, we shall have $1/x^2 > K$ if $x^2 < 1/K$, i.e. for all $x$ such that $|x| < 1/\sqrt{K}$. We say that $1/x^2$ *tends to infinity* when $x$ tends to $0$, and write

$$\frac{1}{x^2} \to \infty \quad \text{when} \quad x \to 0.$$

Graphically, the curve $y = 1/x^2$ has the line $x = 0$ for an asymptote (1.41, Remark ($\alpha$)).

In the general case, the statement

$$f(x) \to \infty \quad \text{when} \quad x \to a$$

means that, given any positive number $K$ however large, a corresponding number $\eta$ can be found so that $f(x) > K$ for all $x$ for which $0 < |x-a| < \eta$. Graphically, the line $x = a$ is an asymptote of the curve $y = f(x)$.

Similarly,

$$f(x) \to -\infty \quad \text{when} \quad x \to a$$

means that, given any negative number $-K$ however large, $\eta$ can be found so that $f(x) < -K$ for all $x$ for which $0 < |x-a| < \eta$.

(2) The function $1/x$ is large and positive when $x$ is small and positive, and can be made as large as we please for all such $x$ sufficiently small. However, if $x$ is small and negative, the function is large and negative. All we can assert in this case is that

$$\frac{1}{x} \to \infty \quad \text{when} \quad x \to 0+ \quad \text{and} \quad \frac{1}{x} \to -\infty \quad \text{when} \quad x \to 0-.$$

In 1.53, (iii) we saw that $\sin(1/x)$ takes all values between $\pm 1$ in any interval however small which encloses the value $x = 0$. Since $1/x$ can be made as numerically large as we please for all $x$ sufficiently small, it follows that the function $(1/x)\sin(1/x)$ can take any value however large, whether positive or negative. However, this function does not *remain* large for *all* values of $x$ near 0; e.g. it is zero when $x = 1/n\pi$ ($n$ being an integer). The condition $(1/x)\sin(1/x) > K$ is not satisfied for *all* $x$ sufficiently near 0, so the function does not tend to infinity when $x \to 0$ (nor even when $x \to 0+$ or $x \to 0-$); similarly it does not tend to minus infinity.

We say that $\sin(1/x)$ *oscillates finitely* (between $+1$ and $-1$) when $x \to 0$, and that $(1/x)\sin(1/x)$ *oscillates infinitely*.

(3) We also say that $1/x^2$, $1/x$, $(1/x)\sin(1/x)$ are *unbounded* near $x = 0$ because they can take arbitrarily large values in this neighbourhood. The function $\sin(1/x)$ is said to be *bounded* near $x = 0$ because it lies between two fixed numbers (namely $+1$ and $-1$) for all small $x$ (and indeed for all $x$). Similarly, the terms 'bounded' or 'unbounded' can be applied to the behaviour of a function in the neighbourhood of any number $x = a$.

*Any function which has a limit when $x \to a$ is certainly bounded near $x = a$; this follows directly from the definition of 'limit'.*

If a function is bounded in the neighbourhood of every $x$ in some closed interval, it is said to be *bounded in this interval*. For example, $\sin(1/x)$ and the functions in 1.41, (i) and (iv), are bounded for all $x$; that in (ii) is bounded in every interval which excludes the number 2, but is unbounded near $x = 2$.

## 2.5 Limits when $x \to \infty$, $x \to -\infty$

If $x$ becomes large and positive, the function $1/x$ becomes small and positive; it can be made as small as we please for all $x$ sufficiently large. Similarly, $\sin x/x$ can be made as small (positive or negative) as we please for all $x$ sufficiently large. We say that $1/x$ and $\sin x/x$ tend to 0 when $x$ *tends to infinity*.

$f(x) \to l$ *when* $x \to \infty$ if, given any positive number $\epsilon$ however small, a positive number $N$ can be found so that $|f(x) - l| < \epsilon$ for all $x$ for which $x > N$. We write $\lim\limits_{x \to \infty} f(x) = l$.

Similar definitions can be given for '$f(x) \to l$ when $x \to -\infty$', '$f(x) \to \infty$ when $x \to \infty$', etc. The reader should formulate them for himself.

## 2.6 Continuity

### 2.61 Definition of 'continuous function'

The intuitive idea of a 'continuous curve' is that of a graph with no breaks ('missing points') in it, so that it could be drawn without raising the pencil from the paper. We should naturally say that the function represented by this curve is a 'continuous function'. Thus the curves and functions in 1.41, (i), (iv) and 1.53, (iii) are 'continuous'; but that in 1.41, (ii) is 'discontinuous' at $x = 2$, that in 1.41, (iii) at $x = +1, -1$, and that in 1.53, (i) has a different sort of 'discontinuity' at $x = 3$.

Taking these notions as a temporary basis, we observe first that 'continuity' is a property of *each point* of the curve rather than of a complete piece of it: the curve in fig. 31 is 'continuous' at most points, but not at $x = a_1$ or $x = a_2$; while that in fig. 30 is 'continuous' even at a point like $Q$ where the *direction* of the curve changes abruptly (cf. 2.15).

If fig. 30 is the graph of $y = f(x)$, we now list some properties of $f(x)$ which must be associated with the point $P$ at which $x = a$ and the curve is 'continuous'.

3

(*a*) $f(x)$ must be *defined at* $x = a$ (otherwise the curve would be broken there).

(*b*) $f(x)$ must be defined *in the neighbourhood of a* (same reason).

(*c*) When $x \to a-$ and when $x \to a+$, we must have $f(x) \to f(a)$.

Fig. 30 suggests other properties which we should expect a 'continuous' function to possess; one of these will be mentioned in 2.65, and others later (6.1) when they are needed, but (*a*)–(*c*) above are sufficient on which to base a precise definition of 'continuity at a point'.

*Definition.* $f(x)$ is *continuous* at $x = a$ if $\lim\limits_{x \to a} f(x) = f(a)$.

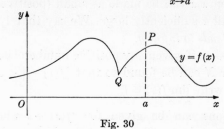

Fig. 30

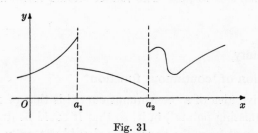

Fig. 31

We can paraphrase this by saying that 'limits coincide with values'. Using the formal definition of 'limit' given in 2.22, the definition above can be restated as follows.

$f(x)$ is continuous at $x = a$ if, given any positive number $\epsilon$ however small, there is a corresponding number $\eta$ such that

$$|f(x) - f(a)| < \epsilon \quad \text{for all } x \text{ for which} \quad |x - a| < \eta.$$

If we write $x = a + h$, we can present these inequalities in the form

$$|f(a + h) - f(a)| < \epsilon \quad \text{for all } h \text{ for which} \quad |h| < \eta.$$

Less precisely, by writing $y = f(x)$ and $h = \delta x$, the condition can be expressed as

$$\delta y \to 0 \quad when \quad \delta x \to 0, \quad \text{or} \quad \lim_{\delta x \to 0} \delta y = 0. \tag{i}$$

If $f(x)$ is continuous at each point for which $a < x < b$, we say that $f(x)$ is *continuous in the open interval* $a < x < b$ (1.12).

*Remark.* To define continuity in the *closed* interval $a \leqslant x \leqslant b$ in this way, a difficulty about the ends $x = a, x = b$ has to be surmounted, because the function may not be defined for $x < a$ or for $x > b$. In order to keep within the interval, the most we can require is that $f(x) \to f(a)$ when $x \to a+$ and that $f(x) \to f(b)$ when $x \to b-$.

## 2.62  Some properties of continuous functions

(1) *The sum, difference, product, and quotient of continuous functions is also a continuous function, provided that in the last case the denominator is non-zero at the point considered.*

The results follow immediately from the definition of 'continuity at a point' and the theorems on limits in 2.3.

(2) *A continuous function of another continuous function is itself continuous.*

We are given that

$$u = g(x) \to g(a) = b \quad \text{when} \quad x \to a,$$

and that $\qquad f(u) \to f(b) \qquad$ when $\quad u \to b$.

We have to prove that

$$f(g(x)) \to f(g(a)) \quad \text{when} \quad x \to a.$$

Given $\epsilon > 0$, there corresponds a number $\delta$ such that

$$|f(u) - f(b)| < \epsilon \quad \text{whenever} \quad |u - b| < \delta.$$

Taking $\delta$ as the '$\epsilon$' in the second hypothesis, there corresponds a number $\eta$ such that

$$|u - b| < \delta \quad \text{whenever} \quad |x - a| < \eta.$$

Hence $\qquad |f(g(x)) - f(g(a))| < \epsilon \quad$ whenever $\quad |x - a| < \eta,$

and the result follows.

## 2.63  Examples of some continuous functions

(1) $x^n$ *is a continuous function for all values of $x$ when $n$ is a positive integer; the same is true except at $x = 0$ when $n$ is a negative integer.*

The function $f(x) = x$ is certainly continuous everywhere. Hence by 2.62 (1) the product $x.x.x \ldots x$ ($n$ factors) is continuous everywhere.

Hence, again by 2.62 (1); the quotient $1/x^n$ is continuous everywhere except where the denominator is zero, viz. at $x = 0$.

It follows by repeated applications of 2.62 (1) that

(a) *all polynomials are continuous everywhere;*

(b) *all rational functions are continuous everywhere except at the points which make the denominator equal to zero.*

(2) $\sin x, \cos x$ *are continuous everywhere.*

For $\quad |\sin(x+h) - \sin x| \; = \; |2\cos(x+\tfrac{1}{2}h)\sin\tfrac{1}{2}h|$

$$\leqslant 2\,|\sin\tfrac{1}{2}h| \quad \text{since} \quad |\cos(x+\tfrac{1}{2}h)| \leqslant 1$$

$$< |h| \quad \text{since} \quad |\sin\tfrac{1}{2}h| < \tfrac{1}{2}|h|.$$

Hence, given $\epsilon$, we have

$$|\sin(x+h) - \sin x| < \epsilon \quad \text{whenever} \quad |h| < \epsilon.$$

Therefore $\sin x$ is continuous for any $x$. A similar proof holds for $\cos x$.

By 2.62 (1), it follows that $\tan x$ *is continuous* except where $\cos x = 0$, i.e. *except when* $x = (k + \tfrac{1}{2})\pi$ where $k$ is any integer (positive, negative, or zero).

## 2.64 Removable discontinuities

In certain cases $\lim\limits_{x \to a} f(x)$ may exist and be equal to $l$, while $f(a)$ is not defined. We can 'remove the discontinuity' at $x = a$ by 'completing the definition' of the function: we may *define* $f(a)$ to be $l$. The new function (defined by means of *two* formulae) is then continuous at $x = a$. For example, we can write

$$f(x) = \frac{x}{\sin x} \quad \text{when} \quad x \neq 0,$$

$$f(0) = 1;$$

then $f(x)$ is continuous at $x = 0$.

On the other hand, no such attempt will make the function $\tan x$ into a continuous function at $x = \tfrac{1}{2}\pi$; for when $x \to \tfrac{1}{2}\pi+$, $\tan x \to -\infty$, and when $x \to \tfrac{1}{2}\pi-$, $\tan x \to +\infty$. Thus $\tan x$ does not approach any limit when $x \to \tfrac{1}{2}\pi$. The same remark applies to $\sin(1/x)$ when $x \to 0$; see 1.53, (iii).

## 2.65 Another property of continuous functions

If $A, B$ are points corresponding to $x = a$, $x = b$ on a continuous curve $y = f(x)$, with $A$ below and $B$ above $Ox$, then geometrical intuition asserts that the curve must cut $Ox$ somewhere (possibly more than once) between $A$ and $B$. We give a formal statement.

*If $f(a)$ and $f(b)$ have opposite signs, then there is at least one number $\xi$ between $a$ and $b$ for which $f(\xi) = 0$.*

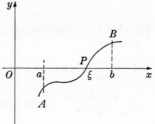

This result can in fact be deduced from the definition in 2.61 with the help of some deeper theorems on 'bounds' of a function. We shall assume it because a rigorous proof would deflect us too far from the course in view.

If we apply the result to the continuous function $g(x) = f(x) - c$, we deduce the

Fig. 32

COROLLARY. *If $f(a) < c < f(b)$, then there is at least one value $x = \xi$ between $a$ and $b$ for which $f(\xi) = c$.*

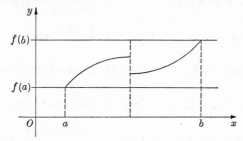

Fig. 33

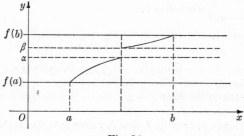

Fig. 34

This can be expressed also as follows: *when $x$ varies from $a$ to $b$, the continuous function $f(x)$ takes every value between $f(a)$ and $f(b)$ at least once.* The reader should illustrate by a sketch.

The converse result, that 'if $f(x)$ takes every value between $f(a)$ and $f(b)$ as $x$ varies from $a$ to $b$, then $f(x)$ is continuous,' is clearly false, as fig. 33 shows.

The result is in general false for a function which is discontinuous in $a \leqslant x \leqslant b$: in fig. 34, the function takes no values between $\alpha$, $\beta$ as $x$ varies from $a$ to $b$.

## Exercise 2(a)

*Find the following limits if they exist; if they do not, state the behaviour of the function.*

1  $\dfrac{x^2 - a^2}{x - a}$ when $x \to a$.

2  $\dfrac{x^2 - 4x + 3}{x^3 - x^2 + x - 1}$ when (i) $x \to 1$; (ii) $x \to \infty$; (iii) $x \to -\infty$; (iv) $x \to 0$.

3  $\dfrac{x^5 - a^5}{x - a}$ when $x \to a$.     4  $\dfrac{1}{x}\{\sqrt{(1 + x)} - \sqrt{(1 - x)}\}$ when $x \to 0$.

5  $\dfrac{\sin 3x}{x}$ when $x \to 0$.     6  $\dfrac{\sin 2x}{\sin 5x}$ when $x \to 0$.

7  $\dfrac{\tan x - \sin x}{\sin^3 x}$ when $x \to 0$.     8  $\dfrac{\cos^{-1} x}{\sqrt{(1 - x^2)}}$ when $x \to 1$.

9  $\dfrac{\tan x}{x}$ when $x \to 0$.     10  $\dfrac{1 - \cos x}{x}$ when $x \to 0$.

11  $\dfrac{1 - \cos x}{x^2}$ when $x \to 0$.     12  $x \sin x$ when (i) $x \to 0$; (ii) $x \to \infty$.

13  $x \sin \dfrac{1}{x}$ when (i) $x \to 0$; (ii) $x \to \infty$.   14  $\cos \dfrac{1}{x}$ when (i) $x \to 0$; (ii) $x \to \infty$.

15  $\dfrac{1}{x} \sin \dfrac{1}{x}$ when (i) $x \to \infty$; (ii) $x \to 0$.  16  $\dfrac{\sin^2 px - \sin^2 qx}{1 - \cos rx}$ when $x \to 0$.

17  $\dfrac{\sec(x + h) - \sec x}{\tan(x + h) - \tan x}$ when $h \to 0$.

18  $\dfrac{\sin(x + h) + \sin(x - h) - 2 \sin x}{h^2}$ when $h \to 0$.

19  If $f(x) = x \sin(1/x)$ $(x \neq 0)$, $f(0) = 0$, *prove that $f(x)$ is continuous at $x = 0$* (and therefore for all values of $x$ by 2.62 (1)). Verify that the graph of $f(x)$ lies in the angles between the lines $y = \pm x$ which contain the $x$-axis. Sketch the graph.

20  If $f(x) = x^2 \sin(1/x)$ $(x \neq 0)$, $f(0) = 0$, prove that $f(x)$ is continuous every-where, and sketch the graph.

## 2.7  Functions of $n$: some important limits

### 2.71  Sequences

Although we have considered functions of a 'continuous' variable $x$ which can range over all values for which the function is defined, we shall also need to consider functions $f(n)$ where $n$ takes only integral values. For example, $f(n)$ may denote 'the sum of the first $n$ positive integers' and is defined only when $n$ is a positive integer.

*Definition (a).* If $f(n)$ is a one-valued function of $n$ which is defined for all positive integers $n$, $f(n)$ is called a *sequence*, and its values

$$f(1), \quad f(2), \quad f(3), \quad \ldots$$

are called *terms* of the sequence. Even if $f(n)$ is defined only for all $n$ greater than some fixed positive integer $m$, it is still called a sequence.

Thus $n^2$, $a^n$, $\sin\{\pi/(n-2)\}$ are sequences, the latter being defined for all $n > 2$; but $\tan \frac{1}{2}n\pi$ is not a sequence because it is undefined for odd values of $n$ however large.

*Remark ($\alpha$).* A sequence is not completely specified when its first few terms are given. For example, although the simplest sequence beginning with

$$1, \quad 4, \quad 9, \quad 16$$

is $n^2$, yet

$$n^2 + (n-1)(n-2)(n-3)(n-4)\,\phi(n)$$

(where $\phi(n)$ is an arbitrary sequence defined for all $n \geqslant 1$) begins in the same way. Sometimes even the 'simplest' sequence is not obvious: thus

$$1, \quad 3, \quad 5, \quad 7$$

could be continued as $9, 11, 13, 15, \ldots$ or as $11, 13, 17, 19, \ldots$ according to whether it is interpreted as the 'arithmetical progression' $2n-1$, or as the sequence of odd prime numbers.

If $f(n)$ has a meaning when $n$ is not a positive integer, we may regard the terms of the sequence as the values of a function $f(x)$ when $x$ takes positive integral values; e.g. $\sin 2n\pi$ could be thought of as the value of $\sin 2\pi x$ when $x = n$. On the other hand, some functions are defined *only* for positive integral values of the variable; $x!$ is an example.

We shall need to know the behaviour of certain sequences when $n$ becomes large. Accordingly we give the general

*Definition (b).* The statements

$$\lim_{n \to \infty} f(n) = l,$$

$$f(n) \to l \quad \text{when} \quad n \to \infty$$

each mean that, given any positive number $\epsilon$ however small, there corresponds a positive number $N$ (which need not be an integer) such that

$$|f(n) - l| < \epsilon \quad \text{for } all \quad n > N.$$

The terms 'tends to infinity', 'tends to minus infinity', 'oscillate' are similarly defined for a sequence when $n \to \infty$. The reader should formulate precise definitions like those in 2.4.

*Remark ($\beta$).* If there is a function $f(x)$ corresponding to the sequence $f(n)$, the fact that $\lim\limits_{n \to \infty} f(n) = l$ does not imply that $f(x)$ approaches a limit. For example, $\lim\limits_{n \to \infty} \sin 2n\pi = 0$ because $\sin 2n\pi = 0$ for all integers $n$, but $\sin 2\pi x$ oscillates between $\pm 1$. However, the converse is true:

$$if \quad \lim_{x \to \infty} f(x) = l, \quad then \quad \lim_{n \to \infty} f(n) = l.$$

For if $|f(x) - l|$ can be made as small as we please for *all* numbers $x > N$, then it will certainly be so for *all integers > N*.

We now consider some important sequences which will be required later.

### 2.72 $a^n$

First suppose $0 < a < 1$. Then

$$a > a^2 > a^3 > \dots > a^n,$$

and hence $\qquad na^n < a + a^2 + \dots + a^n = \dfrac{a - a^{n+1}}{1 - a} < \dfrac{a}{1 - a}$

by summing the geometrical progression. Therefore

$$a^n < \frac{a}{n(1-a)},$$

and the right-hand side can be made as small as we please for all $n$ sufficiently large. Consequently $a^n \to 0$ when $n \to \infty$.

If $-1 < a < 0$, then we have $0 < |a| < 1$ and the above argument shows that $|a|^n \to 0$, and therefore $a^n \to 0$, when $n \to \infty$. If $a = 0$, then $a^n = 0$ for all $n$, and the limit is still zero. Hence

$$\textbf{if} \quad -1 < a < 1, \quad \lim_{n \to \infty} a^n = 0.$$

If $a > 1$, then $a = 1 + b$ where $b > 0$. By Bernoulli's inequality (1.21, ex. (ii)),

$$a^n = (1+b)^n > 1 + nb.$$

Since $nb$ can be made as large as we please for all $n$ sufficiently large, so can $a^n$. Therefore

$$\textbf{if} \quad a > 1, \quad a^n \to \infty \quad \textbf{when} \quad n \to \infty.$$

If $a < -1$, the preceding shows that $a^n$ is large and positive for even values of $n$, and is large and negative for odd values, when $n \to \infty$. Hence $a^n$ *oscillates infinitely* when $a < -1$.

If $a = 1$, then $a^n = 1$ for all $n$, and so the limit when $n \to \infty$ is 1.

If $a = -1$, then $a^n$ is $+1$ for even $n$ and $-1$ for odd $n$. Thus $a^n$ *oscillates finitely* (between $+1, -1$) when $n \to \infty$.

## 2.73 $a^n/n$

If $-1 < a < 1$, then $a^n \to 0$ by 2.72, and hence certainly $a^n/n \to 0$ when $n \to \infty$. If $a = \pm 1$, then $a^n = (\pm 1)/n$, and this also tends to zero when $n \to \infty$. Therefore

$$\text{if} \quad -1 \leqslant a \leqslant 1, \quad \text{then} \quad \lim_{n \to \infty} \frac{a^n}{n} = 0.$$

If $a > 1$, then $a = 1 + b$ where $b > 0$. Writing $f(n) = a^n/n$,

$$\frac{f(n+1)}{f(n)} = \frac{n}{n+1} a = \frac{n}{n+1}(1+b)$$

$$> 1 + \tfrac{1}{2}b, \quad \text{say,}$$

if $\qquad n(1+b) > (n+1)(1+\tfrac{1}{2}b), \quad \text{i.e.} \quad \tfrac{1}{2}nb > 1 + \tfrac{1}{2}b,$

i.e. $\qquad\qquad n > \dfrac{2+b}{b} = N, \quad \text{say.}$

Hence for $n > N$, we have

$$\frac{a^{n+1}}{n+1} > \frac{a^n}{n}(1+\tfrac{1}{2}b),$$

and so

$$\frac{a^n}{n} > \frac{a^{n-1}}{n-1}(1+\tfrac{1}{2}b) > \frac{a^{n-2}}{n-2}(1+\tfrac{1}{2}b)^2 > \dots > \frac{a^N}{N}(1+\tfrac{1}{2}b)^{n-N}.$$

When $n \to \infty$, $(1+\tfrac{1}{2}b)^n \to \infty$ by 2.72; hence†

$$\text{if} \quad a > 1, \quad \frac{a^n}{n} \to \infty \quad \text{when} \quad n \to \infty.$$

If $a < -1$, $a^n/n$ oscillates infinitely because it takes positive values for $n$ even, negative values for $n$ odd, and by the preceding case these can be made as numerically large as we please.

---

† An alternative proof is given in the last part of ex. (vi), 12.52.

**2.74  $a^n/n!$**

First suppose $a > 0$, and choose $N > 2a$; then for $n > N$,

$$\frac{a^n}{n!} = \frac{a}{1} \cdot \frac{a}{2} \cdots \frac{a}{N-1} \cdot \frac{a}{N} \cdots \frac{a}{n}$$

$$< \frac{a}{1} \cdot \frac{a}{2} \cdots \frac{a}{N-1} \cdot \tfrac{1}{2} \cdots \tfrac{1}{2}$$

$$= \left( \frac{a^{N-1}}{(N-1)!} \right) (\tfrac{1}{2})^{n-N+1}.$$

Since $(\tfrac{1}{2})^n \to 0$ when $n \to \infty$, this expression also tends to zero.

If $a < 0$, then $|a^n/n!| = |a|^n/n! \to 0$ when $n \to \infty$ by the preceding case. If $a = 0$, then $a^n/n! = 0$ for all $n$. Hence

$$\textbf{for all values of } \textbf{\textit{a}}, \quad \lim_{n \to \infty} \frac{a^n}{n!} = \mathbf{0}.$$

**2.75  $m(m-1) \dots (m-n+1) \, a^n/n!$, $m$ constant**

If $m$ is a positive integer, then the function is zero for all $n \geqslant m+1$. Ignoring this simple case, write

$$f(n) = \frac{m(m-1) \dots (m-n+1)}{n!} a^n.$$

If $a \neq 0$,      $$\frac{f(n+1)}{f(n)} = \frac{m-n}{n+1} a = -\left( 1 - \frac{m+1}{n+1} \right) a,$$

and we can make the *modulus* of this as close to $|a|$ as we please for all $n$ sufficiently large.

If $|a| < 1$, then clearly $|a| < \tfrac{1}{2}(1+|a|) < 1$, and we can make $|f(n+1)/f(n)|$ less than $\tfrac{1}{2}(1+|a|)$ for all $n > N$, say. Writing

$$k = \tfrac{1}{2}(1+|a|),$$

then for $n > N$ we have

$$|f(n)| < k\,|f(n-1)| < k^2\,|f(n-2)| < \dots < k^{n-N}\,|f(N)|.$$

Since $0 < k < 1$, then $k^n \to 0$ when $n \to \infty$ and so also $|f(n)| \to 0$. Hence

$$\textbf{if } -1 < a < 1, \quad \lim_{n \to \infty} \frac{m(m-1) \dots (m-n+1)}{n!} a^n = \mathbf{0}.$$

We shall not need to consider values of $a$ outside the range $-1 < a < 1$, but a similar argument would show that if $|a| > 1$, then $|f(n)| \to \infty$ when $n \to \infty$.

## 2.76 Further examples

(i) *Prove that for any $a > 0$,* $\lim\limits_{n \to \infty} \sqrt[n]{a} = 1$.

If $a > 1$, then $\sqrt[n]{a} > 1$ (1.21, Corollary II $(c)$); let $\sqrt[n]{a} = 1 + b$, so that $b > 0$. Then

$$a = (1+b)^n > 1 + nb > nb$$

by Bernoulli's inequality (1.21, ex. (ii)). Hence $b < a/n$, and so

$$0 < \sqrt[n]{a} - 1 < \frac{a}{n}.$$

Since $a/n$ can be made as small as we please for all $n$ sufficiently large, hence $\sqrt[n]{a} - 1 \to 0$ when $n \to \infty$.

If $0 < a < 1$, then $1/a > 1$, and so by the preceding case with $1/a$ instead of $a$ we have $1/\sqrt[n]{a} \to 1$ when $n \to \infty$, i.e. $\sqrt[n]{a} \to 1$.

If $a = 1$, then $\sqrt[n]{a} = 1$ for all $n$.

The function is not defined for all $n$ if $a < 0$, and is zero for all $n$ if $a = 0$. The required result has thus been established.

(ii) *If $-1 < a < 1$, prove that* $\lim\limits_{n \to \infty} n^r a^n = 0$ *for any fixed integer $r$.*

If $r$ is negative, the result is evident from 2.72; the particular case $r = -1$ appeared in 2.73. We may assume $0 < a < 1$ in the following (cf. 2.72).

If $r$ is a positive integer, let $b$ denote the positive $(r+1)$th root of $a$, so that $0 < b < 1$ and $b^{r+1} = a$. Then, as in 2.72,

$$nb^n < 1 + b + b^2 + \ldots + b^{n-1} = \frac{1-b^n}{1-b} < \frac{1}{1-b},$$

i.e.

$$na^{n/(r+1)} < \frac{1}{1 - a^{1/(r+1)}}.$$

Hence

$$n^{r+1}a^n < \frac{1}{(1 - a^{1/(r+1)})^{r+1}},$$

and so

$$n^r a^n < \frac{1}{n(1 - a^{1/(r+1)})^{r+1}};$$

and this last expression tends to zero when $n \to \infty$.

## 2.77 Monotonic functions

If $f(n+1) > f(n)$ for all values of $n$, then $f(n)$ is said to be a *steadily increasing function* or to be *monotonic increasing*. If the property holds only for all $n$ beyond a certain integer $N$, $f(n)$ is monotonic increasing for $n > N$. If $\geqslant$ replaces $>$, $f(n)$ is *non-decreasing*.

There are two ways in which such a function can behave when $n$ increases:

(i)  it may increase indefinitely with $n$, i.e. $f(n) \to \infty$ when $n \to \infty$; or

(ii) it may remain less than some fixed number $k$, i.e. $f(n) < k$ for all $n$ (so that the function is *bounded*, 2.4(3)).

In case (ii) it can be shown that $f(n)$ *approaches some limit $l$ when* $n \to \infty$, *where* $l \leqslant k$. Although a strict proof of this assertion cannot be given in this book, the reader may convince himself of its truth by the following geometrical illustration.

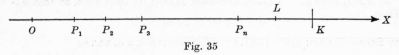

Fig. 35

Let the values $f(1)$, $f(2)$, ..., $f(n)$, ... be represented by lengths $OP_1, OP_2, ..., OP_n, ...$ measured off along a fixed line $OX$. Since the points $P$ move to the right (if $n > N$), then either (i) they will eventually pass beyond any point whatever which one cares to select on the line; or (ii) there will be some point $K$ of the line which they can never pass. Clearly any point to the right of $K$ will have the same property; but a point to the left of $K$ may or may not. The assertion is that there is some point $L$ of the line, either to the left of $K$ or coinciding with $K$, to which the $P$'s approach indefinitely closely, i.e. such that the length $P_n L$ is as small as we please, for all sufficiently large values of $n$.

It should be noticed that although $f(n) < k$ for all $n > N$, we may yet have $\lim\limits_{n \to \infty} f(n) = k$. For example, $1 - 1/n < 1$ for all $n$, but $1 - 1/n \to 1$ when $n \to \infty$. Hence we must allow $l \leqslant k$ and not merely $l < k$ in general.

Similarly, if $f(n+1) < f(n)$ for all $n$ (or perhaps only for all $n > N$), then $f(n)$ is said to be *monotonic decreasing* (perhaps for $n > N$). When $n \to \infty$, *either* $f(n) \to -\infty$; *or* $f(n) > k'$ for some constant $k'$, and in this event $f(n) \to l'$, where $l' \geqslant k'$.

The results stated above are important because they enable us to decide whether a monotonic function has a limit *without our first needing to know what that limit must be.*† If the definition (*b*) in 2.71 were to be used to show directly that $\lim\limits_{n \to \infty} f(n) = l$, we must know $l$ at the start; and this is not always the case (see ex. (ii) below). Rather, we can define a number by the property of being the limit of a bounded monotonic function, as in ex. (ii) and 4.43 (8).

**Examples**

(i) *Discuss* $\lim\limits_{n \to \infty} a^n$ *when* $a > 0$. (Cf. 2.72.)

Taking $f(n) = a^n$, then $f(n+1) = af(n)$.

---

† The first will be basic in convergence tests: see 12.32 (5).

If $a > 1$, then $f(n)$ is steadily increasing for all $n$, and so it either tends to infinity or to some limit $l$. If $f(n) \to l$, then $l > 1$ (since $f(1) = a > 1$ and the function is increasing), and

$$l = \lim f(n+1) = \lim \{af(n)\} = a \lim f(n) = al.$$

The relation $l = al$ is impossible because $a > 1$ and $l > 1$. Hence *when $a > 1$, $a^n \to \infty$ when $n \to \infty$.*

If $0 < a < 1$, then $f(n)$ is steadily decreasing for all $n$, and so either $f(n) \to -\infty$ or $f(n) \to l'$, where as before we find $l' = al'$, so $l' = 0$. Thus *when $0 < a < 1$, $a^n \to 0$ when $n \to \infty$.*

\*(ii) *Prove that $(1 + 1/n)^n$ has a limit when $n \to \infty$. Calling this limit $e$, show that $2\frac{1}{4} < e \leqslant 3$.*

(This example may be postponed until the binomial theorem (12.1) has been revised.)

We first prove that, for all positive integers $n$,

$$\left(1 + \frac{1}{n+1}\right)^{n+1} > \left(1 + \frac{1}{n}\right)^n.$$

The $(r+1)$th term in the expansion of $(1 + 1/n)^n$ is (for $r = 1, 2, ..., n$)

$$\frac{n(n-1)\dots(n-r+1)}{r!} \cdot \frac{1}{n^r} = \frac{1}{r!}\left(1 - \frac{1}{n}\right)\left(1 - \frac{2}{n}\right)\dots\left(1 - \frac{r-1}{n}\right), \qquad (a)$$

which is positive because each factor is positive. Similarly, the $(r+1)$th term in the expansion of $\left(1 + 1/(n+1)\right)^{n+1}$ is (for $r = 1, 2, ..., n+1$)

$$\frac{1}{r!}\left(1 - \frac{1}{n+1}\right)\left(1 - \frac{2}{n+1}\right)\dots\left(1 - \frac{r-1}{n+1}\right) \qquad (b)$$

which is positive. Each factor in $(b)$ which involves $n$ is greater than the corresponding factor in $(a)$.

Hence, apart from the first terms (which are both 1), the expansion of $\left(1 + 1/(n+1)\right)^{n+1}$ is greater term-by-term than the expansion of $(1 + 1/n)^n$; and the $(n+2)$th term of the first, to which there is no corresponding term of the second, is positive. Hence the required inequality follows; it shows that $(1 + 1/n)^n$ *is a monotonic increasing function of $n$.*

When $n = 2$, the function has the value $2\frac{1}{4}$; so for all $n > 2$, we have $2\frac{1}{4} < (1 + 1/n)^n$.

The expression $(a)$ shows that the $(r+1)$th term in $(1 + 1/n)^n$ is less than $1/r!$, and hence

$$\left(1 + \frac{1}{n}\right)^n < 1 + \frac{1}{1!} + \frac{1}{2!} + \frac{1}{3!} + \dots + \frac{1}{n!}.$$

Since $n! = 1.2.3\dots n > 1.2.2\dots 2 = 2^{n-1}$ if $n > 2$, therefore

$$\left(1 + \frac{1}{n}\right)^n < 1 + 1 + \frac{1}{2} + \frac{1}{2^2} + \dots + \frac{1}{2^{n-1}} \quad \text{if} \quad n > 2$$

$$= 1 + \{1 - (\tfrac{1}{2})^n\}/(1 - \tfrac{1}{2}) \quad \text{on summing the G.P.,}$$

$$= 3 - \frac{1}{2^{n-1}}$$

$$< 3.$$

We have now shown that $(1+1/n)^n$ is an increasing function which remains less than 3 for all $n$. Hence $(1+1/n)^n$ tends to a limit, say $e$, when $n \to \infty$, and $2\frac{1}{4} < e \le 3$.

The number $e$, which we shall meet by a different approach in Ch. 4, is very important in mathematics and its applications.

### Exercise 2(b)

*State the behaviour of the following functions when $n \to \infty$.*

**1** $n+(-1)^n$.

**2** $\{n+(-1)^n\}/n$.

**3** $1+(-1)^n$.

**4** $n\{1+(-1)^n\}$.

**5** $n\{2+(-1)^n\}$.

**6** $n\{1+(-2)^n\}$.

**7** $n^2+(-1)^n n$.

**8** $\sin \frac{1}{2}n\pi$.

**9** $\dfrac{\sin \frac{1}{2}n\pi}{n}$.

**10** $\dfrac{\sin n}{n}$.

**11** $\dfrac{\sin n}{\sqrt{n}}$.

**12** $\dfrac{1}{n}+\cos \frac{1}{2}n\pi$.

*Calculate the limits of the following functions when $n \to \infty$.*

**13** $\dfrac{1+2+3+\ldots+n}{n^2}$.

**14** $\dfrac{1+3+5+\ldots+(2n+1)}{n^2}$.

**15** $\dfrac{1}{n}(n^2-1)\sin \dfrac{\pi}{n}$.

**16** $n^2\left\{\cos\left(\dfrac{\theta}{2n}\right)-\cos\left(\dfrac{\theta}{n}\right)\right\}$, where $\theta$ is constant.

**17** $\dfrac{3^n+(-2)^n}{3^n+2^n}$.

**18** $\sqrt[n]{(a^n+b^n)}$, where $a > 0$, $b > 0$.

**\*19** $\dfrac{(n+1)^n}{n^{n+1}}$.

**20** Discuss the behaviour of $n^r a^n$ where $r$ is a fixed positive integer and $|a| > 1$.

**21** Prove that $\sqrt[n]{(n!)} \to \infty$ when $n \to \infty$. [By 2.74, $a^n/n! \to 0$ for any $a$, however large; so $n! > a^n$ for all $n$ sufficiently large. *Alternatively*, use Ex. 1 $(a)$, no. 4.]

**\*22** (i) If $f(n)$ is monotonic increasing or decreasing, prove that

$$g(n) = \frac{f(1)+f(2)+\ldots+f(n)}{n}$$

has the same property.

(ii) If $f(n)$ increases and has limit $l$, prove that $g(n)$ tends to a limit $\le l$. What can be said if $f(n)$ *decreases* and has limit $l$?

### Miscellaneous Exercise 2(c)

**1** Prove $\quad \lim\limits_{x \to 1} \dfrac{x-(n+1)x^{n+1}+nx^{n+2}}{(1-x)^2} = \frac{1}{2}n(n+1)$.

[Use the result $1+x+x^2+\ldots+x^{p-1} = (1-x^p)/(1-x)$, $x \ne 1$.]

**2** If $n > 1$, then $\sqrt[n]{n} = 1+a$ where $a > 0$. Use the binomial theorem to prove that $(1+a)^n > \frac{1}{2}n(n-1)a^2$, and hence show that

$$0 < a < \sqrt{\frac{2}{n-1}}.$$

Deduce that $\lim\limits_{n\to\infty} \sqrt[n]{n} = 1$, and state the value of $\lim\limits_{n\to\infty} \sqrt[n]{(n^r)}$ where $r$ is a fixed integer.

*For convenience we write $f(n) = u_n$ in nos. 3–7.*

**\*3** The values of $u_n$ are defined for $n = 1, 2, 3, \dots$ in succession by the *recurrence formula* $u_{n+1} = \frac{1}{2}(u_n + a/u_n)$. Assuming $a > 0$, prove that
   (i) if $0 < u_n < \sqrt{a}$, then $u_{n+1} > \sqrt{a}$;
   (ii) if $u_n > \sqrt{a}$, then $u_{n+1} > \sqrt{a}$ and also $u_{n+1} < u_n$.
Deduce that if $u_1 > 0$, then
   (iii) $u_n > \sqrt{a}$ for all $n \geqslant 2$, and $u_n$ is monotonic decreasing;
   (iv) $u_n \to \sqrt{a}$ when $n \to \infty$.
What is the conclusion when $u_1 < 0$?

**\*4** Solve no. 3 by showing that

$$\frac{u_{n+1} - \sqrt{a}}{u_{n+1} + \sqrt{a}} = \left(\frac{u_1 - \sqrt{a}}{u_1 + \sqrt{a}}\right)^{2^n}.$$

**\*5** It is given that $4u_{n+2} - 5u_{n+1} + u_n = 0$ for all $n \geqslant 1$, and that $u_1 = 7$, $u_2 = 4$. By writing the relation in the form

$$u_{n+2} - u_{n+1} = \tfrac{1}{4}(u_{n+1} - u_n),$$

prove that $u_{n+1} - u_n = -3(\frac{1}{4})^{n-1}$. Hence express $u_n$ as a function of $n$, and prove $\lim\limits_{n\to\infty} u_n = 3$.

**6** If $u_n > 0$ and $\lim\limits_{n\to\infty}(u_{n+1}/u_n) > 1$, prove $u_n \to \infty$ when $n \to \infty$. [Method of 2.73.]

**7** If $\lim\limits_{n\to\infty}(u_{n+1}/u_n) = l$ where $-1 < l < 1$, prove $\lim\limits_{n\to\infty} u_n = 0$. [Method of 2.75.]

*Obtain results from nos. 6, 7 by taking $u_n$ to be the following functions.*

**8** $\dfrac{a^n}{n!}$ ($a$ constant).     **9** $\dfrac{m(m-1)\dots(m-n+1)}{n!} a^n$ ($a$, $m$ constant).

**10** $n^r a^n$ ($a$, $r$ constant).

**\*11** Find the behaviour of $u = (x^2 + 2y)/(y^2 - 2x)$ when $x$ and $y$ both tend to 0 (i) along the line $y = mx$; (ii) along the curve $x^3 + y^3 = xy$. [Use the parametric equations $x = t/(1+t^3)$, $y = t^2/(1+t^3)$ (Ex. 1 (e), no. 3) to express $u$ in terms of $t$, and observe that both $x$ and $y$ tend to 0 when $t \to 0$, $+\infty$, or $-\infty$.]

**\*12** If $\phi_m(x) = \lim\limits_{n\to\infty} \{\cos(m!\,\pi x)\}^{2n}$, prove that $\phi_m(x)$ is always 0 if $x$ is irrational, and is 1 if $x$ is rational provided that $m$ is sufficiently large. Hence show that $\lim\limits_{m\to\infty} \phi_m(x)$ is 0 if $x$ is irrational and 1 if $x$ is rational. Write down a function $y$ which is 0 when $x$ is rational and 1 when $x$ is irrational.

# 3

# THE DERIVATIVE. SOME APPLICATIONS

## 3.1 The derivative of a function of one variable

### 3.11 Definitions

Let $f(x)$ be a one-valued function defined at and near $x = a$. Then in the process described in 2.11 we can put $\delta x = h$, and express the corresponding change $\delta y$ caused by changing $x$ from $a$ to $a+h$ as

$$\delta y = f(a+h)-f(a),$$

so that
$$\frac{\delta y}{\delta x} = \frac{f(a+h)-f(a)}{h}. \tag{i}$$

The definition given in 2.11 can now be formulated more precisely as follows.

If the ratio (i) tends to a limit when $h \to 0$ in any manner, this limit is called the *derivative* of $f(x)$ at $x = a$, and is written $f'(a)$. The function $f(x)$ is said to be *derivable* at $x = a$.

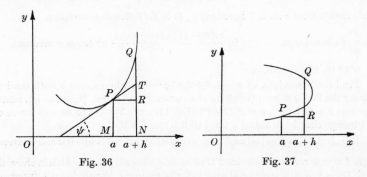

Fig. 36        Fig. 37

Writing $x = a+h$ in (i), so that '$h \to 0$' is equivalent to '$x \to a$', we obtain another form of the definition:

$$\lim_{x \to a} \frac{f(x)-f(a)}{x-a} = f'(a). \tag{ii}$$

The reader will know that the derivative $f'(a)$ is interpreted geometrically as the gradient of the tangent to the curve $y = f(x)$ at the point $\big(a, f(a)\big)$: see fig. 36.

If the function $f(x)$ were not one-valued, the difference $f(a+h)-f(a)$ may approach a limit other than zero when $h \to 0$ (fig. 37), so that the ratio (i) would not tend to a limit. To deal with a many-valued function we must consider its branches separately.

If $f(x)$ has a derivative for all $x$ between $a$ and $b$, we say that $f(x)$ is *derivable in the (open) interval* $a < x < b$. We also say that $f'(x)$ *exists* for $a < x < b$, and call $f'(x)$ the *derived function* of $f(x)$. Other notations have been mentioned in 2.11, the chief ones being $dy/dx$ and $y'$.

*Remark.* If $f(x)$ is defined only for $a \leqslant x \leqslant b$, there is no derivative at $x = a$ or at $x = b$ because $h$ cannot tend to zero *in any manner* without $a+h$ or $b+h$ passing outside the range of definition of $f(x)$.

Given $f(x)$, the process of calculating $f'(x)$ is called *derivation* of $f(x)$ *with respect to x* (abbreviated to 'wo $x$'). The details of this process depend on the function itself.

## Examples

(i) *If* $f(x) = c$, *where c is constant, then* $f'(x) = 0$. For
$$\lim_{h\to0}\frac{f(x+h)-f(x)}{h} = \lim_{h\to0}\frac{c-c}{h} = \lim_{h\to0}\frac{0}{h} = \lim_{h\to0}0 = 0:$$
see the Remark in 2.22.

We are not yet justified in asserting the *converse* of this result; see 3.82.

(ii) *If* $f(x) = x$, *then* $f'(x) = 1$. For
$$\lim_{h\to0}\frac{f(x+h)-f(x)}{h} = \lim_{h\to0}\frac{(x+h)-x}{h} = \lim_{h\to0}\frac{h}{h} = \lim_{h\to0}1 = 1.$$

(iii) *If c is constant*, $d\{cf(x)\}/dx = cf'(x)$. For
$$\lim_{h\to0}\frac{cf(x+h)-cf(x)}{h} = \lim_{h\to0}c\,\frac{f(x+h)-f(x)}{h} = c\lim_{h\to0}\frac{f(x+h)-f(x)}{h}$$

by 2.3, (i), and this last expression is $cf'(x)$.

## 3.12 A derivable function is continuous

If $f(x)$ is derivable at $x = a$, then
$$\frac{f(a+h)-f(a)}{h} \to f'(a) \quad \text{when} \quad h \to 0.$$

We can therefore write
$$\frac{f(a+h)-f(a)}{h} = f'(a)+\eta,$$

where $\eta$ (in general depending on both $h$ and $a$) tends to zero when $h \to 0$. Hence
$$f(a+h)-f(a) = h\{f'(a)+\eta\} \to 0 \quad \text{when} \quad h \to 0,$$

i.e. $$\lim_{h \to 0} f(a+h) = f(a),$$

which shows that $f(x)$ is continuous at $x = a$.

The converse of this result is false, as the following examples show.

## Examples

(i) $f(x) = x^{\frac{2}{3}}$ is continuous at $x = 0$, but $\{f(h) - f(0)\}/h = h^{-\frac{1}{3}}$ which tends to $+\infty$ when $h \to 0+$, and to $-\infty$ when $h \to 0-$. Hence $f(x)$ is not derivable at $x = 0$.

(ii) $f(x) = |x|$ is continuous at $x = 0$; but

$$\frac{f(h) - f(0)}{h} = \frac{|h|}{h} = \begin{cases} +1 & \text{if } h > 0, \\ -1 & \text{if } h < 0, \end{cases}$$

so that this ratio does not approach a limit when $h \to 0$. Consequently $f(x)$ is not derivable at $x = 0$.

(iii) Let $f(x) = x \sin(1/x)$, $x \neq 0$, and $f(0) = 0$. Then (Ex. 2 (a), no. 19) $f(x)$ is continuous at $x = 0$; but

$$\frac{f(h) - f(0)}{h} = \frac{h \sin(1/h)}{h} = \sin(1/h),$$

and when $h \to 0$ this oscillates between $\pm 1$. Hence $f(x)$ is not derivable at $x = 0$.

In all these examples the functions fail to be derivable at a single point (the origin); but continuous functions of $x$ have been constructed which do not possess a derivative for any value of $x$. It should now be clear that *derivability* of a function requires much more than its *continuity*.

We also remark that, even if $f'(x)$ exists everywhere, it may not be a continuous function of $x$. An example is given in Ex. 3 (a), no. 39.

## 3.2   The rules of derivation

The following rules should be familiar to the reader from early work in Calculus, but we formulate and prove them here for completeness.

*Let $u$, $v$ be functions of $x$ which are derivable at $x = a$. Then*

(1) *the sum $y = u + v$ is derivable, and has derivative $u' + v'$;*

(2) *the product $y = uv$ is derivable, and has derivative $vu' + uv'$;* *and*

(3) *if $v \neq 0$ when $x = a$, the quotient $y = u/v$ is derivable, and has derivative $(vu' - uv')/v^2$.*

*Proofs.* A change $\delta x$ in $x$ from $a$ to $a + \delta x$ causes changes $\delta u$, $\delta v$ in $u$, $v$, and these in turn cause a change $\delta y$ in $y$, given by

$$(1) \quad \delta y = (u + \delta u + v + \delta v) - (u + v) = \delta u + \delta v;$$

$$(2) \quad \delta y = (u + \delta u)(v + \delta v) - uv = v\,\delta u + u\,\delta v + \delta u\,\delta v;$$

$$(3) \quad \delta y = \frac{u + \delta u}{v + \delta v} - \frac{u}{v} = \frac{v\,\delta u - u\,\delta v}{v(v + \delta v)},$$

where we assume in (3) that the change $\delta x$ is such that $v + \delta v \neq 0$; by continuity of $v$ and the hypothesis that $v \neq 0$ when $x = a$, this will be the case for all $\delta x$ sufficiently small.

Divide both sides by $\delta x$, and then let $\delta x \to 0$. Since $u$, $v$ are derivable at $x = a$, $\delta u/\delta x \to u'$ and $\delta v/\delta x \to v'$, and by 3.12 $\delta u \to 0$ and $\delta v \to 0$. Hence $\delta y/\delta x$ tends to a limit, given by

$$(1) \quad u' + v', \qquad (2) \quad vu' + uv', \qquad (3) \quad \frac{vu' - uv'}{v^2}.$$

**(4)** *Function of a function.* Suppose $y = f(x)$ and $x = \phi(t)$; then we can express $y$ directly in terms of $t$ as $y = f(\phi(t)) = g(t)$, say.

*If $\phi(t)$ is derivable at $t = t_0$, and if $f(x)$ is derivable at $x = x_0$, where $x_0 = \phi(t_0)$, then we prove that $g(t)$ is derivable at $t = t_0$ and*

$$g'(t_0) = f'(x_0)\,\phi'(t_0).$$

*Proof.* Since $x = \phi(t)$ is derivable at $t = t_0$, then as in 3.12 we can write

$$\frac{\delta x}{\delta t} = \phi'(t_0) + \eta,$$

where $\eta$ (depending in general on $t_0$ and $\delta t$) tends to 0 when $\delta t \to 0$; so

$$\delta x = \{\phi'(t_0) + \eta\}\,\delta t. \tag{i}$$

Similarly, as $y = f(x)$ is derivable at $x = x_0$,

$$\delta y = \{f'(x_0) + \eta_1\}\,\delta x, \tag{ii}$$

where $\eta_1$ (in general a function of $x_0$ and $\delta x$) tends to 0 when $\delta x \to 0$, and hence certainly $\eta_1 \to 0$ when $\delta t \to 0$, by equation (i).†

From (i) and (ii),
$$\delta y = \{f'(x_0) + \eta_1\}\{\phi'(t_0) + \eta\}\,\delta t,$$

---

† $\eta_1$ is undefined if $\delta x = 0$; but (cf. 2.64) we may then define $\eta_1 = 0$. Equation (ii) is now significant even if $\delta x$, given by (i), happens to be zero for some values of $\delta t$.

so that $\qquad \dfrac{\delta y}{\delta t} = \{f'(x_0) + \eta_1\}\{\phi'(t_0) + \eta\}$

$$\to f'(x_0)\,\phi'(t_0) \quad \text{when} \quad \delta t \to 0.$$

Hence $g'(t_0)$ exists and is $f'(x_0)\,\phi'(t_0)$.

The reader may wonder why we give this elaborate proof instead of the usual one in the elementary books, which goes as follows:

By properties of fractions, $\qquad \dfrac{\delta y}{\delta t} = \dfrac{\delta y}{\delta x} \times \dfrac{\delta x}{\delta t}.$

Letting $\delta t \to 0$, then also $\delta x \to 0$ (3.12 and 2.61, (i)) and so, using the limit theorem about products (2.3, (iii)), we have

$$\frac{dy}{dt} = \frac{dy}{dx} \times \frac{dx}{dt}.$$

*This proof fails* if $x$ is a function of $t$ such that $\delta x$ vanishes infinitely many times in the neighbourhood of $t_0$, i.e. as $\delta t \to 0$. For example,

$$x = \phi(t) = \begin{cases} t^2 \sin(1/t) & (t \neq 0), \\ 0 & (t = 0) \end{cases}$$

is derivable at $t = 0$, and $\phi'(0) = 0$; for

$$\frac{\phi(t) - \phi(0)}{t} = t\sin(1/t) \to 0 \quad \text{when} \quad t \to 0 \quad (\text{Ex. 2}\,(a),\ \text{no. 19}).$$

However, there is no interval including $t = 0$ *throughout which* $\delta x \neq 0$; for

$$\delta x = \phi(\delta t) - \phi(0) = (\delta t)^2 \sin\frac{1}{\delta t}$$

vanishes infinitely many times in any interval $-T < \delta t < T$, however small $T$ may be.

(5) *Inverse functions.* The inverse $x = g(y)$ of a given function $y = f(x)$ was defined in 1.51 (2), (3). We suppose that $f(x)$ *is derivable at* $x = x_0$, *with* $f'(x_0) \neq 0$, *and that* $g(y)$ *is continuous at* $y = y_0$, *where* $y_0 = f(x_0)$. *Then* $g(y)$ *is derivable at* $y = y_0$, *and*

$$g'(y_0) = \frac{1}{f'(x_0)}.$$

*Proof.* As in 3.12,

$$\frac{\delta y}{\delta x} = f'(x_0) + \eta \quad \text{where} \quad \eta \to 0 \quad \text{when} \quad \delta x \to 0.$$

Provided that $f'(x_0) + \eta \neq 0$, this gives

$$\frac{\delta x}{\delta y} = \frac{1}{f'(x_0) + \eta}. \tag{iii}$$

As $f'(x_0) \neq 0$, the proviso will be satisfied for all $\eta$ sufficiently small, and this will be the case for all $\delta x$ sufficiently small.

Since $x = g(y)$ is continuous at $y_0$, hence $(2.61, (\mathrm{i}))$ $\delta x \to 0$ when $\delta y \to 0$, and consequently $\eta \to 0$ when $\delta y \to 0$. Letting $\delta y \to 0$ in (iii),

$$\frac{\delta x}{\delta y} \to \frac{1}{f'(x_0)},$$

i.e. $g'(y_0)$ exists and is $1/f'(x_0)$.

## 3.3 Derivatives of some well-known functions

### 3.31 The function $x^m$

*Case 1: m is a positive integer n.*
We have by summing the geometrical progression that

$$a^{n-1} + a^{n-2}b + a^{n-3}b^2 + \ldots + b^{n-1}$$

$$= a^{n-1} \times \frac{1 - \left(\dfrac{b}{a}\right)^n}{1 - \dfrac{b}{a}} \quad \text{if} \quad a \neq b$$

$$= \frac{a^n - b^n}{a - b}.$$

Taking $b = x$, $a = x + h$, we get

$$\frac{(x+h)^n - x^n}{h} = (x+h)^{n-1} + (x+h)^{n-2}x + \ldots + x^{n-1} \quad (n \text{ terms})$$

$$\to x^{n-1} + x^{n-1} + \ldots + x^{n-1} \quad \text{when} \quad h \to 0$$

$$= nx^{n-1}.$$

Hence    $\dfrac{d}{dx}(x^m) = mx^{m-1}$   when $m$ is a positive integer.

*Case 2: m is a negative integer $-n$.*
$y = x^m = x^{-n} = 1/x^n$, and so by the quotient rule and Case 1,

$$\frac{dy}{dx} = \frac{x^n \cdot 0 - nx^{n-1} \cdot 1}{x^{2n}} = -nx^{-n-1}$$

$$= mx^{m-1}.$$

*Case 3: m is a fraction $p/q$.*
From $y = x^{p/q}$ we have $y^q = x^p$. Derive this relation wo $x$, using the 'function of a function' rule on the left-hand side, and Cases 1, 2:

$$qy^{q-1}\frac{dy}{dx} = px^{p-1},$$

so
$$\frac{dy}{dx} = \frac{p}{q}\frac{x^{p-1}}{y^{q-1}} = \frac{p}{q}x^{p-1}(x^{p/q})^{-q+1} = \frac{p}{q}x^{p/q-1}$$

$$= mx^{m-1}.$$

*Conclusion.* For all rational values of $m$,

$$\frac{d}{dx}(x^m) = mx^{m-1}.$$

The function $x^m$ has not yet been defined for $m$ irrational; see 4.41 (6) $(d)$.

### 3.32 The circular functions (angles in *radians*)

If $y = \sin x$, then

$$\delta y = \sin(x+h) - \sin x = 2\cos(x + \tfrac{1}{2}h)\sin\tfrac{1}{2}h,$$

so
$$\frac{\delta y}{\delta x} = 2\cos(x + \tfrac{1}{2}h)\frac{\sin\tfrac{1}{2}h}{h} \to \cos x \times 1 \quad \text{when} \quad h \to 0.$$

Hence
$$\frac{d}{dx}\sin x = \cos x.$$

Similarly,
$$\frac{d}{dx}\cos x = -\sin x.$$

For $y = \tan x = \sin x/\cos x$, the quotient rule gives

$$\frac{dy}{dx} = \frac{\cos x.\cos x - \sin x(-\sin x)}{\cos^2 x} = \frac{\cos^2 x + \sin^2 x}{\cos^2 x} = \sec^2 x.$$

### 3.33 The inverse circular functions

These functions and their graphs were discussed in 1.52 (3), exs. (iv)–(vi).

(1) $y = \operatorname{Sin}^{-1} x$. Since $x = \sin y$,

$$\frac{dx}{dy} = \cos y = \pm\sqrt{(1-x^2)}.$$

Assuming that we are considering one branch of $\operatorname{Sin}^{-1} x$ (say that whose values lie in the range $n\pi - \tfrac{1}{2}\pi \leqslant y \leqslant n\pi + \tfrac{1}{2}\pi$) so that the function is continuous, we have by 3.2 (5) that

$$\frac{dy}{dx} = \pm\frac{1}{\sqrt{(1-x^2)}}.$$

If we consider only the principal branch $y = \sin^{-1}x$, then $\cos y \geqslant 0$, and we must choose the positive sign of the square root:

$$\frac{d}{dx}\sin^{-1}x = +\frac{1}{\sqrt{(1-x^2)}}.$$

(2) $y = \text{Cos}^{-1}x$. Since $x = \cos y$, $dx/dy = -\sin y = \pm\sqrt{(1-x^2)}$, and so on any one branch

$$\frac{dy}{dx} = \pm\frac{1}{\sqrt{(1-x^2)}}.$$

On the principal branch $y = \cos^{-1}x$, $\sin y \geqslant 0$, and hence we choose the negative sign:

$$\frac{d}{dx}\cos^{-1}x = -\frac{1}{\sqrt{(1-x^2)}}.$$

(3) $y = \text{Tan}^{-1}x$. From $x = \tan y$ we have $dx/dy = \sec^2 y = 1+x^2$. Hence, on any one branch,

$$\frac{d}{dx}\text{Tan}^{-1}x = \frac{1}{1+x^2}.$$

## 3.4 Implicit functions and functions defined parametrically

### 3.41 Derivative of a function defined implicitly

In all cases we shall assume that the given equation in $x$ and $y$ does in fact define $y$ as an implicit function of $x$ (see 1.51 (2)) and that this function is derivable.† The method of finding $dy/dx$ is to derive both sides of the equation wo $x$, using the rules of 3.2 and treating functions of $y$ as 'functions of the (implicit) function' $y$ of $x$.

## Examples

(i) If $x^n + y^n = a^n$, then

$$nx^{n-1} + ny^{n-1}\frac{dy}{dx} = 0 \quad \text{and so} \quad \frac{dy}{dx} = -\left(\frac{x}{y}\right)^{n-1}.$$

(ii) If $a\sin mx + b\cos ny = c$, then

$$am\cos mx - bn\sin ny\frac{dy}{dx} = 0, \quad \text{so} \quad \frac{dy}{dx} = \frac{am\cos mx}{bn\sin ny}.$$

(iii) If $x^5 + 6x^3y + 2y^5 = 0$, then

$$5x^4 + 6\left(y\,.\,3x^2 + x^3\frac{dy}{dx}\right) + 2\,.\,5y^4\frac{dy}{dx} = 0,$$

and

$$\frac{dy}{dx} = -\frac{5x^4 + 18x^2y}{6x^3 + 10y^4}.$$

† Sufficient conditions can be given for existence and derivability of implicit functions.

### 3.42 Derivative of a function defined parametrically

Given two functions $x = \phi(t)$, $y = \psi(t)$, we may think of the elimination of $t$ as leading to a relation $F(x, y) = 0$ defining $y$ as an implicit function of $x$. Assuming the existence and derivability of the functions concerned, we require $dy/dx$; it can be found directly in terms of $t$ *without performing the elimination*, and it is this form of the result which is particularly convenient in geometrical problems, where the given equations are the parametric equations of a curve (see 1.61).

By 'function of a function'

$$\frac{dy}{dt} = \frac{dy}{dx} \times \frac{dx}{dt},$$

and so

$$\frac{dy}{dx} = \frac{dy}{dt} \bigg/ \frac{dx}{dt} = \frac{\psi'(t)}{\phi'(t)}.$$

### Example

*If $x = a(\theta - \sin\theta)$ and $y = a(1 - \cos\theta)$, find $dy/dx$.*

$$\frac{dy}{dx} = \frac{dy}{d\theta} \bigg/ \frac{dx}{d\theta} = \frac{a\sin\theta}{a(1-\cos\theta)} = \frac{2\sin\tfrac{1}{2}\theta\cos\tfrac{1}{2}\theta}{2\sin^2\tfrac{1}{2}\theta}$$

$$= \cot\tfrac{1}{2}\theta.$$

This result gives the gradient at $P$ of the cycloid (see 1.61, ex.). If $T$ is the other extremity of the diameter along $NC$ (fig. 22), then

$$\tfrac{1}{2}\theta = \tfrac{1}{2}P\hat{C}N = P\hat{T}N,$$

and hence the gradient at $P$ is $\cot P\hat{T}N = \tan T\hat{P}K$. Therefore $PT$ *is the tangent at $P$.* Since $T\hat{P}N = 1$ right-angle, $PN$ *is the normal at $P$.*

### Exercise 3(a)

1 Find *from first principles* the derivative of (i) $1/x$; (ii) $\sqrt{x}$; (iii) $x^{\frac{3}{2}}$; (iv) $\tan x$.

*Write down the derivatives of the following.*

2 (i) $(x^3 - 3)^{10}$; (ii) $\sqrt{(a^2 + x^2)}$; (iii) $1/(1 - x^2)^4$.

3 (i) $\sin^n x$; (ii) $\sin mx$; (iii) $\sin^n mx$; (iv) $\tan^n x$.

4 If $v = ds/dt$ and $f = dv/dt$, prove $f = v\, dv/ds = \tfrac{1}{2}d(v^2)/ds$.

5 If $x = y^2 + 5y + 2$, find $dy/dx$ (i) in terms of $y$; (ii) in terms of $x$.

6 Obtain the quotient rule (3.2(3)) by applying the product rule to $u = vy$ (assuming that $y$ so defined is derivable).

7 If $u$, $v$, $w$ are derivable functions of $x$, prove

$$\frac{d}{dx}(uvw) = vw\frac{du}{dx} + wu\frac{dv}{dx} + uv\frac{dw}{dx}.$$

8 If $y = f(x)$, $x = \phi(t)$, $t = F(u)$ are derivable functions, express the derivative of $y$ wo $u$ as a function of $u$.

**9** Deduce the derivative of $\cos x$ from that of $\sin x$ by using the relation $\cos x = \sin(x + \tfrac{1}{2}\pi)$ and 'function of a function'.

**10** Calculate $d(\sin x)/dx$ by writing $\delta y$ in the form

$$\delta y = \cos x \sin h - \sin x (1 - \cos h)$$

and using $\qquad \lim \dfrac{\sin h}{h} = 1, \quad \lim \dfrac{1 - \cos h}{h} = 0.$

Treat $\cos x$ similarly.

**11** The following incorrect 'proof' that $d(\sin x)/dx = \cos x$ is sometimes given by students:

'We have $\sin(x+h) = \sin x \cos h + \cos x \sin h$, and we know that when $h \to 0$, $\cos h \to 1$ and $\sin h \to h$; hence $\sin(x+h) \to \sin x + h \cos x$, so

$$\sin(x+h) - \sin x \to h \cos x \quad \text{and} \quad \{\sin(x+h) - \sin x\}/h \to \cos x.'$$

Criticise this argument.

*Verify the following ($x$ being in radians unless otherwise stated).*

**12** $\dfrac{d}{dx} \sec x = \tan x \sec x.$ $\qquad$ **13** $\dfrac{d}{dx} \operatorname{cosec} x = -\cot x \operatorname{cosec} x.$

**14** $\dfrac{d}{dx} \cot x = -\operatorname{cosec}^2 x.$

**15** $\dfrac{d}{dx} \sin x = \dfrac{\pi}{180} \cos x,$ $x$ being in *degrees*.

**16** Deduce the result of 3.33 (2) from the relation $\cos^{-1} x + \sin^{-1} x = \tfrac{1}{2}\pi$.

*Write down the derivatives of*

**17** $\sin^{-1}(5x).$ $\qquad\qquad$ **18** $\tan^{-1}\left(\dfrac{2x}{1 - x^2}\right).$ [Put $x = \tan \tfrac{1}{2}\theta$.]

**19** $x \sin^{-1} x + \sqrt{(1 - x^2)}.$ $\qquad$ **20** $\sin^{-1}\dfrac{x}{a}$ if (i) $a > 0$; (ii) $a < 0$.

**21** $\cos^{-1}\dfrac{x}{a}$ if (i) $a > 0$; (ii) $a < 0$. $\qquad$ **22** $\tan^{-1}\dfrac{x}{a}.$

*By using the relations*

$$\sec^{-1} x = \cos^{-1}\left(\frac{1}{x}\right), \quad \operatorname{cosec}^{-1} x = \sin^{-1}\left(\frac{1}{x}\right), \quad \cot^{-1} x = \tfrac{1}{2}\pi - \tan^{-1} x,$$

*or otherwise, prove that*

**23** $\dfrac{d}{dx} \sec^{-1} x = +\dfrac{1}{x\sqrt{(x^2 - 1)}}$ $\quad (x > 1).$

**24** $\dfrac{d}{dx} \operatorname{cosec}^{-1} x = -\dfrac{1}{x\sqrt{(x^2 - 1)}}$ $\quad (x > 1).$

**25** $\dfrac{d}{dx} \cot^{-1} x = -\dfrac{1}{1 + x^2}.$

**26** If $u$, $v$ are derivable functions of $x$, prove

$$\frac{d}{dx}\left(\tan^{-1}\frac{u}{v}\right) = \frac{vu' - uv'}{u^2 + v^2}.$$

*Find $dy/dx$ from each of the following equations.*

**27** $\dfrac{x^2}{a^2} - \dfrac{y^2}{b^2} = 1.$     **28** $\left(\dfrac{x}{a}\right)^{\frac{1}{2}} + \left(\dfrac{y}{b}\right)^{\frac{1}{2}} = 1.$     **29** $x^3 + y^3 = 3axy.$

**30** If $y$ is a function of $x$, write down in terms of $x$, $y$, $y'$ the derivative of $x^2 y^3$ wo $x$.

*Find $dy/dx$ in terms of the parameter from the following pairs of equations.*

**31** $x = ct$, $y = c/t$.     **32** $x = a\cos\phi$, $y = b\sin\phi$.

**33** $x = \dfrac{3at}{1+t^3}$, $y = \dfrac{3at^2}{1+t^3}.$

**34** Find the equation of the tangent at the point $(at^2, 2at)$ on the curve $y^2 = 4ax$, and interpret $t$ geometrically. Also give the equation of the normal.

**35** Find the equation of the tangent to the curve $x = a\cos^4\theta$, $y = a\sin^4\theta$ at the point $\theta$. Prove that the sum of its intercepts on $Ox$, $Oy$ is always $a$.

\*36 Prove the result of 3.31, Case 1, by using the binomial expansion of $(x+h)^n$.

\*37 Prove that $\lim\limits_{b \to a} (a^m - b^m)/(a-b) = ma^{m-1}$ for all rational values of $m$.

[If $m = p/q$, put $a = x^q$, $b = y^q$:

$$\text{limit} = \lim_{y \to x}\frac{x^p - y^p}{x^q - y^q} = \frac{px^{p-1}}{qx^{q-1}} = \dots.$$

If $m = -s$, then

$$\text{limit} = \lim\frac{b^s - a^s}{a^s b^s}\bigg/(a-b) = -\frac{sa^{s-1}}{a^{2s}} = \dots.]$$

\*38 If $f(x) = x\tan^{-1}(1/x)$ $(x \neq 0)$, $f(0) = 0$, prove that $f(x)$ is continuous at $x = 0$. Also prove that $\{f(h) - f(0)\}/h$ tends to $+\frac{1}{2}\pi$ when $h \to 0+$, and to $-\frac{1}{2}\pi$ when $h \to 0-$, so that $f'(x)$ does not exist at $x = 0$.

\*39 *If* $f(x) = x^2 \sin(1/x)$ $(x \neq 0)$, $f(0) = 0$, *show that*

$$f'(x) = 2x\sin(1/x) - \cos(1/x) \ (x \neq 0) \quad and \quad f'(0) = 0.$$

*Also show that $f'(x)$ is not continuous at $x = 0$ because $\cos(1/x)$ oscillates when $x \to 0$.*

## 3.5 Derivatives of second and higher orders

### 3.51 Notation

The derivative of a derivable function $y = f(x)$ will in general also be a derivable function, and will possess a derivative which is denoted by

$$\frac{d^2y}{dx^2}, \quad \frac{d^2}{dx^2}y, \quad D^2y, \quad D_x^2y, \quad D^2f(x), \quad f''(x), \quad y'',$$

and is called the *second derivative* of $y$ wo $x$.

Similarly $d^2y/dx^2$ may possess a derivative. In general, the result of $n$ repeated derivations of $y = f(x)$ wo $x$ is denoted by

$$\frac{d^n y}{dx^n}, \quad f^{(n)}(x), \quad y^{(n)}, \quad D^n y, \quad \text{etc.}$$

When $t$ is the independent variable (as is frequently the case in mechanics), the first few derivatives wo $t$ are written

$$\dot{y}, \quad \ddot{y}, \quad \dddot{y}, \quad ...,$$

but this notation beyond third order becomes clumsy.

## 3.52 Implications of the existence of $f^{(n)}(a)$, $n > 1$

If $d^n y/dx^n$ exists at $x = a$, then $d^{n-1}y/dx^{n-1}$ must exist at and near $x = a$, and is continuous at $x = a$; for the function $d^{n-1}y/dx^{n-1}$ is derivable at $x = a$ (which implies its existence near $a$ also), and it is continuous at $x = a$ by 3.12. It follows in turn that the lower derivatives $d^{n-2}y/dx^{n-2}$, ..., $dy/dx$ and the function $y$ itself all exist and are continuous *at and near $x = a$.*

## 3.53 Examples

(i) *If $y = \dfrac{1}{x}(a\cos kx + b\sin kx)$, prove $\dfrac{d^2y}{dx^2} + \dfrac{2}{x}\dfrac{dy}{dx} + k^2 y = 0$.*

In cases like this it is easier to work with products than with quotients; we therefore begin by writing the given equation as

$$xy = a\cos kx + b\sin kx,$$

and derive both sides wo $x$:

$$y + x\frac{dy}{dx} = -ak\sin kx + bk\cos kx.$$

Derive again wo $x$:

$$\frac{dy}{dx} + \left(\frac{dy}{dx} + x\frac{d^2y}{dx^2}\right) = -ak^2\cos kx - bk^2\sin kx,$$

i.e.

$$x\frac{d^2y}{dx^2} + 2\frac{dy}{dx} = -k^2 xy,$$

from which the result follows.

This result, which is a relation between $x$, $y$, $dy/dx$, $d^2y/dx^2$, is called a *differential equation*. As it does not contain the constants $a$, $b$, it could be regarded as the result of eliminating these constants from the given equation by use of the derivatives of $y$. The converse problem (of finding the function $y$ in terms of $x$ when the differential equation is given) is more difficult; some manageable cases are discussed in Ch. 5.

(ii) *If $x = a\cos^3 t$, $y = b\sin^3 t$, find $d^2y/dx^2$ in terms of $t$.*

$$\frac{dy}{dx} = \frac{dy}{dt}\bigg/\frac{dx}{dt} = \frac{3b\sin^2 t\cos t}{-3a\cos^2 t\sin t} = -\frac{b}{a}\tan t.$$

$$\frac{d^2y}{dx^2} = \frac{d}{dx}\left(\frac{dy}{dx}\right) = \left\{\frac{d}{dt}\left(\frac{dy}{dx}\right)\right\}\bigg/\frac{dx}{dt}$$

$$= \left(-\frac{b}{a}\sec^2 t\right)\bigg/(-3a\cos^2 t\sin t) = \frac{b}{3a^2}\sec^4 t\operatorname{cosec} t.$$

(iii) *Calculate $d^2y/dx^2$ if $y$ is defined implicitly by the equation*

$$x^3 + 3axy + y^3 = a^3.$$

Derive both sides wo $x$:

$$3x^2 + 3a\left(y + x\frac{dy}{dx}\right) + 3y^2\frac{dy}{dx} = 0,$$

i.e. 　　　　　　$$x^2 + ay + (ax + y^2)\frac{dy}{dx} = 0. \qquad\qquad (a)$$

Derive both sides of $(a)$ wo $x$:

$$2x + a\frac{dy}{dx} + \frac{dy}{dx}\left(a + 2y\frac{dy}{dx}\right) + (ax + y^2)\frac{d^2y}{dx^2} = 0,$$

$$\therefore\quad -(ax + y^2)\frac{d^2y}{dx^2} = 2x + 2a\frac{dy}{dx} + 2y\left(\frac{dy}{dx}\right)^2$$

$$= 2\left\{x - a\frac{x^2 + ay}{ax + y^2} + y\left(\frac{x^2 + ay}{ax + y^2}\right)^2\right\}$$

by using $(a)$. After some calculation, this reduces to

$$2\frac{xy^4 - a^3xy + x^4y + 3ax^2y^2}{(ax + y^2)^2}$$

$$= 2xy\frac{y^3 + 3axy + x^3 - a^3}{(ax + y^2)^2}$$

$$= 0 \quad\text{from the given equation.}$$

Hence 　　　　　　　　$$\frac{d^2y}{dx^2} = 0.$$

The result is explained thus: the given equation can be written

$$(x + y - a)(x^2 + y^2 + a^2 - xy + ax + ay) = 0,$$

and the second bracket is

$$\tfrac{1}{2}\{(x - y)^2 + (x + a)^2 + (y + a)^2\},$$

which is positive for all values of $x$, $y$ except $x = y = -a$, when the expression is zero. Hence either $x = y = -a$, or $x + y - a = 0$. The 'curve' consists of the isolated point $(-a, -a)$ and the straight line $x + y = a$. It is now clear why $d^2y/dx^2 = 0$.

(iv) *Derivatives of the inverse function.* If $y = f(x)$ has derivatives, and there is an inverse function $x = g(y)$ which also has derivatives, express

$$\frac{d^2x}{dy^2},\quad \frac{d^3x}{dy^3}\quad\text{in terms of}\quad \frac{dy}{dx},\quad \frac{d^2y}{dx^2},\quad \frac{d^3y}{dx^3}.$$

We have by 3.2 (5) that

$$\frac{dx}{dy} = \left(\frac{dy}{dx}\right)^{-1}.$$

Hence

$$\frac{d^2x}{dy^2} = \frac{d}{dy}\left\{\left(\frac{dy}{dx}\right)^{-1}\right\} = \frac{d}{dx}\left\{\left(\frac{dy}{dx}\right)^{-1}\right\} \times \frac{dx}{dy}$$

$$= -\left(\frac{dy}{dx}\right)^{-2}\frac{d^2y}{dx^2} \times \left(\frac{dy}{dx}\right)^{-1}$$

$$= -\frac{d^2y}{dx^2}\Big/\left(\frac{dy}{dx}\right)^3.$$

Similarly,

$$\frac{d^3x}{dy^3} = -\frac{d}{dy}\left\{\frac{d^2y}{dx^2}\left(\frac{dy}{dx}\right)^{-3}\right\}$$

$$= -\frac{d}{dx}\left\{\frac{d^2y}{dx^2}\left(\frac{dy}{dx}\right)^{-3}\right\} \times \frac{dx}{dy}$$

$$= -\left\{\left(\frac{dy}{dx}\right)^{-3}\frac{d^3y}{dx^3} - 3\left(\frac{dy}{dx}\right)^{-4}\left(\frac{d^2y}{dx^2}\right)^2\right\} \times \left(\frac{dy}{dx}\right)^{-1}$$

$$= \left\{3\left(\frac{d^2y}{dx^2}\right)^2 - \frac{d^3y}{dx^3}\frac{dy}{dx}\right\}\Big/\left(\frac{dy}{dx}\right)^5.$$

### Exercise 3(b)

*Calculate $d^2y/dx^2$ for the following functions y.*

1 $\dfrac{x-1}{x^2+4}$.

2 $\dfrac{x^3}{(x+1)(x-2)}$.

3 $\dfrac{x}{\sqrt{(1-x^2)}}$.

4 $\cos^2 x \sin x$.

5 $\dfrac{\sin x}{x}$ $(x \neq 0)$.

6 $\cos(m\sin^{-1}x)$.

7 If $y = a\cos mx + b\sin mx$, find $d^2y/dx^2$ in terms of $y$.

8 If $y = x/\sqrt{(a^2-x^2)}$, prove that

$$(a^2-x^2)\frac{d^2y}{dx^2} = 3x\frac{dy}{dx}.$$

9 If $y = \tan^{-1}x$, prove that

$$(1+x^2)\frac{d^2y}{dx^2} + 2x\frac{dy}{dx} = 0.$$

10 If $y = \tan x$, prove that $y' = 1+y^2$. Derive this equation twice wo $x$, and hence obtain the values of $y'$, $y''$, $y'''$ when $x = 0$.

*Write down expressions for the first five derivatives of the following functions, and by inspection try to write down a formula for the nth derivative.*

11 $x^m$ for (i) $m$ a positive integer; (ii) $m$ not a positive integer; (iii) $m = -1$.

12 $\sin x$.     13 $\cos x$.     14 $(ax+b)^m$.     15 $\sin(ax+b)$.

*16 Prove that the nth derivative of $1/(1-x^2)$ is

$$\tfrac{1}{2}(n!)\{(1-x)^{-n-1} + (-1)^n(1+x)^{-n-1}\}.$$

17 If $u$, $v$ are functions of $x$, obtain $d^2(uv)/dx^2$, $d^3(uv)/dx^3$ in terms of $u$, $v$ and their derivatives (supposed to exist). Can you write down $d^4(uv)/dx^4$ without further calculation?

*Find $dy/dx$ and $d^2y/dx^2$ when $y$ is defined implicitly by*

18   $y^2 = 4ax.$          19   $x^3 + y^3 = 3axy.$          \*20   $x^5 + y^5 = 5ax^2y^2.$

*Find $dy/dx$ and $d^2y/dx^2$ in terms of the parameter from the following pairs of equations.*

21   $x = at^2,\ y = 2at.$      22   $x = a\cos\phi,\ y = b\sin\phi.$

23   $x = a(\cos\theta + \theta\sin\theta),\ y = b(\sin\theta - \theta\cos\theta).$      24   $x = x(t),\ y = y(t).$

25   If $y = \sin n\theta$ and $x = \sin\theta$, find $dy/dx$ and $d^2y/dx^2$, and prove that

$$(1 - x^2)\frac{d^2y}{dx^2} - x\frac{dy}{dx} + n^2y = 0.$$

\*26   Writing $t$, $a$, $b$ for $y'$, $y''/2!$, $y'''/3!$ and $\tau$, $\alpha$, $\beta$ for $dx/dy$, $(1/2!)\,d^2x/dy^2$, $(1/3!)\,d^3x/dy^3$, express $bt - a^2$ in terms of $\tau$, $\alpha$, $\beta$.

Some easy work on partial derivatives of first and second orders could now be done, e.g. 9.11–9.24 (1), and Ex. 9 (a), nos. 1–4, 6, 7, 10, 12–15.

### 3.6   Increasing and decreasing functions; maxima and minima

### 3.61   Function increasing or decreasing at a point

If $f(x)$ is defined at and near $x = a$, and if $f(x) < f(a)$ for all values of $x$ just less than $a$, while $f(x) > f(a)$ for all values of $x$ just greater than $a$, then we say that $f(x)$ is *increasing at $x = a$*. A similar definition can be stated for '$f(x)$ is decreasing at $x = a$'.

*If $f'(x)$ exists when $x = a$ and $f'(a) > 0$, then $f(x)$ is increasing at $x = a$.*
*Proof.* We have
$$\frac{f(a+h) - f(a)}{h} = f'(a) + \eta,$$

where $\eta \to 0$ when $h \to 0$. Hence for all $h$ sufficiently small (positive or negative), $f'(a) + \eta$ will have the same sign as $f'(a)$, i.e. it will be positive; and so $f(a+h) - f(a)$ will have the same sign as $h$. Thus, if $h < 0$ then $f(a+h) < f(a)$; and if $h > 0$, $f(a+h) > f(a)$. Therefore $f(x)$ is increasing at $x = a$.

Similarly, *if $f'(a) < 0$, then $f(x)$ is decreasing at $x = a$*.

The converses may be false; e.g. $f(x) = x^3$ is increasing at $x = 0$ according to the above definition, yet $f'(0) = 0$.

### 3.62   Definition of 'maximum', 'minimum'

(a) $f(x)$ has a *maximum value* at $x = a$ if $f(a)$ is the largest value of $f(x)$ in the neighbourhood of $a$; i.e. if for some sufficiently small positive number $\eta$ we have

$$f(a) > f(x) \quad \text{for all } x \text{ for which} \quad 0 < |x - a| < \eta.$$

(b) $f(x)$ has a *minimum value* at $x = a$ if $f(a)$ is the smallest value of $f(x)$ in the neighbourhood of $a$; i.e. if

$$f(a) < f(x) \quad \text{for all } x \text{ for which} \quad 0 < |x - a| < \eta.$$

(c) If $x = a$ gives a maximum or minimum value of $f(x)$, then $a$ is called a *turning point* or *extremum* of $f(x)$, and $f(a)$ is called a *turning value*.

### Remarks

($\alpha$) Although in definition (c) we used 'point' for 'value of $x$', we are not implying any dependence on graphical illustration; but the language is convenient and suggestive.

($\beta$) Since the terms 'maximum', 'minimum' refer to a *neighbourhood* of a 'point' $x = a$, a 'maximum value' is not necessarily the same as the greatest value of the function, because the latter is relative to the values of the function over the whole range of $x$ for which it is defined. A function may have several different maxima, and may or may not possess a 'greatest value'. There is a similar distinction between 'minimum value' and 'least value'. Further, it is no contradiction that a function may have a maximum value which is less than a minimum value; e.g. see 3.66, ex. (i).

($\gamma$) If $f(x)$ is defined only for $a \leqslant x \leqslant b$, then the end-points $x = a$, $x = b$ of this interval are excluded from the title 'maximum' or 'minimum' because they have no 'neighbourhood'. They may of course correspond to greatest or least values of the function.

The above definitions and remarks are given and illustrated graphically in most elementary courses on Calculus, and the following results obtained from graphical considerations.

(A) We *find* possible turning points $x$ by solving $f'(x) = 0$.

(B) We *test* each root $x = a$ (to see whether it actually gives a turning point, and also to distinguish between maxima and minima)

*either* by considering the sign of $f'(x)$ as $x$ increases through $a$,

*or* by finding the sign of $f''(a)$.

It is our purpose in 3.63–3.65 to establish these results directly from the definitions (a), (b) above, independently of graphical considerations. For subsequent illustration we observe that, according to these definitions,

(i) $x^4$ and $x^{\frac{2}{3}}$ have a minimum at $x = 0$;

(ii) $x^3$ has neither maximum nor minimum at $x = 0$.

For in each case $f(0) = 0$; in (i), $f(x) > 0$ for $x \neq 0$, but in (ii) we have $f(x) < 0$ for $x < 0$ and $f(x) > 0$ for $x > 0$. The reader should confirm this graphically.

**3.63** *If $f(x)$ is derivable at a maximum or minimum, then $f'(x) = 0$ there.*

In the proof of this theorem (and later in 6.21) we require the following

LEMMA. *If $\phi(x) > 0$ for all $x$ just greater than $a$, and if $\phi(x)$ approaches a number $l$ when $x \to a+$, then $l \geqslant 0$.*

For if $l < 0$, then since $\phi(x)$ can be made as close to $l$ as we please for all $x$ sufficiently near to (and greater than) $a$, there would be such values of $x$ for which $\phi(x)$ would be negative, contradicting that $\phi(x) > 0$.

The necessity for allowing $l = 0$ is illustrated by the case

$$\phi(x) = (x-1)^{\frac{1}{2}} \quad \text{when} \quad x \to 1+.$$

There are similar results when $\phi(x) < 0$ and $x \to a+$ or $x \to a-$.

*Proof of the theorem*

Suppose $x = a$ gives a *minimum* of $f(x)$. Then $f(x) - f(a) > 0$ for all $x$ sufficiently near $a$. Hence

$$\frac{f(x) - f(a)}{x - a} > 0 \quad \text{when} \quad x > a.$$

From the hypothesis, this fraction has the limit $f'(a)$ when $x \to a$ *in any manner* (see equation (ii) of 3.11). Hence in particular it will approach the value $f'(a)$ when $x \to a+$, and so by the Lemma, $f'(a) \geqslant 0$.

Similarly, since
$$\frac{f(x) - f(a)}{x - a} < 0 \quad \text{when} \quad x < a,$$

we may let $x \to a-$ and conclude that $f'(a) \leqslant 0$. Combining the two conclusions, we have $f'(a) = 0$.

A similar proof and result holds if $x = a$ gives a *maximum*.

*Remarks*

($\alpha$) *The converse of this theorem may be false:* a root of $f'(x) = 0$ may not give either a maximum or a minimum. For example, if $f(x) = x^3$, then $f'(x) = 0$ when $x = 0$, but $x = 0$ is not a turning point.

($\beta$) *Not all maxima or minima may be given by $f'(x) = 0$:* they may occur where $f(x)$ is not derivable. For example, if $f(x) = x^{\frac{2}{3}}$, then $f'(0)$ does not exist (3.12, ex. (i)); yet $x = 0$ gives a minimum.

($\gamma$) In view of ($\alpha$), ($\beta$) we may say that *maxima and minima are to be sought among the values of x for which $f'(x)$ is zero or non-existent.*

A root of $f'(x) = 0$ is called *a stationary point* of $f(x)$.

### 3.64 'Change of sign' test

*If* (i) $f'(x)$ *exists at and near* $x = a$, *and* $f'(a) = 0$,

    (ii) $f'(x) < 0$ *for all* $x$ *less than and sufficiently near to* $a$, *while*

      $f'(x) > 0$ *for all x greater than and sufficiently near to* $a$,

*then* $x = a$ *is a minimum point of* $f(x)$.

*Proof.* (See Note [1] on p. 362.) Hypothesis (i) implies that $f(x)$ is continuous at and near $x = a$ (see 3.52). By this and (ii), there exists a number $\eta > 0$ such that $f'(x) < 0$ for $a - \eta < x < a$ and $f(x)$ is continuous for $a - \eta \leqslant x \leqslant a$. Hence by Corollary 1 in 3.83 we have $f(x) > f(a)$ when $a - \eta < x < a$. Similarly, there is an interval $a < x < a + \eta'$ for which (using 3.83) $f(x) > f(a)$.

Thus in a sufficiently small neighbourhood of $a$, $f(a)$ is the smallest value of $f(x)$; i.e. $x = a$ gives a minimum of $f(x)$.

COROLLARY 1. Replacing condition (ii) by

    (ii)' $f'(x) > 0$ *for all x less than and sufficiently near to* $a$, *and*

      $f'(x) < 0$ *for all x greater than and sufficiently near to* $a$,

we could prove that $x = a$ *gives a maximum of* $f(x)$.

COROLLARY 2. *If* $f'(x)$ *does not change sign as x increases through* $a$, *then* $x = a$ *is not a turning point.*

For if $f'(x) > 0$ on both sides of $a$, then in some neighbourhood of $a$ we have $f(x) < f(a)$ for $x < a$ and $f(x) > f(a)$ for $x > a$; i.e. in this neighbourhood there are some values of $f(x)$ greater and some less than $f(a)$. The definitions ($a$), ($b$) in 3.62 are not satisfied. Similar remarks hold if $f'(x) < 0$ on both sides of $a$.

An example is $f(x) = x^3$: $f'(x) = 3x^2 > 0$ on both sides of $x = 0$, which is therefore not a turning point. (Also see p. 362, Note [2].)

### 3.65 'Second derivative' test

*If* (i) $f'(x)$ *is continuous at* $x = a$,

    (ii) $f'(a) = 0$,             (iii) $f''(a)$ *exists and is not zero*,

*then* $x = a$ *gives a minimum of* $f(x)$ *if* $f''(a) > 0$, *and a maximum if* $f''(a) < 0$.

*Proof.* If $f''(a) > 0$, then by 3.61 applied to the function $f'(x)$ we see that $f'(x)$ is increasing at $x = a$. Since $f'(a) = 0$, $f'(x)$ increases through the value zero: in the neighbourhood of $a$ we have $f'(x) < 0$ for $x < a$ and $f'(x) > 0$ for $x > a$. Hence by 3.64, $x = a$ gives a minimum.

If $f''(a) < 0$, then $f'(x)$ is decreasing through the value zero as $x$ increases through $a$, i.e. it changes from positive to negative. By 3.64, Corollary 1, $x = a$ gives a maximum.

*Remark.* If $f''(a) = 0$, the test is indecisive, for $x = a$ may give a maximum, a minimum, or neither. For example, $f''(0) = 0$ for both $x^4$ and $x^3$; but $x = 0$ gives a minimum of $x^4$, and is not a turning point of $x^3$. When this happens we must revert to the more fundamental test by 'change of sign' (3.64).

### 3.66 Examples

(i) $f(x) = x + \dfrac{1}{x}$.

$$f'(x) = 1 - \frac{1}{x^2} \quad \text{and} \quad f'(x) = 0 \quad \text{when} \quad x = \pm 1.$$

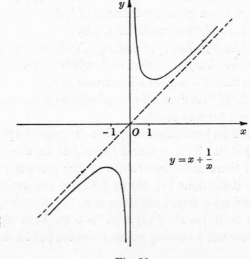

Fig. 38

As $f''(x)$ is easily found, we can test these stationary points by using 3.65. We have

$$f''(x) = \frac{2}{x^3}, \quad \text{and so} \quad f''(1) > 0, \quad f''(-1) < 0.$$

Hence $x = 1$ gives a minimum value of $f(x)$, viz. $f(1) = 2$; and $x = -1$ gives a maximum, viz. $f(-1) = -2$.

*The maximum is thus less than the minimum value.* Further, $f(x)$ has no greatest value because it can be made as large as we please by taking $x$ sufficiently large, or sufficiently small, and positive; similarly $f(x)$ has no least value. Cf. Ex. 1 $(a)$, no. 11.

(ii) $f(x) = \dfrac{13x^2 - 8x + 5}{5x^2 - 16x + 3}$.

It can be verified that
$$f'(x) = \frac{-28(3x-2)(2x+1)}{(5x^2 - 16x + 3)^2},$$

except when $5x^2 - 16x + 3 = 0$, i.e. when $x = \frac{1}{5}$ or 3. For these values the function itself is not defined. We have $f'(x) = 0$ when $x = \frac{2}{3}$ or $-\frac{1}{2}$.

For testing these stationary points the 'change of sign' method of 3.64 is obviously better, owing to the evident complexity of the expression for $f''(x)$. Since the denominator of $f'(x)$ is positive for all $x$ except $\frac{1}{5}$ or 3, we need consider only the change of sign of the numerator.

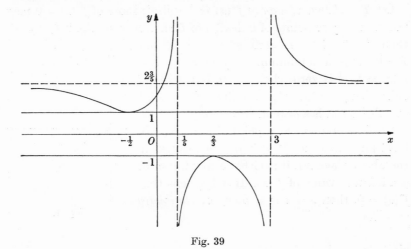

Fig. 39

When $x$ increases through $\frac{2}{3}$, $f'(x)$ changes from positive to negative, so that $x = \frac{2}{3}$ gives a maximum value, viz. $f(\frac{2}{3}) = -1$. When $x$ increases through $-\frac{1}{2}$, $f'(x)$ changes from negative to positive; $x = -\frac{1}{2}$ gives a minimum value $f(-\frac{1}{2}) = 1$.

Again the maximum is less than the minimum; and the function has no greatest or least value, because it can be made as numerically large as we please by taking $x$ sufficiently close to $\frac{1}{5}$ or 3.

This function could also be discussed by an algebraic treatment as in 1.41, ex. (iv). Its graph, sketched by the methods of 1.41, is shown (not to scale) in fig. 39.

### 3.7 Points of inflexion of a curve

### 3.71 Definition and determination

A *point of inflexion* of $f(x)$ is a point where $f'(x)$ exists and has a turning value.

Geometrically, a point of inflexion of the curve $y = f(x)$ is one at which there is a tangent which has maximum or minimum gradient.

Applying 3.64 to the function $f'(x)$, we get the following theorem.

*If* (i) $f''(x)$ *exists at and near* $x = a$, *and* $f''(a) = 0$,

(ii) $f''(x)$ *changes sign as* $x$ *increases through* $a$,

*then* $x = a$ *is a point of inflexion of* $f(x)$.

For, under these conditions, $f'(a)$ is a turning value of $f'(x)$.

*Remarks*

($\alpha$) *The change of sign of* $f''(x)$ *is essential:* roots of $f''(x) = 0$ may not always give points of inflexion of $f(x)$. For example, if $f(x) = x^4$, then $f''(x) = 12x^2$ is zero when $x = 0$, but $x = 0$ gives a minimum. If $f(x) = x^{\frac{5}{2}}$, then $f''(x) = \frac{15}{4}x^{\frac{1}{2}}$ is zero when $x = 0$, but *does not exist* when $x < 0$; thus $x = 0$ cannot be a point of inflexion: it is a *cusp* in this example.

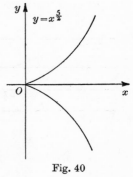

$y = x^{\frac{5}{2}}$

Fig. 40

($\beta$) The gradient at a point of inflexion can have any value, for $f'(a)$ is not mentioned in the above theorem: it is either a maximum or a minimum value of $f'(x)$. If it happens that $f'(a) = 0$, then $x = a$ is a point of *stationary inflexion*.

### Examples

(i) *Find the points of inflexion of* $f(x) = 6x^6 - 25x^4 + 9$.

We find that $f''(x) = 60x^2(3x^2 - 5)$. Hence $f''(x) = 0$ when $x = 0$ or $\pm\sqrt{\frac{5}{3}}$.

Since $f''(x)$ does not change sign as $x$ increases through zero, $x = 0$ is not a point of inflexion. (It is in fact a maximum, by using 3.64.)

$f''(x)$ changes from negative to positive as $x$ increases through $\sqrt{\frac{5}{3}}$, and from positive to negative as $x$ increases through $-\sqrt{\frac{5}{3}}$. Each therefore gives a point of inflexion.

(ii) *If* $f''(x)$ *is continuous, an inflexion occurs between consecutive maxima and minima.*

For at a maximum $x_1$, $f''(x_1) < 0$; and at a minimum $x_2$, $f''(x_2) > 0$. By the continuity of $f''(x)$, it must be zero for at least one point $x_3$ between $x_1$ and $x_2$ (2.65), and changes sign as $x$ increases through $x_3$. Hence $x_3$ is a point of inflexion.

## 3.72 Summary

The points of a continuous curve $y = f(x)$ can be classified as follows.

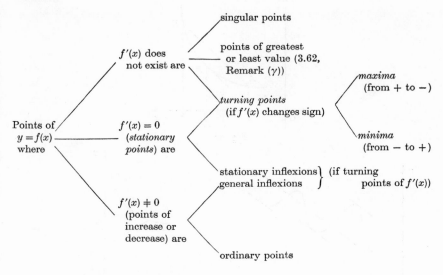

## 3.8    A theorem suggested geometrically

### 3.81 The mean value theorem

Given a continuous curve $AB$, at every point of which there is a tangent (cf. 2.15), fig. 41 suggests that there is a point $P$ on the curve between $A$ and $B$ at which the tangent is parallel to the chord $AB$ (even if $AB$ is itself tangential to the curve at $A$ or $B$ or both). There may be several such points $P$ (fig. 42). If the curve is not continuous the result may clearly be false; e.g. fig. 43 shows the graph of $y = -1/x$ for which every tangent has positive gradient but the gradient of $AB$ is negative.

Let $y = f(x)$ be the equation of the curve, and let $A$, $B$, $P$ correspond to $x = a, b, \xi$. Then we are supposing that $f(x)$ is continuous for $a \leqslant x \leqslant b$, and that $f'(x)$ exists for $a < x < b$. The property can then be expressed as follows.

*If $f(x)$ is continuous for $a \leqslant x \leqslant b$ and if $f'(x)$ exists for $a < x < b$, then there is at least one number $\xi$ such that*

$$\frac{f(b) - f(a)}{b - a} = f'(\xi) \quad (a < \xi < b).$$

This result (the *first mean value theorem*) will be established without appeal to a figure in Ch. 6. Meanwhile we assume it, and use it to make the following important deductions.

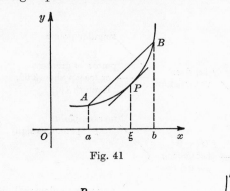

Fig. 41

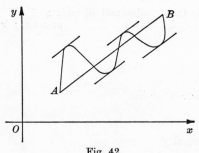

Fig. 42

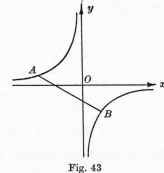

Fig. 43

**3.82** *If $f(x)$ is continuous for $a \leqslant x \leqslant b$ and $f'(x) = 0$ for all $x$ in $a < x < b$, then $f(x)$ is constant for $a \leqslant x \leqslant b$.*

Let $x_1$, $x_2$ be any two numbers such that $a \leqslant x_1 < x_2 \leqslant b$. Then by 3.81 applied to the interval $x_1 \leqslant x \leqslant x_2$,

$$\frac{f(x_2) - f(x_1)}{x_2 - x_1} = f'(\xi)$$

for some number $\xi$ between $x_1$ and $x_2$, i.e. between $a$ and $b$. Since $f'(\xi) = 0$, this shows that $f(x_2) - f(x_1) = 0$, so that $f(x)$ takes the same value for any pair of numbers $x_1 \neq x_2$ satisfying $a \leqslant x \leqslant b$; i.e. $f(x)$ is constant for $a \leqslant x \leqslant b$.

This theorem is the *converse* of 3.11, ex. (i).

The hypothesis that $f'(x)$ is zero for $a < x < b$ implies (3.12) that $f(x)$ is continuous for $a < x < b$. If we do not include continuity at $x = a$ and $x = b$, the theorem may be false. For example, if

$$f(x) = 1 \quad (0 < x < 1), \qquad f(0) = 0 = f(1),$$

then $f'(x) = 0$ for $0 < x < 1$ (3.11, ex. (i)), but $f(x)$ is not constant for $0 \leqslant x \leqslant 1$.

COROLLARY. *If $f(x)$ and $g(x)$ are continuous for $a \leqslant x \leqslant b$ and $f'(x) = g'(x)$ for $a < x < b$, then $f(x) - g(x)$ is constant for $a \leqslant x \leqslant b$.*

This is proved by applying the theorem to $\phi(x) = f(x) - g(x)$. The result, which can be worded 'if two continuous functions have the same derivative then they differ by a constant', is basic for integration (4.11).

### 3.83 Function increasing or decreasing throughout an interval

*Definitions.* (*a*) The function $f(x)$ is *increasing in the interval* $a \leqslant x \leqslant b$ if, for any numbers $x_1, x_2$ such that $a \leqslant x_1 < x_2 \leqslant b$, we have $f(x_1) < f(x_2)$. (*b*) If we have $f(x_1) > f(x_2)$, then $f(x)$ *is decreasing in the interval* $a \leqslant x \leqslant b$. (*c*) In either case $f(x)$ is said to be *monotonic* in $a \leqslant x \leqslant b$. (*d*) If $=$ is included in (*a*), $f(x)$ is *non-decreasing* (cf. 2.77).

The definition (*a*) should be compared with that of 'function increasing at a point' in 3.61. The fact that $f(x)$ is increasing at the point $x_0$ does not necessarily imply that $f(x)$ is increasing *throughout some interval* containing $x_0$: for details and an example the reader may consult Hardy, *Pure Mathematics* (7th–10th ed.), foot of p. 233, and p. 236, penultimate paragraph.

Further, if $f(x)$ is known to be increasing at *each point* of $a < x < b$, one may be tempted to conclude that $f(x)$ is increasing *in the interval* $a \leqslant x \leqslant b$. Although this is true, it is not obvious from the definitions (loc. cit., p. 208, ex. 19). We now give a proof of this property, assuming that $f'(x)$ exists in $a < x < b$. The converse property is of course always true.

*If* (i) $f(x)$ *is continuous for* $a \leqslant x \leqslant b$,

   (ii) $f'(x) > 0$ *throughout* $a < x < b$,

*then $f(x)$ is increasing for $a \leqslant x \leqslant b$.*

*Proof.* If $a \leqslant x_1 < x_2 \leqslant b$, then by 3.81 applied to the interval $x_1 \leqslant x \leqslant x_2$,
$$f(x_2) - f(x_1) = (x_2 - x_1) f'(\xi) \quad (x_1 < \xi < x_2).$$

Since $x_2 - x_1 > 0$, and $f'(\xi) > 0$ by hypothesis (ii), hence $f(x_2) > f(x_1)$. The result follows.

COROLLARY 1. Replacing condition (ii) by

   (ii)' $f'(x) < 0$ *throughout* $a < x < b$,

we can prove that $f(x)$ *decreases for* $a \leqslant x \leqslant b$.

COROLLARY 2. *If $f'(x) > 0$ for $a < x < b$ and $f(a) \geqslant 0$, then $f(x) > 0$ for $a < x \leqslant b$.*

For if $a < x \leqslant b$, then $f(x) > f(a) \geqslant 0$.

COROLLARY 3. Replacing condition (ii) by

   (ii)'' $f'(x) > 0$ for $a < x < b$ *except when* $x = k$ (where $f'(x)$ may be undefined, or zero, or negative), then $f(x)$ *is still increasing for* $a \leqslant x \leqslant b$.

For the conditions of the theorem are satisfied by $f(x)$ in each of the intervals $a \leqslant x \leqslant k$, $k \leqslant x \leqslant b$ (since $f'(x)$ is required to exist only for $a < x < k$, $k < x < b$).

Hence $f(x)$ increases for $a \leqslant x \leqslant k$ and for $k \leqslant x \leqslant b$ and therefore for $a \leqslant x \leqslant b$. For example, $f(x) = x^3$ is increasing for all $x$, although $f'(0) = 0$.

The result can be extended to allow any finite number of points $k$ where $f'(x)$ is not positive.

## Examples

(i) *If* $f(x) = x/\sin x$ $(x \neq 0)$, $f(0) = 1$, *prove that* $f(x)$ *increases for* $0 \leqslant x \leqslant \frac{1}{2}\pi$. *Deduce that* $\sin x \geqslant 2x/\pi$ *for* $0 \leqslant x \leqslant \frac{1}{2}\pi$, *with equality only when* $x = 0$ *or* $\frac{1}{2}\pi$.

$f(x)$ is continuous, and if $x \neq 0$,

$$f'(x) = \frac{\sin x - x \cos x}{\sin^2 x} = \frac{\cos x (\tan x - x)}{\sin^2 x}.$$

If $0 < x < \frac{1}{2}\pi$, then $\cos x > 0$ and $\tan x > x$ (cf. 2.12, inequality (iii)), so that $f'(x) > 0$. Hence $f(x)$ is increasing for $0 \leqslant x \leqslant \frac{1}{2}\pi$.

It follows that, if $0 < x < \frac{1}{2}\pi$, then $f(x) < f(\frac{1}{2}\pi)$, i.e. $x/\sin x < \frac{1}{2}\pi$, and consequently $\sin x > 2x/\pi$. When $x = 0$ or when $x = \frac{1}{2}\pi$, the two sides are equal.

*If* $0 < x < \frac{1}{2}\pi$, *we now have* $2x/\pi < \sin x < x$ (cf. 2.12, inequality (iii)). Also see Ex. 3 (c), no. 19.

(ii) *If* $a$, $b$ *are positive constants and* $f(x) = a + b + x - 3(abx)^{\frac{1}{3}}$ *for* $x > 0$, *prove that the least value of* $f(x)$ *is* $a + b - 2\sqrt{(ab)}$, *and deduce that for* $c \geqslant 0$,

$$a + b + c - 3(abc)^{\frac{1}{3}} \geqslant a + b - 2\sqrt{(ab)}$$

*with equality only when* $c = \sqrt{(ab)}$.

If $x > 0$, $f'(x) = 1 - (ab)^{\frac{1}{3}} x^{-\frac{2}{3}}$. Hence $f'(x) \gtrless 0$ according as $x \gtrless \sqrt{(ab)}$. Hence for $0 \leqslant x < \sqrt{(ab)}$, $f(x)$ is decreasing; and for all $x > \sqrt{(ab)}$, $f(x)$ is increasing. The minimum at $x = \sqrt{(ab)}$ is therefore the *least* value of the function for $x \geqslant 0$, and $f(x) > f(\sqrt{(ab)})$ for $x \geqslant 0$ unless $x = \sqrt{(ab)}$; i.e.

$$a + b + x - 3(abx)^{\frac{1}{3}} > a + b + \sqrt{(ab)} - 3\{(ab)^{\frac{2}{3}}\}^{\frac{1}{3}}$$
$$= a + b - 2\sqrt{(ab)}.$$

Since $a + b - 2\sqrt{(ab)} > 0$ unless $a = b$ (1.22 (1)), we have $a + b + c > 3(abc)^{\frac{1}{3}}$ unless $a = b = c$. The example can be generalised to give a proof of the theorem of the means (1.22 (2)).

## Exercise 3(c)

*Find the stationary points of the following functions, and distinguish between them. Also find any other points of inflexion in nos.* 1–3, 6.

1  $2x^3 + 3x^2 - 12x + 7$.          2  $x^5 + 5x^3$.

3  $\dfrac{9x}{(x-1)^2}$.          4  $\dfrac{(x-1)(x-3)}{5x^2 + 4}$.

5  $a \cos x + b \sin x$ (verify the results trigonometrically).

6  $a \cos^2 x + b \sin^2 x$, $b > a > 0$.          7  $2 \cos x + \sin 2x$.

8  Show that $(x+5)^2 (x^3 - 10)$ has a minimum when $x = 1$. Find its other turning points and its points of inflexion.

9  $f(x)$ is defined by $f(x) = 2x(x-1)$ if $x < 1$, $f(x) = (x-1)(x-2)(x-3)$ if $x \geqslant 1$. Verify that $f(x)$ and $f'(x)$ are continuous for all values of $x$, and find the turning points and inflexions of $f(x)$.

**10** It is sometimes asserted that 'between two consecutive maxima of a continuous function occurs a minimum'. Show that this may not be true by considering the function $f(x)$ in the interval $-3\pi \leqslant x \leqslant 3\pi$, where $f(x) = \cos x$ $(x \geqslant \pi, x \leqslant -\pi), f(x) = -1 \; (-\pi < x < \pi)$.

**11** If $f(\theta) = 4\cos^3 \theta - 3\cos\theta$, we may put $x = \cos\theta$ and consider instead $\phi(x) = 4x^3 - 3x$. Verify that $\phi(x)$ has just two turning points, $x = \pm\frac{1}{2}$; but that $f(\theta) = \cos 3\theta$ has infinitely many turning points, given by $\theta = \frac{1}{3}n\pi$ for all integers $n$. Explain this apparent contradiction.

**12** Show that the stationary points on the curve $x^3 + y^3 = 3xy$ satisfy $x^2 = y$. Hence show that they are $x = 0, \sqrt[3]{2}$.

**13** Post Office regulations require that the length plus girth of a parcel shall not exceed 6 ft. Find the maximum volume if the cross-section is (i) square; (ii) circular.

**14** $AB$, $CD$ are cables of equal length. The strain which any section of $AB$ can stand varies as the cube of its distance from $B$, and the strain that any section of $CD$ can stand varies as the cube of its distance from $D$. The strain that can be carried at $A$ is four times that at $C$. The cables are now woven together so that $D$ coincides with $A$, and $C$ with $B$. How far from $B$ is the weakest point of the composite cable?

**15** A conical tent has a given volume. Find the ratio of height to diameter of base for the area of canvas to be a minimum.

**\*16** Show that $y = (x^2 - 2x + 4)/(x^2 + 2x + 4)$ has three points of inflexion, which lie on the line $x + 3y = 5$. $[y - 1 = -4f(x)$, where $f(x) = x/(x^2 + 2x + 4)$. The condition $f''(x) = 0$ is found to be $x^3 - 12x - 8 = 0$, i.e.

$$(x - 2)(x^2 + 2x + 4) - 12x = 0.$$

The coordinates of the points of inflexion therefore satisfy

$$x - 2 = 12f(x) = -3(y - 1).]$$

**17** If $f(x) = \cos x - 1 + \frac{1}{2}x^2$, prove that $f'(x) > 0$ when $x > 0$. Deduce that for all $x \neq 0$, $\cos x > 1 - \frac{1}{2}x^2$. What happens when $x = 0$?

**18** Using no. 17, show that $\sin x > x - \frac{1}{6}x^3$ when $x > 0$.

**\*19** Extend the result of 3.83, ex. (i) to the interval $\frac{1}{2}\pi \leqslant x \leqslant \pi$. [Put $y = \pi - x$.] Hence prove that

$$\sin x \geqslant \frac{2x(\pi - x)}{\pi^2} \quad \text{for} \quad 0 \leqslant x \leqslant \pi.$$

**20** Prove that the *greatest* value of $x^2 y^3$, where $x$ and $y$ are positive and $x + y = 1$, is $2^2 . 3^3/5^5$. (Cf. method of 1.22, ex. (ii).)

**21** If $f(x) = x^r - 1 - r(x - 1)$ and $r > 1$, prove that for $x > 0$ the least value of $f(x)$ is 0 and occurs when $x = 1$. Show that this is also true when $r < 0$. When $x > 0$, $x \neq 1$, and $r > 1$ or $r < 0$, deduce the inequality $x^r - 1 > r(x - 1)$. Prove a corresponding result when $0 < r < 1$.

## 3.9 Small changes. Differentials

### 3.91 Small changes

If $y = f(x)$ is derivable at $x = a$, then as in 3.12,

$$\frac{f(a + h) - f(a)}{h} = f'(a) + \eta,$$

where $\eta \to 0$ when $h \to 0$. For small values of $h$ we therefore have the approximation

$$\frac{f(a+h)-f(a)}{h} \doteqdot f'(a), \tag{i}$$

i.e.
$$f(a+h)-f(a) \doteqdot hf'(a), \tag{ii}$$

or (less precisely)
$$\delta y \doteqdot \frac{dy}{dx}\,\delta x. \tag{iii}$$

The approximation in (i)–(iii) is in general correct 'to first order in $h$'. For $\eta$ is small when $h$ is small, and so $h\eta$ is small compared with $h$; but $h\eta$ may not be small compared with any power of $h$ higher than the first: see Ex. 3 ($d$), no. 7.

Referring to fig. 36 of 3.11, $PR = \delta x$, $QR = \delta y$ and $\tan \psi = dy/dx$. By (iii), $\delta y \doteqdot PR \tan \psi = RT$. The approximation is therefore equivalent to replacing the step $RQ$ to the curve by the step $RT$ to the tangent; $QT$ represents the error: cf. 6.42, Remark ($\beta$).

Result (i) should be compared with that of 3.81 when $b = a+h$, viz.

$$\frac{f(a+h)-f(a)}{h} = f'(\xi) \quad (a < \xi < a+h). \tag{iv}$$

($a$) The latter is *exact*, although the number $\xi$ is known only to lie *somewhere* between $a$ and $a+h$.

($b$) The approximation (i) assumes only that $f(x)$ is derivable *at* $x = a$, while (iv) requires $f(x)$ to be derivable for $a < x < a+h$.

### Examples

(i) *Calculate* $f(x) = 3x^2 - 7x + 8$ *approximately when* $x = 2 \cdot 015$.

$f'(x) = 6x - 7$, and $f(2) = 6, f'(2) = 5$. Hence by (ii) above,

$$f(2 \cdot 015) \doteqdot f(2) + 0 \cdot 015 \times 5 = 6 + 0 \cdot 075 = 6 \cdot 075.$$

(ii) *Calculate* $\sqrt[3]{126}$ *approximately*.

Take $f(x) = \sqrt[3]{x}$, $a = 125$, $h = 1$. Then $f'(x) = \frac{1}{3}x^{-\frac{2}{3}}$, $f(125) = 5$, and $f'(125) = \frac{1}{75}$. Hence
$$f(126) \doteqdot 5 + \tfrac{1}{75} \doteqdot 5 \cdot 013.$$

(iii) *The area $S$ of a triangle is calculated from the formula* $S = \frac{1}{2}bc \sin A$. *Find the approximate error in $S$ owing to* ($a$) *a small error $\delta A$ in $A$ (measured in radians);* ($b$) *a small error $\delta b$ in $b$.*

($a$) $\quad \delta S \doteqdot \dfrac{dS}{dA}\,\delta A = \tfrac{1}{2}bc \cos A\, \delta A.$

($b$) $\quad \delta S \doteqdot \dfrac{dS}{db}\,\delta b = \tfrac{1}{2}c \sin A\, \delta b.$

(iv) *In ex.* (iii) *suppose there were simultaneous errors* $\delta b$, $\delta c$, $\delta A$ *in* $b$, $c$, $A$. The error in $S$ is then, from first principles,

$$\delta S = \tfrac{1}{2}(b+\delta b)(c+\delta c)\sin(A+\delta A) - \tfrac{1}{2}bc\sin A$$
$$\doteqdot \tfrac{1}{2}(b\,\delta c + c\,\delta b)\sin A + \tfrac{1}{2}bc\cos A\,\delta A$$

if we expand the brackets, use the approximations $\sin\delta A \doteqdot \delta A$ and $\cos\delta A \doteqdot 1$, and neglect all products of the expressions $\delta b$, $\delta c$, $\delta A$.

The result shows that $\delta S$ is approximately what is obtained by adding together the approximate errors in $S$ which would have been caused if the errors $\delta b$, $\delta c$, $\delta A$ had occurred separately. We return to this matter in 9.33.

### Exercise 3(d)

**1** Calculate $\tan 45° 16'$ approximately. [Use $d(\tan x)/dx = \tfrac{1}{180}\pi \sec^2 x$, $x$ in *degrees*.]

**2** Give the approximate error in the volume of a sphere when calculated from the formula $V = \tfrac{4}{3}\pi r^3$, if the radius is in error by $\delta r$.

**3** Taking $f(x) = x^m$, obtain the approximation $(x+h)^m \doteqdot x^m + mx^{m-1}h$ for small $h$.

**4** If $\theta$ is a small angle in radians, prove $\tan(\alpha+\theta) \doteqdot \tan\alpha + \theta\sec^2\alpha$.

**5** If $pv = RT$ where $R$ is constant, find the approximate error in $v$ due to (i) a small error $\delta p$ in $p$ only; (ii) small errors $\delta p$, $\delta T$ in both $p$ and $T$.

**6** If $y = uv/w$, prove

$$\frac{\delta y}{y} \doteqdot \frac{\delta u}{u} + \frac{\delta v}{v} - \frac{\delta w}{w}.$$

[Consider $wy = uv$.]

*7 Find the function $\eta$ in 3.91 when $f(x)$ is (i) $x^2$; (ii) $x^3$; (iii) $1/x$. Verify that $\eta/h$ tends to a limit when $h \to 0$, but that *in general* $\eta/h^m$ does not when $m$ is a constant greater than 1; state the exceptional case in (ii).

### 3.92 Differentials

(1) The symbol $dy/dx$ for the derivative, although of fractional appearance, has been defined as a single indecomposable symbol to represent $\lim_{\delta x \to 0} \delta y/\delta x$ when this limit exists. The parts $dy$, $dx$ cannot be separated, and so far are meaningless alone. The most we have done towards splitting the symbol $dy/dx$ is to write it in the 'operational' form $(d/dx)y$. We now define $dy$, $dx$ separately, in such a way that the *quotient* $dy \div dx$ is equal to the derivative of $y$ wo $x$.

If $y = f(x)$ is derivable at $x$, then

$$\delta y = f'(x)\,\delta x + \eta\,\delta x, \tag{i}$$

where $\eta$ is a function of $\delta x$ (and usually also of $x$) which tends to zero when $\delta x \to 0$. In 3.91 we deduced for small $\delta x$ the approximation

$$\delta y \doteqdot f'(x)\,\delta x. \tag{ii}$$

We now *define dy* by the equation

$$dy = f'(x)\,\delta x. \tag{iii}$$

This equation also defines $dx$; for $dx$ *is dy* when $y$ is the function $x$. In this case we have $f'(x) = 1$ (3.11, ex. (ii)), so that

$$dx = \delta x \tag{iv}$$

and we may rewrite equation (iii) more symmetrically as

$$dy = f'(x)\,dx. \tag{v}$$

(iv) shows that $dx$ *is identical with the arbitrary increment* $\delta x$; but *in general* $dy \neq \delta y$, because by (i)

$$\delta y = dy + \eta\,\delta x \tag{vi}$$

and in general $\eta \neq 0$. The approximation (ii) can thus be restated as $\delta y \doteq dy$. In fig. 36 of 3.11, $PR = \delta x = dx$, $QR = \delta y$, and by (v), $dy = \tan\psi.PR = RT$. Hence $dy$ is represented by the step to the tangent.

On dividing (v) by $dx$, we see that $dy/dx = f'(x)$, where here $dy/dx$ is the *quotient* $dy \div dx$. The symbol $dy/dx$ can therefore be interpreted in two ways: (*a*) as $\lim_{\delta x \to 0} \delta y/\delta x$; (*b*) as $dy \div dx$. No confusion arises, because the results are the same, namely, the derivative $f'(x)$ of $y = f(x)$.

*Definition. dy* is called the *differential* of $y$, and $dx$ is the *differential* of $x$.

In (v), $f'(x)$ is the coefficient of the differential $dx$ in the expression for $dy$. Consequently, the derivative $f'(x)$ is sometimes called the *differential coefficient* of $y$ wo $x$.

In some books and examination papers the approximation (ii) is written $dy \doteq f'(x)\,dx$; that is, differentials are confused with small increments. By definition, relation (iii) and hence (v) are *exact*.

(2) *Invariance property.* Suppose now that $y = f(x)$ and $x = \phi(t)$ are derivable functions; then by 3.2 (4)

$$\frac{dy}{dt} = f'(x)\,\phi'(t),$$

and so

$$dy = f'(x)\,\phi'(t)\,dt.$$

By (v) applied to $\phi(t)$, $dx = \phi'(t)\,dt$. Hence we still have

$$dy = f'(x)\,dx.$$

Thus (v) is more general than it appears to be at first sight, for in the argument leading to (v) $x$ was the independent variable; we have now shown that the same relation is true even when $x$ is not the independent variable in $f(x)$, but is itself a function. We may summarise by saying that *formulae in differentials are valid whether the variable is independent or not*. The technical convenience of differentials arises from this property, especially in geometrical applications of the calculus (Ch. 8).

(3) *Second-order differentials.* Since $d^2y/dx^2$ (as defined in 3.51) is also a composite symbol, we may enquire whether the part $d^2y$ can be defined alone in such a way that $d^2y \div (dx)^2 = f''(x)$. It can, but we shall not do so here because such 'second order' differentials lack the advantage which 'first order' ones possess: there is no invariance property because in general

$$\frac{d^2y}{dt^2} \neq f''(x) \left(\frac{dx}{dt}\right)^2 ;$$

see Ex. 3 (*e*), no. 22.

## 3.93 Differentiable functions

*Definition.* The function $y = f(x)$ is *differentiable* at $x$ if it is defined at and near $x$, and the increment $\delta y$ caused by changing $x$ to $x + \delta x$ can be expressed in the form
$$\delta y = A\,\delta x + \epsilon\,\delta x, \tag{vii}$$
where $A$ is in general a function of $x$, but is independent of $\delta x$, and $\epsilon$ is in general a function of both $x$ and $\delta x$ which tends to zero when $\delta x \to 0$.

In this case, (vii) gives $\delta y/\delta x = A + \epsilon$. Letting $\delta x \to 0$, the right-hand side of this equation tends to $A$. Hence $\lim_{\delta x \to 0} \delta y/\delta x = A$, so that the function $f(x)$ is derivable at $x$ and $A$ is its derivative, $f'(x)$. We have therefore shown that, if $f(x)$ is differentiable at a point, then it is also derivable there.

Conversely, if $f(x)$ is derivable at $x$, then (i) holds, and equation (vii) is satisfied with $A = f'(x)$ and $\epsilon = \eta$. Hence $f(x)$ is also differentiable at $x$.

It appears that, for the function $f(x)$, the properties of being derivable and of being differentiable are equivalent: one implies the other. It is thus customary to speak of *differentiating* $f(x)$ wo $x$ when we mean the process of calculating $f'(x)$, i.e. of *deriving* $f(x)$. Strictly, to *differentiate* $y = f(x)$ is to write down the equation (v) for its differential $dy$; but (v) shows that we may pass directly from differential to derivative on dividing by $dx$, and conversely, any equation involving a derivative can be converted into one between differentials by multiplying by $dx$.

In view of the equivalence of 'derivability' and 'differentiability', the reader may wonder why the two concepts were introduced. When we consider functions of more than one variable in Ch. 9 we shall define 'differentiable function' and 'differential' in essentially the same manner as here, but we shall find that 'derivability' and 'differentiability' are no longer equivalent; and we shall justify the introduction of differentials by their great technical convenience, which is not so well illustrated by functions of a single variable.

## Miscellaneous Exercise 3(e)

*Calculate the derivative of*

1 $\cos(\sin x)$.

2 $(1-x^2)^{\frac{3}{2}} \sin^{-1} x$.

3 $\tan^{-1}(n\tan x)$.

4 $\sin^{-1}\dfrac{a+b\cos x}{b+a\cos x}$ if $0 < a < b, 0 < x < \pi$.

5 $\cos^{-1}\{2x(1-x^2)^{\frac{1}{2}}\}$.

6 $\tan^{-1}\dfrac{2\sqrt{x}}{1-x}$.

7 $\sin^{-1}\sqrt{(1-x^2)}$ if $0 < x < 1$, and explain why the result is also the derivative of $\cos^{-1}x$.

8 Show that each of the functions

$$2\sin^{-1}\sqrt{\frac{x-b}{a-b}}, \quad 2\tan^{-1}\sqrt{\frac{x-b}{a-x}}, \quad \sin^{-1}\left(\frac{2\sqrt{\{(a-x)(x-b)\}}}{a-b}\right)$$

has derivative $1/\sqrt{\{(a-x)(x-b)\}}$.

*Calculate the nth derivative of*

9 $\dfrac{x+1}{x^2-4}$.

10 $\dfrac{x^3}{(x-1)(x-2)}$.

11 $\sin x \sin 3x$.

*Prove the following properties of a polynomial $f(x)$.*

12 If $f(x)$ is divisible by $(x-a)^m$, then $f'(x)$ is divisible by $(x-a)^{m-1}$.

13 *Conversely*, if $f(x)$ is divisible by $x-a$ and $f'(x)$ is divisible by $(x-a)^{m-1}$, then in fact $f(x)$ is divisible by $(x-a)^m$.

14 If $a$, $b$ are roots of $f(x) = 0$, then $f'(x) = 0$ has at least one root between $a$ and $b$ (*Rolle's theorem for polynomials*). [Suppose $a$, $b$ to be *consecutive* roots. Write $f(x) = (x-a)^m(x-b)^n g(x)$, where $g(x)$ has the same sign for $a \leqslant x \leqslant b$. Verify that $f'(x) = (x-a)^{m-1}(x-b)^{n-1}h(x)$ where $h(a)$, $h(b)$ have opposite signs, so that $h(x)$ and therefore $f'(x)$ is zero for some $x$ between $a$ and $b$.]

15 Not more than one root of $f(x) = 0$ can lie between consecutive roots of $f'(x) = 0$. [Suppose $a'$, $b'$ are consecutive roots of $f'(x) = 0$, and if possible let there be two roots $a$, $b$ of $f(x) = 0$ between them. Use no. 17 to show there would be a root $x = c'$ of $f'(x) = 0$ between $a$, $b$, i.e. between $a'$ and $b'$.]

16 There is a root of $f'(x) + \lambda f(x) = 0$ between any pair of roots of $f(x) = 0$. [Begin as in no. 14. Cf. Ex. 6 (a), no. 4.]

*Prove the following properties of a rational function $h(x) = f(x)/g(x)$.*

17 If $g(x)$ has a factor $(x-a)^m$, then the denominator of $h'(x)$ (after $h'(x)$ has been reduced to its lowest terms) is divisible by $(x-a)^{m+1}$ but by no higher power of $x-a$.

18 A rational function whose denominator contains a first degree factor $x-a$ cannot be the derivative of any rational function. In particular, $1/x$ *is not the derivative of a rational function*.

19 If $y = (\tan x + \sec x)^n$, prove $dy/dx = ny \sec x$.

20 If $y = ax \sin(b/x)$, prove

$$x^4\frac{d^2y}{dx^2} + b^2y = 0 \quad \text{and} \quad xy\frac{d^3y}{dx^3} = \frac{d^2y}{dx^2}\left(x\frac{dy}{dx} - 4y\right).$$

**21** Transform the equation

$$4x\frac{d^2y}{dx^2} + 2\frac{dy}{dx} + y = 0$$

into one in which $t$ is the independent variable, where $x = t^2$.

**\*22** Assuming that $d^2y$ has been defined so that $d^2y = f''(x)(dx)^2$, where $y = f(x)$ and $x$ is the independent variable, show that if $x$ is a function of $t$, then

$$\frac{d^2y}{dt^2} = f''(x)\left(\frac{dx}{dt}\right)^2 + f'(x)\frac{d^2x}{dt^2}$$

and so    $d^2y = f''(x)(dx)^2 + f'(x)d^2x \neq f''(x)(dx)^2$
in general.

**23** If $y = \cos(p\sin^{-1}x)$, prove $(1-x^2)y'' - xy' + p^2y = 0$.

**24** If $x = \tan\theta$, $y = \tan k\theta$, prove

$$(1+x^2)\frac{d^2y}{dx^2} = 2(ky-x)\frac{dy}{dx}.$$

**\*25** If $y = \sin n\theta/\sin\theta$ and $x = \cos\theta$, prove that

$$(1-x^2)\frac{dy}{dx} - xy + n\cos n\theta = 0 \quad \text{and} \quad (1-x^2)\frac{d^2y}{dx^2} - 3x\frac{dy}{dx} + (n^2-1)y = 0.$$

Taking $n = 5$, verify that the last equation is satisfied if $y$ is a polynomial of the form $x^4 + ax^2 + b$; and find $a$, $b$.

**26** Find $dy/dx$ if $y = \sin\{(x+y)^2\}$.

*Investigate the turning points (if any) of*

**27** $\dfrac{(1+x)^2}{(1-x)^3}.$      **28** $\dfrac{ax+b}{cx+d}.$      **29** $\dfrac{\sin(x+a)}{\sin(x+b)}.$

**30** $\dfrac{a+b\cos x}{c+d\cos x}.$      **31** $x(x^2+a^2)^{-\frac{1}{2}} - x(x^2+b^2)^{-\frac{1}{2}}$   $(b > a).$

**32** If $(x_0, y_0)$ is a stationary point on the curve $x^3 + y^3 - 9xy + 1 = 0$, prove that at this point $d^2y/dx^2 = 18/(27 - x_0^3)$. Prove also that the stationary points are $x = (27 \pm 3\sqrt{78})^{\frac{1}{3}}$, and determine which is a maximum and which is a minimum.

**33** Prove that $8(\frac{1}{2}x - \sin\frac{1}{2}x) > x - \sin x$ when $x > 0$. [Use 3.83, Corollary 2.]

**34** If $0 < \sin\alpha < 1$, prove that $x\sin\alpha - \sin^{-1}(\sin x\sin\alpha)$ increases as $x$ increases from 0 to $\frac{1}{2}\pi$, $\alpha$ remaining constant. If $0 \leqslant \alpha \leqslant \frac{1}{2}\pi$ and $\alpha$ is varied, show that the expression takes its greatest possible value when $x = \frac{1}{2}\pi$ and $\alpha$ is the positive acute angle such that $\cos\alpha = 2/\pi$.

**\*35** If $f(x) = \sin x + \frac{1}{2}\sin 2x + \frac{1}{3}\sin 3x$, find the stationary points of $f(x)$ in $0 < x < \pi$. Over what part of $0 < x < \pi$ is $f(x)$ (i) increasing; (ii) decreasing? Prove $f(x) > 0$ for $0 < x < \pi$, and find where it attains its greatest value in this interval.

**36** If $f(x) = x^\alpha y^\beta - \alpha x - \beta y$ where $x > 0$, $y > 0$, $\alpha + \beta = 1$ and $0 < \alpha < 1$, and $y$ is a parameter, prove that the *greatest* value of $f(x)$ is 0 and occurs when $x = y$. Deduce that $x^\alpha y^\beta < \alpha x + \beta y$ unless $x = y$.

**37** A horizontal ray of light from a source $A$ meets the vertical plane surface of a block of glass at $P$, and passes inside to a point $B$. If $v_1, v_2$ are the speeds before and after entry, and $\alpha$, $\beta$ are the angles of incidence and refraction, show that the passage from $A$ to $B$ occurs in the least possible time if

$$\frac{\sin \alpha}{\sin \beta} = \frac{v_1}{v_2}.$$

[Let $M$, $N$ be the feet of perpendiculars from $A$, $B$ to the surface; $MP = x$, $AM = a$, $BN = b$, $MN = c$.]

**38** A right circular cone of height $h$, base radius $r$, and slant height $l$ has a constant volume $V$. Show that the combined area $S$ of its base and curved surface is a minimum when $l = 3r$. [Express $S$ in terms of $V$ and the semi-vertical angle $\theta$.]

**39** Find the area of the largest rectangle which can be inscribed in the ellipse $x^2/a^2 + y^2/b^2 = 1$.

**40** Prove that the length intercepted on the tangent to $x^2/a^2 + y^2/b^2 = 1$ by the axes has one stationary value. Find it, and prove it is a minimum.

# 4

## INTEGRATION

### (A) Methods of integration. The logarithmic, exponential and hyperbolic functions

#### 4.1 The process inverse to derivation

#### 4.11 The problem

In Ch. 3 we considered the process of derivation: given a function $y = f(x)$, find $dy/dx$. We now turn to the inverse process: given $f(x)$, find a function $y$ which is such that $dy/dx = f(x)$.

We do not enter here into general considerations of whether the problem always has a solution, i.e. whether a function $y$ having the required property exists; this will depend on $f(x)$, as indicated in 4.16 (2). However, if $y = \phi(x)$ is known to be a solution, then $y = \phi(x) + c$ will also be a solution for any choice of the constant $c$. Further, every solution can then be written in this form; for if $y = \psi(x)$ is another solution, then from $\phi'(x) = f(x)$ and $\psi'(x) = f(x)$ we have $\psi'(x) - \phi'(x) = 0$, so that by the corollary in 3.82, $\psi(x) - \phi(x) = c$ for some constant $c$, i.e. $\psi(x) = \phi(x) + c$.

The general solution $y = \phi(x) + c$ is written

$$y = \int f(x)\, dx$$

and is called an *indefinite integral* of $f(x)$ wo $x$, or a *primitive function* of $f(x)$ wo $x$. In this expression the $dx$ is not a differential, but in 4.21 we shall prove that it behaves like a differential when a substitution is made for $x$ (i.e. a change of variable), and that its inclusion in the symbol is justified. At present, $\int \ldots dx$ is to be regarded as a composite sign for the operation of finding a function having the derivative ..., i.e. the inverse of the operation symbolised by $d/dx$. This inverse process is called *integration* wo $x$, and the function which replaces the dots ... is called the *integrand*.

*We understand always that $x$ is confined to intervals throughout which the integrand is continuous.*

## 4.12 Some standard integrals

*For brevity we shall usually omit the arbitrary constant c in this chapter and in Answers to the Exercises.* The results of 3.3 show that

$$\int x^n dx = \frac{x^{n+1}}{n+1} \quad provided \quad n \neq -1,$$

$$\int \cos x\, dx = \sin x, \quad \int \sin x\, dx = -\cos x,$$

$$\int \sec^2 x\, dx = \tan x, \quad \int \operatorname{cosec}^2 x\, dx = -\cot x,$$

$$\int \frac{1}{1+x^2} dx = \tan^{-1} x,$$

$$\int \frac{1}{\sqrt{(1-x^2)}} dx = \sin^{-1} x \quad \text{or} \quad -\cos^{-1} x.$$

## 4.13 Some properties of indefinite integrals

(1) *If k is constant, then*

$$\int k f(x)\, dx = k \int f(x)\, dx.$$

For both sides have the same derivative $kf(x)$ wo $x$ (using 3.11, ex. (iii) for the right-hand side), and hence differ by a constant at most. As each integral implies the presence of an arbitrary constant, the two sides have the same meaning. Similarly,

$$(2) \qquad \int (u+v)\, dx = \int u\, dx + \int v\, dx,$$

because each side has derivative $u+v$ (3.2 (1)).

(3) *If*
$$\int f(x)\, dx = \phi(x), \quad then \quad \int f(ax+b)\, dx = \frac{1}{a} \phi(ax+b)$$

(where $a \neq 0$, $b$ are constants), because each side of the last equation has derivative $f(ax+b)$, by 3.2 (4).

Property (3) shows that each of the standard integrals in 4.12 can be generalised by replacing $x$ by the *linear* function $ax+b$; (1) and (2) show that if a given function can be split into sums or differences of constant multiples of these standard forms, its integral is found by integrating term-by-term ('integration by decomposition').

**Examples**

(i) $\displaystyle\int \frac{x^3 + 5x + 1}{x^3}\, dx = \int \left(1 + \frac{5}{x^2} + \frac{1}{x^3}\right) dx = x - \frac{5}{x} - \frac{1}{2x^2} + c.$

(ii) $\displaystyle\int \frac{dx}{\sqrt{(1+x)} + \sqrt{x}}$ can be decomposed by first multiplying numerator and

denominator by $\sqrt{(1+x)} - \sqrt{x}$ ('rationalising the denominator'), giving

$$\int \{\sqrt{(1+x)} - \sqrt{x}\}\, dx = \tfrac{2}{3}(1+x)^{\frac{3}{2}} - \tfrac{2}{3}x^{\frac{3}{2}} + c.$$

(iii) $\displaystyle\int \sin^3 x\, dx = \int \tfrac{1}{4}(3\sin x - \sin 3x)\, dx = \tfrac{1}{4}(-3\cos x + \tfrac{1}{3}\cos 3x) + c.$

(iv) $\displaystyle\int \frac{dx}{x^2(1+x^2)} = \int \left(\frac{1}{x^2} - \frac{1}{1+x^2}\right) dx = -\frac{1}{x} - \tan^{-1} x + c.$

(v) $\displaystyle\int \frac{dx}{10 + 12x + 4x^2} = \int \frac{dx}{1 + (2x+3)^2} = \tfrac{1}{2}\tan^{-1}(2x+3) + c.$

The results of Ex. 4 (a), nos. 2, 3 are very useful.

## 4.14 Areas

An elementary account of finding the area under a continuous curve $y = f(x)$ between the ordinates through $x = a$, $x = b$ $(a < b)$ proceeds as follows. (We suppose the part $AB$ of the curve with which we are concerned does not pass below $Ox$.)

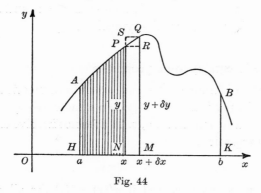

Fig. 44

If $PN$ is the ordinate through the point $P(x, y)$ on the curve, the area $AHNP$ is a function of $x$, say $A(x)$. If $Q$ is the point $(x + \delta x, y + \delta y)$, then area $AHMQ$ is $A + \delta A$, so that the strip $PNMQ$ has area $\delta A$, which lies in value between the areas of the inner and outer rectangles $PNMR$, $SNMQ$. Thus

$$\delta A \text{ lies between } y\,\delta x \text{ and } (y + \delta y)\,\delta x,$$

so
$$\frac{\delta A}{\delta x} \text{ lies between } y \text{ and } y + \delta y.$$

When $\delta x \to 0$, then $\delta y \to 0$ by continuity of $y = f(x)$; so that $\delta A/\delta x$, lying between $y$ (which is fixed) and $y + \delta y$ (which tends to $y$), also tends to $y$ when $\delta x \to 0$. By definition, $\delta A/\delta x \to dA/dx$ when $\delta x \to 0$; hence
$$\frac{dA}{dx} = y = f(x),$$

from which
$$A = \int f(x)\, dx$$

$$= \phi(x) + c, \quad \text{say.}$$

When $PN$ coincides with $AH$, i.e. when $x = a$, the area $AHNP$ is zero. Hence $0 = \phi(a) + c$, $c = -\phi(a)$, and

$$A = \phi(x) - \phi(a).$$

To obtain the complete area, we make $PN$ coincide with $BK$ by putting $x = b$:
$$\text{area } AHKB = \phi(b) - \phi(a).$$

To calculate the required area we therefore
   (a)  write down an indefinite integral $\phi(x)$ of $f(x)$;
   (b)  find its value when $x = b$, then when $x = a$, and subtract.
This process is indicted by the symbol $[\phi(x)]_a^b$.

## 4.15 Definite integrals; some properties

Although the difference $\phi(b) - \phi(a)$ arises in connection with areas, it can be considered independently as a number associated with the function $f(x)$ and two numbers $a, b$. It does not involve any arbitrary constant of integration.

*Definitions*
   (a)  If $f(x)$ is continuous and if $\phi'(x) = f(x)$ for all $x$ for which $a \leqslant x \leqslant b$, the symbol $\int_a^b f(x)\, dx$ is defined to mean $\phi(b) - \phi(a)$ and is called the *definite integral* of $f(x)$ wo $x$ from $a$ to $b$.
   (b)  The numbers $a, b$ are the *limits of integration*,† and the interval $a \leqslant x \leqslant b$ is the *range of integration*.

† Here 'limit' is used in the sense of 'end', not in the technical sense of Ch. 2.

*Remarks*

($\alpha$) The definition

$$\int_a^b f(x)\,dx = \phi(b) - \phi(a)$$

is intended to apply even when $b \leqslant a$, and also when $f(x)$ is negative in all or part of the range of integration, provided that the conditions in ($a$) are satisfied throughout the range.

($\beta$) Although $\int_a^b f(x)\,dx$ depends on the numbers $a$, $b$, *it does not depend on the variable of integration* $x$, which could be replaced by any other letter; e.g.

$$\int_a^b f(t)\,dt = [\phi(t)]_a^b = \phi(b) - \phi(a),$$

so that $\int_a^b f(x)\,dx, \int_a^b f(t)\,dt$ are the same symbol. This is not the case for the corresponding indefinite integral, which is a function of the variable of integration.

The following properties come immediately from the definition ($a$):

**(1)** $\displaystyle\int_a^a f(x)\,dx = 0.$

**(2)** $\displaystyle\int_a^b f(x)\,dx = -\int_b^a f(x)\,dx.$

**(3)** $\displaystyle\int_a^b f(x)\,dx = \int_a^c f(x)\,dx + \int_c^b f(x)\,dx,$ where $c$ *need not lie between* $a$ and $b$.

Using 4.13, we have

**(4)** $\displaystyle\int_a^b kf(x)\,dx = k\int_a^b f(x)\,dx.$

**(5)** $\displaystyle\int_a^b \{f(x)+g(x)\}\,dx = \int_a^b f(x)\,dx + \int_a^b g(x)\,dx.$

For if $\psi(x) = \int g(x)\,dx$, then the left-hand side is

$$[\phi(x)+\psi(x)]_a^b = \phi(b)+\psi(b)-\phi(a)-\psi(a) = \text{right-hand side}.$$

The reader should illustrate properties (1)–(5) by 'areas' under sketch-graphs.

Since the definite integral of $f(x)$ depends solely on the limits $a$, $b$, we may replace $b$ (say) by a variable $x$ and thereby obtain a *function* $\int_a^x f(t)\,dt$ of $x$. (To avoid confusion, the variable of integration has been altered from $x$ to $t$: see Remark $(\beta)$ above.)

(6) $\int_a^x f(t)\,dt$ *is a continuous function of $x$ for $a \leqslant x \leqslant b$.*

For $\int_a^x f(t)\,dt = \phi(x) - \phi(a)$, and $\phi'(x) = f(x)$ for $a \leqslant x \leqslant b$ by the definition $(a)$. Hence $\phi(x)$ is continuous for $a \leqslant x \leqslant b$ because it is derivable (3.12). Thus

(7) $\int_a^x f(t)\,dt$ *is a derivable function of $x$ for $a \leqslant x \leqslant b$, with derivative* $f(x)$.

(8) *If $f(x) > 0$ for $a < x < b$, then $\int_a^x f(t)\,dt$ is an increasing function of $x$ for $a \leqslant x \leqslant b$.*

For it has derivative $f(x)$ which is positive; use 3.83. In particular, from 3.83, Corollary 2:

(9) *If $f(x) > 0$ for $a < x < b$, then $\int_a^b f(x)\,dx > 0$.*

(10) *If $m < f(x) < M$ for $a < x < b$, then*

$$m(b-a) < \int_a^b f(x)\,dx < M(b-a).$$

Replace $f(x)$ by $f(x) - m$ in property (9): then by (4) and (5),

$$0 < \int_a^b \{f(x) - m\}\,dx = \int_a^b f(x)\,dx - \int_a^b m\,dx,$$

i.e. $$m(b-a) < \int_a^b f(x)\,dx.$$

Similarly, writing $M - f(x)$ for $f(x)$, we prove the other inequality.

(11) *If $f(x) > g(x)$ for $a < x < b$, then $\int_a^b f(x)\,dx > \int_a^b g(x)\,dx$.*

Replace $f(x)$ by $f(x) - g(x)$ in (9), and use (4), (5).

## 4.16 Criticism of 4.14, 4.15

(1) The discussion in 4.15 takes for granted that we understand what is meant by 'the area under a curve'. More precisely we are assuming that, given

a function $y = f(x)$ which is continuous for $a \leqslant x \leqslant b$, there is associated with the part of the plane bounded by the curve, the line $Ox$, and the lines $x = a$, $x = b$ a definite number called its 'area'. Thus the work only suggests what the formula for this 'area' should be, when the term 'area under a curve' has been defined to agree as closely as possible with our intuitive ideas of area as obtained from straight-line figures. The matter will be taken up in Ch. 7.

(2) Our definition of $\int_a^b f(x)\,dx$ presupposes that, for all $x$ for which $a \leqslant x \leqslant b$, we can find an indefinite integral $\phi(x)$ of $f(x)$. If we cannot, then the symbol might be meaningless. Also we have required $f(x)$ to be continuous for $a \leqslant x \leqslant b$, but later in this chapter we show how this restriction can be removed in certain circumstances (4.9).

(3) In Ch. 7 we shall approach the definite integral as the limit of a certain summation;† such a limit always exists for suitable classes of functions (e.g. continuous ones). We then deduce that a solution $y$ of $dy/dx = f(x)$ exists for *continuous* functions $f(x)$ by proving that

$$y = \int_a^x f(t)\,dt$$

is such a solution. We should then be assured that an indefinite integral $\phi(x)$ of $f(x)$ exists, even if we cannot perform the inverse operation symbolised by $\int f(x)\,dx$ to express this function in terms of those already known.

    Meanwhile, we advance with the practical technique of finding a formula for $\int f(x)\,dx$ for as many types of function $f(x)$ as possible.

## Exercise 4(a)

*Integrate by decomposition*

1   $1/\{\sqrt{(x+1)} - \sqrt{(x-1)}\}$.      2   $\cos^2 x$.      3   $\sin^2 x$.

4   $\tan^2 x$.      5   $\cos^3 x$.      6   $\dfrac{x^2}{1+x^2}$.      7   $\dfrac{x^4}{1+x^2}$.

8   $\dfrac{x}{\sqrt{(1+x)}}$.      9   $\dfrac{1+x^{-2}}{1+x^2}$.      10   $\sin x \cos 3x$.

*Write down the integrals of*

11   $(ax+b)^n$, $n \neq -1$.      12   $\cos(ax+b)$.

13   $\sin(ax+b)$.      14   $\dfrac{1}{a^2+x^2}$.

15   $\dfrac{1}{\sqrt{(a^2-x^2)}}$.   [In 14, 15 write $x/a$ for $x$ in the standard integrals.]

16   $\dfrac{1}{(1+x^2)(4+x^2)}$.

*By replacing $x$ by a suitable linear function in the standard forms, integrate*

17   $\dfrac{1}{x^2+6x+9}$.      18   $\dfrac{1}{x^2+6x+10}$.      19   $\dfrac{1}{\sqrt{(2x-x^2)}}$.

† This is the origin of the sign $\int$.

*20 Use 4.15(10) to prove $1 \cdot 59 < \int_{\frac{1}{2}}^{1} \sqrt{(10 + x^3)} \, dx < 1 \cdot 66$.

*21 If $n > 1$ and $0 < x < 1$, verify that $\sqrt{(1 - x^2)} < \sqrt{(1 - x^{2n})} < 1$, and deduce

$$0 \cdot 5 < \int_{0}^{\frac{1}{2}} \frac{dx}{\sqrt{(1 - x^{2n})}} < 0 \cdot 524.$$

*22 Use the inequality $\sin x < \sqrt{(\sin x)} < \sqrt{x} \; (0 < x < \frac{1}{2}\pi)$ to prove

$$1 < \int_{0}^{\frac{1}{2}\pi} \sqrt{(\sin x)} \, dx < 1 \cdot 32.$$

## 4.2 Some general methods of integration

The purpose of these methods is to reduce a given integral to another which is already known or can be found easily.

### 4.21 Integration by substitution (change of variable)

To find $\int f(x) \, dx$ we may proceed as follows. Let

$$y = \int f(x) \, dx,$$

so that

$$\frac{dy}{dx} = f(x).$$

Put $x = g(t)$; then by 'function of a function',

$$\frac{dy}{dt} = \frac{dy}{dx}\frac{dx}{dt} = f(x) g'(t) = f\{g(t)\} g'(t).$$

Hence

$$y = \int f\{g(t)\} g'(t) \, dt. \tag{i}$$

For a suitably chosen function $g(t)$ it may happen that $f\{g(t)\} g'(t)$ is a simpler function of $t$ than $f(x)$ is of $x$, and may be recognised as a standard form. We should then have $y$ expressed as a function of $t$, say $y = \psi(t)$. To restore the variable $x$, we should use the original substitution $x = g(t)$ to get $t$ in terms of $x$.

Equation (i) shows that when we transform an integral by putting $x = g(t)$, we may substitute for $x$ in the integrand, and *replace $dx$ by $g'(t) \, dt$*. The $dx$ thus behaves like a differential (cf. 4.11).

### Examples

(i) $y = \int x(3x - 2)^7 \, dx$.

This could be calculated by direct expansion, followed by integration term-by-term. The following is easier.

Put $t = 3x - 2$, so that $x = \frac{1}{3}(t + 2)$. Then

$$\frac{dy}{dt} = \frac{dy}{dx}\frac{dx}{dt} = x(3x - 2)^7 \times \frac{1}{3}$$

$$= \frac{1}{9}(t + 2)\,t^7 = \frac{1}{9}(t^8 + 2t^7).$$

$$\therefore \quad y = \frac{1}{9}(\tfrac{1}{9}t^9 + \tfrac{1}{4}t^8) = \tfrac{1}{324}t^8(4t + 9)$$

$$= \tfrac{1}{324}(3x - 2)^8\,(12x + 1).$$

When surds are involved, we usually choose a substitution which will rationalise them or reduce them to a single term.

(ii) $y = \displaystyle\int \frac{x}{\sqrt{(x - 3)}}\,dx.$

Put $x - 3 = t^2$, so that $x = t^2 + 3$. Then

$$\frac{dy}{dt} = \frac{dy}{dx}\frac{dx}{dt} = \frac{x}{\sqrt{(x - 3)}}\,2t = \frac{t^2 + 3}{t}\,2t = 2t^2 + 6.$$

$$\therefore \quad y = \tfrac{2}{3}t^3 + 6t = \tfrac{2}{3}t(t^2 + 9)$$

$$= \tfrac{2}{3}(x - 3)^{\frac{1}{2}}\,(x + 6).$$

*Alternatively*, using the 'rule' at the end of 4.21,

$$\int \frac{x}{\sqrt{(x - 3)}}\,dx = \int \frac{t^2 + 3}{t}\,2t\,dt = \int (2t^2 + 6)\,dt = \text{etc.}$$

(iii) $\int x^2 \sqrt{(1 + x^3)}\,dx.$

Put $t = 1 + x^3$, so that $dt = 3x^2\,dx$. The integral is

$$\int \tfrac{1}{3}t^{\frac{1}{2}}\,dt = \tfrac{2}{9}t^{\frac{3}{2}} = \tfrac{2}{9}(1 + x^3)^{\frac{3}{2}}.$$

The 'rule' considerably reduces calculations in this example because some of the factors in the integrand cancel out before the substitution is made for $x$ there. The reader should try it by the method of exs. (i), (ii).

The substitution $t^2 = 1 + x^3$ would also rationalise the surd and enable the integral to be found. The one used was preferred because $x^2$ is (apart from a constant factor) the derivative of the expression $1 + x^3$ under the root sign.

The same method can be used for $\int x^{n-1}(ax^n + b)^m\,dx$.

(iv) $\displaystyle\int \frac{x^2}{1 + x^6}\,dx.$

Here the denominator prevents direct integration. Putting $t = 1 + x^6$ would not help because this would give $\dfrac{1}{6}\displaystyle\int \frac{dt}{t\sqrt{(t - 1)}}$, introducing surds.

Observe that $x^2\,dx$ is almost the differential of $x^3$, and put $t = x^3$:

$$\text{integral} = \int \frac{\tfrac{1}{3}dt}{1 + t^2} = \tfrac{1}{3}\tan^{-1} t = \tfrac{1}{3}\tan^{-1}(x^3).$$

Sometimes more than one substitution is needed: the first leads to a result which suggests the second.

(v) $\displaystyle\int \frac{dx}{x^2\sqrt{(x^2+1)}}$.

The reader should verify that putting $t = x^2+1$ or $t^2 = x^2+1$ does not help. Put $x = 1/t$, so that $dx = -(1/t^2)\,dt$:

$$\text{integral} = \int \frac{-(1/t^2)\,dt}{(1/t^2).(1/t)\sqrt{(1+t^2)}} = -\int \frac{t\,dt}{\sqrt{(1+t^2)}}.$$

This suggests putting $u^2 = 1+t^2$, so that $u\,du = t\,dt$:

$$\text{integral} = -\int \frac{u\,du}{u} = -\int du = -u$$

$$= -\sqrt{(1+t^2)} = -\frac{1}{x}\sqrt{(x^2+1)}.$$

It is now clear that the single substitution $x = 1/\sqrt{(u^2-1)}$ would have reduced the integral directly.

Although the attempt to rationalise by the algebraic substitution $t^2 = x^2+1$ at the outset was useless, we remark that *integrals involving $\sqrt{(a^2 \pm x^2)}$ are often reduced by a trigonometrical substitution*. In this example put $x = \tan\theta$, so that $dx = \sec^2\theta\,d\theta$:

$$\text{integral} = \int \frac{\sec^2\theta\,d\theta}{\tan^2\theta\,\sec\theta} = \int \frac{\cos\theta}{\sin^2\theta}\,d\theta.$$

Now put $t = \sin\theta$, $dt = \cos\theta\,d\theta$:

$$\text{integral} = \int \frac{dt}{t^2} = -\frac{1}{t} = -\frac{1}{\sin\theta} = -\frac{1}{x}\sqrt{(x^2+1)},$$

as before.

(vi) $\int \sqrt{(1-x^2)}\,dx$.

The integrand is defined only for $-1 \leqslant x \leqslant 1$, so that an angle $\theta$ between $\pm\frac{1}{2}\pi$ always exists such that $x = \sin\theta$. This substitution is therefore legitimate, and $\theta = \sin^{-1}x$ (*principal* value).

$$\text{Integral} = \int \cos\theta.\cos\theta\,d\theta = \int \cos^2\theta\,d\theta$$

$$= \int \tfrac{1}{2}(1+\cos 2\theta)\,d\theta = \tfrac{1}{2}(\theta + \tfrac{1}{2}\sin 2\theta)$$

$$= \tfrac{1}{2}(\theta + \sin\theta\,\cos\theta)$$

$$= \tfrac{1}{2}\{\sin^{-1}x + x\sqrt{(1-x^2)}\}.$$

## 4.22 Definite integrals by substitution

The method of substitution can also be used to evaluate definite integrals. For example, the result of ex. (vi) gives

$$\int_0^1 \sqrt{(1-x^2)}\,dx = [\tfrac{1}{2}\{\sin^{-1}x + x\sqrt{(1-x^2)}\}]_0^1 = \tfrac{1}{2}(\sin^{-1}1 - \sin^{-1}0) = \tfrac{1}{4}\pi.$$

However, for *definite* integrals it is unnecessary to change back to the original variable of integration. Thus when $x = \sin\theta$, the values of $\sin^{-1}x + x\sqrt{(1-x^2)}$ when $x = 0, 1$ are the same as the values of $\theta + \frac{1}{2}\sin 2\theta$ when $\theta = 0, \frac{1}{2}\pi$, so that

$$[\sin^{-1}x + x\sqrt{(1-x^2)}]_{x=0}^{x=1} = [\theta + \tfrac{1}{2}\sin 2\theta]_{\theta=0}^{\theta=\frac{1}{2}\pi}.$$

In short, when we change the function by the substitution, we also make a corresponding change in the range of integration. The work is set out as follows.

(vii) *Evaluate* $\int_0^1 \sqrt{(1-x^2)}\,dx$.

Putting $x = \sin\theta$, $dx = \cos\theta\,d\theta$; and as $x$ increases from 0 to 1, $\theta$ increases from 0 to $\frac{1}{2}\pi$. Hence

$$\int_0^1 \sqrt{(1-x^2)}\,dx = \int_0^{\frac{1}{2}\pi} \cos\theta . \cos\theta\,d\theta = \int_0^{\frac{1}{2}\pi} \tfrac{1}{2}(1+\cos 2\theta)\,d\theta$$

$$= [\tfrac{1}{2}(\theta + \tfrac{1}{2}\sin 2\theta)]_0^{\frac{1}{2}\pi} = \tfrac{1}{2}(\tfrac{1}{2}\pi - 0) = \tfrac{1}{4}\pi.$$

(viii) $\int_1^{\sqrt{3}} \dfrac{dx}{x^2\sqrt{(x^2+1)}}$.

We saw in ex. (v) that the single substitution $x = (u^2-1)^{-\frac{1}{2}}$ reduces the integral. As $x$ increases from 1 to $\sqrt{3}$, $u = +\sqrt{(1+1/x^2)}$ decreases from $\sqrt{2}$ to $2/\sqrt{3}$. Thus, using the working of ex. (v),

$$\int_1^{\sqrt{3}} \frac{dx}{x^2\sqrt{(x^2+1)}} = [-u]_{\sqrt{2}}^{2/\sqrt{3}} = \sqrt{2} - 2/\sqrt{3}.$$

Write $\phi(x) = \int f(x)\,dx$, and *suppose $a < b$.* Let $g(t)$ be a continuous function of $t$ with a continuous derivative $g'(t)$ in the range between $t_1$ and $t_2$ inclusive, where $a = g(t_1)$ and $b = g(t_2)$.

*Suppose first that $g(t)$ steadily increases from $a$ to $b$ as $t$ increases from $t_1$ to $t_2$.* By equation (i), p. 100,

$$\phi(x) = \int f\{g(t)\}\,g'(t)\,dt = \psi(t), \quad \text{say,}$$

so that $$\phi\{g(t)\} = \psi(t).$$

Then $$\int_a^b f(x)\,dx = \phi(b) - \phi(a) = \phi\{g(t_2)\} - \phi\{g(t_1)\}$$

$$= \psi(t_2) - \psi(t_1) = \int_{t_1}^{t_2} f\{g(t)\}\,g'(t)\,dt.$$

Secondly, suppose $g(t)$ *steadily decreases* from $b$ to $a$ as $t$ increases from $t_2$ to $t_1$; then the above working is unchanged.

*Remark.* If we use the substitution $x = g(t)$ to transform

$$\int_a^b f(x)\,dx \quad \text{into} \quad \int_{t_1}^{t_2} f\{g(t)\}\,g'(t)\,dt,$$

where $t = t_1$ when $x = a$ and $t = t_2$ when $x = b$,

care must be taken to ensure that $g(t)$ *steadily* changes from $a$ to $b$ as $t$ varies from $t_1$ to $t_2$; otherwise $g(t)$ is not a legitimate substitution for $x$, a steadily changing variable. For example, by direct evaluation

$$\int_{-1}^{+1} \frac{dx}{1+x^2} = [\tan^{-1} x]_{-1}^{+1} = \tfrac{1}{2}\pi.$$

The substitution $x = 1/t$, which gives

$$\int_{-1}^{+1} \frac{dx}{1+x^2} = \int_{-1}^{+1} \frac{-dt}{1+t^2} = -\tfrac{1}{2}\pi,$$

is not legitimate because $g(t) = 1/t$ does not *steadily* vary from $-1$ to $+1$ as $t$ increases from $-1$ to $+1$. Also see Ex. 4($b$), nos. 29, 30. This matter will be taken up again in 7.23.

### Exercise 4($b$)

*Integrate each of the following functions by use of a substitution.*

1   $(x+1)(x-5)^6$.

2   $\dfrac{x}{(x+3)^3}$.

3   $\dfrac{x}{\sqrt{(x+3)}}$.

4   $x(1-x^2)^5$.

5   $\dfrac{x}{(4x^2+1)^3}$.

6   $\dfrac{x^3}{\sqrt{(1+x^4)}}$.

7   $\cos x \sin^3 x$.

8   $\tan^3 x \sec^2 x$.

9   $\dfrac{1}{\sqrt{(9-4x^2)}}$.

10   $\dfrac{1}{9+4x^2}$.

11   $(9-x^2)^{-\frac{3}{2}}$.

12   $(9+x^2)^{-\frac{3}{2}}$.

13   $(x^2-9)^{-\frac{3}{2}}$.

14   $\dfrac{x^3}{(x^2+1)^3}$.

15   $\dfrac{(1+x^2)^{\frac{3}{2}}}{x^6}$.

16   $\dfrac{1}{x\sqrt{(x^2-1)}}$.

17   $\dfrac{1}{x^2\sqrt{(x^2-1)}}$.

18   $\dfrac{\tan^{-1}x}{1+x^2}$.

19   $\dfrac{1}{\sqrt{\{x(2-x)\}}}$.

20   $\cos^3 x \sin^4 x$.

21   $\sqrt{\left(\dfrac{1+x}{1-x}\right)}$.   [Rationalise the numerator.]

22   $\dfrac{1}{x^2}\sqrt{\left(\dfrac{x-1}{x+1}\right)}$.   [Put $t = 1/x$.]

*Evaluate the following definite integrals.*

23   $\displaystyle\int_{-1}^{2} \frac{dx}{\sqrt{(3x+5)}}$.

24   $\displaystyle\int_{0}^{1} \sqrt{(x^4+2x^7)}\,dx$.

25   $\displaystyle\int_{0}^{1} x^3\sqrt{(1-x^2)}\,dx$.

26   $\displaystyle\int_{0}^{\frac{1}{2}\pi} \sin x \cos^3 x\,dx$.

27   $\displaystyle\int_{0}^{\frac{1}{4}\pi} \frac{dx}{9\cos^2 x + 25\sin^2 x}$.   [Put $t = \tan x$.]

28   $\displaystyle\int_{\frac{1}{4}\pi}^{\frac{3}{4}\pi} \cos x \operatorname{cosec}^5 x\,dx$.

*Some of the properties in 4.15 will be required in the following.*

*29   The substitution $t = x^{\frac{2}{3}}$ applied to $\displaystyle\int_{-1}^{+1} dx$ appears to give

$$\int_{-1}^{+1} dx = \int_{1}^{1} \tfrac{3}{2}t^{\frac{1}{2}}\,dt = 0.$$

What is the value of the integral? Apply the change of variable process correctly.

**\*30** Show that $\displaystyle\int_0^\pi \cos^2 x \, dx = \frac{1}{2}\pi$ by using the substitution $t = \sin x$.

**31** By putting $x = \frac{1}{2}\pi - t$, show that

$$\int_0^{\frac{1}{2}\pi} x \sin^2 x \, dx = \frac{\pi}{2} \int_0^{\frac{1}{2}\pi} \cos^2 x \, dx - \int_0^{\frac{1}{2}\pi} x \cos^2 x \, dx.$$

Deduce that $$\frac{\pi}{2} \int_0^{\frac{1}{2}\pi} \cos^2 x \, dx = \int_0^{\frac{1}{2}\pi} x \, dx = \frac{\pi^2}{8},$$

and hence evaluate $\displaystyle\int_0^{\frac{1}{2}\pi} \cos^2 x \, dx$.

**\*32** Prove $\displaystyle\int_{-a}^{+a} f(x) \, dx = \int_0^a \{f(x) + f(-x)\} \, dx$. *Deduce that* $(a)$ *if* $f(x)$ *is odd, then*

$\displaystyle\int_{-a}^{+a} f(x) \, dx = 0$; $(b)$ *if* $f(x)$ *is even, then* $\displaystyle\int_{-a}^{+a} f(x) \, dx = 2 \int_0^a f(x) \, dx$.

**\*33** *Prove* $\displaystyle\int_0^a f(x) \, dx = \int_0^a f(a-x) \, dx$. Use this result to show that

$$\int_0^{\frac{1}{2}\pi} \frac{2\sin x + 3\cos x}{\sin x + \cos x} \, dx = \frac{5\pi}{4}.$$

**\*34** Prove $\displaystyle\int_0^\pi \frac{x \sin x}{1 + \cos^2 x} \, dx = \frac{\pi^2}{4}$.

## 4.23 Integration by parts

(1) Just as integration by substitution is the analogue of the 'function of a function' rule, so integration by parts is that of the product formula. From

$$\frac{d}{dx}(uv) = vu' + uv',$$

we have by integrating both sides wo $x$ that

$$uv = \int vu' \, dx + \int uv' \, dx.$$

If one of the integrals on the right is known, then the other can be found. Supposing $\int vu' \, dx$ is known, we may write the formula as

$$\int uv' \, dx = uv - \int vu' \, dx. \tag{ii}$$

## Examples

(i) $\int x \cos x \, dx$.

Take $u = x$, $v' = \cos x$; then $u' = 1$, $v = \sin x$.

$$\int x \cos x \, dx = x \sin x - \int \sin x \cdot 1 \, dx$$

$$= x \sin x + \cos x.$$

Had we taken $u = \cos x$, $v' = x$, the integral on the right would have been $\int -\frac{1}{2}x^2 \sin x\,dx$, which is more complicated than the given integral.

(ii) $\int x^2 \sin 3x\,dx$.

Take $u = x^2$, $v' = \sin 3x$; then $u' = 2x$, $v = -\frac{1}{3}\cos 3x$.

$$\int x^2 \sin 3x\,dx = -\frac{1}{3}x^2 \cos 3x - \int (-\tfrac{2}{3}x \cos 3x)\,dx$$

$$= -\tfrac{1}{3}x^2 \cos 3x + \frac{2}{3}\int x \cos 3x\,dx.$$

Repeating the process on the last integral, take $u = x$, $v' = \cos 3x$; then $u' = 1$, $v = \frac{1}{3}\sin 3x$, and

$$\int x \cos 3x\,dx = \tfrac{1}{3}x \sin 3x - \int \tfrac{1}{3}\sin 3x\,dx$$

$$= \tfrac{1}{3}x \sin 3x + \tfrac{1}{9}\cos 3x.$$

$$\therefore \quad \int x^2 \sin 3x\,dx = \tfrac{1}{27}(2 - 9x^2)\cos 3x + \tfrac{2}{9}x \sin 3x.$$

(iii) $\int x(3x-2)^7\,dx$ (cf. 4.21, ex. (i)).

Take $u = x$, $v' = (3x-2)^7$; then $u' = 1$, $v = \tfrac{1}{24}(3x-2)^8$.

$$\int x(3x-2)^7\,dx = \tfrac{1}{24}x(3x-2)^8 - \int \tfrac{1}{24}(3x-2)^8\,dx$$

$$= \tfrac{1}{24}x(3x-2)^8 - \tfrac{1}{24.27}(3x-2)^9$$

$$= \tfrac{1}{24.27}(3x-2)^8 \{27x - (3x-2)\}$$

$$= \tfrac{1}{324}(3x-2)^8 (12x+1), \quad \text{as before.}$$

(2) Taking $v = x$ in equation (ii), we get

$$\int u\,dx = xu - \int xu'\,dx$$

$$= xu - \int x\,du,$$

where we have used the formula (i) in 4.21 for change of variable, $x$ now being supposed a function $g(u)$ of $u$. Hence, if the integral of $g(u)$ wo $u$ can be found, then so can the integral wo $x$ of the function $u$ inverse to $x$.

(iv) $\int \sin^{-1} x\,dx$.

Take $u = \sin^{-1} x$, $v' = 1$; then $u' = 1/\sqrt{(1-x^2)}$, $v = x$.

$$\int \sin^{-1} x\,dx = x \sin^{-1} x - \int \frac{x}{\sqrt{(1-x^2)}}\,dx$$

$$= x \sin^{-1} x + \sqrt{(1-x^2)}$$

by using the substitution $t = x^2$ or $x = \sin \theta$.

(v) $\int x \tan^{-1} x \, dx$.

Take $u = \tan^{-1} x$, $v' = x$; then $u' = 1/(1+x^2)$, $v = \frac{1}{2}x^2$.

$$\int x \tan^{-1} x \, dx = \frac{1}{2}x^2 \tan^{-1} x - \int \frac{1}{2}x^2 \frac{1}{1+x^2} \, dx$$

$$= \frac{1}{2}x^2 \tan^{-1} x - \frac{1}{2} \int \left(1 - \frac{1}{1+x^2}\right) dx$$

$$= \frac{1}{2}x^2 \tan^{-1} x - \frac{1}{2}x + \frac{1}{2}\tan^{-1} x$$

$$= \frac{1}{2}(x^2 + 1) \tan^{-1} x - \frac{1}{2}x.$$

Examples (iv), (v) indicate that when the integrand involves a transcendental function whose derivative is algebraic, we can get a new integral containing only algebraic functions by taking this function as $u$.

## 4.24 Reduction formulae

If the given integral involves $n$, an integer, then integration by parts may reduce the integral to one of similar form but involving a smaller value of $n$. The relation between these integrals, known as a *reduction formula*, can be used successively until we obtain an integral corresponding to $n = 0$ or $1$ or some other small value, and this last integral may be known.

### Example

Consider $c_n = \int x^n \cos ax \, dx$, $s_n = \int x^n \sin ax \, dx$, where $a$ is constant and $n$ is a positive integer.

In $c_n$ take $u = x^n$, $v' = \cos ax$; then $u' = nx^{n-1}$, $v = (1/a) \sin ax$.

$$c_n = \frac{x^n}{a} \sin ax - \frac{n}{a} \int x^{n-1} \sin ax \, dx$$

$$= \frac{x^n}{a} \sin ax - \frac{n}{a} s_{n-1}.$$

Similarly, $\quad s_{n-1} = -\frac{x^{n-1}}{a} \cos ax + \frac{n-1}{a} \int x^{n-2} \cos ax \, dx$

$$= -\frac{x^{n-1}}{a} \cos ax + \frac{n-1}{a} c_{n-2}.$$

$$\therefore \quad c_n = \frac{x^n}{a} \sin ax - \frac{n}{a} \left\{ -\frac{x^{n-1}}{a} \cos ax + \frac{n-1}{a} c_{n-2} \right\}$$

$$= \frac{x^n}{a} \sin ax + \frac{nx^{n-1}}{a^2} \cos ax - \frac{n(n-1)}{a^2} c_{n-2}.$$

This is a reduction formula for $c_n$ which reduces the value of $n$ by two at each application, until either

$$c_0 = \int \cos ax\, dx \text{ (if } n \text{ is even)} \quad \text{or} \quad c_1 = \int x \cos ax\, dx \text{ (if } n \text{ is odd)}$$

is reached; and these integrals are respectively (see Ex. 4 (c), no. 2 for the latter)

$$\frac{1}{a}\sin ax, \quad \frac{x}{a}\sin ax - \frac{1}{a^2}\cos ax.$$

To calculate $\int x^5 \cos x\, dx$ we should use the formula thus:

$$\begin{aligned}
c_5 = \int x^5 \cos x\, dx &= x^5 \sin x + 5x^4 \cos x - 20c_3 \\
&= x^5 \sin x + 5x^4 \cos x - 20(x^3 \sin x + 3x^2 \cos x - 6c_1) \\
&= x^5 \sin x + 5x^4 \cos x - 20x^3 \sin x - 60x^2 \cos x \\
&\qquad\qquad + 120(x \sin x - \cos x) \\
&= x(x^4 - 20x^2 + 120)\sin x + 5(x^4 - 12x^2 - 24)\cos x.
\end{aligned}$$

A similar reduction formula can be obtained for $s_n$: see Ex. 4 (c), no. 12. Further examples will be met later in this chapter.

## 4.25 Definite integrals by parts and reduction

The methods of integration by parts and by reduction formula can be applied to definite integrals. Thus if $f'(x)$, $g'(x)$ are continuous,

$$\int_a^b f(x)\,g'(x)\, dx = \left[ f(x)\,g(x) - \int f'(x)\,g(x)\, dx \right]_a^b$$

$$= [f(x)\,g(x)]_a^b - \int_a^b f'(x)\,g(x)\, dx.$$

**Examples**

(i)
$$\begin{aligned}
\int_0^{\frac{1}{2}\pi} x \cos x\, dx &= [x \sin x]_0^{\frac{1}{2}\pi} - \int_0^{\frac{1}{2}\pi} \sin x\, dx \\
&= (\tfrac{1}{2}\pi - 0) - [-\cos x]_0^{\frac{1}{2}\pi} \\
&= \tfrac{1}{2}\pi - (0 + 1) \\
&= \tfrac{1}{2}\pi - 1.
\end{aligned}$$

(ii) If $c_n = \displaystyle\int_0^{\frac{1}{2}\pi} x^n \cos x\, dx$, then as in 4.24, example,

$$c_n = [x^n \sin x + nx^{n-1}\cos x]_0^{\frac{1}{2}\pi} - n(n-1)\int_0^{\frac{1}{2}\pi} x^{n-2}\cos x\, dx$$

$$= (\tfrac{1}{2}\pi)^n - n(n-1)c_{n-2} \quad \text{if} \quad n > 1.$$

The case $n = 1$ has just been considered in ex. (i).

## Exercise 4(c)

*Use integration by parts to calculate the integral of the following functions.*

1   $x \sin x$.         2   $x \cos ax$.         3   $x^2 \sin x$.

4   $x \cos^2 x$.         5   $(x+1)(x-5)^6$.

6   $x(1-x)^n$ $(n \neq -1, -2)$.         7   $\cos^{-1} x$.

*8   $\dfrac{\tan^{-1} x}{x^3}$.         9   $x \sin 3x \cos x$. [Convert the product into a sum.]

10   $\sin^3 x$. [Put $u = \sin^2 x = 1 - \cos^2 x$, $v' = \sin x$.] (Cf. 4.13, ex. (iii).)

*11   $x \sec^{-1} x$.

12   Obtain a reduction formula for $s_n = \int x^n \sin ax \, dx$. Hence calculate $\int x^5 \sin x \, dx$.

*Evaluate the following definite integrals.*

13   $\displaystyle\int_0^{\frac{1}{2}\pi} x \sin 5x \, dx$.      14   $\displaystyle\int_0^{\pi} x^2 \sin \tfrac{1}{2}x \, dx$.      15   $\displaystyle\int_0^{\frac{1}{2}\pi} \cos^3 x \, dx$.

16   $\displaystyle\int_0^1 x^2(1-x)^n \, dx$ $(n \neq -1, -2, -3)$.      17   $\displaystyle\int_0^1 x \tan^{-1} x \, dx$.

*The following may require both integration by parts and substitution.*

*18   $\displaystyle\int \frac{x \sin^{-1} x}{\sqrt{(1-x^2)}} \, dx$.      *19   $\displaystyle\int (\sin^{-1} x)^2 \, dx$.      *20   $\displaystyle\int x \sin^{-1} x \, dx$.

*21   Use integration by parts to integrate

$$(1-x^2)\frac{d^2y}{dx^2} - 4x\frac{dy}{dx} - 2y + 6$$

twice wo $x$.

## 4.3   The logarithmic function

### 4.31   The integral $\int dx/x$

Any systematic investigation of what functions can be integrated must begin by considering the case $n = -1$ of $\int x^n dx$, which was excluded from the first entry in the list of standard integrals in 4.12. We now take up this investigation.†

It is more convenient to begin with the *definite* integral

$$\int_1^t \frac{1}{x} dx \quad (t > 0),$$

because (i) no arbitrary constant is involved; (ii) we can visualise this as the 'area' under the curve $y = 1/x$ from $x = 1$ to $x = t$. We are

---

† If there is a function $y$ for which $dy/dx = 1/x$, then Ex. 3(e), no. 18 shows that it cannot be a rational function.

                       

taking for granted the *existence* of the integral (see 4.16 (2)); or equivalently, the existence of a number which measures the 'area' (4.16 (1)) shown in fig. 45.

We do not attempt to consider the case $t \leqslant 0$ because the integrand would then be discontinuous in the range $t \leqslant x \leqslant 1$. The number 1 has been chosen as the lower limit of integration for later convenience, but any positive constant would do.

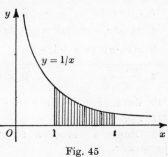

Fig. 45

In view of 4.15, Remark ($\beta$), the definite integral under consideration is a function of $t$ only. We write

$$\phi(t) = \int_1^t \frac{dx}{x} \quad (t > 0), \tag{i}$$

so that $\phi(t)$ represents the 'area' shaded in fig. 45, and is *not defined* for $t \leqslant 0$.

## 4.32 Investigation of $\phi(t)$

We now obtain some properties of $\phi(t)$ which will supply enough information for us to be able to identify the function.

(1) From equation (i) we have $\phi(1) = 0$.

Also, since $1/x$ is positive throughout the range of integration, $\phi(t)$ is continuous (4.15 (6)) and increases as $t$ increases (4.15 (8)) for $t > 0$.

(2) *Functional law for $\phi(t)$.*

We shall prove that, for all rational values of $n$,

$$\phi(t^n) = n\phi(t), \tag{ii}$$

where if $n = p/q$ ($q$ even), $t^n$ denotes the *positive* $q$th root of $t^p$.

*Proof.* By definition,

$$\phi(t^n) = \int_1^{t^n} \frac{dx}{x}.$$

Putting $x = z^n$, we have $dx = nz^{n-1}dz$, so that $dx/x = n\,dz/z$. Also, as $x$ varies from 1 to $t^n$, $z$ varies steadily from 1 to $t$. Hence

$$\phi(t^n) = \int_1^t n\frac{dz}{z} = n\int_1^t \frac{dz}{z} = n\phi(t).$$

Taking $n = -1$, we have $\quad \phi\left(\frac{1}{t}\right) = -\phi(t). \tag{iii}$

**(3)** *Bounds for* $\phi(1+u)$.

If $x$ lies between 1 and $1+u$, where $1+u > 0$ and $u \neq 0$, then $1/x$ lies between 1 and $1/(1+u)$. Hence by 4.15 (10),

$$\int_1^{1+u} \frac{1}{x}\, dx \text{ lies between } u \text{ and } \frac{u}{1+u},$$

i.e.            $\phi(1+u)$ *lies between* $u$ *and* $\dfrac{u}{1+u}$.       (iv)

The argument is easily followed by noticing that the shaded 'area' (fig. 46) lies in value between the areas of the small and large rectangles, whose common base is $u$ and whose respective heights are 1, $1/(1+u)$ (the values of $y$ when $x = 1, 1+u$).

Taking $u = 1$, then by (iv)

$$\tfrac{1}{2} < \phi(2) < 1. \qquad\qquad \text{(v)}$$

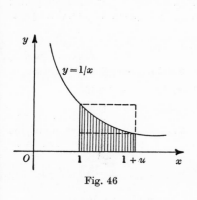

Fig. 46                  Fig. 47

**(4)** *Range of values of* $\phi(x)$.

Given any $x \geqslant 2$, however large, we can find a positive integer $n$ such that $2^n \leqslant x < 2^{n+1}$. Since $\phi(t)$ increases with $t$ (see (1)),

$$\phi(x) \geqslant \phi(2^n),$$
$$= n\phi(2) > \tfrac{1}{2}n$$

by (ii) and (v). When $x \to \infty$, also will $n \to \infty$ (since $2^{n+1} > x$), and the inequality $\phi(x) > \tfrac{1}{2}n$ shows that $\phi(x) \to \infty$. Thus†

$$\phi(x) \to \infty \quad \text{when} \quad x \to \infty. \qquad\qquad \text{(vi)}$$

† Contrast Ex. 4 (*d*), no. 20.

By equation (iii), $\phi(x) = -\phi(1/x)$. When $x \to 0+$, $1/x \to \infty$ and so $\phi(1/x) \to \infty$, by (vi). Hence

$$\phi(x) \to -\infty \quad \text{when} \quad x \to 0+. \tag{vii}$$

Combining these results with those of (1), we have:

$\phi(x)$ *increases continuously from* $-\infty$ *to* $+\infty$ *as* $x$ *increases from* $0$ *to* $+\infty$.

Recalling from (1) that $\phi(1) = 0$, we can now sketch the general form of the graph of $y = \phi(x)$, as in fig. 47.

**(5)** *The number e.*

Since $\phi(x)$ is continuous, steadily increases, and takes all values, therefore it can take any given value just once. In particular, there is a single value of $x$ for which $\phi(x) = 1$. Calling this value $e$, we have effectively defined it by

$$\phi(e) = \int_1^e \frac{dx}{x} = 1. \tag{viii}$$

The reader who has done any work on approximate integration may consider the following way of estimating the numerical value of $e$. By using Simpson's rule for 10 ordinates of the curve $y = 1/x$, equally spaced at intervals of $0.1$, we find (7.33, ex.) that

$$\phi(2) = \int_1^2 \frac{dx}{x} \doteqdot 0.693.$$

Hence $\qquad\qquad \phi(2^n) = n\phi(2) \doteqdot 0.693n.$

We shall have $\phi(2^n) \doteqdot 1$ if $n \doteqdot 1/0.693 \doteqdot 1.44$. Thus $\phi(x) \doteqdot 1$ if

$$x \doteqdot 2^{1.44} \doteqdot 2^{1.5} = \sqrt{8} \doteqdot 2.8.$$

(Use of tables for $2^{1.44}$ would give $x \doteqdot 2.71$. In 6.53, example, we shall find the value more accurately as $2.718282$, and in 12.72 (1) it will be shown that $e$ is an irrational number.)

**(6)** *Identification of* $\phi(x)$.

If $x = e^y$, where $y$ is rational, we have by (ii) and (viii) that

$$\phi(x) = \phi(e^y) = y\phi(e) = y.$$

But if $x = e^y$, then also $y = \log_e x$ by the definition of 'logarithm'. Hence, whenever $x$ is a rational power of $e$,

$$\phi(x) = \log_e x \quad (x > 0). \tag{ix}$$

A difficulty arises if $x$ is not a rational power of $e$. In this case $\log_e x$ is not defined; the elementary definition 'the logarithm of $x$ to base $e$ is the power to which $e$ must be raised to equal $x$' is meaningless because irrational indices have not yet been defined. On the other hand, $\phi(x)$ is defined for all $x > 0$. We may 'complete the definition' of $\log_e x$ when $x$ is not a power of $e$ by *defining* $\log_e x$ to be $\phi(x)$; then (ix) holds for *all* $x > 0$. The properties of logarithms proved in elementary algebra are unchanged: (ii) is the index law; for the addition law, we have by the substitution $t = yu$,

$$\log_e(xy) = \int_1^{xy} \frac{dt}{t} = \int_{1/y}^x \frac{du}{u} = \int_1^x \frac{du}{u} - \int_1^{1/y} \frac{du}{u} \quad (4.15\,(3))$$

$$= \log_e x - \log_e \frac{1}{y} = \log_e x + \log_e y \quad \text{by (iii).}$$

Similarly, $$\log_e \frac{x}{y} = \log_e x - \log_e y.$$

If $x$ is a rational power of 10, say $x = 10^y$, then by the 'elementary' definition, $y = \log_{10} x$. On taking logarithms to base $e$ and using (ii), we get
$$\log_e x = y \log_e 10,$$

i.e. $$\log_{10} x = \frac{\log_e x}{\log_e 10},$$

the usual 'change of base' formula. When $x$ is not a power of 10, this formula can be used to *define* $\log_{10} x$. In particular,

$$\log_{10} e = \frac{\log_e e}{\log_e 10} = \frac{1}{\log_e 10}.$$

Logarithms to base $e$ are called *natural, Napierian*,† or *hyperbolic*‡ *logarithms*. In future we shall write $\log x$ for $\log_e x$; and when any other base (such as 10) is used, we shall indicate this explicitly (as $\log_{10} x$).

**(7)** *Derivative of* $\log x$.
Since $$\phi(x) = \int_1^x \frac{dt}{t},$$
we have by 4.15 (7) that $\phi'(x) = 1/x$, i.e.

$$\frac{d}{dx}(\log x) = \frac{1}{x}. \tag{x}$$

**(8)** $\int \frac{1}{x} dx.$
From equation (x)

$$\int \frac{dx}{x} = \log x + c \quad (x > 0). \tag{xi}$$

---

† After Napier (1550–1617), the inventor of logarithms.
‡ Because associated with the 'area' under the hyperbola $y = 1/x$.

If $x < 0$, put $x = -y$; then

$$\int \frac{dx}{x} = \int \frac{-dy}{-y} = \int \frac{dy}{y} = \log y = \log(-x),$$

so
$$\int \frac{dx}{x} = \log(-x) + c \quad (x < 0).$$
(xi)′

We can combine (xi), (xi)′ into the single equation

$$\int \frac{dx}{x} = \log|x| + c.$$
(xii)

The results of (xi), (xii) conclude the enquiry begun in 4.31, so that the gap in our list of integrals is filled. We shall continue to omit the arbitrary constant $c$ for brevity.

### 4.33 An application to integration

We can generalise (xii) as follows. If $u = f(x)$, then

$$\frac{d}{dx}(\log|u|) = \frac{1}{u}\frac{du}{dx} = \frac{f'(x)}{f(x)}.$$

Hence
$$\int \frac{f'(x)}{f(x)}\,dx = \log|f(x)|.$$
(xiii)

Thus any fraction, rational or not, whose numerator is expressible as a constant multiple of the derivative of the denominator can be integrated. *We usually omit the modulus in Answers to Exercises.*

### Example

Since
$$\tan x = \frac{\sin x}{\cos x} = -\left\{\frac{d}{dx}(\cos x)\right\}\Big/\cos x,$$

$$\int \tan x\,dx = -\log|\cos x| = \log|\sec x|.$$

### Exercise 4(d)

*Write down the derivative of the following (use properties of logarithms to simplify the expression before derivation whenever possible).*

1  $\log(2x)$.  　　2  $\log(x^2)$.  　　3  $\log(1/x)$.  　　4  $\log\sin x$.

5  $\log\sin^2 x$.  　6  $\log\cot x$.  　7  $\log\dfrac{1+x}{1-x}$.  　8  $\log\log x$.

9  $\dfrac{x}{\log x}$.  　　10  $(\log x)^2$.

11  Verify that equation (xii) is equivalent to $\displaystyle\int \frac{dx}{x} = \tfrac{1}{2}\log(x^2)$.

12  Find $d(\log_{10} x)/dx$. [Use the formula for change of base.]

*Obtain and simplify the derivative of*

13   $\log \tan \tfrac{1}{2}x$.        14   $\log (\tan x + \sec x)$.      15   $\log \dfrac{1 + \sin x}{1 - \sin x}$.

16   $\log (1 + \sin 2x) + 2 \log \{\sec (\tfrac{1}{4}\pi - x)\}$.

17   Find the maximum value of $(\log x)/x$.

18   Prove that $\log x - (x - 1)/\sqrt{x}$ is a decreasing function for $x > 1$. Deduce that $\log x < (x - 1)/\sqrt{x}$ for $x > 1$.

*19   Prove that the $n$th derivative of $\log x$ is $(-1)^{n-1} (n - 1)!/x^n$.

*20   *The result* (vi) *of 4.32 is not 'obvious from a figure'.* Verify that $y = 1/x^2$ for $x > 0$ has a graph generally similar to that of $y = 1/x$ for $x > 0$; but that, if

$$\psi(x) = \int_1^x \frac{dt}{t^2}, \quad \text{then } \psi(x) \to 1 \text{ when } x \to \infty.$$

*Integrate the following.*

21   $\dfrac{1}{3x}$.      22   $\dfrac{1}{1 + x}$.      23   $\dfrac{1}{1 - x}$.      24   $\dfrac{x}{1 + x^2}$.

25   $\dfrac{x^3}{1 + x^4}$.      26   $\dfrac{6x - 7}{3x^2 - 7x + 5}$.      27   $\dfrac{2 - x}{3x^2 - 12x + 7}$.      28   $\cot x$.

29   $\tan 3x$.      30   $\dfrac{\sin 2x}{3 + 5 \cos^2 x}$.      31   $\dfrac{1 - \tan x}{1 + \tan x}$.      32   $\dfrac{1}{x \log x}$.

33   $\dfrac{\sec^2 \tfrac{1}{2}x}{\tan \tfrac{1}{2}x}$, and deduce $\int \operatorname{cosec} x \, dx$.

*Using integration by parts, calculate the integral of*

34   $x^m \log x$ $(m \neq -1)$, and deduce $\int \log x \, dx$.      35   $\log (x^{1/x})$, $x > 0$.

36   $\tan^{-1} x$.      37   $(x \log x)^2$.      38   $x \tan^2 x$.

*Evaluate the following definite integrals.*

39   $\displaystyle\int_0^{\frac{1}{8}\pi} \tan 2x \, dx$.      40   $\displaystyle\int_3^8 \frac{dx}{1 + 3x}$.      41   $\displaystyle\int_0^{\frac{1}{2}\pi} \frac{\cos x}{1 + \sin x} dx$.

42   $\displaystyle\int_0^1 \frac{x}{1 + x} dx$.      43   $\displaystyle\int_1^e \log (\sqrt{x}) \, dx$.      44   $\displaystyle\int_0^1 x \log (1 + x) \, dx$.

*45   If $u_n = \int x^m (\log x)^n \, dx$ where $n$ is a positive integer and $m \neq -1$, obtain the reduction formula

$$u_n = \frac{x^{m+1}}{m + 1} (\log x)^n - \frac{n}{m + 1} u_{n-1}.$$

Hence calculate $\displaystyle\int_1^e x^2 (\log x)^3 \, dx$.

*46   If    $C = \displaystyle\int \frac{\cos x \, dx}{a \cos x + b \sin x}$    and    $S = \displaystyle\int \frac{\sin x \, dx}{a \cos x + b \sin x}$,

simplify $aC + bS$ and $bC - aS$. Hence calculate

$$\int_0^{\frac{1}{2}\pi} \frac{\cos x \, dx}{3 \cos x + 4 \sin x}.$$

*47 Find numbers $\lambda$, $\mu$ for which

$$7\sin x + 4\cos x = \lambda(\sin x + 2\cos x) + \mu(\cos x - 2\sin x).$$

Hence calculate $\displaystyle\int \frac{7\sin x + 4\cos x}{\sin x + 2\cos x}\,dx.$

## 4.4 The exponential and hyperbolic functions

### 4.41 The exponential function

(1) If $x = \log y$, then by 4.32 (4) $x$ increases continuously from $-\infty$ to $+\infty$ as $y$ increases from $0$ to $\infty$. Thus,† to any given $x$ corresponds just one value of $y$, and this value is positive. Hence *for all $x$* there is defined a function $y$ which is the inverse of $x = \log y$; it is written $y = e^x$ and called the *exponential function* of $x$.

It obeys the usual laws of indices; this is clear when $x$ is rational, and is true when $x$ is irrational: putting $y_1 = e^{x_1}$ and $y_2 = e^{x_2}$, so that $x_1 = \log y_1$ and $x_2 = \log y_2$, then

$$x_1 + x_2 = \log y_1 + \log y_2 = \log(y_1 y_2) \quad \text{by 4.32 (6),}$$

so $\qquad e^{x_1+x_2} = y_1 y_2$, i.e. $e^{x_1+x_2} = e^{x_1}.e^{x_2}.$

Similarly $\qquad e^{x_1} \div e^{x_2} = e^{x_1-x_2}.$

(2) *$e^x$ is an increasing function of $x$:* if $x_1 < x_2$ and $y_1$, $y_2$ are the corresponding values of $e^x$, then $y_1 < y_2$; for $y_1 \geqslant y_2$ would imply that $x_1 \geqslant x_2$ since $x = \log y$ is an increasing function of $y$. Thus

*$e^x$ increases from $0$ to $\infty$ when $x$ increases from $-\infty$ to $+\infty$.*

In particular we emphasise that $e^x$ *is positive for all $x$*, but that $e^x \to 0+$ when $x \to -\infty$.

(3) *$e^x$ is continuous for all values of $x$.*
*Proof.* Write $y = e^x$ and let $e^{x+h} = y + k$. Then

$$h = \log(y+k) - \log y = \int_1^{y+k} \frac{dt}{t} - \int_1^y \frac{dt}{t} = \int_y^{y+k} \frac{dt}{t}.$$

When $h > 0$, then by (2) also $k > 0$. If $y < t < y+k$ we have

$$\frac{1}{t} > \frac{1}{y+k}, \quad \text{so by 4.15 (10),} \quad h > \frac{k}{y+k}.$$

---

† This assertion is based on the property stated in 2.65, Corollary. We have already made it in 4.32(5) for the case $x = 1$.

When $h < 0$, then also $k < 0$ by (2), and if $y + k < t < y$ we have $1/t > 1/y$, from which

$$|h| = -h = \int_{y+k}^{y} \frac{dt}{t} > \frac{-k}{y} = \frac{|k|}{y}.$$

Hence when $h \to 0$, also $k \to 0$; i.e. $e^x$ is continuous.

(4) The graph of $y = e^x$, i.e. of $x = \log y$, can be obtained from that of $y = \log x$ by interchanging $Ox$, $Oy$ and then reversing the direction of $Ox$ to restore right-handedness.

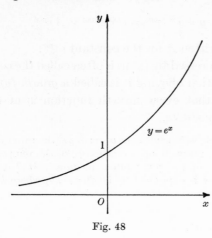

Fig. 48

(5) *Derivative of $e^x$.*

If $y = e^x$, then from $x = \log y$ we have $dx/dy = 1/y$, and hence

$$\frac{dy}{dx} = y; \qquad \text{(xiv)}$$

i.e. $$\frac{d}{dx}(e^x) = e^x, \qquad \text{(xv)}$$

so that $e^x$ *is its own derivative.*

The corresponding result for integration is

$$\int e^x \, dx = e^x. \qquad \text{(xvi)}$$

*Remarks*

($\alpha$) The property (xv) is *characteristic* of $e^x$; i.e. $e^x$ is essentially the *only* function which is equal to its own derivative. This can be shown by starting from equation (xiv), which expresses the property stated, and solving it for $y$ in terms of $x$. More generally, we will consider the equation

$$\frac{dy}{dx} = my, \tag{xvii}$$

where $m$ is constant. To solve this, write it as $dx/dy = 1/(my)$ and integrate both sides wo $y$:

$$x = \frac{1}{m}\int\frac{dy}{y} = \frac{1}{m}\log y + c,$$

hence

$$\log y = m(x-c)$$

and

$$y = e^{mx-mc} = e^{mx}\,e^{-mc} = A\,e^{mx},$$

where we have written $A$ for the constant $e^{-mc}$.

($\beta$) The law expressed by (xvii) is often called the *compound interest law*, and any function obeying it is called a *growth function*. We have therefore shown that every growth function is of the form $A\,e^{mx}$ where $A$, $m$ are constants.

A physical example is *Newton's law of cooling*: 'the rate of decrease of temperature of a body is proportional to the excess of this temperature above that (supposed constant) of the surrounding medium'. If $\theta$ denotes the excess temperature, the law is expressed by $d\theta/dt = -k\theta$, where $k$ is some positive constant.

(6) *The function $a^x$.*

Except for the case when $a = e$ (see (1) above), $a^x$ has been defined only when $x$ is rational. Suppose now that $a$ is any *positive* number, and let $x$ be any rational number $p/q$. (If $q$ is even, there will be two values of $a^{p/q}$; let $y$ denote the *positive* value). Then by 4.32, (ii),

$$\log y = \frac{p}{q}\log a = x\log a,$$

so

$$y = e^{x\log a}.$$

We use this equation to *define* $a^x$ when $x$ is irrational:

$$a^x = e^{x\log a} \quad (a > 0). \tag{xviii}$$

Then

(a) $a^x > 0$ *for all $x$*, by (2).

(b) $a^x$ *satisfies the usual index laws* $a^x \cdot a^y = a^{x+y}$, $(a^x)^y = a^{xy}$.

For

$$a^x \cdot a^y = e^{x\log a}\,e^{y\log a} = e^{(x+y)\log a} = a^{x+y};$$

and, writing $b = e^{x\log a}$,

$$(a^x)^y = (e^{x\log a})^y = b^y = e^{y\log b} = e^{y\cdot x\log a} = e^{(xy)\log a} = a^{xy}.$$

(c) $\dfrac{d}{dx}(a^x) = a^x \log a$.

For $\qquad \dfrac{d}{dx}(a^x) = \dfrac{d}{dx}(e^{x \log a}) = (\log a)\, e^{x \log a} = (\log a)\, a^x$.

Hence also $\qquad\qquad\qquad \displaystyle\int a^x\, dx = \dfrac{a^x}{\log a}$.

(d) We can now extend the formula $d(x^m)/dx = mx^{m-1}$ to irrational $m$ $(x > 0)$.

For $\qquad \dfrac{d}{dx}(x^m) = \dfrac{d}{dx}(e^{m \log x})$ if $\;x > 0$

$\qquad\qquad\qquad\quad = e^{m \log x} \times \dfrac{m}{x}$   by 'function of a function'

$\qquad\qquad\qquad\quad = x^m \times \dfrac{m}{x}$

$\qquad\qquad\qquad\quad = mx^{m-1}$.

### Exercise 4(e)

1 Sketch the graphs of (i) $y = e^{-x}$; (ii) $y = \frac{1}{2}(e^x + e^{-x})$; (iii) $y = \frac{1}{2}(e^x - e^{-x})$.

*Write down the derivative of*

2 $e^{3x}$.        3 $xe^x$.        4 $e^{x^2}$.        5 $e^{3x}\sin 2x$.

6 $e^{\cos x}$.        7 $\log(e^x)$.        8 $e^{\log x}$.        9 $e^{x \log 2}$.

10 $2^x$.        11 $x^x$ $(x > 0)$.

*Write down the integral of*

12 $e^{2x}$.        13 $\dfrac{1}{e^x}$.        14 $\sqrt{e^x}$.        15 $\dfrac{1}{1 + e^{-x}}$.

16 $e^{\sin x}\cos x$.        17 $\log(e^{3x^2})$.

*Using integration by parts, calculate*

18 $\displaystyle\int e^x \cos x\, dx$.        19 $\displaystyle\int e^x \sin x \cos x\, dx$.        20 $\displaystyle\int_0^\pi e^x \sin^2 x\, dx$.

\*21 Writing $u_n = \int x^n e^{ax}\, dx$, where $n$ is a positive integer and $a$ is a non-zero constant, obtain the reduction formula

$$u_n = \frac{1}{a} x^n e^{ax} - \frac{n}{a} u_{n-1}.$$

Hence calculate $\int x^4 e^{2x}\, dx$.

22 Find the constant $m$ if $e^{mx}$ satisfies

$$\frac{d^2 y}{dx^2} - \frac{dy}{dx} - 6y = 0.$$

23 If $y$ is a function of $x$, prove

$$\frac{d^2}{dx^2}(e^{ax}y) = e^{ax}\left(\frac{d^2 y}{dx^2} + 2a\frac{dy}{dx} + a^2 y\right).$$

*Write down the nth derivative of*

**24** $e^{ax}$.                    **25** $xe^x$.                    **26** $a^x$ $(a > 0)$.

***27** Find $dy/dx$ when $y = e^{ax}\sin bx$, and express the result in the form $r e^{ax}\sin(bx+\theta)$. Hence *write down* an expression for the $n$th derivative of $y$.

**28** Find constants $p$, $q$ in terms of $a$ and $b$ so that

$$\frac{d}{dx}\{e^{ax}(p\sin bx + q\cos bx)\} = e^{ax}\sin bx.$$

Hence calculate $\int e^{ax}\sin bx\,dx$.

**29** Calculate $\int e^{ax}\cos bx\,dx$ by integrating by parts twice.

**30** Find the turning point and points of inflexion of $y = e^{-x^2}$. Sketch the curve.

**31** Find the values of $x$ at the maximum, minimum and points of inflexion of $y = x^2 e^{-x}$. Sketch the curve.

**32** What is the maximum *value* of $x^n e^{-x}$ $(n > 0)$? Prove that $n^n e^{-n} > x^n e^{-x}$ when $x > n$, and by putting $x = n+1$ deduce that $(1+1/n)^n < e$.

**33** If $n$ is a positive integer, prove that

$$e^{-x}\left(1 + x + \frac{x^2}{2!} + \frac{x^3}{3!} + \dots + \frac{x^n}{n!}\right)$$

is a decreasing function of $x$ for $x > 0$. Deduce that

$$e^x > 1 + x + \frac{x^2}{2!} + \dots + \frac{x^n}{n!} \quad \text{when} \quad x > 0.$$

***34** If $f(x) = xy - x\log x - e^{y-1}$ $(x > 0)$, show that the greatest value of $f(x)$ is 0, and deduce that for $x > 0$ and $x \neq e^{y-1}$, $xy < x\log x + e^{y-1}$.

***35** If $y = A\sin(\log x) + B\cos(\log x)$, where $A$, $B$ are constants, prove that

$$x^2\frac{d^2y}{dx^2} + x\frac{dy}{dx} + y = 0.$$

***36** By writing $x = e^t$, show that the equation

$$x^3\frac{d^3y}{dx^3} + 3x^2\frac{d^2y}{dx^2} + x\frac{dy}{dx} = 24x^2$$

becomes $d^3y/dt^3 = 24e^{2t}$. Hence express $y$ (i) as a function of $t$; (ii) as a function of $x$.

## 4.42 Logarithmic derivation

The rule for deriving a product $y = uv$ can be proved as follows. First take logarithms: $\log y = \log u + \log v$.

Then derive each side wo $x$, using the rule for 'function of a function':

$$\frac{1}{y}\frac{dy}{dx} = \frac{1}{u}\frac{du}{dx} + \frac{1}{v}\frac{dv}{dx}.$$

Multiply by $y = uv$:                    $\dfrac{dy}{dx} = v\dfrac{du}{dx} + u\dfrac{dv}{dx}.$

A similar method is useful when deriving a continued product of functions. Thus if $y = uvw$,

$$\log y = \log u + \log v + \log w,$$

$$\frac{1}{y}\frac{dy}{dx} = \frac{1}{u}\frac{du}{dx} + \frac{1}{v}\frac{dv}{dx} + \frac{1}{w}\frac{dw}{dx},$$

and

$$\frac{dy}{dx} = vw\frac{du}{dx} + wu\frac{dv}{dx} + uv\frac{dw}{dx}.$$

Cf. Ex. 3 (a), no. 7. Any function of the form

$$y = \frac{u_1 u_2 \dots u_n}{v_1 v_2 \dots v_m}$$

can be dealt with similarly.

### Examples

(i) $y = e^{2x} x^3 \cos x \log x$.

$$\log y = 2x + 3\log x + \log \cos x + \log \log x.$$

$$\therefore \quad \frac{1}{y}\frac{dy}{dx} = 2 + \frac{3}{x} - \frac{\sin x}{\cos x} + \frac{1}{\log x}\frac{1}{x}.$$

$$\therefore \quad \frac{dy}{dx} = e^{2x} x^3 \cos x \log x \left(2 + \frac{3}{x} - \tan x + \frac{1}{x \log x}\right).$$

Although in 4.41 (6) we have defined $a^x = e^{x \log a}$ and thereby written down its derivative, in practice we usually derive this and similar functions involving powers as follows.

(ii) $y = a^x$ $(a > 0)$.

$$\log y = x \log a, \quad \therefore \quad \frac{1}{y}\frac{dy}{dx} = \log a, \quad \therefore \quad \frac{dy}{dx} = a^x \log a.$$

(iii) $y = x^x$ $(x > 0)$; cf. Ex. 4 (e), no. 11.

$$\log y = x \log x, \quad \therefore \quad \frac{1}{y}\frac{dy}{dx} = \log x + 1, \quad \therefore \quad \frac{dy}{dx} = x^x(1 + \log x).$$

### 4.43 The logarithmic inequality. Some important limits

(1) *The logarithmic inequality.*
*If $u \neq 0$ and $u > -1$,*

$$\frac{u}{1+u} < \log(1+u) < u.$$

This follows from statement (iv) of 4.32, since

$$u - \frac{u}{1+u} = \frac{u^2}{1+u} > 0.$$

(2) $\dfrac{\log x}{x} \to 0$ *when* $x \to \infty$.

Put $u = \sqrt{x} - 1$ in (1). Then if $x > 0$ and $x \neq 1$,

$$\log \sqrt{x} < \sqrt{x} - 1 < \sqrt{x},$$

so that $\tfrac{1}{2} \log x < \sqrt{x}$ and $(\log x)/x < 2/\sqrt{x}$. When $x > 1$, $(\log x)/x$ is positive. Hence when $x \to \infty$, $(\log x)/x \to 0$.

(3) $\dfrac{\log x}{x^p} \to 0$ *when* $x \to \infty$, *where $p$ is any fixed positive number.*

Write $x = y^p$ in (2). Then when $y \to \infty$, also $x \to \infty$, and so

$$\frac{p \log y}{y^p} = \frac{\log x}{x} \to 0.$$

Hence $(\log y)/y^p \to 0$ when $y \to \infty$. Thus *log $x$ tends to infinity slower than any positive power of $x$.*

(4) $x \log x \to 0$ *when* $x \to 0+$.

Put $x = 1/y$ in (2). When $y \to 0+$, then $x \to \infty$ and

$$\frac{-\log y}{1/y} = \frac{\log x}{x} \to 0,$$

i.e. $y \log y \to 0$.

(5) $x^p \log x \to 0$ *when* $x \to 0+$, *where $p$ is any fixed positive number.*

Put $x = 1/y^p$ in (2). When $y \to 0+$, then $x \to \infty$ and

$$\frac{-p \log y}{1/y^p} = \frac{\log x}{x} \to 0,$$

i.e. $y^p \log y \to 0$.

(6) $\dfrac{x^m}{e^x} \to 0$ *when* $x \to \infty$, *for any constant $m$.*

This is obvious if $m \leqslant 0$ by 4.41 (2). If $m > 0$, put $p = 1/m$ in (3), which shows that $(\log x)/x^{1/m} \to 0$, and hence $(\log x)^m/x \to 0$, when $x \to \infty$. Put $x = e^y$; then when $y \to \infty$, also $x \to \infty$, and so

$$\frac{y^m}{e^y} = \frac{(\log x)^m}{x} \to 0.$$

Thus $e^x$ *tends to infinity faster than any positive power of $x$.*

(7) *The exponential limit.*

$$\lim_{n \to \infty} \left( 1 + \frac{x}{n} \right)^n = e^x, \quad \text{for all } x.$$

If $n > 0$ and $n + x > 0$, then by (1) with $u = x/n$ and $x \neq 0$,†

$$\frac{x}{n+x} < \log\left(1 + \frac{x}{n}\right) < \frac{x}{n}.$$

On multiplying by $n$ we get

$$\frac{nx}{n+x} < \log\left(1 + \frac{x}{n}\right)^n < x.$$

When $n \to \infty$, $nx/(n+x) \to x$, and hence

$$\log\left(1 + \frac{x}{n}\right)^n \to x.$$

Hence for all $n$ sufficiently large,

$$\left(1 + \frac{x}{n}\right)^n = e^{x+\xi},$$

where $\xi \to 0$ when $n \to \infty$. Since $e^x$ is continuous (4.41 (3)), $e^{x+\xi} \to e^x$, and hence

$$\left(1 + \frac{x}{n}\right)^n \to e^x \quad \text{when} \quad n \to \infty.$$

In particular, with $x = 1$,

$$\lim_{n \to \infty} \left(1 + \frac{1}{n}\right)^n = e.$$

This result is sometimes taken as the definition of $e$ in alternative presentations of the theory. Cf. 2.77, ex. (ii).

(8) $\lim\limits_{n \to \infty} \left(1 + \tfrac{1}{2} + \tfrac{1}{3} + \ldots + \dfrac{1}{n} - \log n\right);$ *Euler's constant.*

Write $\qquad f(n) = 1 + \tfrac{1}{2} + \tfrac{1}{3} + \ldots + \dfrac{1}{n} - \log n.$

Then $f(n)$ *decreases as $n$ increases*, for

$$f(n+1) - f(n) = \frac{1}{n+1} + \log\frac{n}{n+1} = \frac{1}{n+1} + \log\left(1 - \frac{1}{n+1}\right) < 0$$

by (1) with $u = -1/(n+1)$.

Now put $u = 1/(r-1)$, where $r > 1$, in (1):

$$\frac{1}{r} < \log r - \log(r-1) < \frac{1}{r-1}.$$

If we write down inequalities of this type corresponding to $r = 2, 3, \ldots, n$, and then add, we obtain

$$\tfrac{1}{2} + \tfrac{1}{3} + \ldots + \frac{1}{n} < \log n < 1 + \tfrac{1}{2} + \tfrac{1}{3} + \ldots + \frac{1}{n-1},$$

so $\qquad\qquad\qquad\qquad \dfrac{1}{n} < f(n) < 1.$

---

† When $x = 0$ the required result holds trivially.

Hence when $n \to \infty, f(n)$ decreases *but remains positive*. Therefore by 2.77, $f(n)$ tends to some limit $\gamma$, where $0 \leqslant \gamma < 1$. Thus

$$\lim_{n \to \infty} \left( 1 + \tfrac{1}{2} + \tfrac{1}{3} + \ldots + \frac{1}{n} - \log n \right) = \gamma.$$

The number $\gamma$, called *Euler's constant*, is approximately $0 \cdot 577,215,664,9 \ldots$.

## Exercise 4(*f*)

*Find the derivative of*

1   $\dfrac{x}{\sqrt{(1-x^2)}}$.

2   $\dfrac{(x+1)^4 (2x+3)^2}{(3x-5)^3}$.

3   $(x-2)^{\frac{1}{2}} (3x+2)^{\frac{1}{3}} (2x+5)^2$.

4   $x^5 e^{3x} \sin 2x$.

5   $e^{e^x}$.

6   $e^x \sin x (\log x)^2$.

7   $x e^x \tan x$.

8   $(1+x)^{1/x}$ $(x > -1)$.

9   $\log_x 2$ $(x > 0)$.

10   $(\log x)^x$ $(x > e)$.

11   Prove that the derivative of $(x+1)(x+2)\ldots(x+n)$ has the value

$$n! \left( 1 + \tfrac{1}{2} + \tfrac{1}{3} + \ldots + \frac{1}{n} \right)$$

when $x = 0$.

12   If $y^2 = e^{xy}$, prove $\qquad \dfrac{dy}{dx} = \tfrac{1}{2} y^2 \Big/ \log \left( \dfrac{e}{y} \right)$.

*Use the logarithmic inequality to prove the following (nos. 13–17).*

13   $\log x \leqslant n(\sqrt[n]{x} - 1)$ if $x > 0$.

14   $e^x \geqslant 1 + x$ for all $x$, and $e^x \leqslant 1/(1-x)$ if $x < 1$.

15   $\dfrac{1+x}{x} < e^{1/x} < \dfrac{x}{x-1}$ if $x > 1$.

16   $\lim\limits_{x \to 1} \dfrac{\log x}{x - 1} = 1$.

17   $\lim\limits_{h \to 0} \dfrac{\log(x+h) - \log x}{h} = \dfrac{1}{x}$.   [This limit is $d(\log x)/dx$.]

18   Calculate $\lim\limits_{h \to 0} \dfrac{\log(x+h) + \log(x-h) - 2\log x}{h^2}$.

19   Prove $\lim\limits_{x \to 0} \dfrac{e^x - 1}{x} = 1$.   [This limit is the value of $d(e^x)/dx$ when $x = 0$; *or* use no. 14.]

20   Prove $\lim\limits_{x \to 0} \dfrac{a^x - 1}{x} = \log a$ $(a > 0)$.

*Assuming the exponential limit, write down the limit when $n \to \infty$ of*

21   $\left( 1 + \dfrac{2}{n} \right)^{-n}$.

22   $\left( 1 - \dfrac{1}{n^2} \right)^n$.

23   $\dfrac{(n+2)^n}{(n-1)^n}$.

24   Prove $\lim\limits_{n \to \infty}\left(1+\dfrac{1}{n}+\dfrac{1}{n^2}\right)^n = e$. [Use the logarithmic inequality as in 4.43 (7).]

*25   If $f(x) > 0$ and is continuous for $0 \leqslant x \leqslant 1$, prove

$$\int_0^1 \log f(x)\,dx \leqslant \log\left\{\int_0^1 f(x)\,dx\right\}$$

by first considering the case when $\int_0^1 f(x)\,dx = 1$. [In this case, $\log f(x) < f(x) - 1$, so

$$\int_0^1 \log f(x)\,dx \leqslant \int_0^1 f(x)\,dx - 1 = 0 = \log\left\{\int_0^1 f(x)\,dx\right\}.$$

For the general case, when $\int_0^1 f(x)\,dx = c$, apply the special case to $\phi(x) = f(x)/c$.]

*Assuming Euler's limit, calculate* $\lim\limits_{n \to \infty} \phi(n)$ *when* $\phi(n)$ *is*

*26   $\dfrac{1}{n+1}+\dfrac{1}{n+2}+\ldots+\dfrac{1}{2n}.$      *27   $\dfrac{1}{2n+1}+\dfrac{1}{2n+2}+\ldots+\dfrac{1}{3n}.$

*28   $\dfrac{1}{n}+\dfrac{1}{n+1}+\ldots+\dfrac{1}{pn}$, $p$ being any positive integer.

*29   $\dfrac{1}{qn+1}+\dfrac{1}{qn+2}+\ldots+\dfrac{1}{pn}$, $p$ and $q$ being positive integers, and $p > q$.

## 4.44 The hyperbolic functions

**(1)** *Definitions and simple properties.*

In the broad sense, $\log x$ and $e^x$ and any function involving these are 'hyperbolic' functions, because $\log x$ is the measure of an 'area' under the hyperbola $y = 1/x$, and $e^x$ is the inverse function. However, the name is customarily used for two special functions which are defined as follows:

$$\operatorname{ch} x = \tfrac{1}{2}(e^x + e^{-x}), \quad \operatorname{sh} x = \tfrac{1}{2}(e^x - e^{-x}). \tag{i}$$

These, sometimes written $\cosh x$, $\sinh x$, are called the *hyperbolic sine* and *hyperbolic cosine* of $x$.

The name arises from the fact that these functions bear a relation to the hyperbola $x^2 - y^2 = a^2$ resembling that between the trigonometric or *circular functions* $\sin x$, $\cos x$ and the circle $x^2 + y^2 = a^2$. For example, just as the point $(a\cos\theta, a\sin\theta)$ lies on the circle for all values of $\theta$, so does the point $(a\operatorname{ch}u, a\operatorname{sh}u)$ lie on the hyperbola for all $u$; this follows because

$$\operatorname{ch}^2 u - \operatorname{sh}^2 u = \tfrac{1}{4}(e^u + e^{-u})^2 - \tfrac{1}{4}(e^u - e^{-u})^2 = 1$$

by expanding. The hyperbolic functions possess many other properties analogous to those of the trigonometric functions. Thus directly from the definitions we verify that $\text{ch}(-x) = \text{ch}\,x$ (so that $\text{ch}\,x$ is an even function like $\cos x$ is), and $\text{sh}(-x) = -\text{sh}\,x$ (so that $\text{sh}\,x$ is odd). The reason for the similarity of behaviour will be given in 14.68.

By analogy with the remaining trigonometric functions, we define

$$\text{th}\,x \ (\text{or } \tanh x) = \frac{\text{sh}\,x}{\text{ch}\,x}, \quad \coth x = \frac{\text{ch}\,x}{\text{sh}\,x} = \frac{1}{\text{th}\,x}.$$

$$\text{cosech}\,x = \frac{1}{\text{sh}\,x}, \quad \text{sech}\,x = \frac{1}{\text{ch}\,x}.$$

All the properties of these functions can be deduced from their definitions by expressing the functions in terms of powers of $e$. The reader should verify the following results.

(i) $\text{ch}\,0 = 1$,

(ii) $\text{sh}\,0 = 0$,

(iii) $\text{ch}(-x) = \text{ch}\,x$,

(iv) $\text{sh}(-x) = -\text{sh}\,x$,

(v) $\text{th}(-x) = -\text{th}\,x$,

(vi) $\text{ch}^2 x - \text{sh}^2 x = 1$,

(vii) $\text{sech}^2 x = 1 - \text{th}^2 x$,

(viii) $\text{cosech}^2 x = \coth^2 x - 1$,

(ix) $\text{ch}(x+y) = \text{ch}\,x\,\text{ch}\,y + \text{sh}\,x\,\text{sh}\,y$,

(x) $\text{sh}(x+y) = \text{sh}\,x\,\text{ch}\,y + \text{ch}\,x\,\text{sh}\,y$.

From these can be deduced the 'double angle' formulae, the formulae for converting sums into products, etc., exactly as in trigonometry; see Ex. 4 $(g)$, nos. 2–15.

**(2)** *Graphs of* $\text{sh}\,x$, $\text{ch}\,x$, $\text{th}\,x$.

Despite their structural similarity to the trigonometric functions, we shall now see that the range of possible values of the hyperbolic functions is quite different. Observe that

$(a)$ $\text{ch}\,x - 1 = \frac{1}{2}(e^x - 2 + e^{-x}) = \frac{1}{2}(e^{\frac{1}{2}x} - e^{-\frac{1}{2}x})^2 \geqslant 0$,

so that $\text{ch}\,x \geqslant 1$ for all $x$;

$(b)$ $\text{ch}\,x - \text{sh}\,x = e^{-x} > 0$, so that $\text{ch}\,x > \text{sh}\,x$ for all $x$; but when $x \to \infty$, the difference tends to 0;

$(c)$ from the definitions, as $x$ increases from $-\infty$ to 0, $\text{ch}\,x$ decreases from $+\infty$ to 1, and $\text{sh}\,x$ increases from $-\infty$ to 0; while as $x$ increases from 0 to $+\infty$, $\text{ch}\,x$ increases from 1 to $+\infty$ and $\text{sh}\,x$ increases from 0 to $+\infty$.

With this guidance and the results (i)–(iv) above, the graphs of $\text{ch}\,x$ and $\text{sh}\,x$ can be sketched (fig. 49).

For $\text{th}\,x$, we have from the definitions that

$$\text{th}\,x = \frac{\text{sh}\,x}{\text{ch}\,x} = \frac{e^x - e^{-x}}{e^x + e^{-x}} = \frac{e^{2x} - 1}{e^{2x} + 1} = 1 - \frac{2}{e^{2x} + 1}.$$

As $x$ increases from $-\infty$ to $+\infty$, $e^{2x}$ increases from 0 to $+\infty$, so that $\operatorname{th} x$ increases from $-1$ to $+1$, and remains between these numbers for all $x$. Also $\operatorname{th} 0 = 0$, and $\operatorname{th}(-x) = -\operatorname{th} x$, so that the graph is symmetrical about the origin (fig. 50).

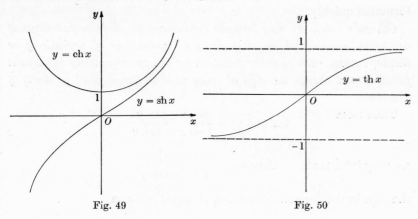

Fig. 49                Fig. 50

**(3)** *Derivatives and integrals.*

From the definitions,

$$\frac{d}{dx}\operatorname{ch} x = \frac{1}{2}\frac{d}{dx}(e^x + e^{-x}) = \tfrac{1}{2}(e^x - e^{-x}) = \operatorname{sh} x,$$

$$\frac{d}{dx}\operatorname{sh} x = \frac{1}{2}\frac{d}{dx}(e^x - e^{-x}) = \tfrac{1}{2}(e^x + e^{-x}) = \operatorname{ch} x,$$

$$\frac{d}{dx}\operatorname{th} x = \frac{d}{dx}\left(\frac{\operatorname{sh} x}{\operatorname{ch} x}\right) = \frac{\operatorname{ch}^2 x - \operatorname{sh}^2 x}{\operatorname{ch}^2 x} = \frac{1}{\operatorname{ch}^2 x} = \operatorname{sech}^2 x,$$

and similarly $$\frac{d}{dx}\operatorname{coth} x = -\operatorname{cosech}^2 x.$$

The corresponding integrals are:

$$\int \operatorname{sh} x\, dx = \operatorname{ch} x, \quad \int \operatorname{ch} x\, dx = \operatorname{sh} x,$$

$$\int \operatorname{sech}^2 x\, dx = \operatorname{th} x, \quad \int \operatorname{cosech}^2 x\, dx = -\operatorname{coth} x.$$

Since $$\operatorname{th} x = \frac{\operatorname{sh} x}{\operatorname{ch} x} = \frac{\dfrac{d}{dx}(\operatorname{ch} x)}{\operatorname{ch} x},$$

$$\int \operatorname{th} x\, dx = \log \operatorname{ch} x;$$

similarly $$\int \operatorname{coth} x\, dx = \log |\operatorname{sh} x|.$$

**(4)** *Relation between formulae for circular and hyperbolic functions.*

Comparison of the results of (1) and Ex. 4 $(g)$, nos. 2–15 with the corresponding trigonometrical formulae will verify the following rule, which can be used as an aid to memory for obtaining hyperbolic formulae quickly.

*Osborn's rule.* *In any formula connecting the circular functions of* GENERAL *angles and not depending on properties of periodicity or limits, replace each circular function by the corresponding hyperbolic function and change the sign of every product (or implied product) of two sines.*

Thus from
$$\tan(x+y) = \frac{\tan x + \tan y}{1 - \tan x \tan y}$$

we may infer that
$$\operatorname{th}(x+y) = \frac{\operatorname{th} x + \operatorname{th} y}{1 + \operatorname{th} x \operatorname{th} y},$$

the sign in the denominator being changed because
$$\tan x \tan y = \frac{\sin x \sin y}{\cos x \cos y}$$

is an implied product of two sines.

The rule excludes the application to periodicity properties, e.g. $\sin(\pi - x) = \sin x$; and to special angles, e.g.
$$\sin(x + \tfrac{1}{4}\pi) = \frac{(\sin x + \cos x)}{\sqrt{2}};$$

and also to limits, e.g. $d(\cos x)/dx = -\sin x$. A complete justification of the rule will be given in 14.68.

Consideration of the integrals in Ex. 4 $(g)$, nos. 40–49, will verify that *the general procedures for integrating circular and hyperbolic functions are the same.*

### Exercise 4(*g*)

*Prove the following formulae.*

1  $\operatorname{ch} x + \operatorname{sh} x = e^{x},\ \operatorname{ch} x - \operatorname{sh} x = e^{-x}.$

2  $\operatorname{sh} 2x = 2 \operatorname{sh} x \operatorname{ch} x.$

3  $\operatorname{ch} 2x = \operatorname{ch}^2 x + \operatorname{sh}^2 x = 2 \operatorname{ch}^2 x - 1 = 1 + 2 \operatorname{sh}^2 x.$

4  $\operatorname{th} 2x = \dfrac{2 \operatorname{th} x}{1 + \operatorname{th}^2 x}.$  5  $\operatorname{sh}(x-y) = \operatorname{sh} x \operatorname{ch} y - \operatorname{ch} x \operatorname{sh} y.$

6  $\operatorname{ch}(x-y) = \operatorname{ch} x \operatorname{ch} y - \operatorname{sh} x \operatorname{sh} y.$  7  $\operatorname{th}(x+y) = \dfrac{\operatorname{th} x + \operatorname{th} y}{1 + \operatorname{th} x \operatorname{th} y}.$

8  Express as sums or differences (i) $\operatorname{sh} x \operatorname{sh} y$; (ii) $\operatorname{sh} x \operatorname{ch} y$; (iii) $\operatorname{ch} x \operatorname{ch} y$.

9 $\operatorname{sh} S + \operatorname{sh} D = 2\operatorname{sh}\tfrac{1}{2}(S+D)\operatorname{ch}\tfrac{1}{2}(S-D).$

10 $\operatorname{sh} S - \operatorname{sh} D = 2\operatorname{sh}\tfrac{1}{2}(S-D)\operatorname{ch}\tfrac{1}{2}(S+D).$

11 $\operatorname{ch} S + \operatorname{ch} D = 2\operatorname{ch}\tfrac{1}{2}(S+D)\operatorname{ch}\tfrac{1}{2}(S-D).$

12 $\operatorname{ch} S - \operatorname{ch} D = 2\operatorname{sh}\tfrac{1}{2}(S+D)\operatorname{sh}\tfrac{1}{2}(S-D).$

13 $\operatorname{sh} 3x = 3\operatorname{sh} x + 4\operatorname{sh}^3 x.$        14 $\operatorname{ch} 3x = 4\operatorname{ch}^3 x - 3\operatorname{ch} x.$

15 If $\tau = \operatorname{th}\tfrac{1}{2}x$, prove (i) $\operatorname{ch} x = \dfrac{1+\tau^2}{1-\tau^2}$; (ii) $\operatorname{sh} x = \dfrac{2\tau}{1-\tau^2}$; (iii) $\operatorname{th} x = \dfrac{2\tau}{1+\tau^2}$.

16 Solve the equation $7\operatorname{sh} x + 20\operatorname{ch} x = 24$ (i) by expressing it as a quadratic in $e^x$; (ii) by using the substitutions in no. 15.

17 Prove          $(\operatorname{ch} x + \operatorname{sh} x)^n = \operatorname{ch} nx + \operatorname{sh} nx,$

and            $(\operatorname{ch} x + \operatorname{sh} x)(\operatorname{ch} y + \operatorname{sh} y) = \operatorname{ch}(x+y) + \operatorname{sh}(x+y).$

18 Prove $\left(\dfrac{1+\operatorname{th} x}{1-\operatorname{th} x}\right)^n = \operatorname{ch} 2nx + \operatorname{sh} 2nx.$

19 If $\operatorname{ch} x \cos x = 1$, prove that $\operatorname{th}\tfrac{1}{2}x = \pm\tan\tfrac{1}{2}x.$ [Express all in half-angles.]

20 Prove $\sin^2 x\,\operatorname{ch}^2 y + \cos^2 x\,\operatorname{sh}^2 y = \tfrac{1}{2}(\operatorname{ch} 2y - \cos 2x).$

21 If $\tan x = \operatorname{th} y$, prove $\sin 2x = \operatorname{th} 2y.$

*Find (and simplify if possible) the derivative of*

22 $\operatorname{sech} x.$      23 $\operatorname{cosech} x.$      24 $\operatorname{coth} x.$      25 $\log\operatorname{sh} x.$

26 $\operatorname{sh}(\log x).$      27 $\log\operatorname{th}\tfrac{1}{2}x.$      28 $\tan^{-1}(\operatorname{coth} x).$    29 $\tan^{-1}(\operatorname{th}\tfrac{1}{2}x).$

30 $\operatorname{ch} x\cos x + \operatorname{sh} x\sin x.$      31 $x^{\operatorname{ch} x}.$      32 $(\operatorname{ch} x)^x.$

33 If $y = A\operatorname{ch} nx + B\operatorname{sh} nx$ where $A$, $B$, $n$ are constants, prove $d^2y/dx^2 = n^2 y.$

34 If $y = \tan^{-1}(\operatorname{sh} x)$, prove $d^2y/dx^2 + (dy/dx)^2\tan y = 0.$

*Write down the integral of*

35 $\operatorname{sh} 2x.$      36 $\operatorname{sech}^2 3x.$      37 $\operatorname{th}\tfrac{1}{2}x.$

38 $e^x(\operatorname{th} x + \operatorname{sech}^2 x).$      39 $e^{ax}\operatorname{ch} bx.$

*Using the formulae of 4.44(1) and nos. 1–15 when necessary, calculate the integral of*

40 $\operatorname{sh}^2 x.$      41 $\operatorname{ch}^2 x.$      42 $\operatorname{th}^2 x.$      43 $\operatorname{sh} x\operatorname{ch} 3x.$

44 $\operatorname{sh}^3 x.$      45 $\operatorname{ch} x\operatorname{ch} 2x\operatorname{ch} 3x.$      46 $\operatorname{ch}(\log x).$

*Using integration by parts, calculate*

47 $\displaystyle\int x\operatorname{ch} x\,dx.$      48 $\displaystyle\int x\operatorname{sech}^2 x\,dx.$      49 $\displaystyle\int x^2\operatorname{sh} x\,dx.$

50 By expressing $\operatorname{sech} x$ in terms of $e$, prove $\int\operatorname{sech} x\,dx = 2\tan^{-1}(e^x) + c.$

51 Reconcile the result of no. 50 with the answer of no. 29 by using

$$\operatorname{Tan}^{-1} a - \operatorname{Tan}^{-1} b = \operatorname{Tan}^{-1}\{(a-b)/(1+ab)\}$$

to show          $\operatorname{Tan}^{-1}(e^x) - \operatorname{Tan}^{-1} 1 = \operatorname{Tan}^{-1}(\operatorname{th}\tfrac{1}{2}x),$

i.e. that the results differ only by a constant.

52 Obtain $\int\operatorname{cosech} x\,dx$ (i) by use of the definition; (ii) from no. 27.

53 Sketch the graph of (i) $y = \coth x$; (ii) $y = \operatorname{sech} x$; (iii) $y = \operatorname{cosech} x$.

54 Prove that $\operatorname{sh} x - x$ and $x - \operatorname{th} x$ are increasing functions for all $x$. Deduce that for $x > 0$, $\operatorname{th} x < x < \operatorname{sh} x < \operatorname{ch} x$.

55 Calculate $\lim_{x \to 0} \operatorname{sh} x / x$ and $\lim_{x \to 0} \operatorname{th} x / x$. [Use no. 54; cf. 2.12.]

### 4.45 The inverse hyperbolic functions

These functions bear the same relation to $\operatorname{sh} x$, $\operatorname{ch} x$, $\operatorname{th} x$, ... that $\sin^{-1} x$, $\cos^{-1} x$, $\tan^{-1} x$, ... bear to $\sin x$, $\cos x$, $\tan x$, ....

(1) Since $x = \operatorname{sh} y$ is a continuous function of $y$ which increases steadily from $-\infty$ to $+\infty$ as $y$ increases from $-\infty$ to $+\infty$ (see 4.44 (2) (c)), hence $\operatorname{sh} y$ takes every value just once; i.e. to any given value of $x$ there corresponds a unique value of $y$, written $y = \operatorname{sh}^{-1} x$ and called the *inverse hyperbolic sine* of $x$.

The graph of $y = \operatorname{sh}^{-1} x$ is obtained from that of $y = \operatorname{sh} x$ as described in 4.41 (4).

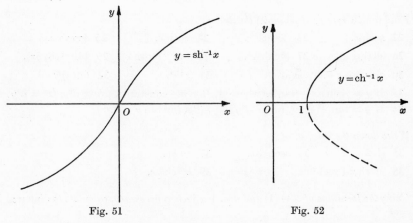

Fig. 51                    Fig. 52

If $y = \operatorname{sh}^{-1} x$, then $\operatorname{sh} y = x$ and so by 4.44 (1), (vi), $\operatorname{ch} y = +\sqrt{(1 + x^2)}$. Also $e^y = \operatorname{sh} y + \operatorname{ch} y = x + \sqrt{(1 + x^2)}$. Hence $y = \log\{x + \sqrt{(x^2 + 1)}\}$, i.e.

$$\operatorname{sh}^{-1} x = \log\{x + \sqrt{(x^2 + 1)}\}.$$

This shows that $\operatorname{sh}^{-1} x$ is continuous for all $x$, since $\log\{x + \sqrt{(x^2 + 1)}\}$ is a continuous function of a continuous function (2.62 (2)).

(2) Since $x = \operatorname{ch} y$ is an even continuous function of $y$ which increases steadily from 1 to $+\infty$ as $y$ increases from 0 to $+\infty$ and also as $y$ decreases from 0 to $-\infty$, hence $\operatorname{ch} y$ has the same value for two (equal but opposite) values of $y$, except when $\operatorname{ch} y = 1$. Thus to any

given value of $x > 1$ correspond two equal and opposite values of $y$; when $x = 1, y = 0$; and for $x < 1$ the function $y$ of $x$ is not defined. The two-valued function $y$ is denoted by $\mathrm{Ch}^{-1}x$, and the *positive* value by $\mathrm{ch}^{-1}x$.

The graph of $y = \mathrm{ch}^{-1}x$ is the 'thick' curve in fig. 52. The complete curve is that of $y = \mathrm{Ch}^{-1}x$.

From $y = \mathrm{Ch}^{-1}x$ we have $x = \mathrm{ch}\,y$, $\mathrm{sh}\,y = \pm\sqrt{(x^2-1)}$, and

$$e^y = \mathrm{ch}\,y + \mathrm{sh}\,y = x \pm \sqrt{(x^2-1)}.$$

Since $\quad \{x + \sqrt{(x^2-1)}\}\{x - \sqrt{(x^2-1)}\} = x^2 - (x^2-1) = 1,$

hence $\quad\quad e^y = x + \sqrt{(x^2-1)} \quad \text{or} \quad \dfrac{1}{x + \sqrt{(x^2-1)}},$

i.e. $\quad\quad \mathrm{Ch}^{-1}x = \pm\log\{x + \sqrt{(x^2-1)}\} \quad (x \geqslant 1),$

and $\quad\quad \mathrm{ch}^{-1}x = +\log\{x + \sqrt{(x^2-1)}\} \quad (x \geqslant 1).$

Thus $\mathrm{ch}^{-1}x$ is continuous for $x \geqslant 1$.

**(3)** Since $x = \mathrm{th}\,y$ is a continuous function of $y$ which increases steadily from $-1$ to $+1$ as $y$ increases from $-\infty$ to $+\infty$, hence $\mathrm{th}\,y$ takes every value between $\pm 1$ just once. Thus, given $x\,(-1 < x < 1)$, there is a unique corresponding $y$, written $y = \mathrm{th}^{-1}x$.

From

$$y = \mathrm{th}^{-1}x, \quad x = \mathrm{th}\,y = \frac{e^{2y}-1}{e^{2y}+1},$$

so

$$e^{2y} = \frac{1+x}{1-x} \quad \text{and} \quad y = \tfrac{1}{2}\log\frac{1+x}{1-x},$$

i.e. $\quad\quad \mathrm{th}^{-1}x = \tfrac{1}{2}\log\dfrac{1+x}{1-x} \quad (|x| < 1),$

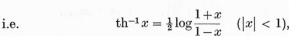

Fig. 53

which is continuous for all $|x| < 1$.

**(4)** The functions $\mathrm{Sech}^{-1}x$, $\mathrm{cosech}^{-1}x$, $\mathrm{coth}^{-1}x$ are seldom used because

$$\mathrm{Sech}^{-1}x = \mathrm{Ch}^{-1}\left(\frac{1}{x}\right), \quad \mathrm{cosech}^{-1}x = \mathrm{sh}^{-1}\left(\frac{1}{x}\right), \quad \mathrm{coth}^{-1}x = \mathrm{th}^{-1}\left(\frac{1}{x}\right).$$

They could be treated similarly; see Ex. 4 (*h*), nos. 38–40.

**(5)** *Derivatives and integrals.*

(i) $y = \text{sh}^{-1} x$.

From $x = \text{sh}\, y$, $dx/dy = \text{ch}\, y = +\sqrt{(1+\text{sh}^2 y)} = +\sqrt{(1+x^2)}$, the positive sign being chosen since $\text{ch}\, y$ is never negative. Hence

$$\frac{d}{dx}\text{sh}^{-1} x = +\frac{1}{\sqrt{(1+x^2)}} \quad (\textit{all } x).$$

(ii) $y = \text{ch}^{-1} x$.

Since $x = \text{ch}\, y$, $dx/dy = \text{sh}\, y = +\sqrt{(\text{ch}^2 y - 1)} = +\sqrt{(x^2 - 1)}$, the positive sign being chosen because the values of $y = \text{ch}^{-1} x$ are not negative, and hence $\text{sh}\, y$ is not negative. Therefore

$$\frac{d}{dx}\text{ch}^{-1} x = +\frac{1}{\sqrt{(x^2 - 1)}} \quad (x > 1).$$

(iii) $y = \text{th}^{-1} x$.

From $x = \text{th}\, y$, $dx/dy = \text{sech}^2 y = 1 - x^2$. So

$$\frac{d}{dx}\text{th}^{-1} x = \frac{1}{1-x^2} \quad (|x| < 1).$$

The corresponding integrals are:

$$\int \frac{dx}{\sqrt{(1+x^2)}} = \text{sh}^{-1} x, \quad \int \frac{dx}{\sqrt{(x^2 - 1)}} = \text{ch}^{-1} x \quad (x > 1),$$

$$\int \frac{dx}{1-x^2} = \text{th}^{-1} x \quad (|x| < 1).$$

These should be compared carefully with the last two results of 4.12.

### Exercise 4(*h*)

**1** Obtain the derivatives of $\text{sh}^{-1} x$, $\text{ch}^{-1} x$, $\text{th}^{-1} x$ by using their logarithmic expressions.

*Write down the derivative of the following, stating any restrictions on the values of $x$.*

**2** $\text{sh}^{-1}(\tfrac{2}{3}x)$.  **3** $\text{ch}^{-1}(2x-1)$.  **4** $\text{th}^{-1}(2x)$.

**5** $\text{th}^{-1}(\tan\tfrac{1}{2}x)$.  **6** $\coth^{-1}\left\{\dfrac{1}{2}\left(x+\dfrac{1}{x}\right)\right\}$.  **7** $\text{sh}^{-1}(\tan x)$.

**8** $\text{th}^{-1}(\sin x)$.  **9** $\text{ch}^{-1}(\sec x)$.

**10** Calculate $\displaystyle\int \frac{dx}{\sqrt{(1+x^2)}}$ by the substitution $x = \text{sh}\, t$.

**11** Calculate $\displaystyle\int \frac{dx}{\sqrt{(x^2 - 1)}}$ $(x > 1)$ by the substitution $x = \text{ch}\, t$.

*By writing $x/a$ $(a > 0)$ for $x$ in the integrals of 4.45(5), prove that*

**12** $\displaystyle\int \frac{dx}{\sqrt{(a^2+x^2)}} = \text{sh}^{-1}\frac{x}{a} + c.$  **13** $\displaystyle\int \frac{dx}{\sqrt{(x^2 - a^2)}} = \text{ch}^{-1}\frac{x}{a} + c \quad (x > a).$

**14** $\int \dfrac{dx}{a^2 - x^2} = \dfrac{1}{a}\,\mathrm{th}^{-1}\dfrac{x}{a} + c \quad (|x| < a).$

**15** Calculate the integrals in nos. 12–14 by using suitable hyperbolic substitutions.

**16** Show that the logarithmic form of no. 12 is

$$\int \frac{dx}{\sqrt{(a^2 + x^2)}} = \log\{x + \sqrt{(a^2 + x^2)}\} + c',$$

and state the relation between the constants $c$, $c'$. Give similar results for nos. 13, 14.

*Write down the integrals of*

**17** $\dfrac{1}{\sqrt{(x^2 + 25)}}.$ **18** $\dfrac{1}{\sqrt{(x^2 - 36)}}.$ **19** $\dfrac{1}{\sqrt{(4x^2 + 25)}}.$ **20** $\dfrac{1}{\sqrt{(36 - x^2)}}.$

**21** $\dfrac{1}{\sqrt{(4x^2 - 9)}}.$ **22** $\dfrac{1}{\sqrt{(2 + 2x + x^2)}}.$ **23** $\dfrac{1}{\sqrt{(2x + x^2)}}.$

**24** By putting $x = a\,\mathrm{sh}\,t$ and using $\mathrm{ch}^2 t = \frac{1}{2}(\mathrm{ch}\,2t + 1)$, prove that if $a > 0$,

$$\int \sqrt{(x^2 + a^2)}\,dx = \tfrac{1}{2}x\sqrt{(x^2 + a^2)} + \tfrac{1}{2}a^2\,\mathrm{sh}^{-1}(x/a) + c.$$

**25** By putting $x = a\,\mathrm{ch}\,t$, calculate $\int\sqrt{(x^2 - a^2)}\,dx \ (a > 0)$. [Contrast 4.21, ex. (vi).]

*By using a suitable hyperbolic substitution, calculate the following.*

**26** $\displaystyle\int \sqrt{(9x^2 - 4)}\,dx.$ **27** $\displaystyle\int_0^2 \sqrt{(9x^2 + 4)}\,dx$ (answer in logarithmic form).

**28** $\displaystyle\int \frac{x^2}{\sqrt{(x^2 - 1)}}\,dx.$ **29** $\displaystyle\int_1^3 \frac{\sqrt{(x^2 + 1)}}{x^2}\,dx.$

**\*30** If $x < -1$, what is the substitution to be used in no. 11? Show that the result is $-\log\{-x + \sqrt{(x^2 - 1)}\} + c$. [As $\mathrm{ch}\,t$ is never negative, the substitution required is $x = -\mathrm{ch}\,t$.]

**\*31** (i) Evaluate $\displaystyle\int_{-13}^{-12} \frac{dx}{\sqrt{(x^2 - 25)}}$ by the substitution $x = -5\,\mathrm{ch}\,t$.

(ii) Also evaluate this integral by first reducing it to $\displaystyle\int_{12}^{13} \frac{dy}{\sqrt{(y^2 - 25)}}$ by the substitution $x = -y$. (This obviates the difficulty of signs indicated in (i).)

*Using integration by parts, calculate the integrals of*

**32** $\mathrm{ch}^{-1}x.$ **33** $\mathrm{th}^{-1}x.$ **34** $x\,\mathrm{sh}^{-1}x.$ **35** $x^2\,\mathrm{th}^{-1}x.$

**\*36** If $c_n = \int x^n\,\mathrm{ch}\,ax\,dx$ and $s_n = \int x^n\,\mathrm{sh}\,ax\,dx$, where $a \neq 0$ and $n$ is a positive integer, prove

$$c_n = \frac{x^n}{a}\,\mathrm{sh}\,ax - \frac{n}{a}s_{n-1}, \quad s_n = \frac{x^n}{a}\,\mathrm{ch}\,ax - \frac{n}{a}c_{n-1}.$$

Hence obtain a reduction formula for $c_n$ and for $s_n$.

**\*37** Using no. 5, show that $\int \sec x\,dx = 2\,\mathrm{th}^{-1}(\tan\frac{1}{2}x)$ if $|\tan\frac{1}{2}x| < 1$. What is the result when $|\tan\frac{1}{2}x| > 1$? Show that both results are given by the logarithmic forms $\log|\tan(\frac{1}{2}x + \frac{1}{4}\pi)| = \log|\sec x + \tan x|$.

**\*38** Verify the results stated in 4.45 (4), and state for what values of $x$ they hold. Use these results to prove that

$$\text{(i)} \quad \frac{d}{dx}\operatorname{sech}^{-1}x = -\frac{1}{x\sqrt{(1-x^2)}} \quad (0 < x < 1);$$

$$\text{(ii)} \quad \frac{d}{dx}\operatorname{cosech}^{-1}x = -\frac{1}{x\sqrt{(1+x^2)}} \quad (x \neq 0);$$

$$\text{(iii)} \quad \frac{d}{dx}\coth^{-1}x = \frac{1}{1-x^2} \quad (|x| > 1).$$

**\*39** Sketch the graphs of (i) $\operatorname{cosech}^{-1}x$; (ii) $\operatorname{Sech}^{-1}x$; (iii) $\coth^{-1}x$.

**\*40** Prove that

$$\text{(i)} \quad \operatorname{sech}^{-1}x = \log\left\{\frac{1+\sqrt{(1-x^2)}}{x}\right\} \quad (0 < x < 1);$$

$$\text{(ii)} \quad \operatorname{cosech}^{-1}x = \log\left\{\frac{1+\sqrt{(1+x^2)}}{x}\right\} \text{ if } x > 0, \quad \log\left\{\frac{1-\sqrt{(1+x^2)}}{x}\right\} \text{ if } x < 0;$$

$$\text{(iii)} \quad \coth^{-1}x = \tfrac{1}{2}\log\left\{\frac{x+1}{x-1}\right\} \quad (|x| > 1).$$

## (B) SYSTEMATIC INTEGRATION

### 4.5 Revised list of standard integrals

In section (A) of this chapter we gave some general methods of integration (decomposition, substitution, parts, reduction formula), and applied them to a variety of functions. In this section we approach the subject from a different point of view by considering various types of function and investigating what methods can be used to integrate them. Although we cannot give rules which will apply for the integration of all functions, we shall see that certain classes of functions can always be integrated by following a well-defined procedure. Our classification of functions will be the 'structural' one outlined in 1.51.

The following results† have been obtained in section (A):

$$\int x^m dx = \frac{x^{m+1}}{m+1} \quad (m \neq -1); \tag{i}$$

$$\int \frac{1}{x}dx = \log|x|; \tag{ii}$$

$$\int e^x dx = e^x; \tag{iii}$$

$$\int \sin x\, dx = -\cos x, \quad \int \cos x\, dx = \sin x; \tag{iv}\, a, b$$

† The arbitrary constant of integration has been omitted for brevity.

$$\int \sec^2 x\,dx = \tan x, \quad \int \operatorname{cosec}^2 x\,dx = -\cot x; \qquad \text{(v)}\,a,b$$

$$\int \operatorname{sh} x\,dx = \operatorname{ch} x, \quad \int \operatorname{ch} x\,dx = \operatorname{sh} x; \qquad \text{(vi)}\,a,b$$

$$\int \operatorname{sech}^2 x\,dx = \operatorname{th} x, \quad \int \operatorname{cosech}^2 x\,dx = -\coth x; \qquad \text{(vii)}\,a,b$$

$$\int \frac{1}{1+x^2}\,dx = \tan^{-1} x; \qquad \text{(viii)}$$

$$\int \frac{1}{\sqrt{(1-x^2)}}\,dx = \sin^{-1} x; \qquad \text{(ix)}$$

$$\int \frac{1}{\sqrt{(1+x^2)}}\,dx = \operatorname{sh}^{-1} x; \qquad \text{(x)}$$

$$\int \frac{1}{\sqrt{(x^2-1)}}\,dx = \operatorname{ch}^{-1} x \quad (x > 1). \qquad \text{(xi)}$$

The result $\displaystyle\int \frac{f'(x)}{f(x)}\,dx = \log|f(x)|$ includes the following:

$$\int \tan x\,dx = -\log|\cos x|, \quad \int \cot x\,dx = \log|\sin x|; \quad \text{(xii)}\,a,b$$

$$\int \operatorname{th} x\,dx = \log \operatorname{ch} x, \quad\quad \int \coth x\,dx = \log|\operatorname{sh} x|. \quad \text{(xiii)}\,a,b$$

We shall show (see 4.81, exs. (i), (ii)) that

$$\int \operatorname{cosec} x\,dx = \log|\tan \tfrac{1}{2}x|; \qquad \text{(xiv)}$$

$$\int \sec x\,dx = \log|\tan(\tfrac{1}{2}x + \tfrac{1}{4}\pi)| \quad or \quad \log|\sec x + \tan x|. \qquad \text{(xv)}\,a,b$$

The reader should know all of these, and should be able to quote the corresponding results when $x$ is replaced by any linear function $px+q$ ($p \neq 0$). Thus by writing $x/a$ for $x$ in nos. (viii)–(xi), we have when $a > 0$:

$$\int \frac{1}{a^2+x^2}\,dx = \frac{1}{a}\tan^{-1}\frac{x}{a}; \qquad \text{(xvi)}$$

$$\int \frac{1}{\sqrt{(a^2-x^2)}}\,dx = \sin^{-1}\frac{x}{a}; \qquad \text{(xvii)}$$

$$\int \frac{1}{\sqrt{(a^2+x^2)}}\,dx = \operatorname{sh}^{-1}\frac{x}{a} \quad or \quad \log\{x + \sqrt{(a^2+x^2)}\}; \quad \text{(xviii)}\,a,b$$

and if $0 < a < x$,

$$\int \frac{1}{\sqrt{(x^2-a^2)}}\,dx = \operatorname{ch}^{-1}\frac{x}{a} \quad or \quad \log\{x+\sqrt{(x^2-a^2)}\}. \quad \text{(xix)}\,a,b$$

Similarly, on replacing $x$ by $px+q$ in (ii),

$$\int \frac{1}{px+q}\,dx = \frac{1}{p}\log|px+q|.$$

Finally,
$$\int \frac{1}{x^2-a^2}\,dx = \frac{1}{2a}\log\left|\frac{x-a}{x+a}\right| \qquad \text{(xx)}$$

because
$$\frac{1}{x^2-a^2} = \frac{1}{2a}\left(\frac{1}{x-a}-\frac{1}{x+a}\right).$$

## 4.6   Integration of rational functions

### 4.61  Preliminary considerations

(1) All rational functions of the form $kf'(x)/f(x)$ (where $k$ is constant) can be integrated immediately as $k\log|f(x)|$, by 4.33.

(2) When the function is not already of this form, we decompose it into simpler rational functions. If the degree of the numerator is not less than that of the denominator, we begin by using 'long division' to reduce the fraction to a polynomial together with a 'proper' fraction.

**Examples**

(i)
$$\frac{x^2}{x+1} = x-1+\frac{1}{x+1},$$

hence
$$\int \frac{x^2}{x+1}\,dx = \tfrac{1}{2}x^2-x+\log|x+1|.$$

(ii)
$$\frac{2x^4-x^3-3x^2+5x-4}{x^3-3x+1} = 2x-1+\frac{3x^2-3}{x^3-3x+1},$$

so the integral of this function is

$$x^2-x+\log|x^3-3x+1|.$$

### 4.62  Digression on partial fractions

We may suppose that, whenever necessary, the above division process has been done. Our problem is then to decompose a 'proper' fraction into a sum of simpler fractions, known as its *partial fractions*. When the denominator can be factorised into a product of linear and irreducible† quadratic factors (repeated or not), the reader will know how this decomposition is made. We give here, for convenience, a summary of the working rules; the theory is in the Algebra section (10.53), where there is a full formulation of the basis of 'equating coefficients' (10.13).

1 (a) For each simple linear factor $x - a$ assume a partial fraction

$$\frac{A}{x-a}.$$

(b) For each squared linear factor $(x-b)^2$ assume a sum

$$\frac{B_1}{x-b} + \frac{B_2}{(x-b)^2}, \quad \text{or equivalently} \quad \frac{Ax+B}{(x-b)^2}.$$

(c) For each cubed linear factor $(x-c)^3$ assume

$$\frac{C_1}{x-c} + \frac{C_2}{(x-c)^2} + \frac{C_3}{(x-c)^3}, \quad \text{or equivalently} \quad \frac{Ax^2+Bx+C}{(x-c)^3};$$

and so on.

2 (a) For each unrepeated irreducible quadratic factor $x^2 + bx + c$ assume a fraction

$$\frac{Ax+B}{x^2+bx+c}.$$

(b) For each squared irreducible quadratic factor $(x^2 + bx + c)^2$ assume a sum

$$\frac{A_1x+B_1}{x^2+bx+c} + \frac{A_2x+B_2}{(x^2+bx+c)^2};$$

and so on.

Thus, in general, for a factor of degree $n$ we assume a partial fraction whose numerator is of degree $n-1$ (cf. the equivalent forms in Rule 1); but in the case of a repeated factor it is usually better to express such a fraction as the sum of $n$ simpler ones, as first stated.

Having followed the above procedure for all factors of the denominator, we convert the relation expressing the equivalence of the given fraction and the sum of partial fractions into a *polynomial* equation, by multiplying both sides by the given denominator. The constants $A, B, C, \ldots$ can then be determined either by (i) equating coefficients, or (ii) substitution of special values for $x$, or (most usually) (iii) a mixture of these methods. The following examples should revise and clarify all that is necessary.

† See the definition in 10.52.

## Examples

(i) *Simple linear factors.*

$$\frac{x^2+1}{x(x^2-4)} = \frac{A}{x} + \frac{B}{x-2} + \frac{C}{x+2}.$$

$$\therefore \quad x^2+1 \equiv A(x^2-4) + Bx(x+2) + Cx(x-2).$$

Put $x = 0$: $\qquad\qquad 1 = -4A$, so $A = -\frac{1}{4}$.

Put $x = 2$: $\qquad\qquad 5 = 8B$, so $B = \frac{5}{8}$.

Put $x = -2$: $\qquad\qquad 5 = 8C$, so $C = \frac{5}{8}$.

Hence $\qquad\qquad \dfrac{x^2+1}{x(x^2-4)} = -\dfrac{1}{4x} + \dfrac{5}{8(x-2)} + \dfrac{5}{8(x+2)}.$

*Repeated linear factor.*

(ii) $\qquad\qquad \dfrac{7x-4}{(x-1)^2(x+2)} = \dfrac{A}{x-1} + \dfrac{B}{(x-1)^2} + \dfrac{C}{x+2}.$

$$\therefore \quad 7x-4 \equiv A(x-1)(x+2) + B(x+2) + C(x-1)^2.$$

Put $x = 1$: $\qquad\qquad 3 = 3B$, so $B = 1$.

Put $x = -2$: $\qquad\qquad -18 = 9C$, so $C = -2$.

Equate coefficients of $x^2$: $\quad 0 = A+C$, so $A = 2$.

Hence $\qquad\qquad \dfrac{7x-4}{(x-1)^2(x+2)} = \dfrac{2}{x-1} + \dfrac{1}{(x-1)^2} - \dfrac{2}{x+2}.$

(iii) $\qquad \dfrac{2x^2+3x+1}{(x-2)^3(x-3)} = \dfrac{A}{x-2} + \dfrac{B}{(x-2)^2} + \dfrac{C}{(x-2)^3} + \dfrac{D}{x-3},$

and we could proceed as before. The following variation of the method should be noticed.

$$2x^2+3x+1 \equiv A(x-2)^2(x-3) + B(x-2)(x-3) + C(x-3) + D(x-2)^3.$$

Put $x = 2$: $\qquad\qquad 15 = -C$, so $C = -15$.

Hence

$$A(x-2)^2(x-3) + B(x-2)(x-3) + D(x-2)^3 \equiv 2x^2+3x+1+15(x-3)$$
$$\equiv 2x^2+18x-44$$
$$\equiv 2(x-2)(x+11).$$

$$\therefore \quad A(x-2)(x-3) + B(x-3) + D(x-2)^2 \equiv 2x+22.$$

Put $x = 2$: $\qquad\qquad -B = 26,$

and hence $\qquad A(x-2)(x-3) + D(x-2)^2 \equiv 2x+22+26(x-3)$
$$\equiv 28(x-2).$$

$$\therefore \quad A(x-3) + D(x-2) \equiv 28.$$

Putting $x = 2$ gives $A = -28$, and $x = 3$ gives $D = 28$.

*When cubic or higher order repetitions occur*, it is shorter to use the following method (equivalent to repeated division).

Putting $y = x - 2$, the fraction given in ex. (iii) becomes

$$\frac{2(y+2)^2 + 3(y+2) + 1}{y^3(y-1)} = \frac{2y^2 + 11y + 15}{y^3(y-1)}.$$

Now† 
$$15 + 11y + 2y^2 = -15(y-1) + 26y + 2y^2$$
$$= -15(y-1) - 26y(y-1) + 28y^2$$
$$= -15(y-1) - 26y(y-1) - 28y^2(y-1) + 28y^3,$$

so the fraction is

$$-\frac{15}{y^3} - \frac{26}{y^2} - \frac{28}{y} + \frac{28}{y-1}$$

$$= \frac{28}{x-3} - \frac{28}{x-2} - \frac{26}{(x-2)^2} - \frac{15}{(x-2)^3}.$$

*Simple quadratic factor.*

(iv) 
$$\frac{x+4}{x(x^2+2)} = \frac{A}{x} + \frac{Bx+C}{x^2+2},$$

$$\therefore \quad x+4 \equiv A(x^2+2) + x(Bx+C).$$

Put $x = 0$:      $4 = 2A$,     so    $A = 2$.

Equate coefficients of $x^2$:    $0 = A + B$,    so    $B = -2$.

Equate coefficients of $x$:    $1 = C$.

Hence 
$$\frac{x+4}{x(x^2+2)} = \frac{2}{x} - \frac{2x-1}{x^2+2}.$$

(v) 
$$\frac{x^2}{(x^2+4)(x^2+9)} = \frac{Ax+B}{x^2+4} + \frac{Cx+D}{x^2+9};$$

but if we notice that the given fraction is a function of $x^2$ and put $y = x^2$, we have (more easily)

$$\frac{y}{(y+4)(y+9)} = \frac{A}{y+4} + \frac{B}{y+9},$$

from which we find $A = -\frac{4}{5}$, $B = \frac{9}{5}$ as in ex. (i). Thus

$$\frac{x^2}{(x^2+4)(x^2+9)} = \frac{9}{5(x^2+9)} - \frac{4}{5(x^2+4)}.$$

(vi) 
$$\frac{x^4 + x^3}{x^3 - 1}.$$

Here the fraction is not 'proper', so that first the division process must be applied. It can be combined with the work for finding the constants in the partial fractions as follows. Since the quotient is of first degree in $x$, it has the form $ax+b$; so, using the factors $(x-1)(x^2+x+1)$ of $x^3-1$,

$$\frac{x^4+x^3}{x^3-1} = ax+b + \frac{A}{x-1} + \frac{Bx+C}{x^2+x+1},$$

$$\therefore \quad x^4 + x^3 \equiv (ax+b)(x^3-1) + A(x^2+x+1) + (Bx+C)(x-1).$$

Equating coefficients of $x^4$ and of $x^3$ gives respectively $1 = a$, $1 = b$. Put $x = 1$:

$$2 = 3A, \quad \text{so} \quad A = \tfrac{2}{3}.$$

† The following can also be obtained by long division of $15 + 11y + 2y^2$ by $-1 + y$ (arranged in *ascending* powers of $y$).

Proceeding as in ex. (iii), first method, we find after a little calculation that

$$3(Bx+C)(x-1) \equiv -(2x+1)(x-1),$$

so that $$3Bx+3C \equiv -2x-1.$$

Equating coefficients of $x$ gives $B = -\tfrac{2}{3}$; and equating constant terms gives $C = -\tfrac{1}{3}$. Hence

$$\frac{x^4+x^3}{x^3-1} = x+1+\frac{2}{3(x-1)} - \frac{2x+1}{3(x^2+x+1)}.$$

*Repeated quadratic factor.*

(vii) $$\frac{x+3}{(x-1)(x^2+1)^2} = \frac{A}{x-1}+\frac{Bx+C}{x^2+1}+\frac{Dx+E}{(x^2+1)^2};$$

but it is easier to proceed as follows. Write

$$\frac{x+3}{(x-1)(x^2+1)^2} = \frac{A}{x-1}+\frac{f(x)}{(x^2+1)^2},$$

$$\therefore \quad x+3 \equiv A(x^2+1)^2+(x-1)f(x).$$

Put $x = 1$: $$4 = 4A, \quad \text{so} \quad A = 1.$$

$$\therefore \quad \frac{f(x)}{(x^2+1)^2} = \frac{x+3}{(x-1)(x^2+1)^2} - \frac{1}{x-1}$$

$$= \frac{-x^3-x^2-3x-2}{(x^2+1)^2}$$

after combining the fractions and then removing the factor $x-1$. We now have

$$f(x) \equiv -x^3-x^2-3x-2,$$

and this can be put in the form $(ax+b)+(cx+d)(x^2+1)$ either by inspection or by division by $x^2+1$. We find

$$f(x) \equiv (-2x-1)+(-x-1)(x^2+1).$$

The given fraction is therefore

$$\frac{1}{x-1}-\frac{x+1}{x^2+1}-\frac{2x+1}{(x^2+1)^2}.$$

(viii) *Case of linear factors with irrational coefficients.*

The method of ex. (i) would apply to a case like

$$\frac{1}{x(x^2-2)} = \frac{1}{x(x+\sqrt{2})(x-\sqrt{2})},$$

and we could express the result in the form

$$\frac{A}{x}+\frac{B}{x+\sqrt{2}}+\frac{C}{x-\sqrt{2}}.$$

For the purpose of integration, *surds can often be avoided by* INCOMPLETELY *decomposing the given fraction.* Thus

$$\frac{1}{x(x^2-2)} = \frac{A}{x}+\frac{Bx+C}{x^2-2},$$

and by the usual methods we should find $A = -\frac{1}{2}$, $B = \frac{1}{2}$, $C = 0$. Then immediately

$$\int \frac{1}{x(x^2-2)}\, dx = -\frac{1}{2}\log|x| + \frac{1}{4}\log|x^2-2|$$

$$= \frac{1}{4}\log\left|\frac{x^2-2}{x^2}\right|.$$

## Exercise 4($i$)

*Using the results of the above worked examples, calculate the integrals of*

1   $\dfrac{x^2+1}{x(x^2-4)}$.

2   $\dfrac{7x-4}{(x-1)^2(x+2)}$.

3   $\dfrac{2x^2+3x+1}{(x-2)^3(x-3)}$.

4   $\dfrac{x+4}{x(x^2+2)}$.

5   $\dfrac{x^2}{(x^2+4)(x^2+9)}$.

6   $\dfrac{x^3(x+1)}{x^3-1}$.

*7   $\dfrac{x+3}{(x-1)(x^2+1)^2}$.

*Express the following completely in partial fractions (using inspection whenever convenient), and hence calculate their integrals.*

8   $\dfrac{1}{x^2-1}$.

9   $\dfrac{x^2+1}{x^2-1}$.

10   $\dfrac{x+1}{(x-1)^2}$.

11   $\dfrac{x^2+1}{(x-2)^3}$.

12   $\dfrac{x-2}{x(x+1)}$.

13   $\dfrac{2x+1}{(x-1)(4x-1)}$.

14   $\dfrac{2x-1}{x^2-x-6}$.

15   $\dfrac{2x^3+x^2+4}{x(x+1)}$.

16   $\dfrac{6x+13}{(x+1)(x+2)(2x+3)}$.

17   $\dfrac{x+18}{x(x-3)^2}$.

18   $\dfrac{2x+7}{(x-1)(x+2)^2}$.

19   $\dfrac{3x^2-3x-2}{x^2(x+1)^2}$.

20   $\dfrac{7x^2+12x+2}{(x+1)^3(x-2)}$.

21   $\dfrac{2x+3}{x(x^2+1)}$.

22   $\dfrac{x^4+4x-16}{(x-2)^2(x^2+4)}$.

*23   $\dfrac{3x^3+22x^2+7x+54}{(1-2x)(x^2+3)^2}$.

24   $\dfrac{1}{(2x^2+1)(3x^2+2)}$.

*Express the following in partial fractions involving only rational coefficients, and hence calculate their integrals.*

25   $\dfrac{x+2}{(2x+3)(x^2-3)}$.

26   $\dfrac{4x+3}{(x-2)(x^2+4x-1)}$.

27   $\dfrac{2x^3+x^2-2}{x^4-4}$.

## 4.63 Summary

It will now be clear that, *provided we can resolve the denominator into linear and quadratic factors*, any given rational function $f(x)/g(x)$ can be expressed as a sum of partial fractions of the types

$$\frac{A}{(x-a)^m}, \quad \frac{Bx+C}{(x^2+bx+c)^m},$$

for $m = 1, 2, 3, \ldots$, where $x^2 + bx + c$ has no factors or else has linear factors with irrational coefficients.

Integration of the first type yields

$$\frac{1}{1-m}\frac{A}{(x-a)^{m-1}} \quad \text{if} \quad m \neq 1, \quad \text{or} \quad A\log|x-a| \quad \text{if} \quad m = 1.$$

We examine the integrals of the second type in 4.64, 4.65.

If the polynomial $g(x)$ cannot be factorised as described above, we remark that the method of partial fractions becomes powerless.† Thus it cannot cope with

$$\int \frac{3x^6 + 5x^4 + x^2 - 2x - 1}{(x^5 - x + 1)^2}\, dx,$$

although the expression for this integral is in fact

$$-\frac{x^2 + 1}{x^5 - x + 1}.$$

**4.64** $\displaystyle\int \frac{Ax+B}{x^2+bx+c}\, dx$

**(1)** *Constant numerator* (case $A = 0$).

Complete the square in the denominator, thus reducing the integral to one of the standard forms (xvi), (xx) in 4.5.

**Examples**

(i) 
$$\int \frac{dx}{x^2 + 6x - 4} = \int \frac{dx}{(x+3)^2 - 13}.$$

Since 
$$\int \frac{dx}{x^2 - a^2} = \int \frac{1}{2a}\left(\frac{1}{x-a} - \frac{1}{x+a}\right) dx = \frac{1}{2a}\log\left|\frac{x-a}{x+a}\right|,$$

therefore 
$$\int \frac{dx}{(x+k)^2 - a^2} = \frac{1}{2a}\log\left|\frac{x+k-a}{x+k+a}\right|.$$

Hence the given integral is

$$\frac{1}{2\sqrt{13}}\log\left|\frac{x+3-\sqrt{13}}{x+3+\sqrt{13}}\right|.$$

(ii) 
$$\int \frac{dx}{4x^2 + 16x + 25} = \frac{1}{4}\int \frac{dx}{x^2 + 4x + 6\frac{1}{4}} = \frac{1}{4}\int \frac{dx}{(x+2)^2 + 2\frac{1}{4}}.$$

Since 
$$\int \frac{dx}{x^2 + a^2} = \frac{1}{a}\tan^{-1}\frac{x}{a}, \quad \int \frac{dx}{(x+k)^2 + a^2} = \frac{1}{a}\tan^{-1}\frac{x+k}{a}.$$

Hence the given integral is

$$\frac{1}{4}\cdot\frac{1}{1\frac{1}{2}}\tan^{-1}\frac{x+2}{1\frac{1}{2}} = \frac{1}{6}\tan^{-1}\frac{2x+4}{3}.$$

† Hermite has shown how to find the *rational part* of any such integral $\displaystyle\int \frac{f(x)}{g(x)}\, dx$ without having to factorise $g(x)$.

*Alternatively*, put $x + 2 = \frac{3}{2}\tan\theta$; the integral becomes

$$\frac{1}{4}\int\frac{\frac{3}{2}\sec^2\theta\,d\theta}{\frac{9}{4}\sec^2\theta} = \frac{1}{6}\int d\theta = \frac{1}{6}\theta = \frac{1}{6}\tan^{-1}\frac{2x+4}{3}.$$

In practice the middle step of the working in each example is done mentally.

(2) *Linear numerator* $(A \neq 0)$.

Express the numerator in the form

$$\lambda\,(\text{derivative of denominator}) + \mu.$$

Hence split the given integral into a sum of two integrals: the first is of the form

$$\int\frac{f'(x)}{f(x)}\,dx = \log|f(x)|;$$

the second is covered by case (1) above.

**Examples**

(iii) $\displaystyle\int\frac{3x-5}{2x^2+2x+41}\,dx.$

Write $\qquad\qquad 3x - 5 \equiv \lambda(4x+2) + \mu.$

Then we find $\lambda = \frac{3}{4}$, $\mu = -\frac{13}{2}$. Hence the given integral is

$$\frac{3}{4}\int\frac{4x+2}{2x^2+2x+41}\,dx - \frac{13}{2}\int\frac{dx}{2x^2+2x+41}.$$

The last integral is

$$\frac{13}{4}\int\frac{dx}{(x+\frac{1}{2})^2+\frac{81}{4}} = \frac{13}{4}\cdot\frac{2}{9}\tan^{-1}\frac{x+\frac{1}{2}}{\frac{9}{2}} = \frac{13}{18}\tan^{-1}\frac{2x+1}{9}.$$

Therefore the given integral is

$$\tfrac{3}{4}\log(2x^2+2x+41) - \tfrac{13}{18}\tan^{-1}\frac{2x+1}{9}.$$

(iv) $\displaystyle\int\frac{2x+7}{11-3x-2x^2}\,dx.$

Write $\qquad\qquad 2x - 7 \equiv \lambda(-3-4x) + \mu;$

then $\lambda = -\frac{1}{2}$, $\mu = \frac{11}{2}$, and the given integral is

$$-\frac{1}{2}\int\frac{-3-4x}{11-3x-2x^2}\,dx + \frac{11}{2}\int\frac{dx}{11-3x-2x^2}.$$

The last integral is

$$-\frac{11}{4}\int\frac{dx}{(x+\frac{3}{4})^2-\frac{97}{16}} = -\frac{11}{4}\cdot\frac{1}{2\cdot\frac{1}{4}\sqrt{97}}\log\left|\frac{x+\frac{3}{4}-\frac{1}{4}\sqrt{97}}{x+\frac{3}{4}+\frac{1}{4}\sqrt{97}}\right|.$$

Therefore the given integral is

$$-\tfrac{1}{2}\log|11-3x-2x^2| - \frac{11}{2\sqrt{97}}\log\left|\frac{4x+3-\sqrt{97}}{4x+3+\sqrt{97}}\right|.$$

## 4.65 Reduction formula for $\displaystyle\int \frac{Ax+B}{(x^2+bx+c)^n}\,dx$

As in 4.64 (2), begin by finding numbers $\lambda$, $\mu$ for which

$$Ax+B \equiv \lambda(2x+b)+\mu.$$

Then $\displaystyle\int \frac{Ax+B}{(x^2+bx+c)^n}\,dx = \lambda \int \frac{2x+b}{(x^2+bx+c)^n}\,dx + \mu \int \frac{dx}{(x^2+bx+c)^n}$

$$= -\frac{\lambda}{n-1}\frac{1}{(x^2+bx+c)^{n-1}}+\mu u_n,$$

where $\displaystyle u_n = \int \frac{dx}{(x^2+bx+c)^n}$

$$= \int \frac{dx}{\{(x+\tfrac12 b)^2+c-\tfrac14 b^2\}^n} \quad \text{by completing the square,}$$

$$= \int \frac{1}{(t^2+k)^n}\,dt$$

on putting $t = x+\tfrac12 b$ and $k = c-\tfrac14 b^2$. Writing the numerator of the integrand as $(1/k)(t^2+k)-(1/k)t^2$, we decompose the integral:

$$u_n = \frac{1}{k}\int \frac{dt}{(t^2+k)^{n-1}} - \frac{1}{k}\int \frac{t^2\,dt}{(t^2+k)^n}$$

$$= \frac{1}{k}u_{n-1} + \frac{1}{2k(n-1)}\int t\,\frac{d}{dt}\left\{\frac{1}{(t^2+k)^{n-1}}\right\}dt$$

$$= \frac{1}{k}u_{n-1} + \frac{1}{2k(n-1)}\frac{t}{(t^2+k)^{n-1}} - \frac{1}{2k(n-1)}u_{n-1}$$

on integrating by parts. Hence

$$u_n = \frac{1}{2(n-1)k}\frac{t}{(t^2+k)^{n-1}} + \frac{2n-3}{2(n-1)k}u_{n-1}.$$

Successive applications of this reduction formula will lead to $u_1$, which is either (xvi) or (xx) of 4.5. In many cases $u_n$ is more easily calculated by means of a substitution, as in the following.

## Example

$$\int \frac{2x-1}{(x^2+2x+5)^2}\,dx.$$

Write $\qquad 2x-1 \equiv \lambda(2x+2)+\mu.$

Then $\lambda = 1$ and $\mu = -3$, so that the integral is

$$\int \frac{2x+2}{(x^2+2x+5)^2}\,dx - 3\int \frac{dx}{(x^2+2x+5)^2}$$

$$= -\frac{1}{x^2+2x+5} - 3\int \frac{dx}{\{(x+1)^2+4\}^2}.$$

Putting $x + 1 = 2\tan\theta$, the last integral becomes

$$\int \frac{2\sec^2\theta}{16\sec^4\theta}\,d\theta = \frac{1}{8}\int\cos^2\theta\,d\theta = \frac{1}{16}\int(1+\cos 2\theta)\,d\theta$$

$$= \tfrac{1}{16}(\theta + \tfrac{1}{2}\sin 2\theta) = \tfrac{1}{16}(\theta + \sin\theta\cos\theta).$$

Since $\qquad \sin\theta = \dfrac{x+1}{\sqrt{\{(x+1)^2+4\}}} \quad$ and $\quad \cos\theta = \dfrac{2}{\sqrt{\{(x+1)^2+4\}}},$

the expression is $\qquad \dfrac{1}{16}\left\{\tan^{-1}\dfrac{x+1}{2} + \dfrac{2(x+1)}{(x+1)^2+4}\right\}.$

Hence the given integral is

$$-\frac{1}{x^2+2x+5} - \tfrac{3}{16}\tan^{-1}\frac{x+1}{2} - \frac{3}{8}\frac{x+1}{x^2+2x+5}$$

$$= -\frac{3x+11}{8(x^2+2x+5)} - \tfrac{3}{16}\tan^{-1}\frac{x+1}{2}.$$

## Exercise 4(*j*)

*Integrate each of the following functions.*

1. $\dfrac{1}{(3x+2)^2+16}.$

2. $\dfrac{1}{(3x+2)^2-16}.$

3. $\dfrac{1}{x^2+2x+2}.$

4. $\dfrac{1}{5+4x-x^2}.$

5. $\dfrac{3x-1}{x^2+2x-10}.$

6. $\dfrac{6x+5}{9x^2-18x+25}.$

7. $\dfrac{x^2+2x+3}{(1+x)(1+x^2)}.$

8. $\dfrac{2x^2-3x-4}{(x^2-1)(x^2+2x+2)}.$

*9. $\dfrac{x^3+7x^2-3x+4}{x^4+x^2+1}.$

10. $\dfrac{1}{(x^2+25)^2}.$

*11. $\dfrac{1}{(x^2+4)^3}.$

*12. $\dfrac{x+2}{(x^2+2x+5)^2}.$

*13. $\dfrac{x^2}{1+x^4}.$

*14. $\dfrac{1}{1+x^4}.$

*15. $\dfrac{1}{1+x^2+x^4}.$

## 4.7 Integration of some algebraical functions

### 4.71 Linear irrationalities

For functions containing roots of linear expressions, a rationalising substitution can always be used.

**Examples**

(i) $\qquad \displaystyle\int \frac{x\,dx}{\sqrt{x+1}} = \int \frac{2t^3\,dt}{t+1} \quad$ by putting $x = t^2$

$$= 2\int\left(t^2-t+1-\frac{1}{t+1}\right)dt$$

$$= \tfrac{2}{3}t^3 - t^2 + 2t - 2\log|t+1|$$

$$= \tfrac{2}{3}x^{\frac{3}{2}} - x + 2x^{\frac{1}{2}} - 2\log|x^{\frac{1}{2}}+1|.$$

(ii) $\displaystyle\int \frac{dx}{x^{\frac{1}{2}}+x^{\frac{1}{3}}}$.

To rationalise both roots, put $x = t^6$ (since 6 is the L.C.M. of 2 and 3). The integral becomes

$$\int \frac{6t^5\,dt}{t^3+t^2} = 6\int \frac{t^3}{t+1}\,dt$$

$$= 6(\tfrac{1}{3}t^3 - \tfrac{1}{2}t^2 + t - \log|t+1|) \quad \text{as in ex. (i)}$$

$$= 2x^{\frac{1}{2}} - 3x^{\frac{1}{3}} + 6x^{\frac{1}{6}} - 6\log|x^{\frac{1}{6}}+1|.$$

(iii) $\displaystyle\int \frac{dx}{x+\surd(2x-1)} = \int \frac{t\,dt}{\tfrac{1}{2}(t^2+1)+t}$, where $2x-1 = t^2$, so that $x = \tfrac{1}{2}(t^2+1)$

$$= \int \frac{2t\,dt}{(t+1)^2}$$

$$= 2\int \left(\frac{1}{t+1} - \frac{1}{(t+1)^2}\right) dt \quad \text{by partial fractions,}$$

$$= 2\left\{\log|t+1| + \frac{1}{t+1}\right\}$$

$$= 2\log|\surd(2x-1)+1| + \frac{2}{\surd(2x-1)+1}.$$

(iv) $\displaystyle\int \frac{dx}{x\,\sqrt[3]{(x-8)}} = \int \frac{3t^2\,dt}{(t^3+8)\,t}$,    where   $x-8 = t^3$

$$= 3\int \frac{t\,dt}{t^3+8} = 3\int \frac{t\,dt}{(t+2)(t^2-2t+4)}$$

$$= 3\int \frac{1}{6}\left(\frac{t+2}{t^2-2t+4} - \frac{1}{t+2}\right) dt \quad \text{by partial fractions,}$$

$$= \frac{1}{2}\int \left(\frac{\tfrac{1}{2}(2t-2)+3}{t^2-2t+4} - \frac{1}{t+2}\right) dt$$

$$= \tfrac{1}{4}\log(t^2-2t+4) + \frac{3}{2}\int \frac{dt}{t^2-2t+4} - \tfrac{1}{2}\log|t+2|.$$

The last integral is

$$\int \frac{dt}{(t-1)^2+3} = \frac{1}{\surd 3}\tan^{-1}\frac{t-1}{\surd 3}.$$

The variable $x$ is now restored by putting $t = (x-8)^{\frac{1}{3}}$.

## 4.72 Quadratic irrationalities

For functions containing $\surd(ax^2+bx+c)$, the substitution

$$ax^2+bx+c = t^2$$

is not usually successful; but 'completing the square' is useful at some stage. We illustrate by a few particular cases; but it can be proved that if $R(x, y)$ is a rational function of $x$ and $y$, where

$$y = \sqrt{(ax^2 + bx + c)},$$

then $\int R(x, y)\, dx$ can always be reduced to the integral of a rational function by means of a substitution.

**4.73** $\displaystyle\int \frac{Ax + B}{\sqrt{(ax^2 + bx + c)}}\, dx$

If we remove the factor $|a|$ from under the root sign, we obtain a multiple of an integral of one of the types

$$\int \frac{Ax + B}{\sqrt{(x^2 + gx + h)}}\, dx, \quad \int \frac{Ax + B}{\sqrt{(-x^2 + gx + h)}}\, dx.$$

**(1)** *Constant numerator* (case $A = 0$).

Completing the square in the denominator reduces either type to one or other of the standard forms (xvii)–(xix) in 4.5.

**Examples**

(i) $\displaystyle\int \frac{dx}{\sqrt{(x^2 + 4x + 13)}} = \int \frac{dx}{\sqrt{\{(x+2)^2 + 9\}}} = \operatorname{sh}^{-1}\frac{x+2}{3}.$

(ii) $\displaystyle\int \frac{dx}{\sqrt{(8 - 5x - 3x^2)}} = \frac{1}{\sqrt{3}}\int \frac{dx}{\sqrt{(\frac{8}{3} - \frac{5}{3}x - x^2)}}$

$$= \frac{1}{\sqrt{3}}\int \frac{dx}{\sqrt{\{\frac{121}{36} - (x + \frac{5}{6})^2\}}} = \frac{1}{\sqrt{3}}\sin^{-1}\frac{6x+5}{11}.$$

(iii) $\displaystyle\int \frac{dx}{\sqrt{\{x(3 + 2x)\}}} = \int \frac{dx}{\sqrt{(2x^2 + 3x)}} = \frac{1}{\sqrt{2}}\int \frac{dx}{\sqrt{(x^2 + \frac{3}{2}x)}}$

$$= \frac{1}{\sqrt{2}}\int \frac{dx}{\sqrt{\{(x + \frac{3}{4})^2 - \frac{9}{16}\}}} = \frac{1}{\sqrt{2}}\operatorname{ch}^{-1}\frac{4x+3}{3}.$$

**(2)** *Linear numerator* ($A \neq 0$).

Since $d\{\sqrt{f(x)}\}/dx = \frac{1}{2}f'(x)/\sqrt{f(x)}$, we begin by writing the numerator in the form

$$\tfrac{1}{2}\lambda \text{ (derivative of expression under the root)} + \mu,$$

where $\lambda$, $\mu$ are constants. The given integral is then split into the sum of two integrals: the first is of the form

$$\int \frac{\frac{1}{2}f'(x)}{\sqrt{f(x)}}\, dx = \sqrt{f(x)};$$

the second is as in case (1).

**Examples**

(iv) $\int \dfrac{1+x}{\sqrt{(1-x+x^2)}}\,dx.$

Write $\qquad 1+x \equiv \tfrac{1}{2}\lambda(-1+2x)+\mu.$

Then $\lambda = 1$ and $\mu = \tfrac{3}{2}$; the integral is

$$\int \frac{\tfrac{1}{2}(2x-1)}{\sqrt{(1-x+x^2)}}\,dx + \frac{3}{2}\int \frac{dx}{\sqrt{(1-x+x^2)}}$$

$$= \sqrt{(1-x+x^2)} + \frac{3}{2}\int \frac{dx}{\sqrt{\{(x-\tfrac{1}{2})^2+\tfrac{3}{4}\}}}$$

$$= \sqrt{(1-x+x^2)} + \tfrac{3}{2}\,\mathrm{sh}^{-1}\frac{2x-1}{\sqrt{3}}.$$

(v) $\displaystyle\int \sqrt{\left(\frac{x-1}{2x+3}\right)}\,dx = \int \frac{x-1}{\sqrt{\{(x-1)(2x+3)\}}}\,dx = \int \frac{x-1}{\sqrt{(2x^2+x-3)}}\,dx,$

by 'rationalising the numerator'. Write

$$x-1 \equiv \tfrac{1}{2}\lambda(4x+1)+\mu.$$

Then $\lambda = \tfrac{1}{2}$ and $\mu = -\tfrac{5}{4}$; the integral is

$$\frac{1}{2}\int \frac{\tfrac{1}{2}(4x+1)}{\sqrt{(2x^2+x-3)}}\,dx - \frac{5}{4}\int \frac{dx}{\sqrt{(2x^2+x-3)}}$$

$$= \tfrac{1}{2}\sqrt{(2x^2+x-3)} - \frac{5}{4\sqrt{2}}\int \frac{dx}{\sqrt{\{(x+\tfrac{1}{4})^2-\tfrac{25}{16}\}}}$$

$$= \tfrac{1}{2}\sqrt{(2x^2+x-3)} - \frac{5}{4\sqrt{2}}\,\mathrm{ch}^{-1}\frac{4x+1}{5}.$$

Examples (i)–(v) could also have been done by using a trigono-metric or hyperbolic substitution. Thus, after the preliminary steps in ex. (v), we could proceed as follows.

Putting $x+\tfrac{1}{4} = \tfrac{5}{4}\,\mathrm{ch}\,u$, the integral is

$$\frac{1}{\sqrt{2}}\int \frac{x-1}{\sqrt{\{(x+\tfrac{1}{4})^2-\tfrac{25}{16}\}}}\,dx = \frac{1}{\sqrt{2}}\int \frac{\tfrac{5}{4}(\mathrm{ch}\,u-1)\cdot\tfrac{5}{4}\,\mathrm{sh}\,u\,du}{\tfrac{5}{4}\,\mathrm{sh}\,u}$$

$$= \frac{5}{4\sqrt{2}}\int (\mathrm{ch}\,u-1)\,du = \frac{5}{4\sqrt{2}}(\mathrm{sh}\,u-u).$$

From $u = \mathrm{ch}^{-1}\tfrac{4}{5}(x+\tfrac{1}{4}) = \mathrm{ch}^{-1}\dfrac{4x+1}{5}$ we have

$$\mathrm{sh}\,u = \sqrt{(\mathrm{ch}^2\,u-1)} = \sqrt{\left\{\left(\frac{4x+1}{5}\right)^2-1\right\}} = \tfrac{1}{5}\sqrt{(16x^2+8x-24)}$$

$$= \frac{2\sqrt{2}}{5}\sqrt{(2x^2+x-3)};$$

this gives the result as before.

**4.74** $\displaystyle\int \frac{dx}{(x-k)\sqrt{(ax^2+bx+c)}}$

This can be reduced by the substitution $x-k = 1/t$ to the form considered in 4.73.

**Examples**

(i) $\displaystyle\int \frac{dx}{(x+1)\sqrt{(1+x-x^2)}}$.

Put $x+1 = 1/t$, so that $dx = -(1/t^2)\,dt$ and

$$1+x-x^2 = -(x+1)^2 + 3x + 2 = -(x+1)^2 + 3(x+1) - 1$$

$$= -\frac{1}{t^2} + \frac{3}{t} - 1 = \frac{-t^2 + 3t - 1}{t^2}.$$

The integral is

$$\int \frac{-(1/t^2)\,dt}{(1/t).(1/t)\sqrt{(-t^2+3t-1)}} = -\int \frac{dt}{\sqrt{\{\frac{5}{4} - (t-\frac{3}{2})^2\}}}$$

$$= -\sin^{-1}\left(\frac{t-\frac{3}{2}}{\frac{1}{2}\sqrt{5}}\right) = \sin^{-1}\frac{3-2t}{\sqrt{5}} = \sin^{-1}\frac{3x+1}{\sqrt{5}(x+1)}$$

by using $t = 1/(x+1)$.

(ii) $\displaystyle\int \frac{x+1}{(x-1)\sqrt{(4x^2+1)}}\,dx = \int \frac{(x-1)+2}{(x-1)\sqrt{(4x^2+1)}}\,dx$

$$= \int \frac{dx}{\sqrt{(4x^2+1)}} + 2\int \frac{dx}{(x-1)\sqrt{(4x^2+1)}}.$$

The first of these integrals is $\frac{1}{2}\text{sh}^{-1}2x$; the second is reduced by the substitution $x-1 = 1/t$ to

$$-\int \frac{dt}{\sqrt{(5t^2+8t+4)}} = -\frac{1}{\sqrt{5}}\int \frac{dt}{\sqrt{\{(t+\frac{4}{5})^2 + \frac{4}{25}\}}}$$

$$= -\frac{1}{\sqrt{5}}\text{sh}^{-1}\frac{5t+4}{2} = -\frac{1}{\sqrt{5}}\text{sh}^{-1}\frac{4x+1}{2x-2}.$$

**4.75** $\displaystyle\int \sqrt{(ax^2+bx+c)}\,dx$

By completing the square, the integrand is reduced to one of the forms

$$\sqrt{(t^2+k^2)}, \quad \sqrt{(k^2-t^2)}, \quad \sqrt{(t^2-k^2)}.$$

The integral can then be found by a trigonometric or hyperbolic substitution. For example, putting $x = a\,\text{sh}\,u$ we have

$$\int \sqrt{(x^2+a^2)}\,dx = \int a\,\text{ch}\,u.a\,\text{ch}\,u\,du$$

$$= a^2\int \text{ch}^2 u\,du = \frac{1}{2}a^2\int(1+\text{ch}\,2u)\,du$$

$$= \frac{1}{2}a^2(u + \frac{1}{2}\text{sh}\,2u) = \frac{1}{2}a^2(u + \text{sh}\,u\,\text{ch}\,u)$$

$$= \frac{1}{2}a^2\,\text{sh}^{-1}(x/a) + \frac{1}{2}x\sqrt{(x^2+a^2)}.$$

We could also rationalise $\sqrt{(x^2+a^2)}$ by putting $x = a\tan\theta$; the integral would become

$$\int a\sec\theta\,.\,a\sec^2\theta\,d\theta = a^2\int\sec^3\theta\,d\theta,$$

but this is less easy to calculate.

Similarly, $\int\sqrt{(a^2-x^2)}\,dx$ is rationalised by $x = a\sin\theta$ (or less conveniently by $x = a\operatorname{th}u$); and $\int\sqrt{(x^2-a^2)}\,dx$ by $x = a\operatorname{ch}u$ (or $x = a\sec\theta$).

## 4.76 Direct use of a trigonometric or hyperbolic substitution

The substitutions mentioned in 4.75 can often be used directly. Others are given in Ex. 4 $(k)$, nos. 32, 33.

### Example

By putting $x = 2\tan\theta$,

$$\int\frac{dx}{x^2\sqrt{(x^2+4)}} = \int\frac{2\sec^2\theta\,d\theta}{4\tan^2\theta\,.\,2\sec\theta} = \frac{1}{4}\int\frac{\cos\theta\,d\theta}{\sin^2\theta}$$

$$= -\frac{1}{4\sin\theta} = -\frac{\sqrt{(x^2+4)}}{4x}.$$

### Exercise 4(k)

*Write down or calculate (if possible by more than one method) the integrals of*

1  $\dfrac{1-\sqrt{x}}{x}$.

2  $\dfrac{2x+1}{\sqrt{(x+1)}}$.

3  $\dfrac{1}{x+\sqrt{x}}$.

4  $\dfrac{1}{1+\sqrt[3]{x}}$.

5  $\dfrac{1}{x\sqrt{(x+4)}}$.

6  $\dfrac{2+\sqrt{x}}{\sqrt[3]{x-1}}$.

7  $\dfrac{1}{\sqrt{(2x+1)}-\sqrt{x}}$.

8  $\dfrac{x}{\sqrt{(1+x)}-\sqrt[3]{(1+x)}}$.

9  $\dfrac{1}{\sqrt{(x^2+6x+13)}}$.

10  $\dfrac{1}{\sqrt{(2x^2+4x-7)}}$.

11  $\dfrac{1}{\sqrt{(6x-7-x^2)}}$.

12  $\dfrac{1}{\sqrt{\{(x+3)(2-x)\}}}$.

13  $\dfrac{x+2}{\sqrt{(x^2+1)}}$.

14  $\dfrac{x}{\sqrt{(x^2+6x+13)}}$.

15  $\dfrac{3x-5}{\sqrt{(2x^2+4x-7)}}$.

16  $\dfrac{2x+7}{\sqrt{(6-x-x^2)}}$.

17  $\sqrt{\left(\dfrac{1+x}{x-1}\right)}$.

18  $\dfrac{x^2}{\sqrt{(x^2+1)}}$.

19  $\dfrac{x^3+1}{\sqrt{(x^2+1)}}$.

20  $\dfrac{1}{x\sqrt{(x^2+4)}}$.

21  $\dfrac{1}{x\sqrt{(2x^2+x)}}$.

22  $\dfrac{1}{x\sqrt{(3x^2+2x-1)}}$.

23  $\dfrac{3x-2}{(x-1)\sqrt{(x^2+9)}}$.

24  $\dfrac{1}{(x+1)\sqrt{(1+2x-x^2)}}$.

25  $\sqrt{(a^2-x^2)}$.

26  $\sqrt{(x^2-a^2)}$.

27  $\sqrt{(3x^2-5x)}$.

28   $\sqrt{(x^2 - 5x + 6)}$.      *29   $(3x+1)\sqrt{(2x^2 - 6x + 1)}$.

*30   $\dfrac{x}{(x^2+4)\sqrt{(x^2+9)}}$.   [Put $t = 1/\sqrt{(x^2+9)}$.]

*31   $\dfrac{1}{(x^2+4)\sqrt{(x^2+9)}}$.   [First put $x = 1/u$.]

32   If $\alpha < \beta$, show that the substitution $x = \alpha\cos^2\theta + \beta\sin^2\theta$ will convert the integrals of $\sqrt{\{(x-\alpha)(\beta-x)\}}$, $1/\sqrt{\{(x-\alpha)(\beta-x)\}}$, and $\sqrt{\{(x-\alpha)/(\beta-x)\}}$ into integrals of rational functions of $\cos\theta$ and $\sin\theta$.

33   If $\alpha > \beta$, show that the substitution $x = \alpha\,\mathrm{ch}^2 u - \beta\,\mathrm{sh}^2 u$ will convert the integrals of $\sqrt{\{(x-\alpha)(x-\beta)\}}$, $1/\sqrt{\{(x-\alpha)(x-\beta)\}}$, and $\sqrt{\{(x-\alpha)/(x-\beta)\}}$ into integrals of rational functions of $\mathrm{ch}\,u$ and $\mathrm{sh}\,u$.

## 4.8   Integration of some transcendental functions

This class of functions is very wide; we concentrate mainly on the circular functions and their hyperbolic analogues, but see 4.85.

### 4.81   Rational functions of sin $x$, cos $x$

**(1)** *The integral of any rational function of* $\sin x$ *and* $\cos x$ *can always be reduced to the integral of a rational function of $t$ by the substitution*

$$t = \tan\tfrac{1}{2}x.$$

For      $\sin x = 2\sin\tfrac{1}{2}x\cos\tfrac{1}{2}x = \dfrac{2\sin\tfrac{1}{2}x\cos\tfrac{1}{2}x}{\sin^2\tfrac{1}{2}x + \cos^2\tfrac{1}{2}x} = \dfrac{2t}{1+t^2},$

$$\cos x = \cos^2\tfrac{1}{2}x - \sin^2\tfrac{1}{2}x = \dfrac{\cos^2\tfrac{1}{2}x - \sin^2\tfrac{1}{2}x}{\cos^2\tfrac{1}{2}x + \sin^2\tfrac{1}{2}x} = \dfrac{1-t^2}{1+t^2},$$

$$\tan x = \dfrac{\sin x}{\cos x} = \dfrac{2t}{1-t^2},$$

and      $\dfrac{dt}{dx} = \tfrac{1}{2}\sec^2\tfrac{1}{2}x = \tfrac{1}{2}(1+t^2),$    so    $dx = \dfrac{2\,dt}{1+t^2}.$

This substitution is especially useful for integrals of the form

$$\int \frac{dx}{a\cos x + b\sin x + c}.$$

## Examples

(i) $\int \operatorname{cosec} x\,dx$.
Putting $t = \tan\tfrac{1}{2}x$, we have

$$\int \frac{dx}{\sin x} = \int \frac{1+t^2}{2t}\frac{2\,dt}{1+t^2} = \int\frac{dt}{t} = \log|t|.$$

Hence      $\int \operatorname{cosec} x\,dx = \log|\tan\tfrac{1}{2}x|.$

(ii) $\int \sec x\, dx$.

Since $\sec x = \operatorname{cosec}(\tfrac{1}{2}\pi + x)$, we have by ex. (i) that

$$\int \sec x\, dx = \log \left| \tan \left(\tfrac{1}{4}\pi + \tfrac{1}{2}x\right)\right|.$$

As $\qquad \tan\left(\tfrac{1}{4}\pi + \tfrac{1}{2}x\right) = \dfrac{\sin\left(\tfrac{1}{4}\pi + \tfrac{1}{2}x\right)}{\cos\left(\tfrac{1}{4}\pi + \tfrac{1}{2}x\right)} = \dfrac{\cos\tfrac{1}{2}x + \sin\tfrac{1}{2}x}{\cos\tfrac{1}{2}x - \sin\tfrac{1}{2}x}$

$$= \dfrac{(\cos\tfrac{1}{2}x + \sin\tfrac{1}{2}x)^2}{\cos^2\tfrac{1}{2}x - \sin^2\tfrac{1}{2}x} = \dfrac{1 + \sin x}{\cos x},$$

and since the last expression can be written in the forms

$$\sec x + \tan x, \qquad \sqrt{\left(\dfrac{1 + \sin x}{1 - \sin x}\right)},$$

the result can also be given as

$$\log\left|\sec x + \tan x\right|, \quad \log\left|\dfrac{1 + \sin x}{\cos x}\right|, \quad \text{or} \quad \tfrac{1}{2}\log\left|\dfrac{1 + \sin x}{1 - \sin x}\right|.$$

(iii) $\qquad \displaystyle\int \frac{dx}{5 + 3\cos x} = \int \frac{2dt}{5(1 + t^2) + 3(1 - t^2)} = \int \frac{dt}{4 + t^2}$

$$= \tfrac{1}{2}\tan^{-1}\tfrac{1}{2}t = \tfrac{1}{2}\tan^{-1}\left(\tfrac{1}{2}\tan\tfrac{1}{2}x\right).$$

(iv) $\qquad \displaystyle\int \frac{dx}{3 + 5\cos x} = \int \frac{2dt}{3(1 + t^2) + 5(1 - t^2)} = \int \frac{dt}{4 - t^2}$

$$= \int \frac{1}{4}\left(\frac{1}{2 - t} + \frac{1}{2 + t}\right) dt$$

$$= \tfrac{1}{4}\{\log\left|2 + t\right| - \log\left|2 - t\right|\} = \tfrac{1}{4}\log\left|\frac{2 + t}{2 - t}\right|$$

$$= \tfrac{1}{4}\log\left|\frac{2 + \tan\tfrac{1}{2}x}{2 - \tan\tfrac{1}{2}x}\right|.$$

(v) $\qquad \displaystyle\int \frac{dx}{\sin x - \cos x + 1} = \int \frac{2dt}{2t - (1 - t^2) + (1 + t^2)} = \int \frac{dt}{t^2 + t}$

$$= \int \left(\frac{1}{t} - \frac{1}{t + 1}\right) dt$$

$$= \log\left|t\right| - \log\left|t + 1\right| = \log\left|\frac{t}{t + 1}\right|$$

$$= \log\left|\frac{\tan\tfrac{1}{2}x}{\tan\tfrac{1}{2}x + 1}\right| = -\log\left|1 + \cot\tfrac{1}{2}x\right|.$$

(vi) The integral

$$\int \frac{a\cos x + b\sin x + c}{A\cos x + B\sin x + C}\, dx$$

is first split into two integrals by writing the numerator in the form

$$\lambda\, (\text{derivative of denominator}) + \mu\, (\text{denominator}) + \nu.$$

It then becomes

$$\lambda \log\left|A\cos x + B\sin x + C\right| + \mu x + \nu \int \frac{dx}{A\cos x + B\sin x + C},$$

and the last integral can be reduced by $t = \tan\tfrac{1}{2}x$.

(2) *Although always available, the substitution* $t = \tan \frac{1}{2}x$ *is sometimes not the best one.*

(a) *If the integrand is an odd function of* $\cos x$, *put* $u = \sin x$.

Let the given integral be $\int R(\sin x, \cos x)\, dx$. Then $R/\cos x$ is an even function of $\cos x$, and will therefore contain only even powers of $\cos x$; using $\cos^2 x = 1 - \sin^2 x$, it can be expressed as a rational function of $\sin x$ only, say $S(\sin x)$, and

$$\int R\, dx = \int S(\sin x) \cos x\, dx = \int S(u)\, du.$$

**Example**

(vii) $$\int \sin^2 x \cos^3 x\, dx = \int \sin^2 x \cos^2 x \cos x\, dx = \int u^2(1 - u^2)\, du$$

$$= \tfrac{1}{3}u^3 - \tfrac{1}{5}u^5, \quad \text{where} \quad u = \sin x,$$

$$= \tfrac{1}{3}\sin^3 x - \tfrac{1}{5}\sin^5 x.$$

(b) *If the integrand is an odd function of* $\sin x$, *put* $u = \cos x$.

**Example**

(viii) $$\int \sin^5 x \sec^6 x\, dx = \int \frac{\sin^4 x}{\cos^6 x} \sin x\, dx = -\int \frac{(1 - u^2)^2}{u^6}\, du$$

$$= -\int \left( \frac{1}{u^6} - \frac{2}{u^4} + \frac{1}{u^2} \right) du, \quad \text{where} \quad u = \cos x,$$

$$= \frac{1}{5u^5} - \frac{2}{3u^3} + \frac{1}{u}$$

$$= \tfrac{1}{5}\sec^5 x - \tfrac{2}{3}\sec^3 x + \sec x.$$

(c) *If the integrand is an even function of both* $\sin x$ *and* $\cos x$, *put* $u = \tan x$.

By writing $\sin x = \cos x \tan x$ the integrand becomes a rational function of $\cos x$ and $\tan x$ which is even in both; using

$$\cos^2 x = \frac{1}{1 + \tan^2 x}$$

we have a rational function of $\tan x$, say $T(\tan x)$, and

$$\int R\, dx = \int T(\tan x)\, dx = \int T(u) \frac{du}{1 + u^2}.$$

**Examples**

(ix) $$\int \frac{dx}{\sin^2 x \cos^2 x} = \int \frac{\sec^4 x}{\tan^2 x}\, dx = \int \frac{1 + u^2}{u^2}\, du$$

$$= \int \left( \frac{1}{u^2} + 1 \right) du = u - \frac{1}{u}$$

$$= \tan x - \cot x.$$

*Alternatively,* the integral is

$$\int \frac{\sin^2 x + \cos^2 x}{\sin^2 x \cos^2 x}\, dx = \int (\sec^2 x + \operatorname{cosec}^2 x)\, dx = \tan x - \cot x;$$

*or* it can be written

$$4 \int \frac{dx}{\sin^2 2x} = 4 \int \operatorname{cosec}^2 2x\, dx = -2 \cot 2x.$$

(x) $\displaystyle \int \frac{dx}{a^2 \cos^2 x + b^2 \sin^2 x} = \int \frac{\sec^2 x\, dx}{a^2 + b^2 \tan^2 x}$    on dividing top

                                                     and bottom by $\cos^2 x$,

$$= \int \frac{du}{a^2 + b^2 u^2} = \frac{1}{ab} \tan^{-1}\left(\frac{bu}{a}\right)$$

$$= \frac{1}{ab} \tan^{-1}\left(\frac{b}{a} \tan x\right).$$

(xi) $\displaystyle \int \frac{dx}{a^2 + b^2 \sin^2 x}$ can be reduced to the form of ex. (x) by first *using* $\cos^2 x + \sin^2 x = 1$ *to make the denominator homogeneous in* $\cos x$ *and* $\sin x$. The integral is

$$\int \frac{dx}{a^2 \cos^2 x + (a^2 + b^2) \sin^2 x}.$$

## 4.82   Circular functions of multiple angles

If the integrand consists of a sum of terms like

$$A \cos^m ax \, \sin^{m'} ax \, \cos^n bx \, \sin^{n'} bx \dots,$$

where the indices $m, m', n, n', \dots$ are positive integers and the multipliers $a, b, \dots$ are any numbers, then by the formulae of elementary trigonometry each such term can be expressed as the sum of a number of terms of the types

$$\lambda \cos\{(pa + qb + \dots)x\}, \quad \mu \sin\{(pa + qb + \dots)x\};$$

these can be integrated at once.

*Alternatively,* if $a, b, \dots$ are *integers,* the functions of the multiple angles can be expanded in powers of $\cos x$ and $\sin x$; here $m, m', n, n', \dots$ may be positive or negative integers. If $a, b, \dots$ are not integers but *rational* numbers, and $k$ is the lowest common multiple of their denominators, the substitution $x = ky$ reduces this case to the one just mentioned. The integral is thus reducible to the form already considered in 4.81.

In practice the elementary trigonometrical formulae can be used for small values of the constants $m, m', \dots, a, b, \dots$; for larger values the help of de Moivre's theorem (Ch. 14) becomes necessary (e.g. see Ex. 14 (b), nos. 6–8), or else a reduction formula is used (see 4.84, and Ex. 4 (m), nos. 28–30).

## Examples

(i)
$$\int \sin 2x \cos x \, dx = \int \tfrac{1}{2}(\sin 3x + \sin x)\, dx$$
$$= -\tfrac{1}{6}\cos 3x - \tfrac{1}{2}\cos x.$$

*Alternatively*, the integral is
$$\int 2\sin x \cos^2 x \, dx = -2\int u^2 du, \quad \text{where} \quad u = \cos x,$$
$$= -\tfrac{2}{3}u^3 = -\tfrac{2}{3}\cos^3 x.$$

This example is easy by either method.

(ii)
$$\int \sin x \cos 2x \sin 3x \, dx = \int \tfrac{1}{2}(\sin 3x - \sin x)\sin 3x \, dx$$
$$= \frac{1}{2}\int \{\sin^2 3x - \tfrac{1}{2}(\cos 2x - \cos 4x)\}\, dx$$
$$= \frac{1}{4}\int \{1 - \cos 6x - \cos 2x + \cos 4x\}\, dx$$
$$= \tfrac{1}{4}(x - \tfrac{1}{6}\sin 6x - \tfrac{1}{2}\sin 2x + \tfrac{1}{4}\sin 4x).$$

The second method leads to complicated calculations.

(iii)
$$\int \cos^2 x \sin^3 2x \, dx = \int \cos^2 x \cdot 8\sin^3 x \cos^3 x \, dx$$
$$= \int 8\sin^3 x \cos^5 x \, dx = -8\int (1 - u^2)\, u^5 du, \quad \text{where} \quad u = \cos x,$$
$$= -\tfrac{4}{3}u^6 + u^8 = \cos^8 x - \tfrac{4}{3}\cos^6 x.$$

The first method is much less direct.

## 4.83  Hyperbolic functions: analogous results

In addition to the standard forms (vi), (vii) and (xiii) of 4.5, we have:

(i)
$$\int \operatorname{cosech} x \, dx = \int \frac{dx}{\operatorname{sh} x} = \int \frac{dx}{2\operatorname{sh}\tfrac{1}{2}x \operatorname{ch}\tfrac{1}{2}x} = \int \frac{\tfrac{1}{2}\operatorname{sech}^2 \tfrac{1}{2}x \, dx}{\operatorname{th}\tfrac{1}{2}x}$$
$$= \int \frac{d\tau}{\tau}, \quad \text{where} \quad \tau = \operatorname{th}\tfrac{1}{2}x,$$
$$= \log\left|\operatorname{th}\tfrac{1}{2}x\right|.$$

(ii)
$$\int \operatorname{sech} x \, dx = \int \frac{dx}{\operatorname{ch} x} = \int \frac{dx}{\operatorname{ch}^2\tfrac{1}{2}x + \operatorname{sh}^2\tfrac{1}{2}x} = \int \frac{\operatorname{sech}^2 \tfrac{1}{2}x}{1 + \operatorname{th}^2\tfrac{1}{2}x}\, dx$$
$$= \int \frac{2d\tau}{1 + \tau^2}, \quad \text{where} \quad \tau = \operatorname{th}\tfrac{1}{2}x,$$
$$= 2\tan^{-1}(\operatorname{th}\tfrac{1}{2}x).$$

*Alternatively,*

$$\int \operatorname{sech} x \, dx = \int \frac{2 \, dx}{e^x + e^{-x}} = \int \frac{2e^x \, dx}{e^{2x} + 1} = \int \frac{2 \, du}{u^2 + 1}, \quad \text{where} \quad u = e^x,$$

$$= 2 \tan^{-1}(e^x).$$

It can be shown (see Ex. 4 (g), no. 51) that these two results differ by a constant.

The substitution $\tau = \operatorname{th} \tfrac{1}{2} x$ (cf. Ex. 4 (g), no. 15) will convert the integral of any rational function of $\operatorname{ch} x$ and $\operatorname{sh} x$ into that of a rational function of $\tau$; but often the substitutions $u = \operatorname{sh} x$, $u = \operatorname{ch} x$, or $u = \operatorname{th} x$ are more convenient, as in 4.81 (2).

Expressions involving powers and products of $\operatorname{ch} ax$, $\operatorname{sh} bx$, ... can be dealt with by the formulae of Ex. 4 (g), nos. 2–4, 8.

In general, *the procedure for integrating hyperbolic functions is similar to that for circular functions.* As the results are less useful, we shall not give further examples.

## Exercise 4(*l*)

*Integrate the following.*

1  $\dfrac{1}{1 - \cos x}$.

2  $\dfrac{1}{1 + \sin x}$.

3  $\dfrac{1}{5 - 3 \cos x}$.

4  $\dfrac{1}{\cos x + \sin x}$.

5  $\dfrac{1}{1 + \cos x + \sin x}$.

6  $\dfrac{1}{2 \cos x + \sin x + 3}$.

7  $\dfrac{\sin x - \cos x}{\sin x + \cos x}$.

8  $\dfrac{\cot x}{1 + \cot x}$.

9  $\dfrac{5 \cos x + 7}{2 \cos x + \sin x + 3}$.

*10  $\dfrac{1}{\sqrt{\{(1 + \sin x)(2 + \sin x)\}}}$.

*11  $\dfrac{x + \sin x}{1 + \cos x}$.

12  $\sin^6 x \cos^3 x$.

13  $\sin^5 x \cos^2 x$.

14  $\dfrac{1}{\sin^4 x \cos^2 x}$.

15  $\cos^3 x \operatorname{cosec}^2 x$.

16  $\tan^3 x$.

17  $\operatorname{cosec}^3 x$.

18  $\dfrac{\cos^2 x}{1 + \sin^2 x}$.

19  $\dfrac{\sin x \cos x}{\sin^2 x - \cos^2 x}$.

20  $\sin 4x \cos 6x$.

21  $\sin x \sin 3x \sin 4x$.

22  $\cos^2 x \sin 5x$.

23  $\cos^3 x \sin^2 2x$.

*24  Find constants $R$, $\alpha$ for which $a \cos x + b \sin x = R \cos (x - \alpha)$. Hence give a method for integrating $1/(a \cos x + b \sin x + c)$. Apply this method to nos. 4, 5 above.

*25  (i) If $a + b > 0$, show that

$$\int \frac{dx}{a + b \cos x} \quad \text{is} \quad \frac{2}{\sqrt{(a^2 - b^2)}} \tan^{-1} \left\{ \sqrt{\left( \frac{a - b}{a + b} \right)} \tan \tfrac{1}{2} x \right\} \quad \text{if} \quad a > b,$$

and is
$$\frac{1}{\sqrt{(b^2-a^2)}} \log \left| \frac{b+a\cos x+\sqrt{(b^2-a^2)}\sin x}{a+b\cos x} \right| \quad \text{if} \quad a < b.$$

(ii) If $a+b < 0$, a change of sign of the whole integral will lead to (i).

(iii) If $a+b = 0$, what is the value of the integral?

*26 Calculate $\displaystyle\int \frac{dx}{a\cos^2 x + b\cos x \sin x + c\sin^2 x}$.

*27 Use integration by parts to calculate
$$\int \frac{x\cos x\sin x}{(a^2\cos^2 x + b^2\sin^2 x)^2}\, dx.$$

*28 (i) If $m \neq n$, prove that
$$\int \cos mx \cos nx\, dx = \frac{1}{m^2 - n^2}(m\sin mx \cos nx - n\cos mx \sin nx),$$
$$\int \sin mx \sin nx\, dx = \frac{1}{m^2 - n^2}(n\sin mx \cos nx - m\cos mx \sin nx),$$
$$\int \sin mx \cos nx\, dx = -\frac{1}{m^2 - n^2}(m\cos mx \cos nx + n\sin mx \sin nx).$$

(ii) What are the values of these integrals when $m = n$?

*29 If $m$ and $n$ are integers, prove that

(i) $\displaystyle\int_0^{2\pi} \sin mx \cos nx\, dx = 0$;

(ii) $\displaystyle\int_0^{2\pi} \cos mx \cos nx\, dx \quad\text{and}\quad \int_0^{2\pi} \sin mx \sin nx\, dx$

are zero if $m \neq n$, and $\pi$ if $m = n$.

*30 If $m$, $n$ are integers, prove that

(i) $\displaystyle\int_0^{\pi} \cos mx \cos nx\, dx \quad\text{and}\quad \int_0^{\pi} \sin mx \sin nx\, dx$

are zero unless $m = n$, when each is $\frac{1}{2}\pi$; or $m = -n$, when they are $\frac{1}{2}\pi$, $-\frac{1}{2}\pi$;

(ii) $\displaystyle\int_0^{\pi} \sin mx \cos nx\, dx$

is $2m/(m^2 - n^2)$ or zero according as $m - n$ is odd or even.

*31 If
$$f(x) = \tfrac{1}{2}a_0 + a_1\cos x + b_1\sin x + a_2\cos 2x + b_2\sin 2x + \dots + a_n\cos nx + b_n\sin nx,$$
and $k$ is a positive integer not exceeding $n$, prove that
$$\int_0^{2\pi} f(x)\, dx = \pi a_0, \quad \int_0^{2\pi} \cos kx\, f(x)\, dx = \pi a_k, \quad \int_0^{2\pi} \sin kx\, f(x)\, dx = \pi b_k.$$

If $k > n$, show that the last two integrals are zero.

**4.84** $\int \sin^m x \cos^n x\,dx$ **by reduction formula**

(1) This integral is an important case of 4.82. It can be treated as described there, using de Moivre's theorem when $m, n$ are not small; *or* by using the substitutions given in 4.81; *or* by a reduction formula. As the last method can be used to simplify the integral even when $m, n$ are not integers, we consider it here.

*First we suppose $m + n \neq 0$.* Write

$$u_{m,n} = \int \sin^m x \cos^n x\,dx.$$

As
$$\int \sin^m x \cos x\,dx = \frac{1}{m+1} \sin^{m+1} x \quad (m \neq -1),$$

we can write     $u = \cos^{n-1} x, \quad v' = \sin^m x \cos x,$

so that     $u' = -(n-1)\cos^{n-2} x \sin x, \quad v = \frac{1}{m+1} \sin^{m+1} x,$

and integrate by parts:

$$u_{m,n} = \frac{1}{m+1} \cos^{n-1} x \sin^{m+1} x + \int \frac{1}{m+1} \sin^{m+1} x \cdot (n-1)\cos^{n-2} x \sin x\,dx$$

$$= \frac{1}{m+1} \cos^{n-1} x \sin^{m+1} x + \frac{n-1}{m+1}\int \sin^{m+2} x \cos^{n-2} x\,dx, \tag{i}$$

provided $m \neq -1$. The last integral is

$$\int \sin^m x(1 - \cos^2 x)\cos^{n-2} x\,dx = \int \sin^m x \cos^{n-2} x\,dx - \int \sin^m x \cos^n x\,dx$$

$$= u_{m,n-2} - u_{m,n}.$$

Putting this in (i) and solving for $u_{m,n}$, we find that, provided $m \neq -1$:

$$u_{m,n} = \frac{1}{m+n} \cos^{n-1} x \sin^{m+1} x + \frac{n-1}{m+n} u_{m,n-2}. \tag{ii}$$

Similarly, taking

$$u = \sin^{m-1} x, \qquad\qquad v' = \cos^n x \sin x,$$

$$u' = (m-1)\sin^{m-2} x \cos x, \quad v = -\frac{1}{n+1}\cos^{n+1} x \quad (n \neq -1),$$

we have, provided that $n \neq -1$:

$$u_{m,n} = -\frac{1}{n+1}\sin^{m-1} x \cos^{n+1} x + \frac{m-1}{n+1} u_{m-2,n+2}, \tag{iii}$$

and
$$u_{m,n} = -\frac{1}{m+n}\sin^{m-1}x\cos^{n+1}x + \frac{m-1}{m+n}u_{m-2,n}. \tag{iv}$$

If both $m$, $n$ are positive, repeated applications of (ii), (iv) will reduce the indices of $\cos x$, $\sin x$ by 2, respectively.

If $m > 0$, $n < 0$, then (iii) numerically reduces *both* indices by 2.
If $m < 0$, $n > 0$, then (i) numerically reduces *both* indices by 2.
If $m < 0$, then $m-2$ is numerically greater than $m$ and the integral on the right of (iv) involves a higher power of $\sin x$ than $u_{m,n}$ does. In this case the formula can be reversed by putting $m+2$ instead of $m$ in (iv), and then solving for $u_{m,n}$ which now appears on the right. We obtain (if $m \neq -1$, $n \neq -1$):

$$u_{m,n} = \frac{1}{m+1}\sin^{m+1}x\cos^{n+1}x + \frac{m+n+2}{m+1}u_{m+2,n}. \tag{v}$$

Similarly, if $n < 0$, we obtain from (ii), if $n \neq -1$, $m \neq -1$:

$$u_{m,n} = -\frac{1}{n+1}\sin^{m+1}x\cos^{n+1}x + \frac{m+n+2}{n+1}u_{m,n+2}. \tag{vi}$$

If $m < 0$ and $n < 0$, then (v), (vi) numerically reduce the powers of $\sin x$, $\cos x$, respectively, by 2.

When $m$ and $n$ are integers (positive or negative), the integral is reduced eventually to one in which the indices of $\sin x$ and $\cos x$ are $-1$, $0$, or $1$; this is easily calculated. (Also see Notes [3], [4] on p. 362.)

The preceding discussion includes the cases

$$u_{m,0} = \int\sin^m x\,dx, \quad u_{0,n} = \int\cos^n x\,dx.$$

*Secondly, suppose $m+n = 0$.* If $m > 0$ the integral is

$$v_m = \int\tan^m x\,dx = \int\tan^{m-2}x(\sec^2 x - 1)\,dx$$

$$= \int\tan^{m-2}x\sec^2 x\,dx - v_{m-2}$$

$$= \frac{1}{m-1}\tan^{m-1}x - v_{m-2} \quad\text{if}\quad m \neq 1.$$

This is the required reduction formula.

If $m < 0$, then $n > 0$; the integral is $\int\cot^n x\,dx$, and we obtain a formula similarly ($n \neq 1$).

**(2)** *Alternative method for getting the reduction formulae.*

To obtain quickly the formula relating $u_{m,n}$ to any specified one of the six integrals $u_{m,n-2}$, $u_{m-2,n}$, $u_{m,n+2}$, $u_{m+2,n}$, $u_{m-2,n+2}$, $u_{m+2,n-2}$, we begin by deriving $\sin^p x\cos^q x$, where $p-1$ is the smaller of the two indices of $\sin x$ and $q-1$ is the smaller of the two indices of $\cos x$. For

$$\frac{d}{dx}(\sin^p x \cos^q x) = p\sin^{p-1}x\cos^{q+1}x - q\sin^{p+1}x\cos^{q-1}x;$$

and after expressing the right-hand side in terms of the required powers of $\sin x$ and $\cos x$, the formula is obtained by integrating the result.

## Example

*Find the formula relating $u_{m,n}$ to $u_{m,n-2}$.*

The index of $\sin x$ in both is $m$, so $p = m+1$. The smaller of the indices of $\cos x$ is $n-2$, so $q = n-1$.

$$\frac{d}{dx}(\sin^{m+1}x\cos^{n-1}x)$$

$$= \cos^{n-1}x.(m+1)\sin^m x\cos x - \sin^{m+1}x.(n-1)\cos^{n-2}x\sin x$$

$$= (m+1)\sin^m x\cos^n x - (n-1)\cos^{n-2}x\sin^m x(1-\cos^2 x)$$

$$= (m+n)\sin^m x\cos^n x - (n-1)\cos^{n-2}x\sin^m x.$$

Integrating,     $\sin^{m+1}x\cos^{n-1}x = (m+n)u_{m,n} - (n-1)u_{m,n-2},$

which gives $u_{m,n}$ in terms of $u_{m,n-2}$.

**(3)** *The definite integral* $\displaystyle\int_0^{\frac{1}{2}\pi}\sin^m x\cos^n x\,dx$ *($m$, $n$ positive integers or zero.*)

By formula (ii),

$$u_{m,n} = \int_0^{\frac{1}{2}\pi}\sin^m x\cos^n x\,dx$$

$$= \left[\frac{1}{m+n}\cos^{n-1}x\sin^{m+1}x\right]_0^{\frac{1}{2}\pi} + \frac{n-1}{m+n}\int_0^{\frac{1}{2}\pi}\sin^m x\cos^{n-2}x\,dx$$

$$= \frac{n-1}{m+n}u_{m,n-2} \quad\text{if}\quad n>1. \tag{vii}$$

Similarly, formula (iv) would give

$$u_{m,n} = \frac{m-1}{m+n}u_{m-2,n}, \quad\text{if}\quad m>1. \tag{viii}$$

In particular, if $m=0$ we have

$$\int_0^{\frac{1}{2}\pi}\cos^n x\,dx = \frac{n-1}{n}\int_0^{\frac{1}{2}\pi}\cos^{n-2}x\,dx \quad (n>1); \tag{ix}$$

and if $n=0$,

$$\int_0^{\frac{1}{2}\pi}\sin^m x\,dx = \frac{m-1}{m}\int_0^{\frac{1}{2}\pi}\sin^{m-2}x\,dx \quad (m>1). \tag{x}$$

By means of these relations the indices $m$, $n$ can be reduced to 0 or 1; the final integral is one of the following:

$$m, n \text{ both odd:} \quad u_{1,1} = \int_0^{\frac{1}{2}\pi} \sin x \cos x \, dx = [\tfrac{1}{2}\sin^2 x]_0^{\frac{1}{2}\pi} = \tfrac{1}{2};$$

$$m \text{ odd, } n \text{ even:} \quad u_{1,0} = \int_0^{\frac{1}{2}\pi} \sin x \, dx = 1;$$

$$m \text{ even, } n \text{ odd:} \quad u_{0,1} = \int_0^{\frac{1}{2}\pi} \cos x \, dx = 1;$$

$$m, n \text{ both even:} \quad u_{0,0} = \int_0^{\frac{1}{2}\pi} dx = \tfrac{1}{2}\pi.$$

If there is an *odd* index, the working is shortened by selecting this (say $n$) for successive reduction: the integral becomes a multiple of

$$\int_0^{\frac{1}{2}\pi} \sin^m x \cos x \, dx = \left[ \frac{1}{m+1} \sin^{m+1} x \right]_0^{\frac{1}{2}\pi} = \frac{1}{m+1},$$

which gives the result *without further formulae*.

Also, it is enough to remember the *single* reduction formula (vii); it includes (ix), and also (viii) and (x) because $u_{m,n} = u_{n,m}$. For, putting $x = \tfrac{1}{2}\pi - y$,

$$u_{m,n} = \int_0^{\frac{1}{2}\pi} \sin^m x \cos^n x \, dx = \int_{\frac{1}{2}\pi}^0 \cos^m y \sin^n y(-dy)$$

$$= \int_0^{\frac{1}{2}\pi} \sin^n y \cos^m y \, dy = u_{n,m}.$$

**Examples**

(i) $\displaystyle\int_0^{\frac{1}{2}\pi} \sin^4 x \cos^7 x \, dx.$

Selecting the odd index 7 for reduction, we have

$$u_{4,7} = \tfrac{6}{11} u_{4,5} = \tfrac{6}{11} \cdot \tfrac{4}{9} u_{4,3} = \tfrac{6}{11} \cdot \tfrac{4}{9} \cdot \tfrac{2}{7} u_{4,1}$$

$$= \frac{6.4.2}{11.9.7} \int_0^{\frac{1}{2}\pi} \sin^4 x \cos x \, dx = \frac{6.4.2}{11.9.7} [\tfrac{1}{5}\sin^5 x]_0^{\frac{1}{2}\pi}$$

$$= \frac{6.4.2}{11.9.7.5} = \frac{16}{1155}.$$

(ii) $\displaystyle\int_0^{\frac{1}{2}\pi} \sin^4 x \cos^6 x \, dx = u_{4,6} = \tfrac{5}{10} u_{4,4} = \tfrac{5}{10} \cdot \tfrac{3}{8} u_{4,2} = \tfrac{5}{10} \cdot \tfrac{3}{8} \cdot \tfrac{1}{6} u_{4,0},$

and $\qquad\qquad u_{4,0} = \tfrac{3}{4} u_{2,0} = \tfrac{3}{4} \cdot \tfrac{1}{2} u_{0,0} = \tfrac{3}{4} \cdot \tfrac{1}{2} \cdot \tfrac{1}{2}\pi.$

$$\therefore \quad u_{4,6} = \frac{5.3.1.3.1}{10.8.6.4.2} \cdot \frac{\pi}{2} = \frac{3\pi}{512}.$$

(iii) $\displaystyle\int_0^{\pi} \sin^3 x \cos^2 x\, dx = \int_0^{\frac{1}{2}\pi} \sin^3 x \cos^2 x\, dx + \int_{\frac{1}{2}\pi}^{\pi} \sin^3 x \cos^2 x\, dx.$

The substitution $x = \pi - y$ in the last integral reduces it to

$$\int_{\frac{1}{2}\pi}^{0} \sin^3 y \cos^2 y(-dy) = \int_0^{\frac{1}{2}\pi} \sin^3 y \cos^2 y\, dy.$$

Therefore the given integral is

$$2\int_0^{\frac{1}{2}\pi} \sin^3 x \cos^2 x\, dx = 2\frac{2.1}{5.3} = \frac{4}{15}.$$

By proceeding in this way for the general case, we deduce the following rule.

*To evaluate* $\displaystyle\int_0^{\frac{1}{2}\pi} \sin^m x \cos^n x\, dx$ *where* $m$, $n$ *are positive integers or zero, write down the expression*

$$\frac{(m-1)(m-3)\dots(n-1)(n-3)\dots}{(m+n)(m+n-2)\dots},$$

*where all three sequences of factors decrease by 2 until either 1 or 2 is reached; if* $m$, $n$ *are both even, multiply the expression by* $\frac{1}{2}\pi$ *(zero counts as an even number).*

(iv) $\displaystyle\int_0^1 x^{2p}(1-x^2)^q\, dx,$ *where* $p$, $q$ *are positive integers.*

Putting $x = \sin\theta$, where $0 \leqslant \theta \leqslant \frac{1}{2}\pi$, the integral becomes

$$\int_0^{\frac{1}{2}\pi} \sin^{2p}\theta \cos^{2q}\theta . \cos\theta\, d\theta = \int_0^{\frac{1}{2}\pi} \sin^{2p}\theta \cos^{2q+1}\theta\, d\theta$$

$$= \frac{(2p-1)(2p-3)\dots 1 . 2q(2q-2)\dots 2}{(2p+2q+1)(2p+2q-1)\dots 1}.$$

## 4.85 Integrals involving other transcendental functions

Integrals of the types

$$\int x^m(\sin^{-1}x)^n\, dx, \quad \int x^m(\cos^{-1}x)^n\, dx, \quad \int x^m(\tan^{-1}x)^n\, dx$$

can be cleared of inverse functions by a substitution. For example, putting $y = \sin^{-1}x$ in the first gives $\int y^n \sin^m y \cos y\, dy$; and since $\sin^m y \cos y$ can be expressed in terms of multiple angles, this integral becomes a sum of integrals like $A\int y^n \sin ay\, dy$, $B\int y^n \cos by\, dy$, each of which can be found by a reduction formula (see 4.24, example).

However, direct integration by parts is often successful: see 4.23 (2).

Ex. 4 $(m)$ includes some integrals of other transcendental functions which can be calculated by a reduction formula. The results of Ex. 4 $(e)$, nos. 28, 29 are useful in applications.

### Exercise 4($m$)

*Calculate the integrals of the following by using a reduction formula.*

1   $\cos^6 x.$              2   $\sin^5 x.$                 3   $\sin^3 x \cos^4 x.$

4   $\sin^4 x \cos^6 x.$       5   $\sec^6 x.$                 6   $\sin^4 x \sec^7 x.$

*Evaluate the following integrals.*

7   $\displaystyle\int_0^{\frac{1}{2}\pi} \cos^5 x\,dx.$           8   $\displaystyle\int_0^{\frac{1}{2}\pi} \sin^6 x\,dx.$          9   $\displaystyle\int_0^{\pi} \cos^4 \tfrac{1}{2}x\,dx.$

10   $\displaystyle\int_0^{\frac{1}{2}\pi} \sin^4 x \cos^5 x\,dx.$    11   $\displaystyle\int_0^{\pi} (1+\cos x)^5\,dx.$    12   $\displaystyle\int_0^{\pi} \sin^3 x(1-\cos x)^4\,dx.$

13   $\displaystyle\int_0^1 x^6 \sqrt{(1-x^2)}\,dx.$      14   $\displaystyle\int_0^1 x^3 \sqrt{(x-x^2)}\,dx.$

15   $\displaystyle\int_0^a x^m(a-x)^n\,dx,$ where $m$ and $n$ are positive integers or zero.

16   $\displaystyle\int_0^{2\pi} \sin^8 x\,dx.$       17   $\displaystyle\int_0^{\frac{3}{2}\pi} \sin^3 x \cos^3 x\,dx.$    18   $\displaystyle\int_0^{\pi} \sin^2 x \cos^3 x\,dx.$

19   If $c_n = \int \cos^n x\,dx$ and $s_n = \int \sin^n x\,dx$, prove that

$$nc_n = \sin x \cos^{n-1} x + (n-1) c_{n-2} \quad \text{and} \quad ns_n = (n-1) s_{n-2} - \cos x \sin^{n-1} x.$$

*Obtain reduction formulae for the following integrals. (Some of these examples have already been given.)*

20   $\int x^n e^{ax}\,dx.$   (Ex. 4 $(e)$, no. 21.)     21   $\int \mathrm{ch}^n x\,dx.$

22   $\int \mathrm{sh}^n x\,dx.$                       23   $\int x^n \sin ax\,dx.$   (Ex. 4 $(c)$, no. 12.)

24   $\int x^n \mathrm{ch}\,ax\,dx.$   (Ex. 4 $(h)$, no. 36.)    25   $\int x^n \mathrm{sh}\,ax\,dx.$

26   $\int e^{ax} \sin^n bx\,dx.$              27   $\int x^m (\log x)^n\,dx.$   (Ex. 4 $(d)$, no. 45.)

28   If $u_{m,\,n} = \int \cos^m x \sin nx\,dx$, prove that

$$(m+n)\,u_{m,\,n} = mu_{m-1,\,n-1} - \cos^m x \cos nx,$$

and calculate              $\displaystyle\int_0^{\frac{1}{4}\pi} \cos^2 x \sin 4x\,dx.$

29   Assuming $m^2 \neq n^2$, obtain a formula relating $u_{m,\,n} = \int \sin^m x \sin nx\,dx$ with $u_{m-2,\,n}.$ Hence prove that

$$(m^2 - n^2) \int_0^{\frac{1}{2}\pi} \sin^m x \sin nx\,dx = m(m-1) \int_0^{\frac{1}{2}\pi} \sin^{m-2} x \sin nx\,dx,$$

and calculate              $\displaystyle\int_0^{\frac{1}{2}\pi} \sin^4 x \sin 5x\,dx.$

30   Prove that

$$\int \cos nx \sec x\,dx = \frac{2}{n-1}\sin(n-1)x - \int \cos(n-2)x \sec x\,dx,$$

and calculate

$$\int_0^{\frac{1}{2}\pi} \frac{\sin 8x \sin x}{\cos x}\, dx.$$

**31** If

$$u_{m,\,n} = \int \frac{x^m\, dx}{(x^2+1)^n},$$

prove that

$$2(n-1)\, u_{m,\,n} = (m-1)\, u_{m-2,\,n-1} - \frac{x^{m-1}}{(x^2+1)^{n-1}}.$$

**\*32** If

$$u_{m,\,n} = \int_0^1 x^m (1-x)^n\, dx,$$

where $m$ and $n$ are positive integers, prove that

(i) $$u_{m,\,n} = u_{n,\,m} \quad [\text{put } y = 1-x];$$

(ii) $$u_{m,\,n} = \frac{n}{m+n+1}\, u_{m,\,n-1}$$

[integrate by parts, and use $x^{m+1}(1-x)^{n-1} \equiv x^m(1-x)^{n-1} - x^m(1-x)^n$].

(iii) Deduce that $$u_{m,\,n} = \frac{m!\,n!}{(m+n+1)!}.$$

**33** Prove $$\int_0^\pi x \sin^4 x \cos^2 x\, dx = \tfrac{1}{32}\pi^2. \quad [\text{Put } y = \pi - x.]$$

**34** Prove $$\int_0^\pi x f(\sin x)\, dx = \tfrac{1}{2}\pi \int_0^\pi f(\sin x)\, dx.$$

**\*35** Write down the derivative of $x^n e^{ax} \cos bx$ and of $x^n e^{ax} \sin bx$. Putting

$$c_n = \int x^n e^{ax} \cos bx\, dx, \qquad s_n = \int x^n e^{ax} \sin bx\, dx,$$

deduce that (for $n > 0$),

$$ac_n - bs_n + nc_{n-1} = x^n e^{ax} \cos bx, \qquad as_n + bc_n + ns_{n-1} = x^n e^{ax} \sin bx.$$

Prove also that

$$ac_0 - bs_0 = e^{ax} \cos bx, \qquad as_0 + bc_0 = e^{ax} \sin bx.$$

[See Ex. 4 (e), nos. 28, 29.] (The *pair* of reduction formulae will determine $c_n$ and $s_n$ when $n$ is a positive integer. Also see 14.66, ex. (iii).)

## 4.9 Generalised integrals

### 4.91 The problem

In our definition of a definite integral (4.15) we required the integrand to be continuous, and we implied that the range of integration was finite. Expressions like the following are at present undefined:

$$\int_{-1}^{+1} \frac{dx}{x^2}, \qquad \int_0^1 \frac{dx}{\sqrt{(1-x^2)}}, \qquad \int_0^\infty \frac{dx}{1+x^2}.$$

For, the integrand of the first is discontinuous at the value $x = 0$ within the range of integration; that in the second is discontinuous at the upper end $x = 1$; and in the third the range of integration is infinite.

We now consider how a meaning may be assigned to such integrals, which are called *improper, infinite,* or *generalised integrals.*

### 4.92 Infinite range ('integrals of the first kind')

(a) Suppose that $f(x)$ is continuous for all $x \geqslant a$, and suppose that

$$\int_a^X f(x)\,dx \to l \quad \text{when} \quad X \to \infty.$$

Then we define $\int_a^\infty f(x)\,dx$ to be $l$; i.e.

$$\int_a^\infty f(x)\,dx = \lim_{X\to\infty} \int_a^X f(x)\,dx$$

provided this limit exists. If the limit does not exist, the infinite integral is not defined.

(b) Similarly, if $f(x)$ is continuous for all $x \leqslant b$, we define

$$\int_{-\infty}^b f(x)\,dx = \lim_{X\to-\infty} \int_X^b f(x)\,dx$$

provided this limit exists.

(c) Finally, if $f(x)$ is continuous for all $x$ and if

$$\int_a^\infty f(x)\,dx, \quad \int_{-\infty}^a f(x)\,dx$$

both exist for some fixed $a$, then we define $\int_{-\infty}^\infty f(x)\,dx$ by

$$\int_{-\infty}^\infty f(x)\,dx = \int_{-\infty}^a f(x)\,dx + \int_a^\infty f(x)\,dx.$$

This definition gives a result independent of $a$; for we have

$$\int_a^b f(x)\,dx + \int_b^X f(x)\,dx = \int_a^X f(x)\,dx,$$

and by letting $X \to \infty$ we obtain

$$\int_a^b f(x)\,dx + \int_b^\infty f(x)\,dx = \int_a^\infty f(x)\,dx;$$

similarly

$$\int_a^b f(x)\,dx + \int_{-\infty}^a f(x)\,dx = \int_{-\infty}^b f(x)\,dx.$$

Hence

$$\int_{-\infty}^b f(x)\,dx + \int_b^\infty f(x)\,dx = \int_a^b f(x)\,dx + \int_{-\infty}^a f(x)\,dx + \int_a^\infty f(x)\,dx - \int_a^b f(x)\,dx$$

$$= \int_{-\infty}^a f(x)\,dx + \int_a^\infty f(x)\,dx.$$

## Examples

(i) If $a > 0$, then $1/x^3$ is continuous for all $x \geqslant a$, and

$$\int_a^X \frac{1}{x^3}\,dx = \left[ -\frac{1}{2x^2} \right]_a^X = \frac{1}{2a^2} - \frac{1}{2X^2}$$

$$\to \frac{1}{2a^2} \quad \text{when} \quad X \to \infty.$$

Hence $\displaystyle\int_a^\infty \frac{1}{x^3}\,dx$ exists and has the value $\dfrac{1}{2a^2}\,(a > 0)$.

(ii) If $a > 0$, then $1/x$ is continuous for all $x \geqslant a$, and

$$\int_a^X \frac{1}{x}\,dx = [\log x]_a^X = \log X - \log a$$

$$\to \infty \quad \text{when} \quad X \to \infty.$$

Hence $\displaystyle\int_a^\infty \frac{1}{x}\,dx$ *does not exist.*

(iii) $\displaystyle\int_0^X \cos x\,dx = [\sin x]_0^X = \sin X$. When $X \to \infty$, $\sin X$ oscillates between

$\pm 1$. Hence $\displaystyle\int_0^\infty \cos x\,dx$ *does not exist.*

(iv) $$\int_1^X \frac{dx}{x(x+1)} = \int_1^X \left( \frac{1}{x} - \frac{1}{x+1} \right) dx = [\log x - \log(x+1)]_1^X \qquad (a)$$

$$= \left[ \log \frac{x}{x+1} \right]_1^X = \log \frac{X}{X+1} - \log \tfrac{1}{2}$$

$$\to \log 1 - \log \tfrac{1}{2} = \log 2 \quad \text{when} \quad X \to \infty.$$

Hence $$\int_1^\infty \frac{dx}{x(x+1)} = \log 2.$$

Had we evaluated the integral at stage $(a)$ as $\log X - \log(X+1) + \log 2$, it would have been necessary to combine the terms containing $X$ into $\log\{X/(X+1)\}$ *before* letting $X \to \infty$, because neither $\log X$ nor $\log(X+1)$ tends to a limit separately.

(v) $$\int_0^X \frac{dx}{a^2 + x^2} = \left[ \frac{1}{a}\tan^{-1}\frac{x}{a} \right]_0^X = \frac{1}{a}\tan^{-1}\frac{X}{a} \to \frac{1}{a}\cdot\frac{\pi}{2}$$

when $X \to \infty$, if $a > 0$. Hence

$$\int_0^\infty \frac{dx}{a^2 + x^2} = \frac{\pi}{2a}.$$

Similarly $$\int_X^0 \frac{dx}{a^2 + x^2} = -\frac{1}{a}\tan^{-1}\frac{X}{a} \to -\frac{1}{a}\left( -\frac{\pi}{2} \right) \quad \text{when} \quad X \to -\infty,$$

and hence $$\int_{-\infty}^0 \frac{dx}{a^2 + x^2} = \frac{\pi}{2a}.$$

Therefore if $a > 0$,

$$\int_{-\infty}^{\infty} \frac{dx}{a^2 + x^2} \quad \text{exists and is} \quad \frac{\pi}{2a} + \frac{\pi}{2a} = \frac{\pi}{a}.$$

(vi)
$$\int_0^X x e^{-x} dx = [-x e^{-x}]_0^X + \int_0^X e^{-x} dx$$
$$= -X e^{-X} + [-e^{-x}]_0^X$$
$$= -X e^{-X} - e^{-X} + 1.$$

When $X \to \infty$, $e^{-X} \to 0$ and $X e^{-X} \to 0$ (4.43 (6)). Hence $\int_0^{\infty} x e^{-x} dx$ exists and is 1.

## 4.93 Discontinuous integrand ('integrals of the second kind')

(a) Suppose $f(x)$ is continuous for $a \leqslant x < b$, but that $f(x) \to \infty$ or $f(x) \to -\infty$ when $x \to b-$. If

$$\int_a^{b-h} f(x)\, dx \to l \quad \text{when} \quad h \to 0+,$$

we define $\int_a^b f(x)\, dx$ to be $l$, i.e.

$$\int_a^b f(x)\, dx = \lim_{h \to 0+} \int_a^{b-h} f(x)\, dx$$

provided this limit exists. Otherwise, the generalised integral does not exist.

(b) Similarly, if $f(x)$ is continuous for $a < x \leqslant b$ and if

$$\lim_{h \to 0+} \int_{a+h}^b f(x)\, dx$$

exists, we define $\int_a^b f(x)\, dx$ to be this limit.

(c) If $f(x)$ is continuous for $a < x < b$, and if for some $c$ satisfying $a < c < b$ the integrals $\int_a^c f(x)\, dx$, $\int_c^b f(x)\, dx$ both exist, we define $\int_a^b f(x)\, dx$ to mean $\int_a^c f(x)\, dx + \int_c^b f(x)\, dx$. As in 4.92 (c), it can be shown that this definition is independent of $c$.

(d) If $f(x)$ is discontinuous for a value $x = c$ within the range $a \leqslant x \leqslant b$, we define $\int_a^b f(x)\, dx$ to mean

$$\int_a^c f(x)\, dx + \int_c^b f(x)\, dx$$

provided that each of these integrals exists.

(e)  Finally, if $f(x)$ is discontinuous for a finite number of values of $x$ in the range of integration, this range can be divided into adjacent intervals for which $f(x)$ is discontinuous at only one of the ends, as in (d). Then $\int_a^b f(x)\,dx$ is defined to be the sum of the integrals over each part, provided that all these exist.

## Examples

(i) $\int_h^1 \frac{1}{\sqrt{x}}\,dx = [2\sqrt{x}]_h^1 = 2 - 2\sqrt{h} \to 2$  when  $h \to 0+$.

Hence $\int_0^1 \frac{1}{\sqrt{x}}\,dx$ exists and has the value 2.

(ii) $\int_0^{2-h} \frac{1}{(x-2)^2}\,dx = \left[-\frac{1}{x-2}\right]_0^{2-h} = \frac{1}{h} - \frac{1}{2} \to \infty$  when  $h \to 0+$.

Hence $\int_0^2 \frac{1}{(x-2)^2}\,dx$ *does not exist.*

(iii) $\int_1^3 \frac{1}{\sqrt[3]{(x-2)}}\,dx$. The integrand is not defined when $x = 2$, which lies within the range of integration. To find whether the integral exists we must consider the integrals

$$\int_1^2 \frac{dx}{\sqrt[3]{(x-2)}}, \quad \int_2^3 \frac{dx}{\sqrt[3]{(x-2)}}.$$

$$\int_1^{2-h} \frac{dx}{\sqrt[3]{(x-2)}} = [\tfrac{3}{2}(x-2)^{\frac{2}{3}}]_1^{2-h} = \tfrac{3}{2}h^{\frac{2}{3}} - \tfrac{3}{2} \to -\tfrac{3}{2} \quad \text{when} \quad h \to 0+$$

and $\int_{2+h}^3 \frac{dx}{\sqrt[3]{(x-2)}} = [\tfrac{3}{2}(x-2)^{\frac{2}{3}}]_{2+h}^3 = \tfrac{3}{2} - \tfrac{3}{2}h^{\frac{2}{3}} \to \tfrac{3}{2} \quad \text{when} \quad h \to 0+$.

Hence $\int_1^3 \frac{dx}{\sqrt[3]{(x-2)}}$ exists and has the value $\tfrac{3}{2} + (-\tfrac{3}{2}) = 0$.

(iv) $\int_0^\infty \frac{1}{x^2}\,dx$. We may consider separately the integrals

$$\int_0^1 \frac{dx}{x^2}, \quad \int_1^\infty \frac{dx}{x^2}.$$

Now $\int_h^1 \frac{dx}{x^2} = \left[-\frac{1}{x}\right]_h^1 = \frac{1}{h} - 1 \to \infty$  as  $h \to 0+$,

so that $\int_0^1 \frac{dx}{x^2}$ does not exist. Hence $\int_0^\infty \frac{dx}{x^2}$ does not exist.

(v) $\int_{-1}^{+1} \frac{dx}{x}$. We consider the integrals $\int_{-1}^0 \frac{dx}{x}$, $\int_0^1 \frac{dx}{x}$.

$$\int_{-1}^{-h} \frac{dx}{x} = [\log|x|]_{-1}^{-h} = \log h,$$

and
$$\int_{h'}^1 \frac{dx}{x} = [\log|x|]_{h'}^1 = -\log h'.$$

Thus neither integral exists separately, and so by $(d)$ $\int_{-1}^{+1} \frac{dx}{x}$ does not exist. However,

$$\int_{-1}^{-h} \frac{dx}{x} + \int_{h'}^1 \frac{dx}{x} = \log h - \log h' = \log \frac{h}{h'},$$

and if a special relation is assumed between $h, h'$, this expression may tend to a limit when $h$ and $h' \to 0+$; e.g. if $h = kh'$ ($k$ being a positive constant), the result is $\log k$. In particular, by taking $h = h'$ we obtain $\log 1 = 0$; Cauchy called this the *principal value* of $\int_{-1}^1 \frac{dx}{x}$, written $P\int_{-1}^1 \frac{dx}{x}$. For $\int_{-1}^1 \frac{dx}{x^2}$, the reader can verify that $P\int_{-1}^1 \frac{dx}{x^2}$ does not exist.

(vi) An integral like $\int_{-5}^4 \frac{dx}{\sqrt{(x^2-9)}}$ is still meaningless because there is an *interval* of values of $x$ in the range of integration for which the integrand is not defined.

## 4.94 The relation $\int_a^b f(x)\,dx = \phi(b) - \phi(a)$

(1) *Application to generalised integrals.*

When we defined the definite integral by this relation in 4.15, we assumed that (i) $f(x)$ is continuous and (ii) $\phi'(x) = f(x)$ for *all* $x$ satisfying $a \leqslant x \leqslant b$. *If these conditions are not satisfied, the relation cannot be used without investigation.*

Thus $\int_{-1}^1 \frac{dx}{x^2}$, which by 4.93, ex. (iv) does not exist, would be 'evaluated' as $[-1/x]_{-1}^1 = -2$ by an incautious application. On the other hand, the relation gives

$$\int_1^3 \frac{dx}{\sqrt[3]{(x-2)}} = [\tfrac{3}{2}(x-2)^{\frac{2}{3}}]_1^3 = \tfrac{3}{2} - \tfrac{3}{2} = 0,$$

which is correct by 4.93, ex. (iii); this is because $\phi(x)$ is continuous at $x = 2$.

*If conditions* (i), (ii) *hold for* $a \leqslant x \leqslant b$ EXCEPT *at* $x = c$ $(a \leqslant c \leqslant b)$, *and if* $\phi(x)$ *is continuous for* $a \leqslant x \leqslant b$, *the relation is valid.*

*Proof.* First supposing $b$ is the exceptional value of $x$, we have by definition (4.93 $(a)$),

$$\int_a^b f(x)\,dx = \lim_{h\to 0+} \{\phi(b-h) - \phi(a)\} = \phi(b) - \phi(a)$$

by continuity of $\phi(x)$ at $x = b$. Similarly, the result follows if $a$ is the exception.

If $a < c < b$, then by definition $(4.93\,(d))$,

$$\int_a^b f(x)\,dx = \int_a^c f(x)\,dx + \int_c^b f(x)\,dx$$

$$= \lim_{h\to 0+} \{\phi(c-h) - \phi(a)\} + \lim_{h'\to 0+} \{\phi(b) - \phi(c+h')\}$$

$$= \{\phi(c) - \phi(a)\} + \{\phi(b) - \phi(c)\}$$

$$= \phi(b) - \phi(a).$$

**(2)** *Definite integrals found by a special method.*

Although most of the definite integrals in this chapter have been obtained directly from the definition in terms of the indefinite integral $\phi(x)$, yet we have given examples where the definite integral has been found independently: Ex. 4 $(b)$, nos. 31, 33, 34 and Ex. 4 $(m)$, nos. 33, 34 depend essentially on the relation

$$\int_0^a f(x)\,dx = \int_0^a f(a-x)\,dx,$$

proved by a substitution; $\displaystyle\int_0^{\frac{1}{2}\pi} \sin^m x \cos^n x\,dx$ $(m,\ n$ positive integers) was found in 4.84 (3) by reduction; also see Ex. 4 $(m)$, no. 32, Ex. 4 $(n)$, nos. 26, 27 and Ex. 4 $(o)$, nos. 79, 80.

Calculation of the definite integral may clearly be possible without knowing $\phi(x)$ explicitly in terms of $x$, because by some device it may be easy to find the *difference* $\phi(b) - \phi(a)$ *between two particular values* of $\phi(x)$.

## 4.95 Integration by parts and by substitution

In suitable circumstances the formulae of 4.22 and 4.25 can be extended to generalised integrals. At this stage it is best to treat each example from first principles: an illustration of integration by parts occurred in 4.92, ex. (vi), and the following will illustrate substitution.

### Example

*Find*

$$\int_a^b \frac{dx}{\sqrt{\{(x-a)\,(b-x)\}}} \quad \text{by using the substitution } x = a + (b-a)\,t^2,\ t \geqslant 0.$$

When $x$ increases from $a$ to $b$, $t$ increases from 0 to 1. Since the integrand is discontinuous at both ends of the range of integration, we consider (see 4.93 $(c)$) the transform of each of the integrals

$$\int_a^c \frac{dx}{\sqrt{\{(x-a)\,(b-x)\}}}, \quad \int_c^b \frac{dx}{\sqrt{\{(x-a)\,(b-x)\}}},$$

where $a < c < b$. Now

$$\int_{a+h}^{c} \frac{dx}{\sqrt{\{(x-a)(b-x)\}}} = \int_{t_1}^{t_2} \frac{2(b-a)\,t\,dt}{(b-a)\,t\sqrt{(1-t^2)}} = 2\int_{t_1}^{t_2} \frac{dt}{\sqrt{(1-t^2)}},$$

where

$$t_1 = \sqrt{\left(\frac{h}{b-a}\right)}, \quad t_2 = \sqrt{\left(\frac{c-a}{b-a}\right)};$$

so the integral is

$$2[\sin^{-1}t]_{t_1}^{t_2} = 2\sin^{-1}t_2 - 2\sin^{-1}t_1 \to 2\sin^{-1}t_2$$

when $h \to 0+$, since then $t_1 \to 0$ also. Hence

$$\int_{a}^{c} \frac{dx}{\sqrt{\{(x-a)(b-x)\}}} = 2\sin^{-1}t_2.$$

Similarly,

$$\int_{c}^{b-h} \frac{dx}{\sqrt{\{(x-a)(b-x)\}}} = 2\int_{t_2}^{t_3} \frac{dt}{\sqrt{(1-t^2)}} = 2\sin^{-1}t_3 - 2\sin^{-1}t_2,$$

where

$$t_3 = \sqrt{\left(\frac{b-a-h}{b-a}\right)}.$$

When $h \to 0+$, the expression tends to

$$2\sin^{-1}1 - 2\sin^{-1}t_2 = \pi - 2\sin^{-1}t_2,$$

since $t_3 \to 1$. Therefore

$$\int_{c}^{b} \frac{dx}{\sqrt{\{(x-a)(b-x)\}}} = \pi - 2\sin^{-1}t_2.$$

Consequently the given integral has the value $\pi$.

## Exercise 4(*n*)

*Discuss the following improper integrals, and evaluate each that exists.*

1   $\displaystyle\int_{1}^{\infty} \frac{dx}{x^2}.$     2   $\displaystyle\int_{1}^{\infty} \frac{dx}{\sqrt{x}}.$     3   $\displaystyle\int_{0}^{\infty} \frac{dx}{x^2+4}.$     4   $\displaystyle\int_{0}^{\infty} x\sin x\,dx.$

5   $\displaystyle\int_{0}^{\infty} e^{-x}\,dx.$    6   $\displaystyle\int_{2}^{\infty} \frac{dx}{x^2-1}.$    7   $\displaystyle\int_{-\infty}^{0} x\,e^{x}\,dx.$    8   $\displaystyle\int_{-\infty}^{\infty} \frac{x^2\,dx}{x^6+1}.$

9   $\displaystyle\int_{0}^{\infty} \operatorname{sech} x\,dx.$   10   $\displaystyle\int_{0}^{\infty} \frac{\cos x}{e^{x}}\,dx.$   11   $\displaystyle\int_{0}^{1} \frac{dx}{\sqrt[3]{x}}.$   12   $\displaystyle\int_{0}^{1} \frac{dx}{x}.$

13   $\displaystyle\int_{-1}^{+1} \frac{dx}{\sqrt{x}}.$   14   $\displaystyle\int_{0}^{1} \frac{dx}{\sqrt{(1-x^2)}}.$   15   $\displaystyle\int_{1}^{2} \frac{dx}{x^2-1}.$   16   $\displaystyle\int_{0}^{e} x\log x\,dx.$

17   $\displaystyle\int_{0}^{\frac{1}{2}\pi} \frac{\sec^{\frac{3}{2}}x}{\sqrt{\sin x}}\,dx.$     18   If $n > -1$, prove $\displaystyle\int_{0}^{1} x^n \log x\,dx = -\frac{1}{(n+1)^2}.$

19   Prove $\displaystyle\int_{0}^{\infty} \frac{dx}{a^2+b^2x^2} = \frac{\pi}{2ab}$ if $a$, $b$ have like signs.

20   Prove $\displaystyle\int_{0}^{\infty} \frac{dx}{(x^2+a^2)(x^2+b^2)} = \frac{\pi}{2ab(a+b)}$ if $ab > 0$, and find

$$\int_{0}^{\infty} \frac{x^2\,dx}{(x^2+a^2)(x^2+b^2)}.$$

**21**   Prove $\displaystyle\int_0^{\frac12\pi} \frac{dx}{a^2\cos^2 x + b^2\sin^2 x} = \frac{\pi}{2ab}$ if $ab > 0$. What is the value if $a$, $b$ have opposite signs?

**\*22**   Prove $\displaystyle\int_0^{\pi} \frac{dx}{a + b\cos x} = \frac{\pi}{\sqrt{(a^2 - b^2)}}$ if $a > b > 0$. [Use Ex. 4 (*l*), no. 25.] What happens if $b \geqslant a > 0$?

**\*23**   Prove $\displaystyle\int_0^{\frac12\pi} \frac{dx}{a + b\cos x} = \frac{1}{\sqrt{(b^2 - a^2)}} \log\left\{\frac{b + \sqrt{(b^2 - a^2)}}{a}\right\}$ if $b > a > 0$.

**24**   If $a > 0$, prove $\displaystyle\int_0^{\infty} e^{-ax}\cos bx\,dx = \frac{a}{a^2 + b^2}$ and $\displaystyle\int_0^{\infty} e^{-ax}\sin bx\,dx = \frac{b}{a^2 + b^2}$. [See Ex. 4 (*e*), nos. 28, 29.]

**25**   If $-\pi < \alpha < \pi$ and $\alpha \neq 0$, prove
$$\int_0^{\infty} \frac{dx}{x^2 + 2x\cos\alpha + 1} = 2\int_0^1 \frac{dx}{x^2 + 2x\cos\alpha + 1} = \frac{\alpha}{\sin\alpha}.$$

If $\alpha = 0$, verify that the value of each expression is 1 (which is the *limit* of $\alpha/\sin\alpha$ when $\alpha \to 0$).

**\*26**   If $u_n = \displaystyle\int_0^{\infty} x^n e^{-ax}\,dx$ and $a > 0$, prove $u_n = (n/a)\,u_{n-1}$. If $n$ is a positive integer, deduce that $u_n = n!/a^{n+1}$.

**\*27**   If $u_{m,n} = \displaystyle\int_0^1 x^{n-1}(\log x)^m\,dx$ where $m$, $n$ are positive integers, prove that $u_{m,n} = -(m/n)\,u_{m-1,n}$. Deduce that $u_{m,n} = (-1)^m m!/n^{m+1}$. [See Ex. 4 (*d*), no. 45.]

**28**   By putting $y = \dfrac{1}{x}$, prove $\displaystyle\int_1^{\infty} \frac{dx}{x^2 + 2x\cos\alpha + 1} = \int_0^1 \frac{dx}{x^2 + 2x\cos\alpha + 1}$.

**29**   Calculate $\displaystyle\int_1^{\infty} \frac{dx}{(1+x)\sqrt{x}}$ by putting $x = t^2$ $(t > 0)$.

**30**   Prove $\displaystyle\int_1^{\infty} \frac{dx}{(x+3)\sqrt{(x-1)}} = \frac12\pi$. [Put $x - 1 = t^2$, $t > 0$.]

**31**   Show $\displaystyle\int_1^{\infty} \frac{\sqrt{x}\,dx}{(1+x)^2} = \frac12 + \frac14\pi$. [Put $x = t^2$, then integrate by parts.]

**32**   Calculate $\displaystyle\int_0^1 \frac{x^m\,dx}{\sqrt{(1-x^2)}}$ if $m$ is a positive integer. [Put $x = \sin\theta$, $0 \leqslant \theta \leqslant \frac12\pi$.]

**33**   Calculate $\displaystyle\int_0^{\infty} \frac{x^m\,dx}{(1+x^2)^n}$ if $n$ and $2n - m - 2$ are positive integers. [Put $x = \tan\theta$, $0 \leqslant \theta \leqslant \frac12\pi$.]

**34**   Prove $\displaystyle\int_a^b \frac{dx}{\sqrt{\{(x-a)(b-x)\}}} = \pi$ by putting $x = a\cos^2\theta + b\sin^2\theta$. (Cf. 4.95. ex.)

**35**   Find $\displaystyle\int_a^b \frac{x\,dx}{\sqrt{\{(x-a)(b-x)\}}}$.

## Miscellaneous Exercise 4(*o*)

*For general practice in integration, the reader may try a random selection from nos. 1–60 following.*

1 $\displaystyle\int \frac{2x^3+7}{(2x+1)(x^2+2)}\,dx.$

2 $\displaystyle\int_2^3 \frac{x^2\,dx}{(x-1)(x+2)}.$

3 $\displaystyle\int \frac{x^2-5x+9}{(x-1)^2(x^2+4)}\,dx.$

4 $\displaystyle\int_0^1 \frac{x\,dx}{(x^2+1)(x+1)^2}.$

5 $\displaystyle\int_0^1 \frac{x^4\,dx}{(1+x^2)^2}.$

6 $\displaystyle\int_1^\infty \frac{x^2+2}{x^4(x^2+1)}\,dx.$

7 $\displaystyle\int \frac{dx}{x^3-1}.$

8 $\displaystyle\int \frac{x^2\,dx}{x^4+x^2-2}.$

9 $\displaystyle\int \frac{dx}{(x^2+1)(x^2+x+1)}.$

10 $\displaystyle\int \frac{dx}{a^4-x^4}.$

11 $\displaystyle\int \frac{dx}{x(1+x+x^2+x^3)}.$

12 $\displaystyle\int \frac{dx}{6-x-4x^2-x^3}.$

13 $\displaystyle\int \frac{5-7x}{2x^3-x^2-2x+1}\,dx.$

14 $\displaystyle\int_1^2 \frac{dx}{x(1+x^4)}.$

*15 $\displaystyle\int \frac{1+x^2}{1+x^4}\,dx.$

16 $\displaystyle\int_0^1 \frac{x\,dx}{1+\sqrt{x}}.$

17 $\displaystyle\int \frac{dx}{\sqrt{(x+a)}-\sqrt{(x-a)}}.$

18 $\displaystyle\int \frac{x^2(x+1)}{(x^2+1)^3}\,dx.$

19 $\displaystyle\int \sqrt{\left(\frac{a+x}{a-x}\right)}\,dx.$

20 $\displaystyle\int_0^a x\sqrt{\left(\frac{a^2-x^2}{a^2+x^2}\right)}\,dx.$

21 $\displaystyle\int_0^a \frac{\sqrt{(a^2-x^2)}}{b^2-x^2}\,dx,\ a^2<b^2.$

22 $\displaystyle\int \sqrt{\left(\frac{a+x}{x}\right)}\,dx.$

23 $\displaystyle\int \frac{dx}{x^4\sqrt{(1+x^2)}}.$

*24 $\displaystyle\int \frac{dx}{\sqrt[3]{(1-x^3)}}.$

*25 $\displaystyle\int \frac{(x^2-1)\,dx}{x\sqrt{(1+3x^2+x^4)}}.$

26 $\displaystyle\int (a^2-x^2)^{-\frac{5}{2}}\,dx.$

27 $\displaystyle\int \frac{\sqrt{(1-x^2)}}{x^4}\sin^{-1}x\,dx.$

28 $\displaystyle\int \frac{\cot^{-1}x\,dx}{x^2(1+x^2)}.$

29 $\displaystyle\int_8^{15} \frac{dx}{(x-3)\sqrt{(x+1)}}.$

30 $\displaystyle\int \frac{dx}{x\sqrt{(1+2x-x^2)}}.$

31 $\displaystyle\int_2^4 \frac{x\,dx}{\sqrt{(6x-8-x^2)}}.$

32 $\displaystyle\int_{\frac{1}{2}}^1 \frac{dx}{x\sqrt{(5x^2-4x+1)}}.$

33 $\displaystyle\int_0^\infty \frac{dx}{(x+1)^2\sqrt{(x^2+1)}}.$

34 $\displaystyle\int_0^1 x^3 e^{x^2}\,dx.$

35 $\displaystyle\int \frac{\cos 2x}{e^x}\,dx.$

36 $\displaystyle\int_0^{\frac{1}{2}\pi} e^{-x}\sin 3x\,dx.$

37 $\displaystyle\int x^3\log(1+x^2)\,dx.$

38 $\displaystyle\int x^{-1}\log(\log x)\,dx.$

39 $\displaystyle\int_0^{\frac{1}{4}\pi} \sqrt{\tan x}\,dx.$

40 $\displaystyle\int_0^{\frac{1}{4}\pi} \sec^3 x\,dx.$

41 $\displaystyle\int \sin x\cos x\cos 2x\,dx.$

42 $\displaystyle\int_{\frac{1}{4}\pi}^{\frac{1}{2}\pi} \cot x\,dx.$

43 (i) $\displaystyle\int \sin x\log(\sin x)\,dx;$ *(ii) $\displaystyle\int_0^{\frac{1}{2}\pi} \sin x\log(\sin x)\,dx.$

44 $\displaystyle\int_0^{\frac{1}{2}\pi} \sin x(1-c^2\sin^2 x)\,dx,\ 0<c^2<1.$

45 $\displaystyle\int_{\frac{1}{4}}^1 \sin^{-1}(\sqrt{x})\,dx.$

46 $\displaystyle\int_0^{\frac{1}{2}\pi} \frac{dx}{3+5\cos x}.$

47 $\int \dfrac{dx}{\sin x - \cos x}.$

48 $\int \dfrac{dx}{1 - \sin x + \cos x}.$

49 $\int_{-\frac{1}{4}\pi}^{+\frac{1}{4}\pi} \dfrac{dx}{5 + 7\cos x + \sin x}.$

50 $\int_a^b \dfrac{dx}{x\sqrt{\{(x-a)(b-x)\}}}, \quad 0 < a < b.$

51 $\int_0^{\frac{1}{2}\pi} \dfrac{\sin \alpha \sin x\,dx}{1 - \sin^2 \alpha \sin^2 x}.$

52 $\int_0^{\frac{1}{2}\pi} \dfrac{1 + 2\cos x}{(2 + \cos x)^2}\,dx.$

53 $\int_0^{\frac{1}{2}\pi} \sin^{\frac{3}{2}} x \cos^3 x\,dx.$

*54 $\int_0^{\frac{1}{4}\pi} \cos^{\frac{3}{2}} 2\phi \sin \phi\,d\phi.$

55 $\int_0^{\log 2} \dfrac{dx}{\operatorname{sh} x + 5\operatorname{ch} x}.$

56 $\int_0^{\frac{1}{4}\pi} \sec^4 x\,dx.$

57 $\int_0^{\frac{1}{2}\pi} \dfrac{\sin^3 x\,dx}{1 + a\cos x}, \quad a > -1.$

58 $\int_0^{\frac{1}{4}\pi} \tan^7 x\,dx.$

59 $\int_0^{2a} x\sqrt{(2ax - x^2)}\,dx.$

60 $\int_0^a x^{\frac{3}{2}}(a - x)^{-\frac{1}{2}}\,dx.$

61 Calculate $\int \log x\,dx$, and deduce

$$\int \sin \theta \log (1 - e\cos \theta)\,d\theta = \left(\dfrac{1}{e} - \cos \theta\right) \log \left(\dfrac{1}{e} - \cos \theta\right).$$

62 If $a\tan \theta = b\tan \phi$, prove that

$$(a^2\sin^2 \theta + b^2\cos^2 \theta)(a^2\cos^2 \phi + b^2\sin^2 \phi) = a^2 b^2$$

and

$$d\theta/(a^2\sin^2 \theta + b^2\cos^2 \theta) = d\phi/ab.$$

Deduce that if $a > 0$, $b > 0$,

$$\int_0^{2\pi} \dfrac{d\theta}{(a^2\sin^2 \theta + b^2\cos^2 \theta)^2} = \dfrac{a^2 + b^2}{a^3 b^3}\,\pi.$$

63 Prove $\int_{-1}^{+1} \dfrac{\sin \alpha\,dx}{1 - 2x\cos \alpha + x^2} = \frac{1}{2}\pi, \ -\frac{1}{2}\pi \text{ or } 0$

according as $2n\pi < \alpha < (2n+1)\pi$, $(2n-1)\pi < \alpha < 2n\pi$, or $\alpha = n\pi$, where $n$ denotes any integer.

64 If $a$ and $b$ are positive, prove $\int_0^\pi \dfrac{(a - b\cos \theta)\,d\theta}{a^2 + b^2 - 2ab\cos \theta} = \dfrac{\pi}{a}, \ 0, \ \dfrac{\pi}{2a}$ according

as $a > b, a < b, a = b.$ $\left[\text{Write the integrand as } \dfrac{1}{2a}\left(1 + \dfrac{a^2 - b^2}{a^2 + b^2 - 2ab\cos \theta}\right).\right]$

65 If $u_n = \int x^{n+\frac{1}{2}}\sqrt{(2a - x)}\,dx$ where $n$ is a positive integer, prove

$$(n + 2)u_n - (2n + 1)au_{n-1} + x^{n-1}(2ax - x^2)^{\frac{3}{2}} = 0.$$

Hence prove $\int_0^{2a} x^2\sqrt{(2ax - x^2)}\,dx = \frac{5}{8}\pi a^4$, and calculate $\int_0^{2a} x^3\sqrt{(2ax - x^2)}\,dx.$

66 If $I(m, n) = \int_0^{\frac{1}{2}\pi} \cos^m \theta \cos n\theta\,d\theta$, prove

$$(m^2 - n^2)I(m, n) = m(m - 1)I(m - 2, n),$$

and calculate $I(4, 5)$.

67  If $I(m,n) = \int_0^{\frac{1}{2}\pi} \sin^m x \cos nx\,dx$ and $J(m,n) = \int_0^{\frac{1}{2}\pi} \sin^m x \sin nx\,dx$, prove

$$(m+n)\,I(m,n) = \sin \tfrac{1}{2}n\pi - mJ(m-1,n-1),$$

and when $m \geqslant 2$ express $I(m,n)$ in terms of $I(m-2,n-2)$.

68  Prove $\int \sin n\theta \sec\theta\,d\theta = -\dfrac{2\cos(n-1)\theta}{n-1} - \int \sin(n-2)\theta \sec\theta\,d\theta$.
and hence evaluate

$$\int_0^{\frac{1}{2}\pi} \frac{\cos 5\theta \sin\theta}{\cos\theta}\,d\theta.$$

69  If $u_n = \int_0^1 x^p(1-x^q)^n\,dx$ where $n$, $p$, $q$ are positive, prove that

$$(nq+p+1)\,u_n = nqu_{n-1}.$$

Evaluate $u_n$ when $n$ is a positive integer.

70  Obtain a reduction formula to express $\int (x^2+a^2)^{\frac{1}{2}n}\,dx$ in terms of $\int (x^2+a^2)^{\frac{1}{2}n-1}\,dx$. Prove $\int_0^a (x^2+a^2)^{\frac{3}{2}}\,dx = \tfrac{1}{8}a^4\{7\sqrt{2}+3\log(1+\sqrt{2})\}$.

*71  If $u_n = \int_0^{\frac{1}{4}\pi} \tan^n x\,dx$, prove $u_n + u_{n-2} = \dfrac{1}{n-1}$, and express $u_n$ as a function of $n$ when $n$ is a positive integer.

*72  Writing $n!\,y_n = \int (x-a)^n \sin x\,dx$, prove that if $n > 1$,

$$y_n + y_{n-2} = \frac{(x-a)^{n-1}}{(n-1)!}\sin x - \frac{(x-a)^n}{n!}\cos x.$$

If $n$ is a positive integer, prove

$$\int_0^a (x-a)^{2n}\sin x\,dx = (-1)^{n-1}(2n)!\left\{\cos a - 1 + \frac{a^2}{2!} - \frac{a^4}{4!} + \dots + (-1)^{n-1}\frac{a^{2n}}{(2n)!}\right\}.$$

73  If $c = \int \mathrm{ch}\,ax \cos bx\,dx$, $s = \int \mathrm{sh}\,ax \sin bx\,dx$, calculate $c$ and $s$ in terms of $x$.

74  Calculate $\int \mathrm{ch}\,ax \sin bx\,dx$, $\int \mathrm{sh}\,ax \cos bx\,dx$.

75  If $f_1(x) = \int_0^x f(t)\,dt$, $f_2(x) = \int_0^x f_1(t)\,dt$, ..., $f_n(x) = \int_0^x f_{n-1}(t)\,dt$, prove that

$$(n-1)!\,f_n(x) = \int_0^x f(t)\,(x-t)^{n-1}\,dt.$$

[Integrate repeatedly by parts. The result expresses repeated integrations as a single integral.]

76  Prove $\int_0^{2a} f(x)\,dx = \int_0^a \{f(2a-x)+f(x)\}\,dx$. Deduce that

(i) if $f(2a-x) = -f(x)$, then $\int_0^{2a} f(x)\,dx = 0$;

(ii) if $f(2a-x) = f(x)$, then $\int_0^{2a} f(x)\,dx = 2\int_0^a f(x)\,dx$;

(iii) $\int_0^\pi f(\sin x)\,dx = 2\int_0^{\frac{1}{2}\pi} f(\sin x)\,dx$;

(iv) $\displaystyle\int_0^\pi f(\cos x)\,dx = 0$ or $2\displaystyle\int_0^{\frac12\pi} f(\cos x)\,dx$ according as $f(t)$ is an odd or even function of $t$.

**77**   Prove $\displaystyle\int_0^{\frac12\pi} f(\sin 2x)\sin x\,dx = \sqrt2\int_0^{\frac14\pi} f(\cos 2x)\cos x\,dx.$

**78**   If $f(a+x) = f(x)$ for all $x$, and $n$ is a positive integer, prove

$$\int_0^{na} f(x)\,dx = n\int_0^a f(x)\,dx.$$

Simplify $$\int_0^{n\pi} f(\cos^2 x)\,dx.$$

**79**   (i) By putting $\theta = \tfrac14\pi - \phi$, prove $\displaystyle\int_0^{\frac14\pi} \log(1+\tan\theta)\,d\theta = \tfrac18\pi\log 2.$

     (ii) Calculate $$\int_0^1 \frac{\log(1+x)}{1+x^2}\,dx.$$

**80**   Prove $$\int_0^{\frac12\pi} \log\left(\frac{1+\sin x}{1+\cos x}\right)dx = 0.$$

# 5

## DIFFERENTIAL EQUATIONS

### 5.1 Construction of differential equations

### 5.11 Elimination of parameters from a function

If $y$ is a function of $x$, an equation involving at least one of the derivatives $dy/dx$, $d^2y/dx^2$, ..., and possibly also $x$, $y$, is called a *differential equation* for $y$. Before considering how to *solve* or *integrate* such an equation (i.e. to find $y$ explicitly or implicitly as a function of $x$), we give some further examples (cf. 3.53, ex. (i)) of differential equations which arise from the process of eliminating parameters from a function. The results will be helpful when we turn to the problem of solving a given differential equation, because they may suggest what sort of solution to expect; see also Ex. 5 (*a*).

### Examples

(i) *Eliminate a from* $y^2 = 4ax$.

To eliminate *one* unknown we must have *two* equations. The second one is obtained by deriving the given equation wo $x$:

$$2yy' = 4a.$$

Eliminating $a$, we have

$$2xyy' = 4ax = y^2,$$

i.e.

$$2xy' = y.$$

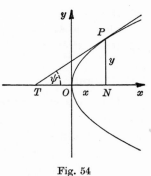

Fig. 54

Geometrically, the given equation represents a parabola having $Ox$ for axis of symmetry and $Oy$ for tangent at the vertex (16.11). The differential equation, which is independent of $a$, thus expresses a geometrical property common to *all* such parabolas. Since $y' = \tan \psi$, we have $NT = y \cot \psi = y/y' = 2x = 2ON$; the property is 'the subtangent = twice the abscissa' (see 5.71).

(ii) *Eliminate A, α from* $x = A \cos(nt + \alpha)$.

To eliminate *two* constants we require *three* equations; hence, deriving twice wo $t$,

$$\frac{dx}{dt} = -An \sin(nt + \alpha), \quad \frac{d^2x}{dt^2} = -An^2 \cos(nt + \alpha).$$

From the last equation and the given one,

$$\frac{d^2x}{dt^2} = -n^2x.$$

(iii) *Eliminate n from the result of ex.* (ii).

Deriving again wo $t$,

$$\frac{d^3x}{dt^3} = -n^2\frac{dx}{dt}.$$

Hence

$$\frac{d^3x}{dt^3} \bigg/ \frac{dx}{dt} = -n^2 = \frac{d^2x}{dt^2} \bigg/ x,$$

so

$$x\frac{d^3x}{dt^3} = \frac{dx}{dt}\frac{d^2x}{dt^2}.$$

Practice in forming differential equations satisfied by given functions has already been offered in Ex. 3 ($b$), nos. 7–10, and Ex. 3 ($e$), nos. 19, 20, 23–25. A few further typical exercises on elimination follow; the reader should do all of nos. 1–10.

## Exercise 5($a$)

1  If $y = a\log x$, eliminate $a$.

2  If $x^2 + y^2 = a^2$, prove $dy/dx = -x/y$, and interpret geometrically.

3  If $y = A\cos nx + B\sin nx$, prove $d^2y/dx^2 = -n^2y$.

4  Eliminate $A$ and $B$ from $y = A\operatorname{ch}px + B\operatorname{sh}px$.

5  If $y = ax + a^2$, prove $y = xy' + y'^2$.

6  Eliminate $c$ from $y^3 = 3cx + c^3$.

7  If $y^2 = 4a(x+b)$, prove $yy'' + y'^2 = 0$; interpret geometrically.

8  If $y = Ae^{kx} + Be^{lx}$, prove $d^2y/dx^2 - (k+l)\,dy/dx + kly = 0$.

9  If $y = (Ax+B)e^{nx}$, prove $d^2y/dx^2 - 2n\,dy/dx + n^2y = 0$.

10  If $y = e^{-ax}(A\cos px + B\sin px)$, prove $d^2y/dx^2 + 2a\,dy/dx + (a^2+p^2)y = 0$.

*11  Eliminate $A$, $B$, $C$ from $y = Ae^x + Be^{2x} + Ce^{3x}$.

*12  If $(x-a)^2 + (y-b)^2 = 1$, prove $(1+y'^2)^3 = y''^2$.

## 5.12 Definitions

All the preceding differential equations involve functions of only one independent variable, and are called *ordinary* differential equations. In Ch. 9 we shall construct differential equations in functions of more than one variable, called *partial* differential equations.

When an ordinary differential equation in $y$ contains $d^ny/dx^n$ but no derivative of higher order, it is said to be of the $n$th *order*. Thus in exs. (i), (ii) and (iii) the equations are of first, second and third order respectively.

If an ordinary differential equation can be expressed as a polynomial in all the derivatives which occur in it (but not necessarily as a polynomial in $x$ or $y$), the *degree* of the equation is the highest power of the highest derivative which occurs. Examples (i)–(iii) are all first-degree equations. In Ex. 5 (a), the equation in no. 5 is of first order, second degree; in nos. 7–10, second order and first degree; in no. 12, second order and second degree. Had the equation in no. 12 been written $(1+y'^2)^{\frac{3}{2}} = y''$ (as it may well have been for geometrical purposes, cf. 8.32 (1)) we should have to rationalise it by squaring both sides before deciding its degree. Finally, $dy/dx = \sqrt{x}/(\sqrt{y}+x)$ is of first order and first degree, although not a polynomial in $x$ or $y$.

## 5.13 Some general conclusions

(1) The above examples and Ex. 5 (a) all suggest that *the result of eliminating n parameters from a function is in general a differential equation of order n*: from the given function we obtain $n$ further equations by deriving $n$ times, and from the total of $n+1$ equations we can in general eliminate the $n$ parameters.

This statement is only true 'in general' because sometimes the elimination leads to an equation of lower order. Thus, elimination of $b$ from $y^2 = 2axy + bx^2$ leads to

$$\left(y - x\frac{dy}{dx}\right)(y - ax) = 0,$$

so that $y - x\,dy/dx = 0$ in either case; and this equation is of order 1 instead of the expected 2. Geometrically, the given equation represents two lines through the origin (see 15.52); each is of the form $y = mx$, and for both it follows that $y = x\,dy/dx$.

(2) Conversely, we should *expect* that *the most general solution of an nth-order differential equation contains n arbitrary constants*. We shall not attempt a general justification of this here, but its truth will be seen in several important cases which follow.

Given a differential equation, the solution which contains the full number of arbitrary constants is called its *general solution* (G.S.) or *complete primitive*. Any relation between $x$ and $y$ which satisfies the equation is called a *particular solution* (P.S.) or *particular integral*. Usually a P.S. is obtained from the G.S. by assigning special values to the arbitrary constants, but exceptionally there are solutions (*singular solutions*) which cannot be so obtained. Thus from Ex. 5 (a), no. 5, the G.S. of $y = xy' + y'^2$ is $y = ax + a^2$; but it is easily verified that $y = -\frac{1}{4}x^2$ also satisfies this differential equation although it cannot be obtained from the G.S. for any particular value of $a$. See also 5.27.

We now give methods for solving the simpler types of ordinary differential equation which arise in geometrical and physical problems.

## 5.2   First-order equations

The standard form is $dy/dx = f(x, y)$, and our problem is to obtain a relation $g(x, y) = 0$ not involving $dy/dx$ but containing one arbitrary constant. If possible we try to give $y$ explicitly in terms of $x$, and free of the sign of integration. We do not discuss whether such a relation *always exists*, but show how to find it in certain special cases.

### 5.21   One variable missing

(1) *If y is missing*, the equation is of the form $dy/dx = f(x)$; the solution is

$$y = \int f(x)\, dx,$$

where the symbol $\int \ldots dx$ implies the presence of one arbitrary constant: cf. 4.11. Thus all the integrations in Ch. 4 could be interpreted as solutions of differential equations having this very simple form.

(2) *If x is missing*, the equation has the form $dy/dx = f(y)$, which can be written $dx/dy = 1/f(y)$. The solution is

$$x = \int \frac{dy}{f(y)}.$$

Equation (xvii) of 4.41 (5) is an example.

### 5.22   Equations whose variables are separable

If the equation can be written in the form

$$f(y)\frac{dy}{dx} + g(x) = 0,$$

then integration wo $x$ gives

$$\int f(y)\frac{dy}{dx}\, dx + \int g(x)\, dx = 0,$$

i.e.
$$\int f(y)\, dy + \int g(x)\, dx = 0.$$

Each of the integrals implies an arbitrary constant, which together give a single such constant.

It is sometimes convenient to obtain this result from a formal† use of differentials by writing the given equation as

$$f(y)\,dy + g(x)\,dx = 0$$

and then integrating throughout. Compare exs. (i), (ii) following; the reader may set out the work in either style, although we shall use the former.

## Examples

(i) $xy^2 \dfrac{dy}{dx} = 1 + y^3$.

Divide both sides by $x(1 + y^3)$:

$$\frac{y^2}{1+y^3}\frac{dy}{dx} = \frac{1}{x}.$$

Integrating wo $x$,      $\displaystyle\int \frac{y^2}{1+y^3}\,dy = \int \frac{1}{x}\,dx,$

$$\therefore \quad \tfrac{1}{3}\log|1+y^3| = \log|x| + c,$$

i.e.           $1 + y^3 = Ax^3,$

since on freeing the equation of logarithms we get $|1+y^3|^{\frac{1}{3}} = |x|\,e^c$, and we may cube both sides, remove moduli, and write $A = \pm\,e^{3c}$.

(ii) $\sin x\,dy/dx - y^2 = 1$.

Rearrange, divide both sides by $(1 + y^2)\sin x$, and multiply by $dx$:

$$\frac{dy}{1+y^2} = \frac{dx}{\sin x},$$

$$\therefore \quad \int \frac{dy}{1+y^2} = \int \operatorname{cosec} x\,dx,$$

$$\therefore \quad \tan^{-1}y = \log|\tan\tfrac{1}{2}x| + c,$$

$$\therefore \quad y = \tan\{\log|\tan\tfrac{1}{2}x| + c\}.$$

## Exercise 5(*b*)

*Solve the following differential equations.*

1   $\dfrac{dy}{dx} = 5x^2$.       2   $\dfrac{dy}{dx} = 3y^5$.       3   $\dfrac{dy}{dx} = \cos^2 y$.

4   $\left(\dfrac{dy}{dx}\right)^2 = x^3$.       5   $\left(\dfrac{dy}{dx}\right)^3 = y$.       6   $\left(\dfrac{dy}{dx}\right)^2 = 1 + y^2$.

7   $x^3\dfrac{dy}{dx} = 1 + x^4$ if $y = 0$ when $x = 1$.       8   $\dfrac{dy}{dx} = \tan x \cot y$.

† Cf. 4.11, where $dx$ is not a differential.

**9** $y\dfrac{dy}{dx}+x = 1$ if $\dfrac{dy}{dx} = -1$ when $x = 2$.   **10** $x\dfrac{dy}{dx}+y = 1$.

**11** $\dfrac{dy}{dx} = y\log x$ if $y = 1$ when $x = e$.

**12** *Show that the general solution of* $dy/dx + P(x)\,y = 0$ *is* $y = e^{-\int P\,dx}$. (A constant is implied in the integral.)

**13** Solve $dy/dx = (x+y)^2$ by putting $z = x+y$.

**14** Solve $(dy/dx + x - 2y)\,dy/dx = 2xy$ by writing $p = dy/dx$ and factorising the quadratic in $p$.

**15** Solve $(3y^2 - x)\,dy/dx = y$. [The equation is $3y^2\,dy/dx = d(xy)/dx$.]

**16** Solve $dy/dx + y/x = 1/x^2$.

\*17 If $dv/dt = (1 - v^2/k^2)\,g$ where $k$, $g$ are constants, and if $v = 0$ when $t = 0$, prove $v = k\,\text{th}\,(gt/k)$. What happens when $t \to \infty$?

## 5.23 Homogeneous equations

If the differential equation can be written

$$f(x, y)\dfrac{dy}{dx}+g(x, y) = 0,$$

where $f(x, y)$, $g(x, y)$ are *homogeneous* and *of the same degree* (1.52 (4)), then the equation can be reduced to the form

$$\dfrac{dy}{dx} = F\!\left(\dfrac{y}{x}\right).$$

For† $\phi(x, y) = -g(x, y)/f(x, y)$ is homogeneous of degree 0, i.e. $\phi(tx, ty) = \phi(x, y)$ for all $t$; taking $t = 1/x$ shows that

$$\phi(x, y) = \phi\!\left(1, \dfrac{y}{x}\right) = F\!\left(\dfrac{y}{x}\right), \quad \text{say.}$$

Putting $y = vx$, then $dy/dx = v + x\,dv/dx$ and the equation becomes

$$v + x\dfrac{dv}{dx} = F(v),$$

in which the variables are separable.

## Example

$$xy\dfrac{dy}{dx} = x^2 - y^2.$$

Since $xy$ and $x^2 - y^2$ are both homogeneous of degree 2, we put $y = vx$ and get

$$x^2v\!\left(v + x\dfrac{dv}{dx}\right) = x^2 - v^2x^2,$$

† Cf. Ex. 1(*d*), nos. 7(i), 8(ii).

i.e. (if $x \neq 0$)
$$v^2 + xv \frac{dv}{dx} = 1 - v^2,$$

$$\therefore \quad \frac{v}{1-2v^2}\frac{dv}{dx} = \frac{1}{x},$$

so
$$-\tfrac{1}{4}\log|1 - 2v^2| = \log|x| + c,$$

i.e.
$$\log|x^4(1 - 2v^2)| = -4c,$$

$$\therefore \quad x^4(1 - 2v^2) = A,$$

and
$$x^4 - 2x^2y^2 = A$$

since $v = y/x$.

## 5.24 Equations reducible to homogeneous type

### Examples

(i) $\dfrac{dy}{dx} = \dfrac{x+y-3}{3x-y-1}$.

This equation is not homogeneous, but by the transformation

$$Y = x+y-3, \quad X = 3x-y-1$$

the given equation becomes
$$\frac{dy}{dx} = \frac{Y}{X}. \tag{$a$}$$

Now
$$\frac{dY}{dX} = \frac{dY}{dx}\bigg/\frac{dX}{dx} = \frac{1+y'}{3-y'}$$

by the formulae of transformation; hence from ($a$),

$$\frac{dY}{dX} = \frac{1+Y/X}{3-Y/X} = \frac{X+Y}{3X-Y}.$$

This is homogeneous; putting $Y = vX$ gives

$$v + X\frac{dv}{dX} = \frac{1+v}{3-v},$$

i.e.
$$X\frac{dv}{dX} = \frac{v^2 - 2v + 1}{3-v},$$

i.e.
$$\frac{3-v}{(v-1)^2}\frac{dv}{dX} = \frac{1}{X},$$

on separating the variables. Integrating wo $X$,

$$\log|X| = \int \frac{3-v}{(v-1)^2}dv$$

$$= \int \left\{ \frac{2}{(v-1)^2} - \frac{1}{v-1} \right\} dv$$

$$= -\frac{2}{v-1} - \log|v-1| + c,$$

hence
$$\log|X(v-1)| + \frac{2}{v-1} = c.$$

Using $v = Y/X$, this is

$$\log|Y-X| + \frac{2X}{Y-X} = c; \tag{b}$$

and from the formulae for $X$, $Y$,

$$\log|2(y-x-1)| + \frac{3x-y-1}{-x+y-1} = c. \tag{c}$$

*Alternatively*, $x+y-3 = 0$ and $3x-y-1 = 0$ represent two straight lines, which intersect at $(1,2)$. If we take new coordinate axes through $(1,2)$, and parallel to the old, we obtain two lines through the new origin: their equations will contain no constant term. To make the change of axes, put (see $15.73\,(2)$)

$$x = X+1, \quad y = Y+2;$$

then $dx = dX$, $dy = dY$ and the given equation becomes

$$\frac{dY}{dX} = \frac{X+Y}{3X-Y},$$

whose solution is $(b)$ above. Using $X = x-1$, $Y = y-2$, $(b)$ becomes

$$\log|y-x-1| + \frac{2(x-1)}{y-x-1} = c'. \tag{d}$$

It is easy to verify that $(c)$, $(d)$ are equivalent, with $c' = 1 - \log 2 + c$.

(ii) $\dfrac{dy}{dx} = \dfrac{2x+y-3}{4x+2y+1}$. *Find the solution for which* $y = 0$ *when* $x = 0$.

The method in ex. (i) fails because, geometrically, the lines $2x+y-3 = 0$, $4x+2y+1 = 0$ are parallel. However, the right-hand side is a function of $2x+y$; therefore put $z = 2x+y$. Then

$$\frac{dy}{dx} = \frac{dz}{dx} - 2,$$

and the equation becomes
$$\frac{dz}{dx} - 2 = \frac{z-3}{2z+1},$$

i.e.
$$\frac{dz}{dx} = \frac{5z-1}{2z+1}.$$

$$\therefore \quad \frac{dx}{dz} = \frac{2z+1}{5z-1} = \frac{2}{5} + \frac{7}{5(5z-1)},$$

$$\therefore \quad x+c = \tfrac{2}{5}z + \tfrac{7}{25}\log|5z-1|,$$

i.e. $\qquad 25x + 25c = 10(2x+y) + 7\log|10x+5y-1|,$

or $\qquad 7\log|10x+5y-1| = 5x - 10y + 25c.$

The condition that $y = 0$ when $x = 0$ enables us to determine $c$:

$$7\log|-1| = 25c,$$

hence $c = 0$. The required solution is

$$7\log|10x+5y-1| = 5x - 10y.$$

*Remark.* Had we written our solution less precisely as

$$7\log(10x+5y-1) = 5x-10y+25c,$$

substitution of the values $x = 0$, $y = 0$ would have given the meaningless equation $7\log(-1) = 25c$, since $\log t$ is not defined for $t \leqslant 0$.

### Exercise 5(c)

*Solve the following.*

1   $x\dfrac{dy}{dx} = x-y.$

2   $(x+y)\dfrac{dy}{dx} = x-y$ if $y = 0$ when $x = 3$.

3   $xy\dfrac{dy}{dx} = x^2+y^2.$          4   $2\dfrac{dy}{dx} = \dfrac{y}{x}+\dfrac{y^2}{x^2}.$

5   $\dfrac{dy}{dx} = \dfrac{y}{x}+\tan\dfrac{y}{x}$ if $y = \tfrac{1}{6}\pi$ when $x = 1$.

6   $x\dfrac{dy}{dx} = y+\sqrt{(x^2+y^2)}.$          7   $\dfrac{dy}{dx} = \dfrac{2x^2+xy+y^2}{x^2+xy+2y^2}.$

8   $\dfrac{dy}{dx} = \dfrac{2x+y-1}{x+2y+1}.$          9   $\dfrac{dy}{dx} = \dfrac{x+y-1}{x+y+1}.$

10   $\dfrac{dy}{dx} = \dfrac{(x+y-1)^2}{4(x-2)^2}.$

11   $\{(x+1)^2+(y-2)^2\}\dfrac{dy}{dx}+2(x+1)^2+2(x+1)(y-2) = 0.$

## 5.25 Linear equations of first order

If $y$ and its derivatives occur to the first power only, and in separate terms, the differential equation is said to be *linear* in $y$.

The most general first-order linear equation in $y$ can therefore be expressed in the form

$$\frac{dy}{dx}+P(x)\,y = Q(x), \tag{i}$$

where $P(x)$, $Q(x)$ are functions of $x$ only, or possibly constants. Ex. 5(b), no. 16 is an example, which happened to be integrable because on clearing the left side of fractions the equation becomes $d(xy)/dx = 1/x$. Similarly, no. 12 is the case when $Q = 0$; the solution can be written

$$y\,e^{\int P\,dx} = 1,$$

and its correctness verified since

$$\frac{d}{dx}(y\,e^{\int P\,dx}) = e^{\int P\,dx}\left(\frac{dy}{dx}+Py\right).$$

These considerations suggest a method of solving the general linear equation:† *we try to make the left side into an exact derivative by multiplying the equation by a suitable function* $R(x)$. We obtain

$$R\frac{dy}{dx} + RPy = RQ,$$

and if the left side is now an exact derivative, it must be the derivative of $Ry$, as the first term indicates. Since

$$\frac{d}{dx}(Ry) = R\frac{dy}{dx} + y\frac{dR}{dx},$$

we see by comparing the terms in $y$ that $R$ must satisfy

$$RP = \frac{dR}{dx},$$

i.e.

$$\frac{1}{R}\frac{dR}{dx} = P,$$

so that

$$\log|R| = \int P\,dx,$$

and hence $R = e^{\int P\,dx}$.

Thus we can begin to solve (i) by multiplying both sides by $e^{\int P\,dx}$. The equation is then

$$\frac{d}{dx}(y\,e^{\int P\,dx}) = Q\,e^{\int P\,dx}, \tag{ii}$$

and the right-hand side is a function of $x$ only.

The function $e^{\int P\,dx}$ which makes the left-hand side into an exact derivative is called an *integrating factor* of (i). Notice that there is no need to include an arbitrary constant in the expression $e^{\int P\,dx}$, provided that a constant is introduced when integrating (ii): any particular function will serve as an integrating factor, without any loss of generality. Often an integrating factor can be seen by inspection; if not, it is best to obtain it from first principles as in the examples following.

### Examples

(i) $dy/dx + 2y = e^{3x}$.

If $R$ is an integrating factor, the left-hand side of

$$R\frac{dy}{dx} + 2Ry = R\,e^{3x}$$

must be identical with     $\dfrac{d}{dx}(Ry) = R\dfrac{dy}{dx} + y\dfrac{dR}{dx}.$

† And others: see Ex. 5 (*l*), no. 58.

Hence
$$2R = \frac{dR}{dx}, \quad \frac{1}{R}\frac{dR}{dx} = 2 \quad \text{and} \quad \log|R| = 2x,$$

so that $R = e^{2x}$. The equation becomes

$$\frac{d}{dx}(y\,e^{2x}) = e^{5x},$$

$$\therefore \quad y\,e^{2x} = \tfrac{1}{5}e^{5x} + c,$$

$$\therefore \quad y = \tfrac{1}{5}e^{3x} + c\,e^{-2x}.$$

*Remark.* Similar work will show that, if $k$ is constant, an integrating factor of

$$\frac{dy}{dx} + ky = Q(x)$$

is always $e^{kx}$. In future this case can be done by inspection.

(ii) $dy/dx + 2y\tan x = \sin^3 x$.

For a suitable $R$, the left-hand side of

$$R\frac{dy}{dx} + 2Ry\tan x = R\sin^3 x$$

must be identical with $\qquad \dfrac{d}{dx}(Ry) = R\dfrac{dy}{dx} + y\dfrac{dR}{dx}.$

Hence
$$2R\tan x = \frac{dR}{dx},$$

$$\frac{1}{R}\frac{dR}{dx} = 2\tan x, \quad \log|R| = -2\log|\cos x|,$$

and $R = \sec^2 x$. The equation becomes

$$\frac{d}{dx}(y\sec^2 x) = \frac{\sin^3 x}{\cos^2 x},$$

$$\therefore \quad y\sec^2 x = \int \frac{\sin^3 x}{\cos^2 x}\,dx$$

$$= \int \frac{u^2 - 1}{u^2}\,du, \quad \text{where} \quad u = \cos x,$$

$$= u + \frac{1}{u} + c = \cos x + \sec x + c.$$

$$\therefore \quad y = \cos^3 x + \cos x + c\cos^2 x.$$

## 5.26 Equations reducible to linear form

**(1)** *Bernoulli's equation*

$$\frac{dy}{dx} + P(x)\,y = Q(x)\,y^n.$$

As a first step towards the standard linear equation, divide both sides by $y^n$:

$$\frac{1}{y^n}\frac{dy}{dx} + P(x)\,y^{1-n} = Q(x).$$

As a second step to standard form put $z = y^{1-n}$, so that

$$\frac{dz}{dx} = (1-n)\,y^{-n}\frac{dy}{dx}.$$

The equation becomes

$$\frac{dz}{dx} + (1-n)\,P(x)\,z = (1-n)\,Q(x),$$

which is linear in $z$.

### Example

(i) $\cos x\,dy/dx + y\sin x + 2y^3 = 0$.

This is

$$\frac{dy}{dx} + y\tan x = -2y^3\sec x,$$

i.e.

$$\frac{1}{y^3}\frac{dy}{dx} + \frac{1}{y^2}\tan x = -2\sec x.$$

Put $z = 1/y^2$, so that

$$\frac{dz}{dx} = -\frac{2}{y^3}\frac{dy}{dx}.$$

Then

$$-\frac{1}{2}\frac{dz}{dx} + z\tan x = -2\sec x,$$

i.e.

$$\frac{dz}{dx} - 2z\tan x = 4\sec x.$$

The reader should verify that $\cos^2 x$ is an integrating factor; so

$$\frac{d}{dx}(z\cos^2 x) = 4\cos x,$$

$$z\cos^2 x = 4\sin x + c,$$

and since $z = 1/y^2$,

$$y^2 = \frac{\cos^2 x}{4\sin x + c}.$$

**(2)** *Change to the inverse function.*

### Example

(ii) $(x + 2y^3)\,dy/dx = y$ is not linear but, regarding $x$ as a function of $y$, we have

$$x + 2y^3 = y\frac{dx}{dy},$$

i.e.

$$\frac{dx}{dy} - \frac{x}{y} = 2y^2,$$

which is linear in $x$ and has integrating factor $1/y$ (seen by inspection, or otherwise). Hence

$$\frac{d}{dy}\left(\frac{x}{y}\right) = 2y,$$

$$\frac{x}{y} = y^2 + c,$$

and

$$x = y^3 + cy.$$

## 5.27 Clairaut's equation

All the standard first-order equations so far considered in 5.2 have also been of first degree. Clairaut's equation

$$y = px + f(p),\qquad\qquad\text{(i)}$$

where $p$ denotes $dy/dx$, is not of first degree unless $f(p)$ is linear in $p$. To solve (i), *derive both sides wo x*:

$$p = p + x\frac{dp}{dx} + f'(p)\frac{dp}{dx},$$

$$\therefore \ \ \left(x + f'(p)\right)\frac{dp}{dx} = 0,$$

and so

$$\frac{dp}{dx} = 0 \quad\text{or}\quad x + f'(p) = 0.$$

The first alternative gives $p = c$, so that from (i), $y = cx + f(c)$. This, containing one arbitrary constant, is the G.S.

From the second alternative and (i),

$$x = -f'(p), \quad y = f(p) - pf'(p),$$

and elimination of $p$ gives another solution. Since these equations do not involve an arbitrary constant, nor will the solution obtained from them.

Nos. 16–19 of Ex. 5(d) illustrate that:

(a) the second solution cannot be obtained from the G.S. by assigning a special value to $c$; it is consequently a *singular solution* (s.s.) (see 5.13(2));

(b) the G.S. represents a family of lines which are tangents to the curve represented by the s.s. (see no. 20). This fact is confirmed geometrically: for at the point of contact of any tangent to the s.s., the curve and tangent have the same $x$, $y$ *and* $dy/dx$; hence the tangent satisfies the same differential equation (i) as the s.s., so this tangent must be included among the lines in the G.S. of (i).†

## Exercise 5(d)

*Solve the following, finding an integrating factor by inspection whenever possible.*

1   $\dfrac{dy}{dx} - y = e^x.$          2   $\dfrac{dy}{dx} + y\cot x = \sin 2x.$

3   $\sin x\dfrac{dy}{dx} - y\cos x = \sin 2x.$     4   $\dfrac{dy}{dx} + 2xy = 2x$ if $y = 2$ when $x = 1.$

5   $x\log x\dfrac{dy}{dx} + y = 2\log x.$

† This fact alone does not show that *every* line of the G.S. touches the s.s.; but see 8.52, ex. (vi).

*6  If $L\dfrac{dx}{dt} + Rx = E\cos pt$, where $L$, $R$, $E$, $p$ are constants, prove

$$x = c\,e^{-Rt/L} + \frac{E}{R^2 + L^2 p^2}(Lp\sin pt + R\cos pt).$$

If $p = 0$ and $R/L$ is positive, prove $x \to E/R$ when $t \to \infty$.

7  $\dfrac{dy}{dx} + y = y^3.$          8  $y\dfrac{dy}{dx} = y^2 + x$ if $y = 0$ when $x = -\frac{1}{2}$.

9  $xy\dfrac{dy}{dx} - 3y^2 = 3x^3.$      10  $2x\dfrac{dy}{dx} = \dfrac{\sin x}{y} - y.$

11  $\dfrac{dy}{dx} + y\tan x = y^3\sec^4 x.$      12  $x^2\dfrac{dy}{dx} = xy + y.$

13  $(2x - 5y^3)\dfrac{dy}{dx} = y.$      14  $(1 + x\,e^y)\dfrac{dy}{dx} = e^y.$

15  $\dfrac{d^2 y}{dx^2} + \dfrac{dy}{dx} = 1.$

*Find the* G.S. *and* S.S. *of the following.*

*16  $y = px + \dfrac{1}{p}.$             *17  $(y - px)^2 = p^2 + 1.$

*18  $y = px + p^3$. Also find all solutions for which $y = 6$ when $x = -7$.

*19  $y = px + \sin p.$

*20  Verify that, in nos. 16–19, the G.S. represents a family of straight lines which are tangents to the curve represented by the S.S.

*21  Find the differential equation of a curve such that any tangent makes with the coordinate axes a triangle of constant area $2k^2$. Hence show that the curves are rectangular hyperbolas $xy = k^2$.

## 5.3   Second-order equations

These involve $d^2 y/dx^2$ and some or all of $dy/dx$, $y$, $x$.

### 5.31  Some simple special types

(1) *$dy/dx$ and $y$ missing.* If the equation can be put in the form $d^2 y/dx^2 = f(x)$, it is solved by integrating twice wo $x$; two arbitrary constants are introduced.

(2) *$dy/dx$ and $x$ missing.* If the equation is $d^2 y/dx^2 = f(y)$, put $p = dy/dx$, so that

$$\frac{d^2 y}{dx^2} = \frac{d}{dx}\left(\frac{dy}{dx}\right) = \frac{dp}{dx} = \frac{dp}{dy}\frac{dy}{dx} = \frac{dp}{dy}p.$$

Then                        $p\dfrac{dp}{dy} = f(y),$

and by integrating wo $y$,

$$\tfrac{1}{2}p^2 = \int f(y)\,dy = \phi(y) + c,$$

say. Thus

$$\frac{dy}{dx} = \{2\phi(y) + 2c\}^{\frac{1}{2}} \quad \text{and} \quad x = \int \{2\phi(y) + 2c\}^{-\frac{1}{2}}\,dy.$$

### Example

$d^2y/dx^2 + n^2y = 0$ (the equation of *simple harmonic motion*).

Putting $p = dy/dx$, the equation becomes

$$p\,\frac{dp}{dy} + n^2y = 0,$$

i.e.

$$\frac{1}{2}\frac{d}{dy}(p^2) + \tfrac{1}{2}n^2\frac{d}{dy}(y^2) = 0.$$

$$\therefore \quad \tfrac{1}{2}p^2 + \tfrac{1}{2}n^2y^2 = \text{constant} = \tfrac{1}{2}n^2a^2, \quad \text{say.}$$

$$\therefore \quad \left(\frac{dy}{dx}\right)^2 = p^2 = n^2(a^2 - y^2),$$

$$\therefore \quad \frac{dx}{dy} = \pm\frac{1}{n\sqrt{(a^2 - y^2)}}.$$

$$\therefore \quad nx + b = \pm\int\frac{dy}{\sqrt{(a^2 - y^2)}} = \pm\sin^{-1}\frac{y}{a},$$

$$\therefore \quad y = \pm a\sin(nx + b).$$

Since the constants $a$, $b$ are arbitrary, there is no loss of generality in taking the ambiguous sign as $+$. The general solution is therefore

$$y = a\sin(nx + b).$$

It can also be written in the forms (cf. 5.11, ex. (ii), and Ex. 5(a), no. 3)

$$y = a\cos(nx + b'), \quad y = A\cos nx + B\sin nx.$$

The above differential equation will be met frequently when solving others, and for convenience we will quote the solutions when needed; the last is usually the most suitable when finding a general solution.

**(3)** *y missing.* The equation has the form†

$$f\!\left(\frac{d^2y}{dx^2}, \frac{dy}{dx}, x\right) = 0.$$

The substitution $p = dy/dx$ immediately reduces it to a first-order equation in $p$, viz. $f(dp/dx, p, x) = 0$.

---

† This includes (1).

**(4)** *x missing*. The equation is†

$$f\left(\frac{d^2y}{dx^2}, \frac{dy}{dx}, y\right) = 0.$$

Putting $p = dy/dx$ as in (2), it becomes $f(p\,dp/dy, p, y) = 0$, a first-order equation in $p$.

### Exercise 5(e)

*Solve the following.*

1   $x^3y'' = 1$.       2   $y^3y'' = a^2$.       3   $y'' = n^2y$.

4   $y'' = 3y'$.       5   $xy'' = 3y'$ if $y = \frac{3}{2}$ and $y' = 2$ when $x = 1$.

6   $y'' = 6yy'^3$.       7   $y''^2 = 1 + y'^2$.

8   Solve $2y'' + y'^2 = 4y$, where $y = 1$ and $dy/dx = 0$ when $x = 0$.

9   Solve $d(y\,dy/dx)/dx = 6y$ with the conditions that $y = 0$ and $dy/dx = 0$ when $x = 0$.

*Given that $u = 1/c$ and $du/d\theta = 0$ when $\theta = 0$, solve $d^2u/d\theta^2 + u = P/(h^2u^2)$ (the equation for central orbits) when $h$, $\mu$ are constants and*

10   $P = \mu u^2$.       11   $P = \mu u^3$.       12   $P = \mu/u$.

### 5.32 Linear second-order equations

The general *linear* equation (see 5.25) of second order is

$$p(x)\frac{d^2y}{dx^2} + q(x)\frac{dy}{dx} + r(x)\,y = g(x);$$

$p(x)$, $q(x)$, $r(x)$ are called its *coefficients*. It can always be reduced to the standard form

$$\frac{d^2y}{dx^2} + a(x)\frac{dy}{dx} + b(x)\,y = f(x), \tag{i}$$

where $f(x)$ and the coefficients $a(x)$, $b(x)$ are functions of $x$ or possibly constants.

**(1)** *The complementary function.*

Let $y = u$ be any particular solution of (i), however simple and however obtained. Then

$$\frac{d^2u}{dx^2} + a(x)\frac{du}{dx} + b(x)\,u = f(x).$$

The function $y = z + u$ will satisfy (i) if and only if

$$\frac{d^2z}{dx^2} + \frac{d^2u}{dx^2} + a(x)\frac{dz}{dx} + a(x)\frac{du}{dx} + b(x)\,z + b(x)\,u = f(x),$$

† This includes (2).

i.e. if and only if $z$ satisfies the equation

$$\frac{d^2z}{dx^2} + a(x)\frac{dz}{dx} + b(x)z = 0, \tag{ii}$$

which is simpler than (i).

The general solution of (ii) will contain† two arbitrary constants. When this solution is known, the general solution of (i) is given by $y = z + u$, also involving two arbitrary constants. Thus

$$\text{G.S. of (i)} = \text{G.S. of (ii)} + \text{P.S. of (i)}.$$

The general solution of (ii) is called the *complementary function* (C.F.) of equation (i). Hence

$$\text{G.S. of (i)} = \text{C.F. of (i)} + \text{P.S. of (i)}.$$

### Example

$$\frac{d^2y}{dx^2} + 9y = 5e^x.$$

To discover a P.S., try $y = ke^x$, where $k$ is some constant, to be found by substitution in the given equation. (This trial is reasonable, since the term in $e^x$ on the right can arise only from derivatives of functions containing $e^x$.) Then

$$ke^x + 9ke^x = 5e^x, \quad \therefore \ k = \tfrac{1}{2}.$$

Hence a P.S. is $y = \tfrac{1}{2}e^x$.

The C.F. is the solution of $d^2y/dx^2 + 9y = 0$; by 5.31 (2), example, this is

$$y = A\cos 3x + B\sin 3x.$$

The G.S. of the given equation is therefore

$$y = A\cos 3x + B\sin 3x + \tfrac{1}{2}e^x.$$

(2) C.F. *as a linear combination of independent particular solutions of* (ii).

I. *If* $z = \phi(x)$, $z = \psi(x)$ *are two solutions of* (ii), *then so also is*

$$z = A\phi(x) + B\psi(x)$$

*for arbitrary constants* $A, B$.

*Proof.* We have

$$\phi'' + a\phi' + b\phi = 0 \quad \text{and} \quad \psi'' + a\psi' + b\psi = 0, \tag{ii} a, b$$

so that

$$\frac{d^2}{dx^2}(A\phi + B\psi) + a\frac{d}{dx}(A\phi + B\psi) + b(A\phi + B\psi)$$
$$= A(\phi'' + a\phi' + b\phi) + B(\psi'' + a\psi' + b\psi)$$
$$= 0;$$

therefore $A\phi + B\psi$ satisfies (ii).

If $\phi(x)/\psi(x)$ is not constant, i.e. if $\phi(x)$, $\psi(x)$ are 'independent' solutions, then the solution $z = A\phi + B\psi$ involves two arbitrary constants $A, B$. Conversely—

II. *If* $\phi$, $\psi$ *are independent solutions of* (ii), *then any solution* $z$ *is of the form* $A\phi + B\psi$, *where* $A, B$ *are constants.*

† From the considerations in 5.13 (2); but see 5.32 (2) for a proof.

*Lemma.* If $y_1$, $y_2$ are any two solutions of $y' + Py = 0$, then $y_2 = Ay_1$.
For by elimination of $P$ from

$$y_1' + Py_1 = 0, \quad y_2' + Py_2 = 0$$

we have $y_1 y_2' - y_2 y_1' = 0$, i.e. $d(y_2/y_1)/dx = 0$, i.e. $y_2/y_1 = A$.

*Proof of* II.

By elimination of $b$ from (ii) and (ii) $b$, we get

$$\psi z'' - z\psi'' + a(\psi z' - z\psi') = 0,$$

i.e.
$$\frac{d}{dx}(\psi z' - z\psi') + a(\psi z' - z\psi') = 0.$$

Similarly from (ii) $a$ and (ii) $b$,

$$\frac{d}{dx}(\psi \phi' - \phi \psi') + a(\psi \phi' - \phi \psi') = 0.$$

Hence, by the lemma,
$$\psi z' - z\psi' = A(\psi \phi' - \phi \psi'),$$

i.e.
$$\frac{d}{dx}\left(\frac{z}{\psi}\right) = A\frac{d}{dx}\left(\frac{\phi}{\psi}\right),$$

so that
$$\frac{z}{\psi} = A\frac{\phi}{\psi} + B \quad \text{and} \quad z = A\phi + B\psi.$$

*Hence the* G.S. *of* (ii) *is a linear combination of any two independent particular solutions.*

If $\phi$ and $\psi$ are not independent, i.e. $\phi = k\psi$ for all $x$, then $A\phi + B\psi = C\psi$ where $C = Ak + B$, which involves effectively only one constant $C$. Thus a knowledge of a single particular solution will not give the G.S. in this way; but see 5.64 for a method applicable in this case.

From the preceding account it appears that, to solve equations like (i), our first problem is to solve the simpler equation (ii). We do not attempt this in general here, but confine ourselves mainly to the case when $a$, $b$ in (ii) are *constants*; see 5.6, however.

### 5.33 Linear second-order equations with constant coefficients

Our standard equation now is

$$\frac{d^2y}{dx^2} + a\frac{dy}{dx} + by = 0, \tag{iii}$$

where $a$ and $b$ are constants.

**(1)** *Direct solution.*

The equation
$$\frac{d}{dx}(y' - ky) = l(y' - ky), \tag{iv}$$

i.e.
$$y'' - (k+l)y' + kly = 0,$$

will be the same as the given equation (iii) if and only if

$$-(k+l) = a \quad \text{and} \quad kl = b, \tag{v}$$

i.e. if and only if $k, l$ are the roots of the quadratic in $m$:

$$m^2 + am + b = 0.$$

*Case (i): suppose $k, l$ exist and $k \neq l$.*

Put $z = y' - ky$, so that (iv) becomes $dz/dx = lz$. This has solution $z = C\,e^{lx}$, so that

$$y' - ky = C\,e^{lx}. \tag{vi}$$

This can be solved EITHER by use of the integrating factor $e^{-kx}$:

$$\frac{d}{dx}(y\,e^{-kx}) = C\,e^{(l-k)x},$$

$$\therefore \quad y\,e^{-kx} = \frac{C}{l-k}\,e^{(l-k)x} + B,$$

$$\therefore \quad y = A\,e^{lx} + B\,e^{kx},$$

where $A = C/(l-k)$; OR by observing that, since the relations (v) are symmetrical in $k$ and $l$, the given equation (iii) is also equivalent to

$$\frac{d}{dx}(y' - ly) = k(y' - ly),$$

from which we deduce that ($C'$ being another arbitrary constant)

$$y' - ly = C'\,e^{kx}. \tag{vii}$$

Elimination of $y'$ from (vi), (vii) gives

$$(k-l)\,y = C'\,e^{kx} - C\,e^{lx},$$

so that $y$ has the form $A\,e^{lx} + B\,e^{kx}$ as before. Cf. Ex. 5 (a), no. 8.

*Case (ii): suppose $k = l$.*

Proceeding as before, the given equation is equivalent to

$$y' - ky = A\,e^{kx}. \tag{vi}'$$

Using the integrating factor $e^{-kx}$, it becomes

$$\frac{d}{dx}(y\,e^{-kx}) = A,$$

$$\therefore \quad y\,e^{-kx} = Ax + B,$$

$$\therefore \quad y = (Ax + B)\,e^{kx}.$$

Cf. Ex. 5 (a), no. 9.

*Case (iii): suppose the m-quadratic has no roots.*

The preceding method then breaks down because (iii) cannot be expressed in the form (iv). However, taking our clue from Ex. 5 (*a*), no. 10, we write equation (iii) as

$$y'' + ay' + (\tfrac{1}{4}a^2 + p^2) y = 0. \tag{viii}$$

This is justified because the *m*-quadratic has no roots, and so $a^2 - 4b < 0$, i.e. $b - \tfrac{1}{4}a^2$ is positive and equal to $p^2$, say.

If we now put† $y = z e^{-\frac{1}{2}ax}$, i.e. $z = y e^{\frac{1}{2}ax}$, we have

$$z' = (y' + \tfrac{1}{2}ay) e^{\frac{1}{2}ax} \quad \text{and} \quad z'' = (y'' + ay' + \tfrac{1}{4}a^2 y) e^{\frac{1}{2}ax}.$$

Hence (viii) becomes         $z'' + p^2 z = 0.$

$$\therefore \quad z = A \cos px + B \sin px$$

and         $y = e^{-\frac{1}{2}ax} (A \cos px + B \sin px).$

*Conclusions.* We have now *proved* that, in the case of linear second-order equations with constant coefficients, the G.S. always involves two arbitrary constants; and that the form of the solution depends on the nature of the roots of the quadratic $m^2 + am + b = 0$, called the *auxiliary equation*.

## Examples

(i) $y'' - 6y' + 8y = 0.$

This is equivalent to

$$\frac{d}{dx}(y' - ky) = l(y' - ky),$$

i.e.         $y'' - (k+l) y' + kly = 0,$

if         $k + l = 6 \quad \text{and} \quad kl = 8.$

Hence $k, l$ are the roots of $m^2 - 6m + 8 = 0$, viz. 2 and 4. Taking $k = 2, l = 4$, we have $dz/dx = 4z$ where $z = y' - 2y$. Therefore $z = C e^{4x}$, and

$$y' - 2y = C e^{4x},$$

$$\therefore \quad \frac{d}{dx}(y e^{-2x}) = C e^{2x},$$

$$\therefore \quad y e^{-2x} = \tfrac{1}{2}C e^{2x} + B,$$

$$\therefore \quad y = A e^{4x} + B e^{2x},$$

where $A = \tfrac{1}{2}C$.

*Alternatively*, by the symmetry in $k, l$ we also have

$$y' - 4y = C' e^{2x}.$$

Elimination of $y'$ gives         $2y = C e^{4x} - C' e^{2x},$

† Cf. Ex. 4 (*e*), no. 23.

so
$$y = A\,e^{4x} + B\,e^{2x}$$
as before, where $A = \frac{1}{2}C$ and $B = -\frac{1}{2}C'$.

(ii) $y'' - 6y' + 9y = 0$.

Proceeding as before, the auxiliary equation is found to be
$$m^2 - 6m + 9 = 0,$$
which has the repeated root $m = 3$. The given equation is equivalent to
$$\frac{dz}{dx} = 3z,$$
where $z = y' - 3y$. Hence $z = A\,e^{3x}$, and so
$$y' - 3y = A\,e^{3x},$$
$$\frac{d}{dx}(y\,e^{-3x}) = A,$$
$$y\,e^{-3x} = Ax + B,$$
and
$$y = (Ax + B)\,e^{3x}.$$

(iii) $y'' - 6y' + 25y = 0$.

The auxiliary equation is $m^2 - 6m + 25 = 0$, i.e. $(m-3)^2 + 16 = 0$, which has no roots. We therefore write the given equation as
$$(y'' - 6y' + 9y) + 16y = 0,$$
and put $y = z\,e^{3x}$, i.e. $z = y\,e^{-3x}$, so that
$$z' = e^{-3x}(y' - 3y), \quad z'' = e^{-3x}(y'' - 6y' + 9y^2).$$
Hence
$$z'' + 16z = 0,$$
$$z = A\cos 4x + B\sin 4x,$$
and
$$y = e^{3x}(A\cos 4x + B\sin 4x).$$

**(2)** *Solution by trial exponentials.*

Since the linear *first*-order equation $dy/dx + Py = 0$ with constant coefficient $P$ has solution $A\,e^{Px}$ (Ex. 5 (*b*), no. 12), we may try to find a P.S. of this type for equation (iii). The function $y = e^{mx}$ will satisfy (iii) if and only if
$$e^{mx}(m^2 + am + b) = 0,$$
i.e.
$$m^2 + am + b = 0.$$

If this quadratic has distinct roots $m = k$, $m = l$, then $e^{kx}$ and $e^{lx}$ are independent solutions of (iii). Hence by 5.32 (2), the G.S. is
$$y = A\,e^{kx} + B\,e^{lx}.$$

If the roots are equal, say $m = k$, then we seek a more general solution than $y = e^{kx}$ by putting $y = v\,e^{kx}$. On substituting in (iii) we find that $e^{kx}v'' = 0$, so that $v'' = 0$ and $v = Ax + B$. The G.S. is thus $y = (Ax + B)\,e^{kx}$.

If the quadratic has no roots, the trial method fails without the aid of complex numbers: see 14.66, ex. (ii).

The reader may feel that, when it is applicable, the trial method is quicker than that in (1); but he may wonder why only the exponential function is selected for a trial P.S. The direct calculations in (1) make clear why exponentials are involved, without *a priori* assumptions about the form of the solution. Both methods can be applied to linear equations with constant coefficients and order greater than 2: see Ex. 5 (*l*), no. 56.

We have now discussed completely the equation (ii) for the case of *constant coefficients*. With the same restriction we next consider the P.S. of (i); Ex. 5 (*f*), nos. 9–28, provides essential information, and the reader should consider these examples carefully, especially nos. 9–18.

### Exercise 5(*f*)

*Find the general solution of the following.*

1  $y'' - 4y' + 3y = 0.$     2  $y'' - y' - 12y = 0.$     3  $y'' + 10y' + 25y = 0.$

4  $y'' - 4y = 0.$     5  $y'' + 4y = 0.$     6  $y'' + 4y' + 13y = 0.$

7  $y'' - y' + y = 0.$     8  $y'' + 3y' = 0.$

*By substitution find the values of the constants A, B, C for which the following equations have the particular solutions indicated.*

9  $y'' + 3y' + 2y = 5; y = A.$

10  $y'' - 2y' + 3y = 6x; y = Ax + B.$

11  $y'' + 5y' - 2y = 4x^2; y = Ax^2 + Bx + C.$

12  $y'' - 3y' + y = 7e^{2x}; y = A e^{2x}$ (cf. 5.32, ex.).

13  $y'' - y' + 7y = 8 \sin 3x + \cos 3x; y = A \cos 3x + B \sin 3x.$ (This is reasonable because terms in $\sin 3x$ and $\cos 3x$ can arise by derivation only from similar terms.)

14  $y'' + 4y' = 8x.$ If we try $y = Ax + B$, as suggested by no. 10, we find that there are no values of $A, B$ for which this can satisfy the equation. Integrating once (without arbitrary constant, since only a P.S. is required) gives $y' + 4y = 4x^2$ and no. 11 suggests trying $y = Ax^2 + Bx + C$. Values for $A, B$ can now be found. Hence a suitable trial for the given equation is $y = Ax^2 + Bx.$

15  $y'' + 5y' = 3x^2; y = Ax^3 + Bx^2 + Cx.$

16  $y'' - 2y' - 3y = 4e^{3x}.$ The trial $y = A e^{3x}$ suggested by no. 12 fails, so put $y = v e^{3x}$; we find that $v$ satisfies $v'' + 4v' = 4$, and to solve this we try $v = Ax$ as in no. 14. Hence $y = Ax e^{3x}$ would be a suitable trial for the original equation. (The trial $y = A e^{3x}$ for a P.S. is futile since the C.F. of the equation is $y = A e^{-x} + B e^{3x}$, which *already* involves $e^{3x}$.)

17  $y'' - 6y' + 9y = 2e^{3x}; y = v e^{3x}$ gives $v'' = 2$, whence $v = x^2$, and so $y = Ax^2 e^{3x}$ would be a suitable trial. (Here the C.F. is $y = (Ax + B) e^{3x}$, so the trials $y = A e^{3x}$ and $y = Ax e^{3x}$ would both be futile.)

18  $y'' + 9y = 3 \cos 3x - 2 \sin 3x.$ The C.F. is $y = A \cos 3x + B \sin 3x$, so we may try $y = x(A \cos 3x + B \sin 3x).$

*19  $y'' - 5y' + 6y = x e^x; y = v e^x$ gives an equation for $v$ like no. 10, hence $y = (Ax + B) e^x$ would be a suitable trial.

*20  $y'' - 5y' + 6y = x e^{3x}; y = v e^{3x}$ gives type of no. 14; hence $y = (Ax^2 + Bx) e^{3x}.$

*21  $y'' - 6y' + 9y = x e^{3x}; y = v e^{3x}$, i.e. $v = y e^{-3x}$, gives $v'' = x$; so $y = Ax^3 e^{3x}.$

*22   $y'' - 5y' + 6y = x^2 e^x$;   $y = (Ax^2 + Bx + C) e^x$.

*23   $y'' - 5y' + 6y = x^2 e^{2x}$;   $y = (Ax^3 + Bx^2 + Cx) e^{2x}$.

*24   $y'' - 6y' + 9y = x^2 e^{3x}$;   $y = Ax^4 e^{3x}$.

*25   $y'' + 3y' + 2y = 10e^x \sin 2x$;   $y = v e^x$ gives type like no. 13.

*26   $y'' + 3y' + 2y = 10e^{-x} \sin 2x$.       *27   $y'' - 6y' + 9y = e^{3x} \sin x$.

*28   $y'' - 4y' + 13y = 6e^{2x} \sin 3x$;   $y = v e^{2x}$ gives type of no. 18.

## 5.34 Particular solution in the case of constant coefficients

(1) The results of Ex. 5 $(f)$, nos. 9–28 suggest useful trial methods for finding a P.S. of

$$y'' + ay' + by = f(x)$$

when $f(x)$ is a polynomial, an exponential function, a sine or cosine, or certain products of these. Our conclusions are summarised as follows:

(i) *If $f(x)$ is a polynomial of degree $n$,* a P.S. is also a polynomial of degree $n$ unless $b = 0$, in which case a P.S. is a polynomial of degree $n + 1$.

(ii) *If $f(x) = k e^{px}$,* a P.S. is usually $A e^{px}$; but if $e^{px}$ belongs to the C.F., a P.S. is $Ax e^{px}$ unless $x e^{px}$ also belongs to the C.F., in which event a P.S. is $Ax^2 e^{px}$.

(iii) *If $f(x) = \lambda \cos px + \mu \sin px$,* a P.S. is usually $A \cos px + B \sin px$; but if these terms occur in the C.F., a P.S. is $x(A \cos px + B \sin px)$.

(iv) *If $f(x) = kx^n e^{px}$,* a P.S. is usually $e^{px} \times$ (polynomial of degree $n$); but if $e^{px}$ belongs to the C.F., a P.S. is $e^{px} \times$ (polynomial of degree $n + 1$), unless $x e^{px}$ also belongs to the C.F., in which case a P.S. is $Ax^{n+2} e^{px}$.

(v) *If $f(x) = e^{px} (\lambda \cos qx + \mu \sin qx)$,* a P.S. is usually $e^{px} (A \cos qx + B \sin qx)$; but if these terms appear in the C.F., a P.S. is $x e^{px} (A \cos qx + B \sin qx)$.

(2) Nos. 14–28 also suggest that in solving a differential equation it is desirable to find the C.F. first, so that when we seek a P.S. we shall be aware of what trials to avoid.

Sometimes we are asked for the special solution in which $y$, $y'$ have prescribed values for given $x$. This information (called *initial* or *boundary conditions*) enables us to determine $A$, $B$ from the G.S. (not, of course, from the C.F. alone). Boundary conditions are important in applications; e.g. see Ex. 5 $(h)$, nos. 1 (ii), 2 (ii).

Hence the *order of procedure* is:

($a$) Find the C.F.

($b$) Inspect it, and then try a suitable P.S. as indicated in (i)–(v) above.

(c) G.S. = C.F. + P.S.

(d) If required, find the solution which satisfies given boundary conditions.

(3) When $f(x)$ consists of a sum of terms $f_1(x)$, $f_2(x)$ of the above types (i)–(v), we can use the following theorem.

*If*       $y = u_1$ *is a* P.S. *of* $y'' + ay' + by = f_1(x)$,

*and*      $y = u_2$ *is a* P.S. *of* $y'' + ay' + by = f_2(x)$,

*then*     *a* P.S. *of* $y'' + ay' + by = f_1(x) + f_2(x)$ *is* $y = u_1 + u_2$.

*Proof.* If $u = u_1 + u_2$, then

$$u'' + au' + bu = (u_1'' + u_2'') + a(u_1' + u_2') + b(u_1 + u_2)$$
$$= (u_1'' + au_1' + bu_1) + (u_2'' + au_2' + bu_2)$$
$$= f_1(x) + f_2(x).$$

Therefore $y = u$ satisfies the original equation, i.e. $u_1 + u_2$ is a P.S.

If $f(x) = 2x - x^2$, we should not apply this theorem with $f_1(x) = 2x$ and $f_2(x) = -x^2$ because it is easier to make the trial $y = Ax^2 + Bx + C$ (or whatever is appropriate) for the *whole* P.S. directly. The theorem is used when $f_1(x)$, $f_2(x)$ are functions of different 'types', as in the following example.

### Example

*Solve* $y'' - 6y' + 8y = e^{2x} + \sin 2x$, *and find the solution for which* $y = 0$ *and* $y' = 0$ *when* $x = 0$.

First consider the C.F., which is the general solution of

$$y'' - 6y' + 8y = 0;$$

it is (see **5.33**, ex. (i)) $y = A e^{4x} + B e^{2x}$.

Next, seek a P.S. of $y'' - 6y' + 8y = e^{2x}$. Since $e^{2x}$ occurs in the C.F. but $x e^{2x}$ does not, we try $y = Ax e^{2x}$; this is found to give $A = -\frac{1}{2}$, so a P.S. for this equation is $-\frac{1}{2}x e^{2x}$.

Finally, seek a P.S. of $y'' - 6y' + 8y = \sin 2x$; we may try $y = A \cos 2x + B \sin 2x$, for which $y' = -2A \sin 2x + 2B \cos 2x$ and $y'' = -4y$, so that $A$, $B$ must be chosen to satisfy

$$-4y - 6y' + 8y = \sin 2x,$$

i.e.       $4(A \cos 2x + B \sin 2x) - 6(-2A \sin 2x + 2B \cos 2x) = \sin 2x.$

This requires $4A - 12B = 0$ and $4B + 12A = 1$, whence $A = \frac{3}{40}$, $B = \frac{1}{40}$. A P.S. of the last equation is thus $\frac{1}{40}(3 \cos 2x + \sin 2x)$.

Hence a P.S. of the given equation is

$$-\tfrac{1}{2}x e^{2x} + \tfrac{1}{40}(3 \cos 2x + \sin 2x),$$

and the G.S. is

$$y = A e^{4x} + (B - \tfrac{1}{2}x) e^{2x} + \tfrac{1}{40}(3 \cos 2x + \sin 2x).$$

From this

$$y' = 4A e^{4x} + (2B - x - \tfrac{1}{2}) e^{2x} + \tfrac{1}{20}(-3 \sin 2x + \cos 2x);$$

hence for the special solution required we have, when $x = 0$:

$$0 = A + B + \tfrac{3}{40}, \quad 0 = 4A + (2B - \tfrac{1}{2}) + \tfrac{1}{20},$$

from which $A = \tfrac{3}{10}$, $B = -\tfrac{3}{8}$. This solution is therefore

$$y = \tfrac{3}{10} e^{4x} - (\tfrac{3}{8} + \tfrac{1}{2}x) e^{2x} + \tfrac{1}{40}(3 \cos 2x + \sin 2x).$$

### Exercise 5(g)

*Find the* G.S. *of the following.*

1   $y'' - 3y' + 2y = 6.$        2   $y'' - 2y' + 3y = 6x - 1.$

3   $y'' - y' - 2y = -\tfrac{1}{2}x^2.$

4   $2y'' - 5y' + 3y = 4e^{2x}.$ Also find the solution for which $y = 0$ and $y' = 0$ when $x = 0$.

5   $y'' - y' - 12y = 2e^{4x}.$        6   $y'' - 2y' + 2y = \sin 3x.$

7   $y'' - 14y' + 50y = 2\cos x.$

8   $y'' + 16y = \cos 4x.$ Also give the most general solution which vanishes when $x = \tfrac{1}{4}\pi$.

9   $y'' + 8y' + 16y = 6e^{-4x}.$      10   $y'' + 6y' = 4x.$

11   $y'' + 4y = \sin x \sin 3x.$      12   $y'' - 3y' + 18y = \operatorname{sh} 2x.$

13   $y'' - 6y' + 8y = e^{4x} - \cos 2x.$      *14   $y'' + 3y' + 2y = e^{-x} \sin x.$

*15   $3y'' - 5y' + 2y = x^2 e^x.$      *16   $y'' + 6y' + 9y = (1+x)e^{-3x}.$

*17   $y'' + y' + y = e^x(x + \cos x).$      *18   $y'' - 4y' + 4y = 8x^2 e^{2x} \sin 2x.$

## 5.4   The operator $D$; calculation of a P.S.

In 5.34 we gave some trial methods for finding a P.S. of a linear second-order equation with constant coefficients. We now show how it can be found more systematically (and often more easily) by formal calculation. The discussion, given here for second-order equations only, extends naturally to higher orders.

### 5.41   Algebraic properties of $D$

The notations $Dy$, $D^2 y$ for first and second derivatives of $y$ were mentioned in 2.11, 3.51 respectively. We now write the standard linear equation

$$\frac{d^2 y}{dx^2} + a\frac{dy}{dx} + by = f(x)$$

as

$$D^2 y + aDy + by = f(x),$$

or

$$(D^2 + aD + b)y = f(x),$$

or briefly

$$F(D)y = f(x),$$

where

$$F(D) = D^2 + aD + b.$$

When the coefficients $a$, $b$ are *constants* the operator $F(D)$ can be treated for certain purposes as if it were an algebraical polynomial in a variable $D$, because $D$ obeys the same laws of addition and multiplication as do algebraical symbols. For, by the rules of derivation,

$$D(u+v) = Du + Dv \qquad \text{(distributive law)}$$

$$D^m(D^n u) = D^{m+n} u = D^n(D^m u) \qquad \text{(index law)}$$

where $m$, $n$ are *positive integers*; and if $k$ *is constant*,

$$D(ku) = kDu \qquad \text{(commutative law)}.$$

**Example**

If $x^2 + ax + b$ has factors $(x-k)(x-l)$, then $k+l = -a$ and $kl = b$. By the preceding laws,

$$(D-k)(D-l)y = (D-k)(Dy - ly)\dagger$$

$$= D(Dy - ly) - k(Dy - ly)$$

$$= D^2 y - (k+l)Dy + kly$$

$$= D^2 y + aDy + by$$

$$= F(D)y.$$

Hence the operator $F(D)$ can be 'factorised' as $(D-k)(D-l)$, (or, if we wish, as $(D-l)(D-k)$ by a similar argument), just as if it were a quadratic in $D$.

For the same reason, *polynomials in $D$ with constant coefficients are added, subtracted and multiplied just like algebraic polynomials.*

If $D^n y = \phi(x)$, we write $y = D^{-n}\phi(x)$, so that $D^{-n}$ denotes $n$ *integrations* wo $x$, thus introducing $n$ arbitrary constants. Observe that $D^n(D^{-n})y = y$, but that $D^{-n}(D^n y) = y + $ a polynomial of degree $n-1$ in $x$ with $n$ arbitrary constants for coefficients; the above index law fails for negative indices.

### 5.42 Shift theorem

*If $u = u(x)$, then $F(D)\{e^{px}u\} = e^{px}F(D+p)u$, where $p$ is constant.*

*Proof.* By the product rule,

$$D\{e^{px}u\} = e^{px}Du + p e^{px}u = e^{px}(Du + pu) = e^{px}(D+p)u.$$

Repeating this, with $u$ replaced by $(D+p)u$, we get

$$D^2\{e^{px}u\} = e^{px}(D+p)\{(D+p)u\} = e^{px}(D+p)^2 u.$$

---

† An operator is understood to act on the function placed immediately after it.

Hence     $F(D)\{e^{px}u\} = (D^2+aD+b)\{e^{px}u\}$

$$= e^{px}\{(D+p)^2+a(D+p)+b\}u$$

$$= e^{px}F(D+p)u.$$

## 5.43 Calculation of P.S. by symbolic methods

If $f(x) = f_1(x)+f_2(x)+...+f_n(x)$, then by 5.34 (3) we can seek a P.S. of each of the equations

$$\frac{d^2y}{dx^2}+a\frac{dy}{dx}+by = f_r(x) \quad (r = 1, 2, ..., n),$$

and add the results. We therefore suppose $f(x)$ to be one of the following simple types.

*Case* (i): $f(x)$ *is a polynomial in $x$ of degree $m$*, say $P$. The equation to be considered is then $F(D)y = P$.

*Method.* Write     $y_1 = \dfrac{1}{F(D)}P,$

$$= D^{-\mu}(c_0+c_1D+c_2D^2+...)P$$

by algebraic long division (in practice by resolution of $1/F(D)$ into partial fractions, followed by use of a relation like

$$\frac{1}{\alpha-D} = \frac{1}{\alpha}\left(1+\frac{1}{\alpha}D+\frac{1}{\alpha^2}D^2+...\right),$$

which is verified by summing the geometrical progression). Here $0 \leqslant \mu \leqslant 2$, which covers the cases when

$$F(D) = \begin{cases} D^2+aD+b & (\mu = 0), \\ D(D+a) & (b = 0, \mu = 1), \\ D^2 & (a = b = 0, \mu = 2). \end{cases}$$

Hence     $y_1 = D^{-\mu}(c_0+c_1D+...+c_mD^m)P,$

since $P$ is a polynomial *of degree $m$*.

*Justification.* We have to show that $y_1$ constructed above is a P.S. of $F(D)y = P$. Write

$$F(D) = D^\mu G(D), \quad \text{where} \quad 0 \leqslant \mu \leqslant 2,$$

and     $G(D) = b_0+b_1D+...+b_{2-\mu}D^{2-\mu},$

and consider the polynomial in $x$,

$$G(x) = b_0+b_1x+...+b_{2-\mu}x^{2-\mu}.$$

By long division, or otherwise,

$$\frac{1}{G(x)} = c_0 + c_1 x + \ldots + c_m x^m + \frac{R(x)}{G(x)},$$

where $R(x)$ is a polynomial with all terms *of degree greater than m.* Clearing of fractions,

$$1 = G(x) (c_0 + c_1 x + \ldots + c_m x^m) + R(x).$$

This result holds for all values of $x$: it is an *identity* between two polynomials. On replacing $x$ by $D$ and recalling the principle in 5.41 that polynomials in $D$ are manipulated like algebraic polynomials, we obtain the following *equivalence of operators* (cf. 5.62):

$$1 = G(D) (c_0 + c_1 D + \ldots + c_m D^m) + R(D).$$

Operating on $P$ with both gives

$$P = G(D) (c_0 + c_1 D + \ldots + c_m D^m) P + R(D) P$$

$$= G(D) (c_0 + c_1 D + \ldots + c_m D^m) P \quad \text{since} \quad R(D) P = 0$$

$$= G(D) D^\mu \{ D^{-\mu} (c_0 + c_1 D + \ldots + c_m D^m) \} P$$

$$= F(D) \{ D^{-\mu} (c_0 + c_1 D + \ldots + c_m D^m) \} P$$

$$= F(D) y_1.$$

Hence $y_1$ does satisfy the given equation.

### Examples

(i) *Find a* P.S. *of* $y'' + 4y = x^3$.

Rewriting the equation as $(D^2 + 4) y = x^3$, a P.S. is given symbolically by

$$y = \frac{1}{D^2 + 4} x^3 = \frac{1}{4} \frac{1}{1 + \frac{1}{4} D^2} x^3$$

$$= \frac{1}{4} (1 - \frac{1}{4} D^2 + \frac{1}{16} D^4 - \ldots) x^3$$

$$= \frac{1}{4} (x^3 - \frac{3}{2} x).$$

(ii) $(D^2 + 3D - 4) y = x^3$.

Here $F(D) = -(1 - D)(4 + D)$, so by partial fractions

$$\text{P.S.} = \frac{1}{D^2 + 3D - 4} x^3 = -\frac{1}{5} \left( \frac{1}{1 - D} + \frac{1}{4 + D} \right) x^3$$

$$= -\frac{1}{5} \{ (1 + D + D^2 + D^3 + \ldots) + \frac{1}{4} (1 - \frac{1}{4} D + \frac{1}{16} D^2 - \frac{1}{64} D^3 + \ldots) \} x^3$$

$$= -\frac{1}{5} \{ \frac{5}{4} + \frac{15}{16} D + \frac{65}{64} D^2 + \frac{255}{256} D^3 + \ldots \} x^3$$

$$= -\{ \frac{1}{4} x^3 + \frac{3}{16} 3x^2 + \frac{13}{64} 6x + \frac{51}{256} 6 \}$$

$$= -\frac{1}{4} x^3 - \frac{9}{16} x^2 - \frac{39}{32} x - \frac{153}{128}.$$

*Case* (ii): $f(x) = e^{px} P$, *where $P$ is a polynomial.*

*Method.* From $F(D) y = e^{px} P$ we have, proceeding symbolically:

$$y_1 = \frac{1}{F(D)} \{e^{px} P\}$$

$$= e^{px} \frac{1}{F(D+p)} P$$

by an extension, at present unjustified, of the shift theorem to rational functions of $D$. The calculation is now formally reduced to Case (i).

*Justification.* Suppose that work as in Case (i) gives

$$\frac{1}{F(D+p)} P = u, \quad \text{so that} \quad P = F(D+p) u.$$

Then we have to show that $y_1 = e^{px} u$ satisfies $F(D) y = e^{px} P$.

By the shift theorem,

$$F(D) \{e^{px} u\} = e^{px} F(D+p) u = e^{px} P;$$

so $y_1$ does satisfy the equation.

*Short method when $F(p) \neq 0$.*

If $F(p) \neq 0$, we have

$$\frac{1}{F(x+p)} = \frac{1}{F(p)} + c_1 x + \dots$$

since the two sides must agree when $x = 0$. Hence if $f(x) = k e^{px}$ (i.e. if $P = k$), the P.S. is

$$e^{px} \frac{1}{F(D+p)} k = e^{px} \left( \frac{1}{F(p)} + c_1 D + \dots \right) k = \frac{k e^{px}}{F(p)}.$$

## Examples

(iii) $(D^2 - 3D + 2) y = 8 e^{4x}$.

A P.S. is

$$\frac{1}{D^2 - 3D + 2} 8 e^{4x}$$

$$= e^{4x} \frac{1}{(D+4)^2 - 3(D+4) + 2} 8 \quad \text{by 'extension' of 5.42,}$$

$$= e^{4x} \frac{1}{D^2 + 5D + 6} 8$$

$$= e^{4x} \frac{1}{6} \frac{1}{1 + \frac{1}{6}(D^2 + 5D)} 8$$

$$= \tfrac{1}{6} e^{4x} \{1 - \tfrac{1}{6}(D^2 + 5D) + \dots\} 8$$

$$= \tfrac{1}{6} e^{4x} . 8$$

$$= \tfrac{4}{3} e^{4x}.$$

*Alternatively*, since $F(4) = 16 - 12 + 2 = 6$, the 'short method' at once gives

$$\text{P.S.} = \frac{8e^{4x}}{F(4)} = \tfrac{4}{3}e^{4x}.$$

(iv) $(D^2 - 3D + 2)y = 3e^{2x}$.

Here $F(D) = (D-1)(D-2)$, and $F(2) = 0$, so the 'short method' is not applicable. We have

$$\text{P.S.} = \frac{1}{D^2 - 3D + 2}\, 3e^{2x}$$

$$= \frac{1}{(D-1)(D-2)}\, 3e^{2x}$$

$$= e^{2x}\frac{1}{(D+1)D}\, 3 \quad \text{by 'extension' of the shift theorem (5.42),}$$

$$= e^{2x}\frac{1}{D}\{1 - D + \ldots\}\, 3$$

$$= e^{2x}D^{-1}3 = 3x\, e^{2x},$$

omitting the arbitrary constant of integration since only a P.S. is being sought.

(v) $y'' - 6y' + 9y = 20x^3 e^{3x}$.

$$F(D) = D^2 - 6D + 9 = (D-3)^2,$$

so $\qquad \text{P.S.} = \dfrac{1}{(D-3)^2}\, 20x^3 e^{3x}$

$$= e^{3x}\frac{1}{D^2}\, 20x^3 \quad \text{by 'extension' of 5.42,}$$

$$= x^5 e^{3x} \quad \text{on integrating twice and omitting constants.}$$

*Case (iii)*. If $f(x)$ involves $\cos qx$ and $\sin qx$ linearly, we use

$$e^{i\theta} = \cos\theta + i\sin\theta, \quad e^{-i\theta} = \cos\theta - i\sin\theta$$

(equivalent to Euler's exponential forms, 14.65) to express such trigonometric terms as complex exponentials, thus reducing the work formally to Cases (i), (ii). In practice it is easier to select the real or imaginary part at the *end* of the work, as in the examples below.

### Examples

(vi) $y'' + 4y = \cos 2x$.

Consider instead the equation $(D^2 + 4)y = e^{2ix}$, of which the given equation is the real part.

$$\text{P.S.} = \frac{1}{D^2 + 4}\, e^{2ix}$$

$$= e^{2ix}\frac{1}{(D + 2i)^2 + 4}\, 1$$

$$= e^{2ix}\frac{1}{D(D + 4i)}\, 1$$

$$= e^{2ix} \frac{1}{4i} D^{-1} \left( 1 - \frac{D}{4i} + \ldots \right) 1$$

$$= \frac{1}{4i} e^{2ix} D^{-1} 1 = \frac{1}{4i} e^{2ix} x$$

$$= -\tfrac{1}{4} ix(\cos 2x + i \sin 2x).$$

The P.S. of the given equation is the real part of this result, viz. $\tfrac{1}{4} x \sin 2x$.
*Alternatively*, using complex factors of the operator,

$$\text{P.S.} = \frac{1}{D^2 + 4} e^{2ix}$$

$$= \frac{1}{D - 2i} \left( \frac{1}{D + 2i} e^{2ix} \right)$$

$$= \frac{1}{D - 2i} \left( \frac{1}{4i} e^{2ix} \right) \quad \text{by the 'short method',}$$

$$= e^{2ix} \frac{1}{D} \left( \frac{1}{4i} \right) \quad \text{by the 'extension' of the shift theorem (5.42),}$$

$$= \frac{x}{4i} e^{2ix} \quad \text{as before.}$$

(vii) $(D^2 - 6D + 13) y = e^{3x} \cos 2x$.

$$\text{P.S.} = \frac{1}{D^2 - 6D + 13} e^{3x} \cos 2x$$

$$= e^{3x} \frac{1}{(D+3)^2 - 6(D+3) + 13} \cos 2x \quad (u = \cos 2x \text{ in the shift theorem})$$

$$= e^{3x} \frac{1}{D^2 + 4} \cos 2x$$

$$= e^{3x} \tfrac{1}{4} x \sin 2x \quad \text{as in ex. (vi).}$$

(viii)* $(D^2 - 3D + 4) y = x \sin x$.
Considering $(D^2 - 3D + 4) y = x e^{ix}$,

$$\text{P.S.} = \frac{1}{D^2 - 3D + 4} x e^{ix}$$

$$= e^{ix} \frac{1}{(D+i)^2 - 3(D+i) + 4} x$$

$$= e^{ix} \frac{1}{D^2 + (2i - 3) D + 3(1 - i)} x$$

$$= e^{ix} \frac{1}{3(1 - i)} \left\{ 1 + \frac{D^2 + (2i - 3) D}{3(1 - i)} \right\}^{-1} x$$

$$= \frac{e^{ix}}{3(1 - i)} \left\{ 1 - \frac{D^2 + (2i - 3) D}{3(1 - i)} + \ldots \right\} x$$

$$= \frac{e^{ix}}{3(1 - i)} \left\{ x - \frac{2i - 3}{3(1 - i)} \right\}$$

$$= \tfrac{1}{6} e^{ix} (1 + i) \{ x + \tfrac{1}{6}(i + 5) \}$$

$$= \tfrac{1}{6} e^{ix} \{ (1 + i) x + \tfrac{1}{3}(3i + 2) \}.$$

The required P.S. is the imaginary part of this, viz.

$$\tfrac{1}{6}\{(x+1)\cos x + (x+\tfrac{2}{3})\sin x\}.$$

*The reader should now find by the symbolic method the* P.S. *of some of the differential equations in Exs.* 5 (f), (g); e.g. Ex. 5 (f), nos. 16–28 and Ex. 5 (g), nos. 13–18 *are best done this way.* Ex. 5 (h) *includes some general examples from Mechanics and Electricity.*

### Exercise 5(h)

*Forced harmonic motion:* $\qquad \dfrac{d^2x}{dt^2} + n^2x = a\cos pt.$

**1** (i) If $p \neq n$, show that $x = A\cos nt + B\sin nt + \{a/(n^2-p^2)\}\cos pt$. (The C.F. represents the *free oscillations*, and the P.S. represents the *forced oscillations*. When $p \doteqdot n$, the amplitude of the forced oscillations becomes large—a phenomenon known as *resonance*.)

(ii) Find $x$ if $x = 0$ and $dx/dt = 0$ when $t = 0$.

**2** (i) If $p = n$, prove $x = A\cos nt + B\sin nt + (a/2n)\,t\sin nt$. (The amplitude of the forced oscillations therefore increases without limit when $t \to \infty$.)

(ii) Find $x$ if $x = 0$ and $dx/dt = 0$ when $t = 0$, and verify that the result is the *limit* when $p \to n$ of that in no. 1 (ii).

*Damped harmonic motion:* $\qquad \dfrac{d^2x}{dt^2} + 2k\dfrac{dx}{dt} + n^2x = 0.$

**3** If $n^2 < k^2$, prove $x = A\exp\{-kt + \surd(k^2-n^2)\,t\} + B\exp\{-kt - \surd(k^2-n^2)\,t\}$ (which represents a non-oscillatory motion). (Here $\exp u$ denotes $e^u$.)

**4** If $n^2 = k^2$, prove $x = e^{-kt}(At + B)$ (also non-oscillatory).

**5** If $n^2 > k^2$, prove $x = e^{-kt}\{A\cos(t\surd(n^2-k^2)) + B\sin(t\surd(n^2-k^2))\}$ (representing *damped oscillations*). If $k$ is small compared with $n$, prove that

$$x \doteqdot e^{-kt} \times \text{(the solution which would be obtained if } k = 0\text{)}.$$

(Hence small damping changes the period only slightly, but diminishes the amplitude of successive vibrations in geometrical progression.)

*Damped forced harmonic motion:*

$$\frac{d^2x}{dt^2} + 2k\frac{dx}{dt} + n^2x = a\cos pt.$$

**6** (i) Prove that the P.S. can be written $b\cos(pt + \alpha)$, where

$$b = a\{(n^2-p^2)^2 + 4k^2p^2\}^{-\frac{1}{2}} \quad \text{and} \quad \tan\alpha = 2kp/(p^2-n^2).$$

(This represents the forced oscillations; if $n^2 > k^2$, the free oscillations are given by no. 5.)

(ii) If $a$, $k$, $n$ are constants and $p$ varies, prove that $b$ is greatest when $p = \surd(n^2-2k^2)$. If also $k$ is small compared with $n$, show $\alpha \doteqdot -\tfrac{1}{2}\pi$, $b \doteqdot a/(2kn)$ (*resonance* again).

7   If the free vibrations (no. 5) are in resonance with the applied force, and both are damped, then $\sqrt{(n^2 - k^2)} = p$ and the equation becomes

$$\frac{d^2x}{dt^2} + 2k\frac{dx}{dt} + (k^2 + p^2)\,x = a\,e^{-kt}\cos pt.$$

Show that the P.S. represents an oscillation of amplitude $(a/2p)\,t\,e^{-kt}$, and that this has a maximum value $a/(2pke)$ when $t = 1/k$ and tends to 0 when $t \to \infty$. (Thus if the damping is small, the forced vibrations become large but remain bounded.)

*Electric circuit with self-inductance, capacitance, resistance, and applied electromotive force:*

$$L\frac{d^2x}{dt^2} + R\frac{dx}{dt} + \frac{x}{C} = E\cos pt.$$

(Observe the analogy with no. 6.)

8   Find the C.F. when $CR^2 < 4L$.

9   Find the P.S. in the form $x = b\cos(pt + \alpha)$.

10   When $t \to \infty$, no. 8 shows that the free oscillations die away. Ignoring these, prove that for given $E$, $L$, $p$, $R$, the greatest values of $x$ occur when $\sqrt{(LC)} = 1/p$.

11   If $F(x)$ is a polynomial in $x$, prove $F(D^2)\cos ax = F(-a^2)\cos ax$.

12   Assuming that this result can be used when $F(x)$ is a rational function, find a P.S. of $(D^2 + D + 1)\,y = \cos 2x$ as follows:

$$\frac{1}{D^2 + D + 1}\cos 2x = \frac{1}{-4 + D + 1}\cos 2x = \frac{1}{D - 3}\cos 2x = \frac{D + 3}{D^2 - 9}\cos 2x$$

$$= \frac{D + 3}{-4 - 9}\cos 2x = -\tfrac{1}{13}(3\cos 2x - 2\sin 2x).$$

Verify this solution directly.

*13   Generalise this method to $\{\phi(D^2) + D\psi(D^2)\}\,y = \lambda\cos ax + \mu\sin ax$, where $\phi(x)$, $\psi(x)$ are polynomials in $x$.

## 5.5   Simultaneous linear first-order equations with constant coefficients

We restrict ourselves to the case of two functions $x$, $y$ of $t$ related by two linear equations of first order with constant coefficients.

### Examples

(i) *Solve*     $\dfrac{dx}{dt} + x - y = 25t\,e^t, \quad 2y - \dfrac{dx}{dt} + \dfrac{dy}{dt} = 25e^t.$

Using $D$ for $d/dt$ here, these equations can be written

$$(D + 1)\,x - y \qquad = 25t\,e^t, \tag{i}$$

$$-Dx + (D + 2)\,y = 25e^t. \tag{ii}$$

As in algebraic simultaneous equations, we eliminate one unknown. We choose $y$ for elimination because it is more simply involved in the equations than $x$. First operate on (i) by† $D+2$:

$$(D+2)(D+1)x - (D+2)y = 25(D+2)te^t,$$

i.e. $(D^2+3D+2)x - (D+2)y = 25(3t+1)e^t.$ (iii)

Now add (ii) and (iii):
$$(D^2+2D+2)x = 25(3t+2)e^t,$$

which is a linear equation for $x$. The C.F. is $e^{-t}(A\cos t + B\sin t)$, and a P.S. is found to be $(15t-2)e^t$. Hence

$$x = e^{-t}(A\cos t + B\sin t) + e^t(15t-2).$$

From (i), $y = (D+1)x - 25te^t$;
and since

$$(D+1)x = e^{-t}\{(-A\sin t + B\cos t) - (A\cos t + B\sin t) + (A\cos t + B\sin t)\}$$

$$+ e^t\{(15t-2) + 15 + (15t-2)\}$$

$$= e^{-t}(B\cos t - A\sin t) + e^t(30t+11),$$

$$\therefore \quad y = e^{-t}(B\cos t - A\sin t) + e^t(5t+11).$$

We could find $y$ independently of $x$ by operating on (i) with $D$, on (ii) with $D+1$, and adding. This would lead to a linear second-order equation for $y$, whose solution would introduce two further constants of integration, say $A'$, $B'$. The relations between $A, B, A', B'$ would then have to be found by substituting the solutions for $x$, $y$ in both of the equations (i), (ii). This method, clearly involving more work, should be avoided.

(ii) *Solve* $\quad 2\dfrac{dx}{dt} + \dfrac{dy}{dt} = x, \quad \dfrac{dx}{dt} + \dfrac{dy}{dt} = 5x + y + e^{-t}.$

The equations are $\quad (2D-1)x + Dy = 0,$ (i)

$$(D-5)x + (D-1)y = e^{-t}.$$ (ii)

To eliminate $y$, operate on (i) with $D-1$, and on (ii) with $D$:

$$(2D^2 - 3D + 1)x + D(D-1)y = 0,$$

$$(D^2 - 5D)x + D(D-1)y = -e^{-t}.$$

Subtract: $\quad (D^2 + 2D + 1)x = e^{-t}.$

The C.F. is $x = (A+Bt)e^{-t}$, and the P.S. is $\frac{1}{2}t^2 e^{-t}$; hence

$$x = e^{-t}(A + Bt + \tfrac{1}{2}t^2).$$

Before finding $y$, *eliminate $dy/dt$ from the given equations* by subtracting:

$$\frac{dx}{dt} = -4x - y - e^{-t},$$

$$\therefore \quad y = -\frac{dx}{dt} - 4x - e^{-t}.$$

Now substitute for $x$:

$$y = -e^{-t}[\{B - A + (1-B)t - \tfrac{1}{2}t^2\} + \{4A + 4Bt + 2t^2\} + 1]$$

$$= -e^{-t}\{(3A + B + 1) + (3B+1)t + \tfrac{3}{2}t^2\}.$$

† In full this means deriving (i) wo $t$, and adding twice (i) to the result. Thus symbolic methods are convenient but not essential in examples like this.

Had we substituted directly for $x$ in either of the given equations, we should have obtained a *differential* equation for $y$, and hence a redundant constant of integration. The method used avoids this, and involves no more than derivations wo $t$.

## Exercise 5(*i*)

*Solve the following pairs of simultaneous differential equations.*

1 $\dfrac{dx}{dt} + y = \sin t,$     2 $\dfrac{dx}{dt} + x - y = e^t,$     3 $\dfrac{dx}{dt} + 2x + y = 0,$

  $\dfrac{dy}{dt} + x = \cos t.$       $\dfrac{dy}{dt} + y - x = 0.$      $\dfrac{dy}{dt} + x + 2y = 0.$

4 $\left.\begin{array}{l}\dfrac{dx}{dt} + 3x - 2y = 1, \\[3mm] \dfrac{dy}{dt} - 2x + 3y = e^t\end{array}\right\}$ if $x = 0$ and $y = 0$ when $t = 0$.

5 $5\dfrac{dx}{dt} + 3\dfrac{dy}{dt} - 11x - 7y = e^t,$

  $3\dfrac{dx}{dt} + 2\dfrac{dy}{dt} - 7x - 5y = e^{2t}.$

6 $\left.\begin{array}{l}3\dfrac{dx}{dt} + \dfrac{dy}{dt} + 2x = 3\cos t, \\[3mm] \dfrac{dx}{dt} + 2\dfrac{dy}{dt} + 3y = 7\cos t - 4\sin t\end{array}\right\}$ if $x = 0$ and $y = 0$ when $t = 0$.

7 Writing $u = x + y$ and $v = x - y$, show by adding and subtracting the equations in no. 2 that $du/dt = e^t$ and $dv/dt + 2v = e^t$, and solve each of these. Hence find $x$ and $y$.

8 Solve no. 3 by the method of no. 7.

9 Solve     $5\dfrac{d^2x}{dt^2} + \dfrac{dy}{dt} + 2x = 4\cos t,$    $3\dfrac{dx}{dt} + y = 4t\cos t.$

*10 Solve     $m\dfrac{du}{dt} = eE - evH,$    $m\dfrac{dv}{dt} = euH,$

where $m, e, E, H$ are constants (the equations of motion of a particle of mass $m$ and charge $e$ in perpendicular electric and magnetic fields of strengths $E, H$). If also $u = \dot{x}$, $v = \dot{y}$, and $x = y = u = v = 0$ when $t = 0$, prove

$$x = \frac{E}{\omega H}(1 - \cos \omega t) \quad \text{and} \quad y = \frac{E}{\omega H}(\omega t - \sin \omega t),$$

where $\omega = eH/m$. (Hence the particle moves along a cycloid: 1.61, ex.)

## 5.6 Some linear second-order equations with variable coefficients

### 5.61 Euler's 'homogeneous' equation

This is of the form
$$x^2\frac{d^2y}{dx^2} + ax\frac{dy}{dx} + by = f(x),$$

where $a$, $b$ are constants. It can be reduced to an equation with constant coefficients by the substitution $x = e^t$; for then

$$\frac{dy}{dx} = \frac{dy}{dt} \Big/ \frac{dx}{dt} = e^{-t}\frac{dy}{dt},$$

and

$$\frac{d^2y}{dx^2} = \frac{d}{dx}\left(\frac{dy}{dx}\right) = \left\{\frac{d}{dt}\left(e^{-t}\frac{dy}{dt}\right)\right\} \Big/ \frac{dx}{dt} = e^{-t}\left\{e^{-t}\frac{d^2y}{dt^2} - e^{-t}\frac{dy}{dt}\right\}$$

$$= e^{-2t}\left\{\frac{d^2y}{dt^2} - \frac{dy}{dt}\right\},$$

and the equation becomes

$$\left(\frac{d^2y}{dt^2} - \frac{dy}{dt}\right) + a\frac{dy}{dt} + by = f(e^t),$$

i.e.

$$\frac{d^2y}{dt^2} + (a-1)\frac{dy}{dt} + by = f(e^t),$$

which gives $y$ as a function of $t$.

### Examples

(i)  $4x^2\dfrac{d^2y}{dx^2} + x\dfrac{dy}{dx} - y = x + \log x.$

Put $x = e^t$; then as above the equation becomes

$$4\left(\frac{d^2y}{dt^2} - \frac{dy}{dt}\right) + \frac{dy}{dt} - y = e^t + t,$$

i.e.

$$4\frac{d^2y}{dt^2} - 3\frac{dy}{dt} - y = e^t + t.$$

The C.F. is $A e^t + B e^{-\frac{1}{4}t}$, and the P.S. is found to be $\frac{1}{5}t e^t - t + 3$. Hence

$$y = A e^t + B e^{-\frac{1}{4}t} + \tfrac{1}{5}t e^t - t + 3$$

$$= Ax + Bx^{-\frac{1}{4}} + \tfrac{1}{5}x\log x - \log x + 3.$$

(ii) The equation

$$(px+q)^2\frac{d^2y}{dx^2} + a(px+q)\frac{dy}{dx} + by = f(x)$$

can be reduced to the standard homogeneous form by putting $z = px + q$, for then

$$\frac{dy}{dx} = \frac{dy}{dz}\frac{dz}{dx} = p\frac{dy}{dz} \quad \text{and} \quad \frac{d^2y}{dx^2} = p^2\frac{d^2y}{dz^2}.$$

### 5.62 Remarks on the use of equivalent operators

If $y$ is any derivable function of $x$, and $x = e^t$, then by 3.2 (4) $y$ is a derivable function of $t$ and we showed above that

$$\frac{dy}{dx} = e^{-t}\frac{dy}{dt}.$$

The operations $d/dx$ on $y$ (as a function of $x$) and $e^{-t}d/dt$ on $y$ (as a function of $t$) therefore have the same effect. We say that these operators are *equivalent*, and write

$$\frac{d}{dx} = e^{-t}\frac{d}{dt}.$$

Since the operator acts on the function which follows it, we must avoid writing the second operator as $(d/dt)\,e^{-t}$; this expression already denotes the *function* which is the derivative of $e^{-t}$.

By applying these operators to $dy/dx$ we have

$$\frac{d^2y}{dx^2} = e^{-t}\frac{d}{dt}\left\{e^{-t}\frac{dy}{dt}\right\} = e^{-2t}\left\{\frac{d^2y}{dt^2} - \frac{dy}{dt}\right\},$$

as before. The use of other equivalent operators will be illustrated in 5.63.

## 5.63 Solution of other equations by a given substitution

### Examples

(i) *Solve*     $(1+x^2)^3\dfrac{d^2y}{dx^2} + 2x(1+x^2)^2\dfrac{dy}{dx} - 9(1+x^2)\,y = x^2 - 1$

*by putting* $x = \tan\theta$.

We have     $\dfrac{dy}{dx} = \dfrac{dy}{d\theta}\bigg/\dfrac{dx}{d\theta} = \dfrac{dy}{d\theta}\bigg/\sec^2\theta = \dfrac{1}{1+x^2}\dfrac{dy}{d\theta},$

so                   $(1+x^2)\dfrac{dy}{dx} = \dfrac{dy}{d\theta}.$            (a)

Deriving both sides of this wo $x$,

$$(1+x^2)\frac{d^2y}{dx^2} + 2x\frac{dy}{dx} = \frac{d^2y}{d\theta^2}\bigg/\frac{dx}{d\theta} = \frac{1}{1+x^2}\frac{d^2y}{d\theta^2}. \qquad (b)$$

*Alternatively*, applying the equivalent operators

$$(1+x^2)\frac{d}{dx} = \frac{d}{d\theta},$$

obtained from (a), to both sides of (a), we get

$$(1+x^2)\frac{d}{dx}\left\{(1+x^2)\frac{dy}{dx}\right\} = \frac{d^2y}{d\theta^2},$$

i.e.         $(1+x^2)^2\dfrac{d^2y}{dx^2} + 2x(1+x^2)\dfrac{dy}{dx} = \dfrac{d^2y}{d\theta^2}.$       (c)

Hence from either (b) or (c) the given equation becomes

$$(1+x^2)\frac{d^2y}{d\theta^2} - 9(1+x^2)\,y = x^2 - 1;$$

and since $x = \tan \theta$, we have

$$\frac{d^2y}{d\theta^2} - 9y = \frac{\tan^2\theta - 1}{\sec^2\theta} = \sin^2\theta - \cos^2\theta = -\cos 2\theta.$$

The C.F. is $y = A e^{3\theta} + B e^{-3\theta}$, and the P.S. is found to be $\frac{1}{13}\cos 2\theta$.

$$\therefore \quad y = A e^{3\theta} + B e^{-3\theta} + \tfrac{1}{13}\cos 2\theta$$

$$= A e^{3\tan^{-1}x} + B e^{-3\tan^{-1}x} + \frac{1-x^2}{13(1+x^2)},$$

since $\qquad \cos 2\theta = 2\cos^2\theta - 1 = \dfrac{2}{\sec^2\theta} - 1 = \dfrac{2}{1+x^2} - 1.$

(ii) *Solve* $\qquad \cos x \dfrac{d^2y}{dx^2} + \sin x \dfrac{dy}{dx} + 4y \cos^3 x = 2\cos^5 x$

*by putting* $x = \sin^{-1} u.$

$$\frac{dy}{dx} = \frac{dy}{du}\frac{du}{dx} = \frac{dy}{du}\cos x = \frac{dy}{du}\sqrt{(1-u^2)};$$

so, by using equivalent operators,

$$\frac{d^2y}{dx^2} = \sqrt{(1-u^2)}\frac{d}{du}\left\{\sqrt{(1-u^2)}\frac{dy}{du}\right\}$$

$$= (1-u^2)\frac{d^2y}{du^2} - u\frac{dy}{du}.$$

The equation becomes

$$\sqrt{(1-u^2)}\left\{(1-u^2)\frac{d^2y}{du^2} - u\frac{dy}{du}\right\} + u\left\{\frac{dy}{du}\sqrt{(1-u^2)}\right\} + 4(1-u^2)^{\frac{3}{2}}y = 2(1-u^2)^{\frac{5}{2}},$$

i.e. $\qquad\qquad\qquad \dfrac{d^2y}{du^2} + 4y = 2(1-u^2).$

We find as usual that the G.S. is

$$y = A\cos 2u + B\sin 2u + \tfrac{1}{4}(3 - 2u^2)$$

$$= A\cos(2\sin x) + B\sin(2\sin x) + \tfrac{1}{4}(3 - 2\sin^2 x).$$

(iii) *Transform the equation*

$$x^2y'' + (3x^2 + 4x)y' + (2x^2 + 6x + 2)y = e^x$$

*by the substitution* $x^2y = z$. *Hence solve the equation.*
Deriving the relation $x^2y = z$ twice wo $x$,

$$x^2y' + 2xy = z' \quad \text{and} \quad x^2y'' + 4xy' + 2y = z''.$$

Hence the equation becomes

$$z'' + 3z' + 2z = e^x,$$

and the G.S. of this is $z = A e^{-x} + B e^{-2x} + \tfrac{1}{6}e^x$. Hence

$$y = \frac{1}{x^2}\left(A e^{-x} + B e^{-2x} + \tfrac{1}{6}e^x\right)$$

is the G.S. of the given equation.

(iv) *Express the equation*

$$\frac{d^2y}{dx^2} + 3\left(\frac{dy}{dx}\right)^2 = (2x - y)\left(\frac{dy}{dx}\right)^3$$

*as a differential equation for $x$ as a function of $y$, and hence solve it.*

From 3.53, ex. (iv),

$$\frac{d^2y}{dx^2} = -\frac{d^2x}{dy^2}\bigg/\left(\frac{dx}{dy}\right)^3.$$

Hence the given equation becomes

$$-\frac{d^2x}{dy^2}\bigg/\left(\frac{dx}{dy}\right)^3 + 3\bigg/\left(\frac{dx}{dy}\right)^2 = (2x - y)\bigg/\left(\frac{dx}{dy}\right)^3,$$

i.e.

$$\frac{d^2x}{dy^2} - 3\frac{dx}{dy} + 2x = y,$$

which is a linear second-order equation with constant coefficients whose solution is found to be

$$x = A\,e^y + B\,e^{2y} + \tfrac{1}{2}y + \tfrac{3}{4}.$$

## 5.64 General case: one integral belonging to the C.F. known

Consider the general linear second-order equation

$$\frac{d^2y}{dx^2} + a(x)\frac{dy}{dx} + b(x)\,y = f(x),$$

and suppose that a solution $y = u(x)$ of

$$\frac{d^2y}{dx^2} + a(x)\frac{dy}{dx} + b(x)\,y = 0$$

is known. Then the given equation can be reduced to a linear one of first order in $v'$ by putting $y = vu$; for by hypothesis

$$u'' + a(x)\,u' + b(x)\,u = 0,$$

and

$$y' = vu' + v'u, \quad y'' = vu'' + 2v'u' + v''u,$$

so the given equation becomes

$$(vu'' + 2v'u' + v''u) + a(x)\,(vu' + v'u) + b(x)\,vu = f(x),$$

i.e.

$$v(u'' + a(x)\,u' + b(x)\,u) + v'(2u' + a(x)\,u) + v''u = f(x).$$

Hence

$$uv'' + \left(2u' + a(x)\,u\right)v' = f(x),$$

i.e.

$$u\frac{dw}{dx} + \left(2u' + a(x)\,u\right)w = f(x),$$

where $w = v'$; this can be solved by using an integrating factor, or otherwise.

## Example

$$x^2y'' - (x^2 + 2x)\,y' + (x + 2)\,y = x^3\,e^x.$$

We easily verify that $y = x$ satisfies

$$x^2y'' - (x^2 + 2x)\,y' + (x + 2)\,y = 0.$$

Putting $y = vx$, so that $y' = v + v'x$ and $y'' = 2v' + v''x$, we have

$$x^2(2v' + v''x) - (x^2 + 2x)\,(v + v'x) + (x + 2)\,vx = x^3\,e^x,$$

which reduces to             $x^3v'' - x^3v' = x^3\,e^x,$

i.e.                          $v'' - v' = e^x.$

Integrating,     $v' - v = e^x + c,$   i.e. $\dfrac{d}{dx}(v\,e^{-x}) = 1 + c\,e^{-x},$

$$\therefore \quad e^{-x}v = x - c\,e^{-x} + A, \quad v = (x + A)\,e^x - c,$$

and hence             $y = e^x(Ax + x^2) + Bx,$

where $B = -c$.

## Exercise 5($j$)

*Solve the following homogeneous equations.*

1  $x^2\dfrac{d^2y}{dx^2} - 3x\dfrac{dy}{dx} + 4y = x^2\log x.$          2  $x^2\dfrac{d^2y}{dx^2} + x\dfrac{dy}{dx} - 4y = x^2.$

3  $9x^2\dfrac{d^2y}{dx^2} + 3x\dfrac{dy}{dx} + y = \log x$   if $y = 0$ and $\dfrac{dy}{dx} = 0$ when $x = 1.$

4  $x^2\dfrac{d^2y}{dx^2} - 2x\dfrac{dy}{dx} - 4y = x^3 + 2\log x.$

5  $(1 + 2x)^2\dfrac{d^2y}{dx^2} - 6(1 + 2x)\dfrac{dy}{dx} + 16y = 8\log(1 + 2x).$

*6  Solve the simultaneous equations

$$t^2\dfrac{d^2x}{dt^2} + t\dfrac{dx}{dt} - 4y = 0, \quad t^2\dfrac{d^2y}{dt^2} + t\dfrac{dy}{dt} - 4x = 0.$$

*Solve the following by using the substitution given.*

7  $(1 - x^2)\dfrac{d^2y}{dx^2} - x\dfrac{dy}{dx} + y = x;\; x = \sin\theta.$

8  $(1 + x^2)\dfrac{d^2y}{dx^2} + x\dfrac{dy}{dx} = 3(\text{sh}^{-1}x)^2;\; x = \text{sh}\,\theta.$

9  $(1 + x^2)^2\dfrac{d^2y}{dx^2} + 2x(1 + x^2)\dfrac{dy}{dx} + 4y = 0;\; x = \tan\tfrac{1}{2}\theta.$

10  $x^2\dfrac{d^2y}{dx^2} + 2x(2x + 3)\dfrac{dy}{dx} + 3(x^2 + 4x + 2)\,y = 0;\; y = zx^{-3}.$

11  Show that the constant $n$ can be chosen so that the substitution $y = x^n z$ reduces the equation $x^2y'' + 4x(x + 1)\,y' + 2(4x + 1)\,y = \cos x$ to the form $z'' + az' + bz = \cos x$, where $a$, $b$ are constants. Hence solve the given equation.

*Solve the following, in which an integral belonging to the* c.f. *is given.*

12  $x^3y'' - 2xy' + 2y = 0$; $y = x$.

13  $(x+1)y'' - 2xy' + (x-1)y = 0$; $y = e^x$.

14  $xy'' - 2(x+1)y' + (x+2)y = (x-2)e^{2x}$; $y = e^x$.

15  $x^2y'' + xy' - 9y = x^3$; try $y = kx^3$, *or* treat as 'homogeneous' type.

16  $x^2y'' + x^2y' + (x-2)y = x^3 e^x$; $y = x^{-1}$.

17  Prove that $ay'' + by' + cy = 0$ is satisfied by $y = e^x$ if and only if $a + b + c \equiv 0$, and by $y = x$ if and only if $b + cx \equiv 0$.

*18  *Riccati's equation* $dy/dx = P(x) + Q(x)y + R(x)y^2$. Prove that the substitution $y = -u'/Ru$ reduces this to the *linear* second-order equation

$$Ru'' - (QR + R')u' + PR^2u = 0.$$

*19  Solve $x^2y' + 3 - 3xy + x^2y^2 = 0$.

## 5.7  Some geometrical applications

We now illustrate how differential equations arise geometrically.

### 5.71  Definitions

Given a plane curve, let $P$ be a point at which there is a definite tangent (2.15); $PN$ is the ordinate of $P$. If the tangent and normal at $P$ cut $Ox$ at $T$, $G$, then $NT$ is the *subtangent* and $NG$ is the *subnormal* to the curve at $P$. Also $PT$, $PG$ are called the *lengths* of the tangent and normal at $P$.

Since $\tan\psi = y'$, it is clear from fig. 55 (where $x$, $y$, $y'$ are positive) that

$$NT = y/y', \quad NG = yy',$$
$$PT = y\sqrt{(1 + y'^2)}/y',$$
$$PG = y\sqrt{(1 + y'^2)}.$$

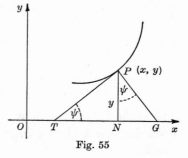

Fig. 55

In the general case these results are obtained from the equations of the tangent and normal at $P$ by finding the intercepts on $Ox$; the lengths are *signed* magnitudes. Thus, the tangent at $(x_1, y_1)$ has equation

$$y - y_1 = y_1'(x - x_1),$$

and cuts $Ox$ where $y = 0$ and $-y_1 = y_1'(x - x_1)$, i.e. $x_1 - x = y_1/y_1'$; and $TN = x_1 - x$.

### Example

For the parabola $y^2 = 4ax$ (16.11) we have $2yy' = 4a$, so that the subnormal $NG$ is $2a$.

It is often stated that 'the subnormal of *any* parabola is constant'; but this is true only if $N$, $G$ are understood to be points on the axis of symmetry.

## 5.72 Orthogonal families of curves

For a fixed value of $\alpha$ (supposed independent of $x$ and $y$), the equation $F(x, y, \alpha) = 0$ represents a plane curve (1.61). For different values of $\alpha$ we thus obtain a *family* or *system* of curves; $\alpha$ 'labels' the individual members of the family. For example, $x^2 + y^2 = \alpha^2$ represents a family of concentric circles of various radii; $y^2 = 4ax$ represents a family of parabolas situated as in 5.11, ex. (i). The curves of the family may not all be of the same geometrical 'type': thus $x^2 + \alpha y^2 = 1$ represents ellipses for $\alpha > 0$, hyperbolas for $\alpha < 0$, and a pair of lines for $\alpha = 0$, as will be shown later.

The *angle of intersection* of two curves is the angle between the tangents at the common point.

Two intersecting families of curves are *orthogonal*, or *orthogonal trajectories* of one another, if each member of one cuts each member of the other at right-angles.

Let $F(x, y, \alpha) = 0$ be the equation of the given family. We first get its differential equation by eliminating $\alpha$ (5.1); suppose the result is $f(x, y, y') = 0$. When $P(x, y)$ is given, this equation determines the gradient $y'$ at $P$ of the member(s) of the family through $P$, say $y' = \phi(x, y)$. The gradient of the orthogonal curve through $P$ will therefore be $-1/\phi(x, y)$; hence on this curve,

$$y' = -\frac{1}{\phi(x, y)}, \quad \text{i.e.} \quad -\frac{1}{y'} = \phi(x, y), \quad \text{so} \quad f\left(x, y, -\frac{1}{y'}\right) = 0.$$

This is the differential equation of the orthogonal family; its solution, involving one arbitrary constant, is the equation of the orthogonal family.

### Examples

(i) *Find the orthogonal trajectories of the family of concentric circles* $x^2 + y^2 = a^2$.

The differential equation of the family is $x + yy' = 0$. The gradient of the circle through $P(x, y)$ is therefore $y' = -x/y$. Hence the gradient of the orthogonal curve through $P$ is $y/x$, so that on this curve we have

$$y' = \frac{y}{x}, \quad \text{i.e.} \quad xy' - y = 0, \quad \text{i.e.} \quad \frac{d}{dx}\left(\frac{y}{x}\right) = 0.$$

The orthogonal family is $y/x = c$, a system of straight lines through $O$.

The result is evident geometrically: the diameters of a circle cut the curve orthogonally.

(ii) *Prove that the system*

$$\frac{x^2}{a^2+\lambda}+\frac{y^2}{b^2+\lambda}=1,$$

*where $\lambda$ is the parameter, is identical with the orthogonal system.* (For a geometrical interpretation see Ex. 18 $(d)$, nos. 18, 19.)

To form the differential equation of the family we must eliminate $\lambda$. First clearing of fractions,

$$\lambda^2+\lambda(a^2+b^2-x^2-y^2)+(a^2b^2-b^2x^2-a^2y^2)=0.$$

Derive wo $x$:     $\lambda(x+yy')+(b^2x+a^2yy')=0,$

so     $$\lambda=-\frac{b^2x+a^2yy'}{x+yy'}$$

and     $$a^2+\lambda=\frac{(a^2-b^2)x}{x+yy'},\quad b^2+\lambda=\frac{(b^2-a^2)yy'}{x+yy'}.$$

The required differential equation is therefore

$$\frac{x(x+yy')}{a^2-b^2}+\frac{y(x+yy')}{(b^2-a^2)y'}=1,$$

i.e.     $$(x+yy')\left(x-\frac{y}{y'}\right)=a^2-b^2.$$

Since this equation is unaltered by replacing $y'$ by $-1/y'$, the orthogonal family is the same as the given family.

## Exercise 5(k)

*Find the differential equation of curves having the following properties, and hence find these curves by solving (notation of 5.71).*

1 Subnormal is constant $k$.     2 Subtangent is constant $k$.

3 Projection of the ordinate on the normal is constant $k$.

4 $OP=PT$.    5 $OG=OP$.    6 $OT=OP$.    7 $ON=NG$.

8 $OT=kPN$.    9 $PG=k$.    10 $PG=kPN^2$.

*Find the orthogonal trajectories of the following families, a being the parameter.*

11 $xy=a^2$.     12 $y^2=4ax$.     13 $ay^2=x^3$.

14 $x^{\frac{2}{3}}+y^{\frac{2}{3}}=a^{\frac{2}{3}}$.     15 $y^3-3x^2y=a$.     16 $y=axe^x$.

17 (i) Prove that the system of parabolas $y^2=4a(x+a)$ is identical with the orthogonal system.

*(ii) Show that the curves corresponding to $a=a_1,a_2$ will intersect only if $a_1,a_2$ have opposite signs. [Consider $y^2=4a(x+a)$ as a quadratic in $a$.]

18 Prove that in general there are two curves which satisfy the differential equation

$$y\left(\frac{dy}{dx}\right)^2-2x\frac{dy}{dx}+1=0$$

and pass through a given point. Prove that these curves coincide if and only if the given point lies on the parabola $y=x^2$.

**19** Find the inclinations to $Ox$ of the two curves through $(3, 5)$ which satisfy the equation in no. 18.

**20** Find the locus of a point such that the two curves through it in no. 18 cut at right-angles.

## Miscellaneous Exercise 5($l$)

*Eliminate the parameters from the following functions.*

**1** $y = c \operatorname{ch}(x/c)$.

**2** $y = A \cos^{-1} x + B$.

**3** $y = A \cos^{-1} x + B \sin^{-1} x$, and explain why the result is the same as for no. 2.

**4** $y = Ax \sin(x^{-1} + B)$.

*Obtain the differential equation of*

**5** all straight lines.

**6** all circles which touch both coordinate axes.

**7** all tangents to the rectangular hyperbola $xy = 1$. [The general point is $(t, 1/t)$.]

*For general practice the reader may solve a random selection from the following differential equations (nos. 8–52).*

**8** $\dfrac{dy}{dx} = \dfrac{1+y^2}{1+x^2}$.

**9** $x\dfrac{dy}{dx} - xy = y$.

**10** $(x+y)^2 \dfrac{dy}{dx} = 1$.

**11** $(x-y)\dfrac{dy}{dx} = x+y$.

**12** $(1-4y')y = (4+y')x$.

**13** $(x-y)^2 \dfrac{dy}{dx} = (x-y-1)^2$.

**14** $(x^2-y^2)\dfrac{dy}{dx} = 2xy$.

**15** $(3x+5y+1)\dfrac{dy}{dx} = 7x-3y+2$.

**16** $x\dfrac{dy}{dx} = y - x\cos^2\dfrac{y}{x}$.

**17** $\dfrac{dy}{dx} - x = xy$.

**18** $\dfrac{dy}{dx} + \dfrac{y}{x} = x^2$.

**19** $(x^2+1)\dfrac{dy}{dx} - xy = 1$.

**20** $\dfrac{dy}{dx} + 2xy = 2x^3 y^3$.

**21** $(xy^3+y)\dfrac{dy}{dx} = \dfrac{1}{x}$.

**22** $x\dfrac{dy}{dx} + y = y^2 \log x$.

**23** $\dfrac{dy}{dx}\sin x - y\cos x = e^x \sin^2 x$.

**\*24** $y = px + \tfrac{1}{3}p^3$.

**\*25** $y = px + e^p$.

**26** $y'' + y'^2 = 0$.

**27** $(x^2+1)y'' + xy' = 0$.

**28** $yy'' + y'^2 = 1$.

**29** $xyy'' + xy'^2 + yy' = 0$.

**30** $y'' - y' - 2y = 0$.

**31** $y'' - 2y' + y = 0$.

**32** $y''' - 3y'' = 0$.

**\*33** $y''' + 3y'' + 3y' + y = 0$ [put $u = y e^x$ and calculate $u'''$].

**34** $y'' - 3y' - 4y = 10\cos 2x$.

**35** $y'' - 5y' + 6y = 4x^2 e^x$.

**36** $y'' - 4y' + 3y = x^3$.

**37** $y'' - 10y' + 29y = e^{5x}\sin 2x$.

**38** $x^2 y'' - 20y = 7x^3$.

**39** $x^2 y'' - xy' + 2y = x\log x$.

**40** $x^2 y'' - 2xy' + 2y = x^3\cos(\log x)$.

**41** $y'' + (2/x)y' = 0$.

**\*42** $x^2 y''' + 3xy'' + y' = \log x$.

**43** $y'' + y'\tan x - y\cos^2 x = 0$ [put $u = \sin x$].

**44** $(1 + x^2)^3 y'' + 2x(1 + x^2)^2 y' + (1 + x^2) y = x$ [put $x = \tan\theta$].

**45** $\dfrac{dx}{dt} + 2x = 2y$,

**46** $\dfrac{dx}{dt} - 3\dfrac{dy}{dt} + 2x + 6y = 0$,

$\dfrac{dy}{dt} + y = 3x$.

$\dfrac{dx}{dt} + 3\dfrac{dy}{dt} - 2x + 6y = 0$.

**47** $\dfrac{dx}{dt} + 5x - 2y = 40e^t$,

$\dfrac{dy}{dt} - x + 6y = 27e^{2t}$.

**48** $t\dfrac{dx}{dt} + y = 0$, $\quad t\dfrac{dy}{dt} + x = \log t$.

**\*49** $\dfrac{dx}{dt} = y$, $\quad \dfrac{dy}{dt} = z$, $\quad \dfrac{dz}{dt} = x$.

**\*50** $\dfrac{d^2 x}{dt^2} + 3\dfrac{dy}{dt} - 4x + 6y = 0$, $\quad \dfrac{d^2 y}{dt^2} + \dfrac{dx}{dt} - 2x + 4y = 0$.

[For nos. 49, 50 see no. 56.]

**51** $xy'' - (2x + 1)y' + (x + 1)y = 0$; particular solution $y = e^x$.

**52** $x(1 + x)y'' = 2y' + 2y$; particular solution $y = 1/(1 + x)$.

**53** Transform $4x^2 y'' + 4xy' + (4x^2 - 1)y = 0$ by the substitution $y = zx^{-\frac{1}{2}}$, and hence solve the equation.

**54** Show that the constant $n$ can be chosen so that the substitution $y = x^n z$ transforms the equation $x^2 y'' + 2x(x + 2)y' + 2(x + 1)^2 y = e^{-x}\cos x$ into one with constant coefficients. Hence solve the given equation.

**55** Prove that a differential equation of the form $p^2 + p\phi(x, y) - 1 = 0$, where $p = dy/dx$, represents a system of plane curves such that in general two pass through every point and intersect at right-angles. Find the system of curves when $\phi(x, y) = -2y/x$.

**\*56** Show that $y''' + ay'' + by' + cy = 0$ is the same equation as

$$\frac{d}{dx}\{y'' - (k + l)y' + kly\} = m\{y'' - (k + l)y' + kly\}$$

if and only if $k + l + m = -a$, $lm + mk + kl = b$ and $klm = -c$, i.e. $k, l, m$ are the roots of $t^3 + at^2 + bt + c = 0$. Putting $z = y'' - (k + l)y' + kly$, show that the solution of the given equation is reduced to solving one of second order in $y$.

**\*57** If $\int f(x)\,dx = \log\{1 + f(x)\}$, find $f(x)$.

**\*58** Verify that $\cos nx$ and $\sin nx$ are both integrating factors of $y'' + n^2 y = f(x)$ and obtain two first integrals. By eliminating $y'$, obtain the G.s. and express it in the form

$$y = A\cos nx + B\sin nx + \frac{1}{n}\int_0^x f(t)\sin n(x - t)\,dt.$$

*59  Solve $y'' + y = \sec x$ by the method of no. 58, and show that it will succeed also when $f(x) = \operatorname{cosec} x$, $\tan x$, or $\cot x$.

*60  (i) If $u$, $v$, $w$ are functions of $x$, and constants $a$, $b$, $c$ not all zero can be found such that $au + bv + cw \equiv 0$ for all $x$, prove that the determinant

$$W = \begin{vmatrix} u & v & w \\ u' & v' & w' \\ u'' & v'' & w'' \end{vmatrix}$$

(the *Wronskian* of $u$, $v$, $w$) is zero for all $x$.

(ii) *Conversely*, if $W \equiv 0$, prove that $u$, $v$, $w$ are connected by a linear relation. [Consider the linear second-order equation in $y$ obtained by replacing $u, u', u''$ by $y, y', y''$ in $W$: it has solutions $y = u, v, w$; use 5.32 (2), II.]

# 6

## SOME THEOREMS OF THE DIFFERENTIAL CALCULUS

### 6.1 Two properties of continuous functions

This chapter contains a sequence of theorems, some of which are 'obvious geometrically'; they are of considerable importance. All of them depend for their proof on the following two properties of continuous functions, which we will state and then explain.

(1) *If $f(x)$ is continuous in $a \leqslant x \leqslant b$, then it is bounded in this range.*

(2) *If $f(x)$ is continuous in $a \leqslant x \leqslant b$, then it possesses a greatest and a least value when $x$ varies in this range.*

Property (1) means that, for all values of $x$ satisfying $a \leqslant x \leqslant b$, there exists a positive number $K$ such that $|f(x)| \leqslant K$. In other words (cf. 2.4(3)), $f(x)$ cannot tend to $+\infty$ or to $-\infty$ or oscillate infinitely when $x$ varies in $a \leqslant x \leqslant b$: it is 'bounded' by the number $K$ in that its numerical value cannot exceed $K$; we can write $-K \leqslant f(x) \leqslant K$ and call $K$ an *upper bound* for $f(x)$, and $-K$ a *lower bound*. Clearly any fixed positive number greater than $K$ would also serve as a bound for $f(x)$ in $a \leqslant x \leqslant b$.

It may happen that some numbers less than $K$ would serve as bounds for $f(x)$. It can be proved that, of all numbers which serve as upper bounds for $f(x)$, there is a *smallest* such number, say $M$; i.e. $f(x) \leqslant M$ for all $x$ for which $a \leqslant x \leqslant b$, but if $H$ is any number just less than $M$, there is at least one value $x_1$ of $x$ in the interval such that $f(x_1) > H$. Similarly, of all lower bounds for $f(x)$ there is a largest, say $m$, which has the property that $f(x) \geqslant m$ for $a \leqslant x \leqslant b$, while if $h$ is any number just greater than $m$, there is at least one value $x = x_2$ in the interval for which $f(x_2) < h$. These numbers $M$, $m$ are called *the* upper, lower bounds of $f(x)$ in $a \leqslant x \leqslant b$. Functions other than continuous ones may possess upper and lower bounds in a given interval; on the other hand they may not. Property (1) implies that for a continuous function these numbers *always* exist.

Property (2) asserts that, for continuous functions, these bounds are actually *values* taken by the function for suitable values of $x$ in the interval: $M$ is the greatest and $m$ the least value of $f(x)$ for $a \leqslant x \leqslant b$.

The property is usually expressed by saying that 'a continuous function attains its bounds in any closed interval'. This may or may not be true for a discontinuous function: even if it possesses upper and lower bounds $M$, $m$, there may be no $x$ in the range for which $f(x) = M$ or for which $f(x) = m$.

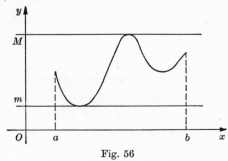

Fig. 56

The properties (1), (2) and others can be deduced from the definition of 'continuity' given in 2.61 with the help of some fundamental theorems on bounds of a function. We shall assume them, because to give rigorous proofs would deflect us too far from the course in view.

In the following theorems we use the *closed* interval $a \leqslant x \leqslant b$ for continuity of $f(x)$ (see 2.61, Remark), and the *open* interval $a < x < b$ for derivability. This is reasonable because (3.11, Remark), if $f(x)$ is defined only in $a \leqslant x \leqslant b$, then $f'(x)$ will not be defined at $x = a$ or at $x = b$; hence existence in the open interval is as much as we can demand of $f'(x)$.

## 6.2   Rolle's theorem

**6.21**  *If* (i) $f(x)$ *is continuous for* $a \leqslant x \leqslant b$, (ii) $f'(x)$ *exists for* $a < x < b$, (iii) $f(a) = f(b)$, *then there is at least one number* $\xi$ *for which* $a < \xi < b$ *and* $f'(\xi) = 0$.

Geometrically the result is obvious (figs. 57, 58): a continuous curve with a definite tangent at every point *between* $A$ and $B$, where it meets the line $y = k$, must have a tangent parallel to $y = k$ at some intermediate point or points. The reader who is willing to accept the theorem on these grounds may omit the following proof.

*First suppose* $f(a) = f(b) = 0$ *in hypothesis* (iii). Since $f(x)$ is continuous, then by 6.1, Property (1), the function has upper and lower bounds $M$, $m$ in $a \leqslant x \leqslant b$. As $f(a) = 0 = f(b)$, therefore $M \geqslant 0$ and $m \leqslant 0$; we consider the following three cases.

(a) $M > 0$. By Property (2) of 6.1, there is a number $\xi$ such that $a \leqslant \xi \leqslant b$ and $f(\xi) = M$. Since $f(a) = f(b) = 0$ and $M > 0$, $\xi$ must be different from $a$ and $b$, so that $a < \xi < b$. We now prove that $f'(\xi) = 0$.

Let $h$ be any positive number such that $\xi + h$ lies in $a \leqslant x \leqslant b$. Then since $f(\xi + h) \leqslant M$, it follows that $f(\xi + h) \leqslant f(\xi)$, and so

$$\frac{f(\xi + h) - f(\xi)}{h} \leqslant 0.$$

Letting $h \to 0+$, we have $f'(\xi) \leqslant 0$ by using hypothesis (ii) of the theorem together with a variant of the lemma in 3.63.

Similarly, if $h > 0$ and $\xi - h$ lies in $a \leqslant x \leqslant b$, then $f(\xi - h) \leqslant f(\xi)$, so that

$$\frac{f(\xi - h) - f(\xi)}{-h} \geqslant 0.$$

Letting $h \to 0+$ gives $f'(\xi) \geqslant 0$. Comparing the two conclusions shows $f'(\xi) = 0$.

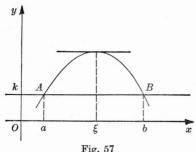

Fig. 57

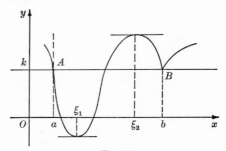

Fig. 58

(b) $M = 0$, *but* $m < 0$. There is a number $\xi$ for which $f(\xi) = m$, and as before, $a < \xi < b$; $f(\xi + h) \geqslant m = f(\xi)$, from which $f'(\xi) = 0$ by an argument similar to the above.

(c) If $M = 0 = m$, then $f(x) = 0$ everywhere in $a \leqslant x \leqslant b$, so $f'(x) = 0$ everywhere in $a < x < b$; any number in $a < x < b$ would do for $\xi$.

*Secondly, suppose* $f(a) = f(b) = k$ *in* (iii). Apply the result just proved to the function $\phi(x) = f(x) - k$, which clearly satisfies hypotheses (i) and (ii) and also $\phi(a) = \phi(b) = 0$. Then Rolle's theorem follows.

*Remark.* The number $\xi$ depends on the function $f(x)$ and on the interval; thus for a given $f(x)$, $\xi$ is in general a function of $a$ and $b$. See Ex. 6 (a), no. 3.

*Definition.* If $f(x)$ is continuous for $a \leqslant x \leqslant b$ and derivable for $a < x < b$, we say that $f(x)$ *satisfies the Rolle conditions* in $a \leqslant x \leqslant b$.

**6.22** If the Rolle conditions are not satisfied, there may not be a number $\xi$ with the required property. Figs. 59, 60 illustrate cases in which hypotheses (i), (ii) respectively are not satisfied; but fig. 61, in which neither is satisfied, shows that the result of the theorem may still be true—the conditions are sufficient but not necessary.

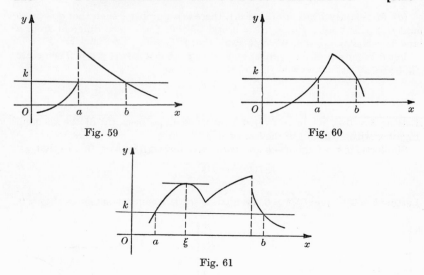

Fig. 59                              Fig. 60

Fig. 61

## 6.23  Application to algebraic equations

If $f(x)$ is a polynomial, then $f(x)$ and $f'(x)$ certainly satisfy the Rolle conditions in any interval. Taking $a$, $b$ of 6.21 to be roots of $f(x) = 0$, we have (cf. Ex. 3 $(e)$, no. 14):

    **(1)** *Between two roots of $f(x) = 0$ lies at least one root of $f'(x) = 0$.*

Further (cf. Ex. 3 $(e)$, no. 15)

    **(2)** *Not more than one root of $f(x) = 0$ can lie between consecutive roots $\alpha$, $\beta$ of $f'(x) = 0$.*

If there were two, then by (1) there would be a root of $f'(x) = 0$ between them, so that $\alpha$, $\beta$ would not be *consecutive* roots of $f'(x) = 0$. (There will be *exactly* one root of $f(x) = 0$ between $\alpha$ and $\beta$ if and only if $f(\alpha), f(\beta)$ have opposite signs, by 2.65.)

The result (2) is helpful in locating the roots of an equation $f(x) = 0$ when those of $f'(x) = 0$ are easily found; see 13.62 (3) for some worked examples.  Finally

    **(3)** *If $f'(x) = 0$ has $s$ roots, then $f(x) = 0$ cannot have more than $s+1$ roots.*

For the roots of $f'(x) = 0$ must separate those of $f(x) = 0$, by (1).

## 6.3  Lagrange's mean value theorem

In 3.81 we conjectured on geometrical grounds the following *mean value theorem.*

*If $f(x)$ satisfies the Rolle conditions in $a \leqslant x \leqslant b$, then there is at least one number $\xi$ for which $a < \xi < b$ and*

$$f(b) - f(a) = (b-a) f'(\xi).$$

## 6.31 Linear approximation to $f(x)$

We may prove this theorem by applying Rolle's theorem to the error obtained by making a linear approximation to $f(x)$ in $a \leqslant x \leqslant b$.

When $a$ is given, any linear function $g(x)$ can be put in the form

$$g(x) = p_0 + p_1(x-a).$$

We choose the coefficients $p_0$, $p_1$ so that it takes the same values as $f(x)$ at the ends of the interval: $g(a) = f(a)$ and $g(b) = f(b)$, so that

$$p_0 = f(a) \quad \text{and} \quad p_0 + p_1(b-a) = f(b).$$

Hence 
$$g(x) = f(a) + \frac{f(b) - f(a)}{b - a}(x-a).$$

(Geometrically, $g(x) = 0$ is the equation of the chord $AB$ in fig. 41 of 3.81; we now consider the error caused by replacing the arc $AB$ by the chord $AB$.)

At any point $x$ in $a < x < b$, the error involved in the approximation $f(x) \doteqdot g(x)$ is $\phi(x) = f(x) - g(x)$. Now $\phi(a) = 0$ and $\phi(b) = 0$, and $\phi(x)$ satisfies the Rolle conditions in $a \leqslant x \leqslant b$ because $f(x)$ and $g(x)$ do. Hence there is a number $\xi$ for which $a < \xi < b$ and $\phi'(\xi) = 0$, i.e.

$$0 = f'(\xi) - \frac{f(b) - f(a)}{b - a},$$

from which the mean value theorem follows. Incidentally we have justified the geometrical conjecture in 3.81; Rolle's theorem is the case when $AB$ is parallel to $Ox$.

**6.32** *We may set out the preceding proof as follows.* Consider the function

$$\phi(x) = f(x) - f(a) + A(x-a).$$

Clearly $\phi(a) = 0$; we choose $A$ so that also $\phi(b) = 0$:

$$0 = f(b) - f(a) + A(b-a). \tag{i}$$

The hypotheses of Rolle's theorem are now satisfied by $\phi(x)$, so that there is a number $\xi$ for which $a < \xi < b$ and $\phi'(\xi) = 0$, i.e. $0 = f'(\xi) + A$. Hence from (i), 
$$f(b) = f(a) + (b-a) f'(\xi). \tag{ii}$$

## 6.33 Alternative versions of the theorem

(1) If $0 < \theta < 1$, then $a + \theta(b - a)$ lies between $a$ and $b$, so that for a suitable $\theta$ we can write $\xi = a + \theta(b - a)$. Putting $b = a + h$, the result (ii) takes the form

$$f(a + h) = f(a) + hf'(a + \theta h) \quad (0 < \theta < 1). \tag{iii}$$

*Remark.* Since $\xi$ depends on $a$ and $b$ (6.21, Remark), hence $\theta$ in (iii) will be some function of $a$ and $h$.

(2) Since $b - h = a$, we also have from (ii) that

$$f(b) = f(b - h) + hf'(b - (1 - \theta)h).$$

Putting $h = -h_1$, $1 - \theta = \theta_1$ (so that $0 < \theta_1 < 1$) and rearranging,

$$f(b + h_1) = f(b) + h_1 f'(b + \theta_1 h_1),$$

which is the mean value theorem for the interval $b + h_1 \leqslant x \leqslant b$ where $h_1 < 0$. Hence *the result* (iii) *remains valid when* $h < 0$; in the statement of the theorem the only modification required is the replacement of $a \leqslant x \leqslant a + h$ by $a + h \leqslant x \leqslant a$.

## Example

Take $f(x) = \log x$, $a > 0$, $a + h > 0$. Then (iii) becomes

$$\log(a + h) = \log a + \frac{h}{a + \theta h},$$

i.e.

$$\log\left(1 + \frac{h}{a}\right) = \frac{h}{a + \theta h},$$

from which we find

$$\theta = \frac{1}{\log\left(1 + \dfrac{h}{a}\right)} - \frac{a}{h}.$$

Put $u = h/a$; then since $a + h > 0$ and $h \neq 0$, also $u > -1$ and $u \neq 0$, and

$$\theta = \frac{1}{\log(1 + u)} - \frac{1}{u}.$$

Since $0 < \theta < 1$,

$$\frac{1}{u} < \frac{1}{\log(1 + u)} < 1 + \frac{1}{u},$$

i.e.

$$\frac{u}{1 + u} < \log(1 + u) < u,$$

the logarithmic inequality (4.43 (1)).

This example illustrates how the mean value theorem can be regarded as a *disguised inequality* owing to the condition $0 < \theta < 1$.

Applications of the theorem to increasing functions and to inequalities were given in 3.83. A few further examples appear in Ex. 6 (a).

### Exercise 6(a)

*Find ξ when Rolle's theorem is applied to*

1 $x(1-x)^n$ in $0 \leqslant x \leqslant 1$, $n$ being a positive integer.

2 $e^x \sin x$ in $0 \leqslant x \leqslant 2\pi$.        3 $(x-a)(b-x)^2$ in $a \leqslant x \leqslant b$.

4 If $f(x)$ satisfies the Rolle conditions in $a \leqslant x \leqslant b$ and $f(a) = 0 = f(b)$, prove that for any given number $\lambda$ the equation $f'(x) + \lambda f(x) = 0$ has at least one root between $a$ and $b$. [Consider $e^{\lambda x} f(x)$. Cf. Ex. 3 (e), no. 16.]

*Find ξ when the mean value theorem is applied to*

5 $\cos x$ in $0 \leqslant x \leqslant \frac{1}{2}\pi$.    6 $x^n$ in $0 \leqslant x \leqslant 1$, $n$ being an integer $> 1$.

*Find θ when the mean value theorem is applied for the interval $a \leqslant x \leqslant a+h$ to*

7 $x^2$.             8 $x^3$, and verify that $\theta \to \frac{1}{2}$ when $h \to 0$.

9 $e^x$.            10 Deduce an inequality from the result of no. 9.

11 Explain why the mean value theorem fails when $f(x) = x^{-1}$ and $a < 0 < a+h$.

12 *Mean value theorem for integrals.* If $f(x)$ is continuous in $a \leqslant x \leqslant b$, apply the mean value theorem to $\phi(x) = \int_a^x f(t)\,dt$ to show that there is a number $\xi$ for which $a < \xi < b$ and $\int_a^b f(x)\,dx = (b-a)f(\xi)$. [Use 4.15(7).]

13 If $f(x)$ and $g(x)$ satisfy the Rolle conditions in $a \leqslant x \leqslant b$, prove that the functions $Af(x) + Bg(x)$ ($A, B$ constant), $f(x)g(x)$, $f(x)/g(x)$ do also, provided that in the last case $g(x)$ is never zero for $a \leqslant x \leqslant b$.

14 If $f'(x)$ satisfies the Rolle conditions in $a \leqslant x \leqslant b$, prove that $f(x)$ does also. [See 3.52.]

*15 If $f(a) = f(c) = f(b)$, where $a < c < b$, and if $f'(x)$ satisfies the Rolle conditions in $a \leqslant x \leqslant b$, prove there is a number $\xi$ for which $a < \xi < b$ and $f''(\xi) = 0$. [Use Rolle's theorem three times.]

*16 If $\{f(b)-f(c)\}/(b-c) = \{f(c)-f(a)\}/(c-a)$, where $a < c < b$, and $f'(x)$ satisfies the Rolle conditions in $a \leqslant x \leqslant b$, prove there is a number $\xi$ for which $a < \xi < b$ and $f''(\xi) = 0$. Interpret geometrically.

*17 If $x > 1$, prove $x + 2 + \log x > \sqrt{(x^2 + 10x - 2)}$. [Use 3.83, Corollary 2.]

*18 If $0 < x < y$, prove $\dfrac{\log(1+y)}{\log y} < \dfrac{\log(1+x)}{\log x}$. $\left[ \text{Prove } \dfrac{d}{dt}\left\{\dfrac{\log(1+t)}{\log t}\right\} < 0. \right]$

*19 If $f(0) = 0$ and $f'(x)$ is an increasing function for $x > 0$, prove that when $x > 0$, (i) $f(x) = xf'(\theta x)$, $0 < \theta < 1$; (ii) $f(x) < xf'(x)$; (iii) $d\{f(x)/x\}/dx > 0$, and deduce that $f(x)/x$ is increasing for $x > 0$.

### 6.4  The second mean value theorem

### 6.41  An algebraic lemma

In 6.31 we proved the first mean value theorem by considering a linear approximation to $f(x)$. By using other polynomials, we obtain mean value theorems of higher orders.

*Any polynomial of degree $n$ in $x$ can be written as a polynomial of degree $n$ in $(x-a)$.*

If the given polynomial is

$$g(x) = a_0 + a_1 x + a_2 x^2 + \dots + a_n x^n,$$

put† $x = a + y$ and expand each term. The term of highest degree in $y$ will arise only from $a_n(a+y)^n$, and hence is $a_n y^n$. We obtain an expression of the form

$$p_0 + p_1 y + p_2 y^2 + \dots + p_n y^n,$$

which is a polynomial in $x - a = y$ of degree $n$.

### 6.42  Quadratic approximation to $f(x)$

For given $a$, every quadratic can be written

$$g(x) = p_0 + p_1(x-a) + p_2(x-a)^2.$$

We choose the coefficients $p_0$, $p_1$, $p_2$ so that $g(a) = f(a)$, $g(b) = f(b)$, and $g'(a) = f'(a)$. (Geometrically, the quadratic curve passes through the extremities $A, B$ of the given arc and has the same gradient at $A$.) Then

$$p_0 = f(a),$$

$$p_0 + p_1(b-a) + p_2(b-a)^2 = f(b), \quad \text{(i)}$$

and since

$$g'(x) = p_1 + 2p_2(x-a), \quad p_1 = f'(a).$$

The 'error' function $\phi(x) = f(x) - g(x)$ clearly satisfies

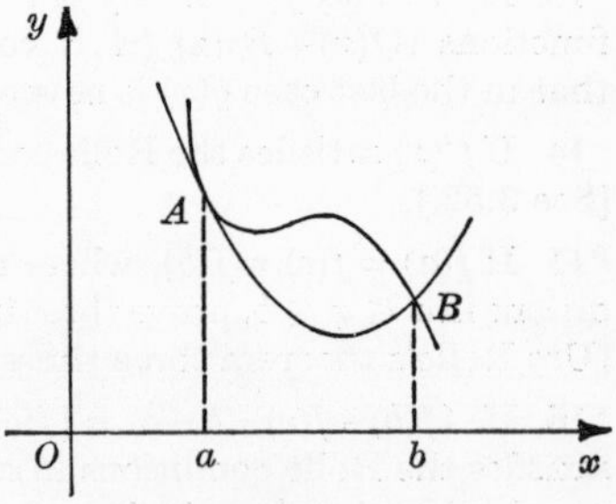

$$\phi(a) = 0 = \phi(b).$$

Fig. 6 2

Hence by Rolle's theorem there is a number $\xi_1$ for which $a < \xi_1 < b$ and $\phi'(\xi_1) = 0$.

Since also $\phi'(a) = 0$, we may apply Rolle's theorem to $\phi'(x)$ for $a \leqslant x \leqslant \xi_1$ to show that there is a number $\xi$ for which $a < \xi < \xi_1$ and

† Another proof is indicated in the Remark of 10.53.

$\phi''(\xi) = 0$. Now $\phi''(x) = f''(x) - 2p_2$, and hence $p_2 = \frac{1}{2}f''(\xi)$; therefore by (i),

$$f(b) = f(a) + (b-a)f'(a) + \tfrac{1}{2}(b-a)^2 f''(\xi). \tag{ii}$$

For the preceding applications of 6.21 to be valid, the Rolle conditions must be satisfied by $\phi(x)$ in $a \leqslant x \leqslant b$ and by $\phi'(x)$ in $a \leqslant x \leqslant \xi_1$. Since we know only that $\xi_1$ lies *somewhere* in $a < x < b$, we shall require $\phi'(x)$, and consequently $f'(x)$, to satisfy the conditions in $a \leqslant x \leqslant b$. As in 6.33(1), we may write $b = a+h$ and $\xi = a+\theta h$ $(0 < \theta < 1)$ since $\xi$ certainly lies between $a$ and $b$. Hence—

*If $f'(x)$ satisfies the Rolle conditions in $a \leqslant x \leqslant a+h$, then*

$$f(a+h) = f(a) + hf'(a) + \tfrac{1}{2}h^2 f''(a+\theta h), \tag{iii}$$

*where $0 < \theta < 1$, and in general $\theta$ depends on $a$ and $h$.*

*Remarks*

($\alpha$) The result (iii) remains true when $h < 0$ if we write the interval as $a+h \leqslant x \leqslant a$. For, the same proof now shows that $\xi_1, \xi$ satisfy $a+h < \xi_1 < a$, $\xi_1 < \xi < a$ respectively, and therefore $a+h < \xi < a$. We can still write $\xi = a+\theta h$ where $0 < \theta < 1$, since $h < 0$.

($\beta$) From fig. 36 in 3.11 we see that

$$f(a+h) - f(a) - hf'(a) = NQ - MP - PR \tan TPR$$
$$= RQ - RT = TQ.$$

Hence $TQ = \tfrac{1}{2}h^2 f''(a+\theta h)$. Compare the comments on the approximations (i)–(iii) of 3.91.

We give some applications of (iii) in 6.7.

## 6.5   Theorems of Taylor and Maclaurin

### 6.51   Approximation to $f(x)$ by a polynomial of degree $n$

The first and second mean value theorems are particular cases of the following mean value theorem of order $n$, called *Taylor's theorem*.

*If $f^{(n-1)}(x)$ satisfies the Rolle conditions for $a \leqslant x \leqslant a+h$, then*

$$f(a+h) = f(a) + hf'(a) + \frac{h^2}{2!}f''(a) + \ldots + \frac{h^{n-1}}{(n-1)!}f^{(n-1)}(a) + \frac{h^n}{n!}f^{(n)}(a+\theta h),$$

*where $0 < \theta < 1$ and $\theta$ depends in general on $a$, $h$ and $n$.*

*Proof.* Choose the coefficients in the polynomial

$$g(x) = p_0 + p_1(x-a) + p_2(x-a)^2 + \ldots + p_n(x-a)^n$$

so that
$$g(a) = f(a), \quad g'(a) = f'(a), \quad \ldots, \quad g^{(n-1)}(a) = f^{(n-1)}(a),$$
$$g(a+h) = f(a+h). \quad \text{(i)}$$

If $\psi(x) = (x-a)^m$ where $m$ is a positive integer, then it is easily verified that
$$\psi^{(r)}(a) = \begin{cases} 0 & \text{if } r \neq m, \\ m! & \text{if } r = m. \end{cases}$$

It follows that
$$g^{(r)}(a) = r! \, p_r,$$

and hence $\quad p_0 = f(a), \quad p_1 = f'(a), \quad 2! \, p_2 = f''(a),$
$$\ldots, \quad (n-1)! \, p_{n-1} = f^{(n-1)}(a), \quad \text{(ii)}$$

and
$$p_0 + p_1 h + p_2 h^2 + \ldots + p_n h^n = f(a+h). \quad \text{(iii)}$$

Write $\phi(x) = f(x) - g(x)$; then by (i)
$$\phi(a+h) = \phi(a) = \phi'(a) = \ldots = \phi^{(n-1)}(a) = 0.$$

Also $\phi(x)$ and its first $n-1$ derivatives satisfy the Rolle conditions in $a \leqslant x \leqslant a+h$ since by the hypothesis and 3.52 this is true for $f^{(n-1)}(x)$, $f^{(n-2)}(x), \ldots, f'(x), f(x)$.

Since $\phi(a+h) = 0 = \phi(a)$, there is a number $\xi_1$ for which
$$a < \xi_1 < a+h \quad \text{and} \quad \phi'(\xi_1) = 0.$$

Since also $\phi'(a) = 0$, there is a number $\xi_2$ such that $a < \xi_2 < \xi_1$ and $\phi''(\xi_2) = 0$. Proceeding similarly, there is a number $\xi_{n-1}$ for which $\phi^{(n-1)}(\xi_{n-1}) = 0$. Lastly, since also $\phi^{(n-1)}(a) = 0$, there is a number $\xi$ for which $a < \xi < \xi_{n-1}$ and $\phi^{(n)}(\xi) = 0$. We know that
$$a < \xi < \xi_{n-1} < \ldots < \xi_2 < \xi_1 < a+h,$$

so we may put $\xi = a + \theta h$, where $0 < \theta < 1$.

Now
$$\phi^{(n)}(x) = f^{(n)}(x) - g^{(n)}(x) = f^{(n)}(x) - n! \, p_n,$$

hence
$$f^{(n)}(\xi) = n! \, p_n.$$

From this with equations (ii), the result follows by using (iii).

With a modification as in Remark ($\alpha$) of 6.42, the result holds for $h < 0$.

## 6.52 Maclaurin's form of the theorem

If we replace $a$ by $0$ and $h$ by $x$ in Taylor's theorem, we get the following.

*If $f^{(n-1)}(t)$ satisfies the Rolle conditions in $0 \leqslant t \leqslant x$, then*

$$f(x) = f(0) + xf'(0) + \frac{x^2}{2!}f''(0) + \ldots + \frac{x^{n-1}}{(n-1)!}f^{(n-1)}(0) + \frac{x^n}{n!}f^{(n)}(\theta x),$$

*where $0 < \theta < 1$ and in general $\theta$ depends on $x$ and $n$.*

Conversely, starting from Maclaurin's form, we could apply it to the function $g(x) = f(a+x)$ in $0 \leqslant t \leqslant h$ and arrive at Taylor's theorem:

$$g(x) = g(0) + xg'(0) + \frac{x^2}{2!}g''(0) + \ldots + \frac{x^{n-1}}{(n-1)!}g^{(n-1)}(0) + \frac{x^n}{n!}g^{(n)}(\theta x)$$

$$= f(a) + xf'(a) + \frac{x^2}{2!}f''(a) + \ldots + \frac{x^{n-1}}{(n-1)!}f^{(n-1)}(a) + \frac{x^n}{n!}f^{(n)}(a+\theta x),$$

since
$$g^{(r)}(x) = \frac{d^r}{dx^r}f(a+x) = f^{(r)}(a+x).$$

Thus the two forms are equivalent.

## 6.53 Closeness of the polynomial approximation

When the conditions of Taylor's (or Maclaurin's) theorem are satisfied, we can apply the theorem to approximate to the given function by a polynomial of degree $n-1$. The closeness of the approximation is measured by the size of the 'remainder term'

$$R = \frac{h^n}{n!}f^{(n)}(a+\theta h), \quad \text{or} \quad \frac{x^n}{n!}f^{(n)}(\theta x).$$

Considering the Maclaurin form, we see that $R$ depends on $n$ and $x$; we may be able to make $R$ small ($a$) by increasing $n$ when $x$ is given (i.e. taking more terms); ($b$) by decreasing $x$ when $n$ is given (i.e. narrowing the range).

Keeping $n$ fixed, suppose $f^{(n)}(\theta x)$ is bounded for all $x$ sufficiently small, say for $|x| < \eta$. Then $|R/x^n| \leqslant K$, where $K$ is independent of $x$; we say that $R$ is *of order* $x^n$ when $x$ is small, and write $R = O(x^n)$. Standing alone, $O(x^n)$ means 'any function which is of order $x^n$'. We can now write *either*

$$f(x) = f(0) + xf'(0) + \ldots + \frac{x^{n-1}}{(n-1)!}f^{(n-1)}(0) + O(x^n),$$

*or*

$$f(x) \doteqdot f(0) + xf'(0) + \ldots + \frac{x^{n-1}}{(n-1)!}f^{(n-1)}(0) \quad \text{correct to order } n-1.$$

## Example

If $f(x) = e^x$, then $f^{(r)}(x) = e^x$ and $f^{(r)}(0) = 1$ for each $r$. Maclaurin's theorem gives

$$e^x = 1 + x + \frac{x^2}{2!} + \ldots + \frac{x^{n-1}}{(n-1)!} + \frac{x^n}{n!} e^{\theta x} \quad (0 < \theta < 1).$$

Hence

$$e^x \doteqdot 1 + x + \frac{x^2}{2!} + \ldots + \frac{x^{n-1}}{(n-1)!}. \tag{i}$$

Since $e^x$ possesses derivatives of all orders, $n$ can be as large as we please; and the conditions of the theorem are satisfied for all $x$.

(a) When $x$ is given, we know by 2.74 that $x^n/n! \to 0$ when $n \to \infty$, which shows that the more terms we take, the better does the polynomial (i) approximate to $e^x$; and this happens to be true here for any $x$. (In general it will be true of $f(x)$ only for $x$ within some definite range.) In particular, by taking $x = 1$ we get

$$e \doteqdot 1 + 1 + \frac{1}{2!} + \frac{1}{3!} + \ldots + \frac{1}{(n-1)!}.$$

Successive terms in this expression are easy to calculate, because the $(r+1)$th term is $1/r! = (1/r)\{1/(r-1)!\} = (1/r) \times r$th term. With $n = 11$ we find that $e \doteqdot 2 \cdot 718282$, correct to 6 places of decimals. A much rougher result was obtained in 4.32 (5).

(b) When $n$ is given, the approximation (i) is *correct to order* $x^{n-1}$, and improves the smaller $|x|$ becomes; for $e^{\theta x} < e^{\eta}$ when $|x| < \eta$.

In order to use Taylor's or Maclaurin's theorem directly in this way, it is essential to be able to find $f^{(n)}(a)$ or $f^{(n)}(0)$ explicitly in terms of $n$. Consequently, in 6.6 we obtain formulae for the $n$th derivative of some elementary functions.† For most functions $f(x)$ the formula for $f^{(n)}(x)$ cannot be found explicitly; but we may still employ the theorem to infer approximations of a given order, expecting these to hold 'for all $x$ sufficiently small' (see Ex. 6 (b), nos. 26–30).

We return to Maclaurin's theorem in Ch. 12 when we discuss infinite series; there we shall sometimes require a different expression for the remainder term.

## 6.54 Other expressions for the remainder term

(1) *Alternative proof of Taylor's theorem.*

For a proof involving only one application of Rolle's theorem like that in 6.32, we could begin by considering the function

$$\psi(x) = F(x) - A(b-x)^p \quad (p > 0),$$

where $F(x) = f(b) - f(x) - (b-x)f'(x) - \ldots - \frac{(b-x)^{n-1}}{(n-1)!} f^{(n-1)}(x)$

$$= f(b) - f(x) - \sum_{r=1}^{n-1} \frac{(b-x)^r}{r!} f^{(r)}(x).$$

† In Ch. 3 this would have seemed an academic exercise; we now see its significance.

Clearly $\psi(b) = 0$; we choose $A$ so that $\psi(a) = 0$:

$$0 = F(a) - A(b-a)^p.$$

By the hypothesis of Taylor's theorem and 3.52, $\psi(x)$ satisfies the Rolle conditions in $a \leqslant x \leqslant b$. Hence there is a number $\xi$ for which $a < \xi < b$ and $\psi'(\xi) = 0$. Now

$$\psi'(x) = -f'(x) + \sum_{r=1}^{n-1} \frac{(b-x)^{r-1}}{(r-1)!} f^{(r)}(x) - \sum_{r=1}^{n-1} \frac{(b-x)^r}{r!} f^{(r+1)}(x) + pA(b-x)^{p-1}$$

$$= \sum_{r=1}^{n-2} \frac{(b-x)^r}{r!} f^{(r+1)}(x) - \sum_{r=1}^{n-1} \frac{(b-x)^r}{r!} f^{(r+1)}(x) + pA(b-x)^{p-1}$$

$$= pA(b-x)^{p-1} - \frac{(b-x)^{n-1}}{(n-1)!} f^{(n)}(x)$$

$$= (b-x)^{p-1} \left\{ pA - \frac{(b-x)^{n-p} f^{(n)}(x)}{(n-1)!} \right\}.$$

From $\psi'(\xi) = 0$ and the fact that $b - \xi \neq 0$ we now have

$$A = \frac{(b-\xi)^{n-p} f^{(n)}(\xi)}{(n-1)! \, p},$$

and so
$$F(a) = \frac{(b-a)^p (b-\xi)^{n-p} f^{(n)}(\xi)}{(n-1)! \, p}.$$

Writing $b = a+h$ and $\xi = a+\theta h$ (where in general $\theta$ will depend on $a, h, n, p$), we find

$$F(a) = \frac{h^n (1-\theta)^{n-p}}{(n-1)! \, p} f^{(n)}(a+\theta h),$$

so that from our definition of $F(x)$ we have

$$f(a+h) = f(a) + hf'(a) + \ldots + \frac{h^{n-1}}{(n-1)!} f^{(n-1)}(a) + \frac{h^n (1-\theta)^{n-p}}{(n-1)! \, p} f^{(n)}(a+\theta h).$$

The last term is known as *Schlömilch's remainder*. In particular, by taking $p = n$ we recover that given in 6.51 (*Lagrange's remainder*); and by taking $p = 1$ we have *Cauchy's remainder*

$$\frac{h^n (1-\theta)^{n-1}}{(n-1)!} f^{(n)}(a+\theta h),$$

which will be used in Ch. 12.

**(2)** *Remainder as a definite integral.*

From (1) we have $F(b) = 0$ and

$$F'(x) = -\frac{(b-x)^{n-1}}{(n-1)!} f^{(n)}(x).$$

If we suppose $f^{(n)}(x)$ is *continuous* for $a \leqslant x \leqslant b$, then

$$F(a) = F(b) - \int_a^b F'(x)\,dx$$

$$= \frac{1}{(n-1)!} \int_a^b (b-x)^{n-1} f^{(n)}(x)\,dx.$$

Putting $b = a+h$ and $x = a+th$, we find

$$F(a) = \frac{h^n}{(n-1)!} \int_0^1 (1-t)^{n-1} f^{(n)}(a+th)\,dt.$$

This form of the remainder term does not involve an undetermined function $\theta$; but we have assumed more about $f^{(n)}(x)$ than in 6.51.

### 6.6    Calculation of some $n$th derivatives

### 6.61 Elementary functions

(i) $x^m$. Here $\quad f'(x) = mx^{m-1}, \quad f''(x) = m(m-1)x^{m-2}, \quad \dots,$

$$f^{(n)}(x) = m(m-1)\dots(m-n+1)x^{m-n}.$$

(a) In particular, *if $m$ is a positive integer*,

$$f^{(n)}(x) = \begin{cases} \dfrac{m!}{(m-n)!}\,x^{m-n} & (n \leqslant m), \\[2mm] m! & (n = m), \\[2mm] 0 & (n > m). \end{cases}$$

(b) When $m = -1, f(x) = 1/x$ and $f^{(n)}(x) = (-1)^n\,n!/x^{n+1}$.

(c) If $f(x) = (ax+b)^m$, then similarly

$$f^{(n)}(x) = a^n m(m-1)\dots(m-n+1)\,(ax+b)^{m-n}.$$

(ii) $\log x$. Since $f'(x) = x^{-1}$, we have by (i) (b)

$$f^{(n)}(x) = \frac{d^{n-1}}{dx^{n-1}}\left(\frac{1}{x}\right) = (-1)^{n-1}\frac{(n-1)!}{x^n}.$$

(iii) $e^{ax}$. Clearly $\qquad f^{(n)}(x) = a^n e^{ax}$.

(iv) $\sin(ax+b)$.

$$f'(x) = a\cos(ax+b) = a\sin(ax+b+\tfrac{1}{2}\pi).$$

Thus each successive derivation will add $\tfrac{1}{2}\pi$ to the 'angle'; hence

$$f^{(n)}(x) = a^n \sin(ax+b+\tfrac{1}{2}n\pi).$$

(v) $\cos(ax+b)$. Similarly

$$f^{(n)}(x) = a^n \cos(ax+b+\tfrac{1}{2}n\pi).$$

(vi) $e^{ax}\sin bx$.

$$f'(x) = e^{ax}(a\sin bx + b\cos bx).$$

This can be written in the form $e^{ax} R \sin(bx+\theta)$, i.e.

$$e^{ax}(R\cos\theta\sin bx + R\sin\theta\cos bx)$$

if we choose $R$ and $\theta$ so that $R\cos\theta = a$ and $R\sin\theta = b$. These give $R = \sqrt{(a^2+b^2)}$, and $\theta$ is determined from

$$\cos\theta : \sin\theta : 1 = a : b : R.$$

The next derivation gives

$$f''(x) = R\frac{d}{dx}\{e^{ax}\sin(bx+\theta)\} = R^2 e^{ax}\sin(bx+2\theta),$$

with the *same* $R$, $\theta$ as before since the equations for determining these are independent of $x$. Hence we see that

$$f^{(n)}(x) = R^n e^{ax}\sin(bx+n\theta).$$

Similarly, $\qquad \dfrac{d^n}{dx^n}\{e^{ax}\cos bx\} = R^n e^{ax}\cos(bx+n\theta),$

where $R$, $\theta$ have the *same* values as before.

*(vii) $\tan^{-1}x$. (This example requires complex numbers and de Moivre's theorem.)

Since $\qquad f'(x) = \dfrac{1}{1+x^2}, \quad$ we have $\quad f^{(n)}(x) = \dfrac{d^{n-1}}{dx^{n-1}}\left(\dfrac{1}{1+x^2}\right).$

Now $\qquad\qquad\qquad \dfrac{1}{1+x^2} = \dfrac{1}{2i}\left(\dfrac{1}{x-i} - \dfrac{1}{x+i}\right),$

hence by (i)(b)

$$\frac{d^{n-1}}{dx^{n-1}}\frac{1}{1+x^2} = (-1)^{n-1}(n-1)!\,\frac{1}{2i}\left\{\frac{1}{(x-i)^n} - \frac{1}{(x+i)^n}\right\}.$$

Write $\qquad\qquad\qquad x+i = \rho(\cos\phi + i\sin\phi),$

so that $x = \rho\cos\phi$, $1 = \rho\sin\phi$ and $\rho = \sqrt{(1+x^2)}$, $\phi = \cot^{-1}x$. By de Moivre's theorem,

$$\frac{1}{(x+i)^n} = \rho^{-n}(\cos\phi + i\sin\phi)^{-n} = \rho^{-n}(\cos n\phi - i\sin n\phi)$$

and $\qquad \dfrac{1}{(x-i)^n} = \rho^{-n}(\cos\phi - i\sin\phi)^{-n} = \rho^{-n}(\cos n\phi + i\sin n\phi).$

$$\therefore \quad \frac{d^{n-1}}{dx^{n-1}}\frac{1}{1+x^2} = (-1)^{n-1}\frac{(n-1)!}{\rho^n}\sin n\phi,$$

i.e. $\qquad f^{(n)}(x) = (-1)^{n-1}(n-1)!\,(1+x^2)^{-\frac{1}{2}n}\sin(n\cot^{-1}x).$

## 6.62 Theorem of Leibniz on the $n$th derivative of a product

If $f(x) = uv$, where $u$ and $v$ are functions of $x$ whose successive derivatives are known, then $f^{(n)}(x)$ can be found by the following theorem. Write

$$u_r = \frac{d^r u}{dx^r}, \quad v_r = \frac{d^r v}{dx^r}, \quad (uv)_r = \frac{d^r}{dx^r}(uv).$$

Then

$$(uv)_n = u_n v + {}^nC_1 u_{n-1}v_1 + {}^nC_2 u_{n-2}v_2 + \dots + {}^nC_r u_{n-r}v_r + \dots + uv_n.$$

This result has already been verified when $n = 2, 3$ in Ex. 3(b), no. 17.

9

*Proof by Mathematical Induction.* (The general principle is explained in 12.28.)

Suppose the result is true for some particular value of $n$, say $n = m$. Then

$$(uv)_m = u_m v + {}^mC_1 u_{m-1} v_1 + \ldots + {}^mC_r u_{m-r} v_r + \ldots + uv_m,$$

and $\quad (uv)_{m+1} = \dfrac{d}{dx}(uv)_m$

$$= (u_m v_1 + vu_{m+1}) + {}^mC_1(u_{m-1}v_2 + u_m v_1) + \ldots$$
$$+ {}^mC_{r-1}(u_{m-r+1}v_r + u_{m-r+2}v_{r-1})$$
$$+ {}^mC_r(u_{m-r}v_{r+1} + u_{m-r+1}v_r) + \ldots + (uv_{m+1} + u_1 v_m),$$
$$= u_{m+1}v + (1 + {}^mC_1)u_m v_1 + \ldots + ({}^mC_{r-1} + {}^mC_r)u_{m-r+1}v_r$$
$$+ \ldots + uv_{m+1}$$
$$= u_{m+1}v + {}^{m+1}C_1 u_m v_1 + \ldots + {}^{m+1}C_r u_{m-r+1}v_r + \ldots + uv_{m+1}$$

since $\qquad\qquad {}^mC_{r-1} + {}^mC_r = {}^{m+1}C_r \quad (r \geqslant 1).$

Hence, *if* the theorem holds for $n = m$, *then* it also holds for $n = m+1$. It does hold when $n = 1$ (it is then the product rule of 3.2), therefore when $n = 2$, therefore when $n = 3, \ldots$, and so for all positive integers $n$.

*Alternative proof.* Successive derivation of $uv$ (as in Ex. 3 (*b*), no. 17) shows that the expression for the $n$th derivative will be of the form

$$(uv)_n = c_0 u_n v + c_1 u_{n-1}v_1 + c_2 u_{n-2}v_2 + \ldots + c_r u_{n-r}v_r + \ldots + c_n uv_n,$$

where the coefficients $c_0, c_1, \ldots, c_n$ are numbers *independent of u, v*. We can therefore determine these coefficients from the convenient special case when $u = e^{ax}$, $v = e^{bx}$ and hence $uv = e^{(a+b)x}$; for then

$$u_r = a^r e^{ax}, \quad v_r = b^r e^{bx}, \quad (uv)_n = (a+b)^n e^{(a+b)x},$$

and the above formula becomes

$$(a+b)^n e^{(a+b)x} = \sum_{r=0}^{n} (c_r a^{n-r} e^{ax} \cdot b^r e^{bx}) = e^{(a+b)x} \sum_{r=0}^{n} (c_r a^{n-r} b^r).$$
$$\therefore \quad (a+b)^n = \sum_{r=0}^{n} (c_r a^{n-r} b^r).$$

Hence $c_r$ is the coefficient of $a^{n-r}b^r$ in the expansion of $(a+b)^n$, viz. ${}^nC_r$, and the result follows.

Leibniz's theorem is easily remembered by analogy with the binomial expansion of $(u+v)^n$.

## Examples

(i) *Find the nth derivative of $x^2 \sin x$.*

Taking $u = \sin x$ and $v = x^2$ we have

$$u_r = \sin\left(x + \tfrac{1}{2}r\pi\right), \quad v_1 = 2x, \quad v_2 = 2, \quad v_r = 0 \text{ for } r > 2.$$

Hence

$$\frac{d^n}{dx^n}(x^2 \sin x) = x^2 \sin\left(x + \tfrac{1}{2}n\pi\right) + 2nx \sin\left(x + \tfrac{1}{2}(n-1)\pi\right)$$

$$+ n(n-1)\sin\left(x + \tfrac{1}{2}(n-2)\pi\right).$$

(ii) *Find a relation between any three consecutive derivatives of $\sin\left(p \sin^{-1} x\right)$.*

We first find a relation between $y = \sin\left(p\sin^{-1}x\right)$, $y_1$, and $y_2$. By direct derivation,

$$y_1 = \frac{p}{\sqrt{(1-x^2)}}\cos\left(p\sin^{-1}x\right),$$

i.e.

$$y_1\sqrt{(1-x^2)} = p\cos\left(p\sin^{-1}x\right).$$

Deriving again,

$$y_2\sqrt{(1-x^2)} - \frac{x}{\sqrt{(1-x^2)}}y_1 = -\frac{p^2}{\sqrt{(1-x^2)}}\sin\left(p\sin^{-1}x\right),$$

$$\therefore \quad (1-x^2)y_2 - xy_1 + p^2 y = 0. \tag{a}$$

Now derive this equation $n$ times,† using Leibniz's theorem:

$$(1-x^2)y_{n+2} + n(-2x)y_{n+1} + \tfrac{1}{2}n(n-1)(-2)y_n$$

$$-xy_{n+1} \quad -ny_n$$

$$+p^2 y_n = 0,$$

i.e. $\qquad (1-x^2)y_{n+2} - (2n+1)xy_{n+1} + (p^2 - n^2)y_n = 0. \tag{b}$

This is the required relation, and it holds for $n \geqslant 0$.

## 6.63 Maclaurin coefficients from a recurrence relation

In Maclaurin's theorem it is only the values of the successive derivatives of $f(x)$ *at* $x = 0$ which are required. A relation between any three consecutive coefficients can be obtained as in ex. (ii) above.

## Example

*Obtain the first 4 terms of the Maclaurin expansion of $\sin\left(p\sin^{-1}x\right)$.*

In equation (b) of ex. (ii), put $x = 0$, and denote $y_r$ when $x = 0$ by $a_r$; then

$$a_{n+2} + (p^2 - n^2)a_n = 0.$$

Hence if $a_n$ is known, $a_{n+2}$ can be found.

Now $a_0 = \sin\left(p\sin^{-1}0\right) = 0$, so that $a_2 = 0$, $a_4 = 0$, ...; i.e. all coefficients with even suffix are zero.

---

† Had the given function been $\sin\left(n\sin^{-1}x\right)$, we should say 'derive $k$ times', to avoid using $n$ in two senses.

Also $\qquad a_1 = p\cos(p\sin^{-1}0) = p;$

therefore $\qquad a_3 = (1^2 - p^2)\,a_1 = (1^2 - p^2)\,p,$

$$a_5 = (3^2 - p^2)\,a_3 = (3^2 - p^2)\,(1^2 - p^2)\,p,$$

$$a_7 = (5^2 - p^2)\,a_5 = (5^2 - p^2)\,(3^2 - p^2)\,(1^2 - p^2)\,p,$$

and so on. The required expansion is

$$\sin(p\sin^{-1}x) = a_0 + a_1 x + \frac{a_2}{2!}x^2 + \frac{a_3}{3!}x^3 + \ldots + \frac{x^n}{n!}f^{(n)}(\theta x)$$

$$= px + (1^2 - p^2)\,p\,\frac{x^3}{3!} + (3^2 - p^2)\,(1^2 - p^2)\,p\,\frac{x^5}{5!}$$

$$+ (5^2 - p^2)\,(3^2 - p^2)\,(1^2 - p^2)\,p\,\frac{x^7}{7!} + \ldots + \frac{x^n}{n!}f^{(n)}(\theta x).$$

*This method only shows what the Maclaurin expansion would be if the conditions of the theorem were satisfied:* it does not discuss the existence and continuity of the successive derivatives, nor does it help us to calculate the remainder term. The method is formal, and approximations so established may only be expected to hold 'for $x$ sufficiently small'.

### Exercise 6(*b*)

*Calculate the nth derivative of the following functions.*

1  $\cos^2 x$ [first express in terms of $2x$].

2  $\sin^3 x$.  $\qquad$ 3  $\operatorname{ch}x$.  $\qquad$ 4  $\dfrac{x}{x^2 + x - 2}$.

5  $x^4 e^x$.  $\qquad$ 6  $x^3/e^x$.  $\qquad$ 7  $x^4 \log x$, $n > 4$.

8  $x^3 \sin^2 x$, $n > 3$.  $\qquad$ 9  $\sin x \sin 2x \sin 3x$.

10  If $f(x) = e^x \sin x$, prove $f^{(n)}(0) = 2^{\frac{1}{2}n}\sin(\frac{1}{4}n\pi)$.

11  If $f(x) = \log x/x$, prove $f^{(n)}(x) = (-1)^n\,n!\{\log x - 1 - \frac{1}{2} - \ldots - 1/n\}\,x^{-n-1}$.

12  If $y = f(x) = (\sin^{-1}x)/\sqrt{(1 - x^2)}$, prove that

$$(1 - x^2)\,y_1 - xy = 1 \quad \text{and} \quad f^{(n+1)}(0) = n^2 f^{(n-1)}(0).$$

13  If $y = \sin(\log x)$, prove $x^2 y_2 + xy_1 + y = 0$ and deduce that

$$x^2 y_{n+2} + (2n + 1)\,xy_{n+1} + (n^2 + 1)\,y_n = 0.$$

14  If $y = e^{\tan^{-1}x}$, prove $(1 + x^2)\,y_{n+2} + \{(2n + 2)\,x - 1\}\,y_{n+1} + n(n + 1)\,y_n = 0$.

15  *If* $y = \tan^{-1}x$, *prove* $(1 + x^2)\,y_2 + 2xy_1 = 0$ *and deduce that*

$$(1 + x^2)\,y_{n+2} + 2(n + 1)\,xy_{n+1} + n(n + 1)\,y_n = 0.$$

*Hence calculate the first four non-zero terms of the Maclaurin expansion of* $\tan^{-1}x$, *assuming that an expansion of* $\tan^{-1}x$ *is possible.*

16  If $y = (\sin^{-1}x)^2$, prove $(1 - x^2)\,y_{n+2} - (2n + 1)\,xy_{n+1} - n^2 y_n = 0$  $(n \geqslant 1)$. Hence find the values of all the derivatives of $y$ when $x = 0$, and write down a few terms of the Maclaurin expansion (assuming this exists).

17 Explain why Maclaurin's theorem cannot be applied to $\log x$ or to $\sin(\log x)$.

*Verify the following expansions.*

18 $e^{mx} = 1 + mx + \dfrac{m^2 x^2}{2!} + \ldots + \dfrac{m^{n-1}x^{n-1}}{(n-1)!} + \dfrac{m^n x^n}{n!} e^{m\theta x}.$

19 $\log(1+x) = x - \frac{1}{2}x^2 + \frac{1}{3}x^3 - \ldots + (-1)^n \dfrac{x^{n-1}}{n-1} + (-1)^{n+1}\dfrac{x^n}{n}(1+\theta x)^{-n}.$

20 $\sin x = x - \dfrac{x^3}{3!} + \dfrac{x^5}{5!} - \ldots + (-1)^{k-1}\dfrac{x^{2k-1}}{(2k-1)!} + (-1)^k \dfrac{x^{2k+1}}{(2k+1)!}\cos(\theta x).$

21 $\cos x = 1 - \dfrac{x^2}{2!} + \dfrac{x^4}{4!} - \ldots + (-1)^{k-1}\dfrac{x^{2k-2}}{(2k-2)!} + (-1)^k \dfrac{x^{2k}}{(2k)!}\cos(\theta x).$

22 $(1+x)^m = 1 + mx + \dfrac{m(m-1)}{2!}x^2 + \ldots + \dfrac{m(m-1)\ldots(m-n+2)}{(n-1)!}x^{n-1}$

$$+ \dfrac{m(m-1)\ldots(m-n+1)}{n!}x^n(1+\theta x)^{m-n}.$$

*23 $\sin(x+h) = \sin x + h\cos x - \dfrac{h^2}{2!}\sin x - \dfrac{h^3}{3!}\cos x + \ldots$

$$+ (-1)^{n-1}\dfrac{h^{2n-1}}{(2n-1)!}\cos x + (-1)^n \dfrac{h^{2n}}{(2n)!}\sin(x+\theta h).$$

Give the corresponding formula up to $h^{2n+1}$.

*24 $\cos(x+h) = \cos x - h\sin x - \dfrac{h^2}{2!}\cos x + \dfrac{h^3}{3!}\sin x - \ldots$

$$+ (-1)^n \dfrac{h^{2n-1}}{(2n-1)!}\sin x + (-1)^n \dfrac{h^{2n}}{(2n)!}\cos(x+\theta h).$$

Give the corresponding formula up to $h^{2n+1}$.

25 Writing $a^x = e^{x\log a}$ $(a > 0)$, obtain an expansion for $a^x$.

*Verify the following approximations 'for $x$ sufficiently small'.*

26 $\tan x \doteqdot x + \frac{1}{3}x^3 + \frac{2}{15}x^5.$ [See Ex. 3(b), no. 10.]

27 $\sin^{-1}x \doteqdot x + \frac{1}{6}x^3 + \frac{3}{40}x^5.$     28 $e^{\cos x} \doteqdot e(1 - \frac{1}{2}x^2 + \frac{1}{6}x^4).$

29 $\sec x \doteqdot 1 + \frac{1}{2}x^2 + \frac{5}{24}x^4.$     30 $\dfrac{x}{\sin x} \doteqdot 1 + \frac{1}{6}x^2 + \frac{7}{360}x^4.$

*31 Prove the second mean value theorem by one application of Rolle's theorem to the function $\psi(x) = f(b) - f(x) - (b-x)f'(x) - A(b-x)^2$, choosing $A$ so that $\psi(a) = 0$.

## 6.7  Further applications of the mean value theorems

### 6.71  Turning points; concavity, inflexions

(1) The second mean value theorem can be written

$$f(a+h) - f(a) = hf'(a) + \tfrac{1}{2}h^2 f''(a+\theta h) \quad (0 < \theta < 1).$$

For $x = a$ to give a turning value of $f(x)$, $f(a+h)-f(a)$ must have the same sign for all $h$ sufficiently small, positive or negative (3.62). This can be so only if $f'(a) = 0$, otherwise the right-hand side would change sign with $h$. Hence

$$f(a+h)-f(a) = \tfrac{1}{2}h^2 f''(a+\theta h).$$

If $f''(a+\theta h)$ is positive for all $h$ sufficiently small, then $f(a+h)-f(a) > 0$ for all such $h$, and hence $x = a$ gives a minimum. If $f''(a+\theta h)$ is negative, we similarly find that $x = a$ gives a maximum.

If $f''(x)$ is *continuous* at $x = a$, then $f''(a+\theta h) \to f''(a)$ as $h \to 0$, and hence $f''(a+\theta h)$ has the same sign as $f''(a)$ for all $h$ sufficiently small (assuming $f''(a) \neq 0$). Hence if $f''(a) > 0$, $x = a$ gives a minimum; if $f''(a) < 0$, it gives a maximum. These results agree with 3.65, but here we have assumed more about $f''(x)$.

**(2)** *Definition.* A curve is *concave upwards* at $P$ if, in the neighbourhood of $P$, it lies above the tangent at $P$ (fig. 63).

*If $f''(x)$ is continuous at $x = a$, the curve $y = f(x)$ is concave upwards or downwards at $x = a$ according as $f''(a) \gtrless 0$.*

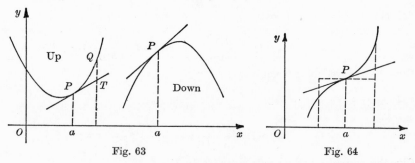

Fig. 63                  Fig. 64

Near $P$ we have by 6.42, Remark $(\beta)$ that

$$\text{ordinate to curve} - \text{ordinate to tangent} = \tfrac{1}{2}h^2 f''(a+\theta h), \qquad \text{(i)}$$

and this expression has the same sign as $f''(a)$ for all $h$ sufficiently small since $f''(x)$ is continuous at $x = a$. The result follows.

COROLLARY 1. *At a point of inflexion the concavity changes sense.*

For by 3.71, $f''(x)$ changes sign as $x$ increases through a point of inflexion.

COROLLARY 2. *At a point of inflexion the curve crosses its tangent.*

For $f''(a+\theta h)$ has opposite signs when $h \lessgtr 0$, so that the difference (i) changes sign as $x$ increases through $a$ (fig. 64).

**(3)** The preceding results can be generalised by using Taylor's theorem.

I. *If $f'(a) = 0$, and $f^{(n)}(x)$ is continuous at $x = a$ and is the first of the derivatives of $f(x)$ which is non-zero at $x = a$, then:*

(i) *if $n$ is odd, there is no turning point at $x = a$;*

(ii) *if $n$ is even, there is a maximum at $x = a$ if $f^{(n)}(a) < 0$, and a minimum if $f^{(n)}(a) > 0$.*

For from the hypothesis,

$$f(a+h) - f(a) = \frac{h^n}{n!} f^{(n)}(a + \theta h),$$

and the results follow immediately from the definitions in 3.62.

II. *If $f''(a) = 0$, and $f^{(n)}(x)$ is continuous at $x = a$ and is the first of the derivatives of $f(x)$ of order greater than 2 which is non-zero at $x = a$, then $x = a$ is a point of inflexion of $f(x)$ if and only if $n$ is odd.*

Applying Taylor's theorem to $f'(x)$, we have by hypothesis:

$$f'(a+h) - f'(a) = \frac{h^{n-1}}{(n-1)!} f^{(n)}(a + \theta h).$$

Hence $f'(x)$ has a turning point at $x = a$ if and only if $n - 1$ is even, i.e. $n$ is odd. The result follows from the definition in 3.71.

## 6.72 Closeness of contact of two curves

Geometrically, 6.53 shows that in the neighbourhood of $x = a$, the closeness of the curve $y = f(x)$ to the polynomial approximating curve

$$y = f(a) + (x-a)f'(a) + \ldots + \frac{(x-a)^{n-1}}{(n-1)!} f^{(n-1)}(a) \tag{ii}$$

is measured by $(x-a)^n f^{(n)}(\xi)/n!$; in particular, $\frac{1}{2}(x-a)^2 f''(\xi)$ measures the closeness of the curve to its tangent (cf. equation (i) above).

If two curves $y = f(x)$, $y = g(x)$ have a point $P$ in common where $x = a$, we now consider how their closeness near $P$ can be defined. If they merely *intersect* at $P$, then $f(a) = g(a)$. If also $f'(a) = g'(a)$, they have the same tangent and are said to *touch* at $P$. If further $f''(a) = g''(a)$, they have the same quadratic approximating curve. In general—

*Definition.* If $m$ is the largest integer for which the two polynomial approximating curves of degree $m$ are the same near $P$, then the given curves are said to have *$m$th-order contact* at $P$.

The necessary and sufficient conditions† for this are that

$$f(a) = g(a), \quad f'(a) = g'(a), \quad \ldots, \quad f^{(m)}(a) = g^{(m)}(a),$$
$$\text{but} \qquad f^{(m+1)}(a) \neq g^{(m+1)}(a). \tag{iii}$$

*Remarks*

($\alpha$) If $f(x)$, $g(x)$ were *polynomials*, then by Ex. 3 ($e$), no. 13 (or 10.43) the conditions (iii) show in turn that the equation $f^{(m)}(x) = g^{(m)}(x)$ has a simple root $x = a$, $f^{(m-1)}(x) = g^{(m-1)}(x)$ has a double root $x = a$, ...,

---

† Subject, of course, to the validity of Taylor's theorem for $n = m + 2$.

and finally $f(x) = g(x)$ has an $(m+1)$-fold root $x = a$. On account of this it is customary to say that the curves $y = f(x)$, $y = g(x)$ possess '$m+1$ *coincident points*' in common at $P$.

($\beta$) The above definition fails if the tangent at $P$ is parallel to $Oy$, since then $f'(a)$, $g'(a)$ would not exist and the conditions of Taylor's theorem would be violated; but the curves could be taken in the form $x = F(y)$, $x = G(y)$, and approximating polynomials in $y$ considered similarly.

THEOREM. *If two curves have mth-order contact at $P$, they cross there if m is even, and do not cross if m is odd.*

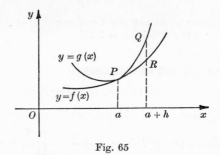

Fig. 65

For, near $P$, the difference of corresponding ordinates is

$$QR = f(a+h) - g(a+h)$$

$$= \sum_{r=0}^{m} \frac{h^r}{r!} \{f^{(r)}(a) - g^{(r)}(a)\} + \frac{h^{m+1}}{(m+1)!} \{f^{(m+1)}(a+\theta h) - g^{(m+1)}(a+\theta h)\}$$

$$= \frac{h^{m+1}}{(m+1)!} \{f^{(m+1)}(a+\theta h) - g^{(m+1)}(a+\theta h)\}$$

by Taylor's theorem with $n = m+1$, and the conditions (iii). Assuming that $f^{(m+1)}(x)$, $g^{(m+1)}(x)$ are *continuous* at $x = a$, the content of the braces has the same sign as $f^{(m+1)}(a) - g^{(m+1)}(a)$ for all $h$ sufficiently small, positive or negative. Hence $f(a+h) - g(a+h)$ has the same sign for all small $h$ if $m$ is odd, so that the curves do not cross at $P$. If $m$ is even, the expression takes opposite signs for $h < 0$ and $h > 0$, i.e. the curves cross.

## 6.73 Approximate solution of equations by Newton's method

(1) *Outline of the method.* Let $f(x)$ be such that $f''(x)$ (and hence by 3.52 also $f'(x)$, $f(x)$) is continuous in $a \leqslant x \leqslant b$, and suppose $f(a)$, $f(b)$ have opposite signs. Then (2.65) there is a root of $f(x) = 0$ in $a < x < b$.

Let $x = c$ be an approximation to this root, and suppose the actual value of the root is $c + h$ ($h$ may be negative). If $c$ is a fair approximation, $h$ will be small. Since $f(c + h) = 0$, the second mean value theorem gives

$$0 = f(c) + hf'(c) + \tfrac{1}{2}h^2 f''(c + \theta h) \quad (0 < \theta < 1).$$

Assuming that $f''(c + \theta h)$ is not large, and that $f'(c) \neq 0$, we may neglect the term in $h^2$ and get

$$0 \doteqdot f(c) + hf'(c),$$

so that $h \doteqdot -f(c)/f'(c)$. Thus, in general, a closer approximation to the root is $c_1 = c - f(c)/f'(c)$.

This process can now be repeated, starting with the approximation $x = c_1$, and leads (in general) to a closer approximation

$$c_2 = c_1 - \frac{f(c_1)}{f'(c_1)};$$

and so on. It is an example of an *iterative process* or *iteration*; i.e. a process by which the accuracy of an approximation is improved by repeating the calculation on successive previous approximations.

(2) *Geometrical interpretation.* The equation of the tangent to the curve $y = f(x)$ at the point $x = c$ is

$$y - f(c) = f'(c)\,(x - c).$$

It meets $Ox$ where $y = 0$ and $x = c - f(c)/f'(c)$, i.e. at $x = c_1$. Hence the approximation is geometrically equivalent to replacing the arc of the curve for which $a \leqslant x \leqslant b$ by the tangent line at $x = c$.

For the process to be successful, the situation must be as indicated in fig. 66, where the numbers $c_1, c_2, \dots$ approach $c + h$. In fig. 67, the approximation gets worse: the trial $c$ is too rough.

(3) *Estimation of the error at any stage.* Let $c_1, c_2, c_3$ be successive approximations. Then

$$c_2 - c_1 = -\frac{f(c_1)}{f'(c_1)}, \quad c_3 - c_2 = -\frac{f(c_2)}{f'(c_2)},$$

and

$$c_3 - c_2 = -\frac{f(c_1 + h)}{f'(c_1 + h)},$$

where $h = -f(c_1)/f'(c_1)$ and is supposed small. Then

$$c_3 - c_2 = -\frac{f(c_1) + hf'(c_1) + \frac{1}{2}h^2 f''(c_1 + \theta h)}{f'(c_1 + h)}$$

$$= -\frac{\frac{1}{2}h^2 f''(c_1 + \theta h)}{f'(c_1 + h)}$$

$$\doteq -\frac{\frac{1}{2}h^2 f''(c_1)}{f'(c_1)}$$

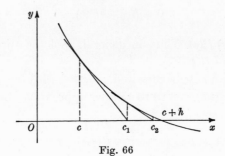

Fig. 66

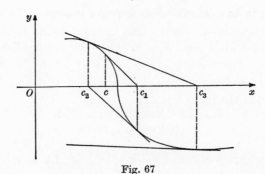

Fig. 67

by the assumed continuity of $f'$ and $f''$; hence, substituting for $h$,

$$c_3 - c_2 \doteq -\frac{\{f(c_1)\}^2 f''(c_1)}{2\{f'(c_1)\}^3}. \tag{i}$$

The expression (i) will be small if

(a) $f(c_1)$ is small, i.e. if the first approximation is good;

(b) $f''(c_1)$ is small, i.e. if the direction of the curve is changing slowly—the curve is approximately straight; and

(c) $f'(c_1)$ is large, i.e. the curve is steeply inclined to $Ox$.

The method is most effective when these conditions are satisfied near the required root. Figures 66, 67 confirm this.

When a specified degree of accuracy is required, and (i) is small enough not to affect the result, then $c_2$ is the approximation sought, and $c_3$ need not be calculated at all.†

### Example

*Solve $x + \sin x = 1\cdot5$ correct to four places of decimals ($x$ in radians).*
Trying $x = 1$ as a first approximation, we find

$$\sin 1 = 0\cdot8415 \quad \text{and} \quad \cos 1 = 0\cdot5402,$$

so
$$\frac{f(1)}{f'(1)} = \frac{1 + 0\cdot8415 - 1\cdot5}{1 + 0\cdot5402} = \frac{0\cdot3415}{1\cdot5402} \doteqdot 0\cdot2217.$$

A second approximation is therefore $1 - 0\cdot2217 = 0\cdot7783$. Similarly,

$$\frac{f(0\cdot7783)}{f'(0\cdot7783)} = \frac{0\cdot7783 + 0\cdot7022 - 1\cdot5}{1\cdot7120} \doteqdot -0\cdot0114,$$

and a third approximation is $0\cdot7783 + 0\cdot0114 = 0\cdot7897$.

This result is in fact correct to four places of decimals; for by taking $c_1 = 0\cdot7783$ in the expression (i) above, it can be shown to be of order $+2\cdot7 \times 10^{-5}$, and hence $c_2 = 0\cdot7897$ is the required value.

**(4)** *Refinement of Newton's method.*‡ By using a *quadratic* approximation for $h$ we can improve the linear one $0 \doteqdot f(c) + hf'(c)$. For

$$0 \doteqdot f(c) + hf'(c) + \tfrac{1}{2}h^2 f''(c),$$

and since $h^2 \doteqdot -hf(c)/f'(c)$, we have

$$0 \doteqdot f(c) + h\left\{ f'(c) - \frac{\tfrac{1}{2}f(c)f''(c)}{f'(c)} \right\}.$$

The next approximation is therefore

$$c_1 = c - f(c) \Big/ \left\{ f'(c) - \frac{\tfrac{1}{2}f(c)f''(c)}{f'(c)} \right\}.$$

It can be shown‡ that the error diminishes very rapidly. Taking $c = 1$ in the above example, we should find $c_1 \doteqdot 0\cdot7909$, which is very near the correct value. Taking $c = 0\cdot7909$ in Newton's process, we should reach the result.

### Exercise 6(c)

*Test the stationary point $x = 0$ for each of the following functions.*

*1   $\cos x - 1$.        *2   $\sin x - x$.        *3   $\cos x - 1 + \dfrac{x^2}{2!}$.

---

† Without an error estimate we should have to calculate $c_3$ and then observe that, to the required degree of accuracy, it does not differ from $c_2$. Our estimate involves only the *previous* approximation.

‡ E. H. Bateman, *Mathematical Gazette*, xxxvii (1953), p. 96.

$*4 \quad \sin x - x + \dfrac{x^3}{3!}.$     $*5 \quad \cos x - 1 + \dfrac{x^2}{2!} - \dfrac{x^4}{4!}.$     $*6 \quad \sin x - x + \dfrac{x^3}{3!} - \dfrac{x^5}{5!}.$

7 Prove that at a point of inflexion, a curve and its tangent have contact of at least the second order.

8 Show that a root of $x^4 + 5x - 15 = 0$ is approximately $1 \cdot 6206$.

9 Show that the smallest positive root of $2 \sin x = 1/x$ is about $0 \cdot 741$.

10 Find correct to three places of decimals the positive root of $\sin x = \frac{1}{2}x$.

*Further examples on approximate solution will be found in Ex.* 13 $(f)$.

## 6.8 Cauchy's mean value theorem

**6.81** For a curve given parametrically by $x = g(t)$, $y = f(t)$, the gradient of the chord joining the points given by $t = a$, $t = b$ is $\{f(b) - f(a)\}/\{g(b) - g(a)\}$, and the gradient of a tangent is $f'(t)/g'(t)$. The geometrical principle in 3.81 suggests that there is a value $t = \xi$ for which $\{f(b) - f(a)\}/\{g(b) - g(a)\} = f'(\xi)/g'(\xi)$. We now prove this under suitable conditions.

**6.82** *If* (i) $f(x)$ *and* $g(x)$ *satisfy the Rolle conditions in* $a \leqslant x \leqslant b$,

     (ii) $f'(x)$, $g'(x)$ *are not zero for the same value of* $x$ *in* $a < x < b$,

and   (iii) $g(b) \neq g(a)$,

*then there is at least one number* $\xi$ *for which* $a < \xi < b$ *and*

$$\frac{f(b) - f(a)}{g(b) - g(a)} = \frac{f'(\xi)}{g'(\xi)}.$$

*Proof.* Consider the function

$$\phi(x) = \{f(x) - f(a)\} + A\{g(x) - g(a)\}.$$

Clearly $\phi(a) = 0$; let us choose $A$ so that $\phi(b) = 0$:

$$0 = \{f(b) - f(a)\} + A\{g(b) - g(a)\},$$

and hence by hypothesis (iii),

$$-A = \frac{f(b) - f(a)}{g(b) - g(a)}.$$

By hypothesis (i), $\phi(x)$ now satisfies the conditions of Rolle's theorem, so that there is at least one number $\xi$ for which $a < \xi < b$ and $\phi'(\xi) = 0$. Since $\phi'(x) = f'(x) + Ag'(x)$, we have

$$f'(\xi) = \frac{f(b) - f(a)}{g(b) - g(a)} g'(\xi).$$

We cannot have $g'(\xi) = 0$, otherwise the last equation would imply $f'(\xi) = 0$ and hypothesis (ii) would be contradicted. Hence we can divide both sides by $g'(\xi)$, and obtain the result.

COROLLARY. *If $g'(x)$ never vanishes in $a < x < b$, then hypothesis (ii) is satisfied; and so is (iii), for if $g(b) = g(a)$ then Rolle's theorem shows that $g'(x)$ would be zero between $a$ and $b$.*

*Remarks*

($\alpha$) The mean value theorem (6.3) is the case when $g(x) = x$.

($\beta$) The result is more general than that obtained by applying the mean value theorem to $f(x)$, $g(x)$ separately and then dividing. For we should have

$$f(b) - f(a) = (b-a)f'(\xi_1) \quad (a < \xi_1 < b),$$

and

$$g(b) - g(a) = (b-a)g'(\xi_2) \quad (a < \xi_2 < b),$$

from which

$$\frac{f(b)-f(a)}{g(b)-g(a)} = \frac{f'(\xi_1)}{g'(\xi_2)}.$$

There is no reason why $\xi_1$ and $\xi_2$ should be equal, and in general they will be different (see Ex. 6 (*d*), no. 1).

## 6.9 'Indeterminate forms': l'Hospital's rules

**6.91** Suppose $f(a) = 0$, $g(a) = 0$; then the function $\phi(x) = f(x)/g(x)$ is not defined when $x = a$ because it takes the meaningless or 'indeterminate' form $0/0$. Yet $\lim\limits_{x \to a} \phi(x)$ may exist: e.g. when $f(x) = \sin x$, $g(x) = x$, $a = 0$. The following two rules give means of calculating this limit when certain conditions are satisfied.

### 6.92 First rule

*If* (i) $f(a) = 0$, $g(a) = 0$, (ii) $f'(a)$ *and* $g'(a)$ *exist and* $g'(a) \ne 0$, *then*

$$\lim_{x \to a} \frac{f(x)}{g(x)} = \frac{f'(a)}{g'(a)}.$$

*Proof.* If $x \ne a$, then by hypothesis (i)

$$\frac{f(x)}{g(x)} = \frac{f(x)-f(a)}{g(x)-g(a)}$$

$$= \frac{f(x)-f(a)}{x-a} \bigg/ \frac{g(x)-g(a)}{x-a}$$

$$\to f'(a)/g'(a) \quad \text{when} \quad x \to a$$

by hypothesis (ii) and the definition of 'derivative' in 3.11, (ii).

COROLLARY. If $f'(a) \neq 0$ and $g'(a) = 0$, the conclusion is that $|f(x)/g(x)| \to \infty$ when $x \to a$.

## Example

$$\lim_{x \to 0} \frac{1 - \cos x}{x} = \frac{\sin 0}{1} = 0.$$

## 6.93  Second rule

*If* (i) $f(a) = 0$, $g(a) = 0$, (ii) $f(x)$, $g(x)$ *are continuous at* $x = a$, (iii) $\lim\limits_{x \to a} \dfrac{f'(x)}{g'(x)}$ *exists, then*

$$\lim_{x \to a} \frac{f(x)}{g(x)} = \lim_{x \to a} \frac{f'(x)}{g'(x)}.$$

*Proof.* Hypothesis (iii) implies that $f'(x)/g'(x)$ exists throughout some sufficiently small interval $a < x < a + H$, i.e. that $f'(x)$ and $g'(x)$ exist and $g'(x) \neq 0$ in this range (otherwise $f'(x)/g'(x)$ would be meaningless for an infinity of values of $x$ just greater than $a$). If $0 < h < H$, the conditions of Cauchy's mean value theorem are satisfied for $a \leqslant x \leqslant a + h$: condition (i) since $f'(x)$, $g'(x)$ exist in $a < x \leqslant a + h$, and $f(x)$, $g(x)$ are therefore continuous in $a \leqslant x \leqslant a + h$, using our hypothesis (ii) for $x = a$; conditions (ii), (iii) by 6.82, Corollary, since $g'(x) \neq 0$ for $a < x \leqslant a + h$.

Hence by Cauchy's theorem and hypothesis (i)

$$\frac{f(a+h)}{g(a+h)} = \frac{f'(\xi)}{g'(\xi)} \quad (a < \xi < a + h).$$

When $h \to 0+$, also $\xi \to a+$ and $f'(\xi)/g'(\xi)$ tends to its limit, say $l$. Hence $f(a+h)/g(a+h) \to l$ when $h \to 0+$, i.e. $f(x)/g(x) \to l$ when $x \to a+$.

Similarly, hypothesis (iii) implies that there is an interval $a - k \leqslant x \leqslant a$ for which Cauchy's formula holds, and hence that $f(x)/g(x) \to l$ when $x \to a-$. Consequently,

$$f(x)/g(x) \to l \quad \text{when} \quad x \to a,$$

and the result follows.

COROLLARY. If $f'(x)/g'(x) \to \infty$ when $x \to a$, the same argument shows that also $f(x)/g(x) \to \infty$.

## Remarks

($\alpha$) *The conditions of the rule are sufficient but not necessary*, i.e. $\lim\{f(x)/g(x)\}$ may exist when $\lim\{f'(x)/g'(x)\}$ does not. For example, if $f(x) = x^2 \sin(1/x)$, $g(x) = x$, then $f(x)/g(x) = x \sin(1/x) \to 0$ when $x \to 0$; but $f'(x) = 2x \sin(1/x) - \cos(1/x)$ $(x \neq 0)$ and $g'(x) = 1$, so

$$\frac{f'(x)}{g'(x)} = 2x \sin\frac{1}{x} - \cos\frac{1}{x},$$

and this does not approach a limit when $x \to 0$ because $\cos(1/x)$ oscillates.

($\beta$) The second rule can be used repeatedly, provided the conditions are satisfied; and it can be used in combination with the first.

### Examples

(i) $\qquad \lim\limits_{x \to 0} \dfrac{1 - \cos x}{x^2} = \lim\limits_{x \to 0} \dfrac{\sin x}{2x} = \tfrac{1}{2}$  since  $\lim\limits_{x \to 0} \dfrac{\sin x}{x} = 1.$

(We cannot use the first rule to show $\lim\limits_{x \to 0} \sin x/x = 1$ because this result was assumed when proving that the derivative of $\sin x$ is $\cos x$ (3.32): we should be 'arguing in a circle'. We could use the first rule to prove $\lim\limits_{x \to 0} \tan x/x = 1$ because this last limit has not been employed fundamentally in finding $d(\tan x)/dx$.)

(ii) $\qquad \lim\limits_{x \to \pi} \dfrac{x \cos x + \pi}{\sin x} = \lim\limits_{x \to \pi} \dfrac{\cos x - x \sin x}{\cos x} = \dfrac{-1}{-1} = 1.$

This could also be done by the first rule.

(iii) $\quad \lim\limits_{x \to 0} \dfrac{2 \log \sec x - x^2}{x^4} = \lim\limits_{x \to 0} \dfrac{2 \tan x - 2x}{4x^3}$

$\qquad\qquad\qquad\qquad = \lim\limits_{x \to 0} \dfrac{\sec^2 x - 1}{6x^2} \quad$ by the second rule again,

$\qquad\qquad\qquad\qquad = \lim\limits_{x \to 0} \dfrac{1}{6}\left(\dfrac{\tan x}{x}\right)^2 \quad$ by trigonometry,

$\qquad\qquad\qquad\qquad = \tfrac{1}{6} \qquad\qquad\quad$ since $\lim\limits_{x \to 0} \dfrac{\tan x}{x} = 1.$

The expression should be simplified at each stage of the work whenever possible: the trigonometrical reduction eases the calculation here, as the reader may verify by proceeding directly by rule to

$$\lim\limits_{x \to 0} \frac{\sec^2 x - 1}{6x^2} = \lim\limits_{x \to 0} \frac{2 \tan x \sec^2 x}{12x} = \ldots.$$

(iv) $\lim\limits_{x \to 1} (1 - x) \tan \tfrac{1}{2}\pi x \quad$ (the form $0 \times \infty$)

$\qquad\qquad = \lim\limits_{x \to 1} \dfrac{1 - x}{\cot \tfrac{1}{2}\pi x} \quad$ (the form $0/0$)

$\qquad\qquad = \lim\limits_{x \to 1} \dfrac{-1}{-\tfrac{1}{2}\pi \operatorname{cosec}^2 \tfrac{1}{2}\pi x} = \dfrac{2}{\pi}.$

To find limits when $x \to \infty$ we may put $y = 1/x$ and consider the corresponding limit when $y \to 0+$.

(v) $\displaystyle \lim_{x \to \infty} \left( x \tan \frac{1}{x} \right)^{x^2} = \lim_{y \to 0+} \left( \frac{\tan y}{y} \right)^{1/y^2}$, where $y = 1/x$.

Put $\quad u = \left( \dfrac{\tan y}{y} \right)^{1/y^2} \quad$ and consider $\quad \log u = \log \left( \dfrac{\tan y}{y} \right) \Big/ y^2$.

$$\lim_{y \to 0} \log u = \lim_{y \to 0} \left\{ \frac{y}{\tan y} \left( \frac{y \sec^2 y - \tan y}{y^2} \right) \Big/ 2y \right\}$$

$$= \lim_{y \to 0} \frac{y \sec^2 y - \tan y}{2y^2 \tan y}$$

$$= \lim_{y \to 0} \frac{\sec^2 y + 2y \sec^3 y \sin y - \sec^2 y}{4y \tan y + 2y^2 \sec^2 y}$$

$$= \lim_{y \to 0} \frac{\tan y}{\sin 2y + y} \quad \text{after reduction,}$$

$$= \lim_{y \to 0} \frac{\sec^2 y}{2 \cos 2y + 1} = \tfrac{1}{3}.$$

Hence $u \to e^{\frac{1}{3}}$ when $y \to 0$, i.e. when $x \to \infty$.

### Exercise 6(d)

**1** Find $\xi$, $\xi_1$, $\xi_2$ in 6.82, Remark $(\beta)$, when $f(x) = x^3$, $g(x) = x^2$ and the interval is $1 \leqslant x \leqslant 2$.

*Calculate the limits, taken when $x \to 0$ unless otherwise stated, of the following.*

**2** $\dfrac{1 - \cos 5x}{x^2}$.

**3** $2x \tan x - \pi \sec x \ (x \to \tfrac{1}{2}\pi)$.

**4** $\dfrac{\log (1 + x)}{\sin x}$.

**5** $\dfrac{10^x - e^x}{x}$.

**6** $\dfrac{e^{ax} - e^{-ax}}{\log (1 + bx)}$.

**7** $\dfrac{\cos^2 \frac{1}{2}\pi x}{ex - e^x} \ (x \to 1)$.

**8** $\dfrac{\tan x - x}{x - \sin x}$.

**9** $\dfrac{e^x - e^{\sin x}}{x - \sin x}$.

**10** $\dfrac{\log (1 + x^2 + x^4)}{x(e^x - 1)}$.

**11** $(\cos x)^{1/x}$.

**12** $|\tan x|^{\sin 2x} \ (x \to \tfrac{1}{2}\pi)$.

**13** $(\cos x + 2 \sin x)^{\cot x}$.

**14** $\cot x - 1/x$.

**\*15** $1/x^2 - \operatorname{cosec}^2 x$.

### Miscellaneous Exercise 6(e)

*Verify the following approximations 'for x sufficiently small'.*

**1** $x \cot x \doteqdot 1 - \tfrac{1}{3}x^2 - \tfrac{1}{45}x^4$.

**2** $\dfrac{x}{e^x - 1} \doteqdot 1 - \tfrac{1}{2}x + \tfrac{1}{12}x^2$.

**3** $\log \left( \dfrac{\sin x}{x} \right) \doteqdot -\tfrac{1}{6}x^2 - \tfrac{1}{180}x^4$.

**4** Prove that the $n$th derivative of $\cos \pi x$ is $\pi^n \cos(\pi x + \tfrac{1}{2}n\pi)$. Show that the $2m$th derivative of $x^2 \cos \pi x$ has the value $(-1)^{m+1} \pi^{2m-2}(\pi^2 + 2m - 4m^2)$ when $x = 1$.

**5** If $m_0 = 1$ and $m_r = m(m-1)(m-2)\ldots(m-r+1)/r!$ $(r \neq 0)$, prove that

$$m_r n_0 + m_{r-1}n_1 + m_{r-2}n_2 + \ldots + m_0 n_r = (m+n)_r$$

for all values of $m$, $n$ and any positive integer $r$ by deriving the identity $x^m . x^n = x^{m+n}$ $r$ times by Leibniz's theorem.

*Calculate the limit when $x \to 0$ of each of the following.*

**6** $\dfrac{e^{\sin x} - 1 - x}{x^2}$.

**7** $\dfrac{x^2 \operatorname{sh} x}{(1+x^3)^4 - (1-x^3)^4}$.

**8** $\dfrac{2x - 2x^2 - \log(1+2x)}{x^2 \tan^{-1} x}$.

**9** $\dfrac{1}{x}\left\{\operatorname{cosec} \pi x - \dfrac{1}{\pi x}\right\}$.

**10** $(\cos x)^{\cot^2 x}$.

**11** Apply the second mean value theorem to $e^x$ to prove that $e^x > 1 + x$ $(x \neq 0)$.

**12** Apply Maclaurin's theorem to prove that, when $n$ is *odd*,

$$e^x > 1 + x + \frac{x^2}{2!} + \ldots + \frac{x^n}{n!} \quad (x \neq 0);$$

and that if $n$ is *even* this result holds for $x > 0$, while for $x < 0$ the inequality is reversed.

*In nos. 13 and 14* $\qquad \phi(x) = \begin{vmatrix} f(a) & g(a) & h(a) \\ f(b) & g(b) & h(b) \\ f(x) & g(x) & h(x) \end{vmatrix}.$

**\*13** If $f(x)$, $g(x)$, $h(x)$ satisfy the Rolle conditions for $a \leqslant x \leqslant b$, apply Rolle's theorem to $\phi(x)$ to prove that there is a number $\xi$ for which $a < \xi < b$ and

$$\begin{vmatrix} f(a) & g(a) & h(a) \\ f(b) & g(b) & h(b) \\ f'(\xi) & g'(\xi) & h'(\xi) \end{vmatrix} = 0.$$

Deduce (i) the first mean value theorem; (ii) Cauchy's mean value theorem.

**\*14** If $f'(x), g'(x), h'(x)$ satisfy the Rolle conditions for $a \leqslant x \leqslant b$, and $a < c < b$, prove by considering the function

$$F(x) = \phi(x) - \frac{(x-a)(x-b)}{(c-a)(c-b)}\phi(c)$$

that there is a number $\xi$ for which $a < \xi < b$ and

$$\phi(c) = \tfrac{1}{2}(c-a)(c-b)\phi''(\xi).$$

**\*15** By taking $g(x) = x$ and $h(x) = 1$ in no. 14, prove

$$\frac{f(c) - f(a)}{c-a} = \frac{f(b) - f(a)}{b-a} - \tfrac{1}{2}(b-c)f''(\xi).$$

The value of $f(x)$ is known when $x = a$ and when $x = b$, and the value at an intermediate point $x = c$ is calculated approximately by the 'rule of proportional parts' (see 13.71). Prove that the error in the value thus found does not exceed $\tfrac{1}{2}(b-c)(c-a)M$ numerically, where $M$ is the upper bound of $|f''(x)|$ for $a \leqslant x \leqslant b$.

**\*16** (i) If $f'(x)$ satisfies the Rolle conditions in $a - h \leqslant x \leqslant a + h$, prove by considering the function $\phi(x) = f(x) - Ax - Bx^2$ that there is a number $\xi$ for which $a - h < \xi < a + h$ and

$$\frac{f(a+h) - 2f(a) + f(a-h)}{h^2} = f''(\xi).$$

[Choose $A$, $B$ so that $\phi(a - h) = \phi(a) = \phi(a + h)$, and apply Rolle's theorem for $a - h \leqslant x \leqslant a$ and for $a \leqslant x \leqslant a + h$.]

(ii) If also $f''(x)$ is *continuous* at $x = a$, deduce that

$$\lim_{h \to 0} \frac{f(a+h) - 2f(a) + f(a-h)}{h^2} = f''(a).$$

**\*17** Apply Cauchy's mean value theorem to the function $F(x)$ defined in 6.54(1) and $G(x) = (b-x)^p$, $0 < p \leqslant n$, to obtain Taylor's theorem with Schlömilch's remainder.

# 7

## INTEGRATION AS A SUMMATION PROCESS

### 7.1 Theory of the definite integral

### 7.11 'Area under a curve'

In 4.14 we gave an account of the process for calculating the area under a continuous curve $y = f(x)$. It is now desirable to re-cast the argument.

Suppose that the area under $y = f(x)$ between $x = a$ and $x = b$ is divided into $n$ strips by ordinates through the points

$$x = x_1, x_2, ..., x_{n-1},$$

where $\qquad a < x_1 < x_2 < ... < x_{n-1} < b.$

(The points need not be at equal distances apart.) It will be convenient to write $x_0$ for $a$, $x_n$ for $b$.

Referring to fig. 68, suppose $M$ corresponds to $x = x_r$ and $N$ to $x = x_{r+1}$. Then the area of the strip $PMNQ$ lies in value between the areas of the rectangles $PMNR$ and $SMNQ$, viz.

$$f(x_r)(x_{r+1} - x_r) \quad \text{and} \quad f(x_{r+1})(x_{r+1} - x_r).$$

Hence the total area under the curve lies between the sums of the areas of all such 'inner' and 'outer' rectangles, viz.

$$\sum_{r=0}^{n-1} f(x_r)(x_{r+1} - x_r) \quad \text{and} \quad \sum_{r=0}^{n-1} f(x_{r+1})(x_{r+1} - x_r). \tag{i}$$

Geometrical intuition leads us to expect that, as we make the division of the given area gradually finer by increasing the number of ordinates and consequently decreasing the width of each strip, the sum of the corresponding outer rectangles will approximate to the required area 'from above', and the sum of the inner rectangles will approximate to it 'from below'. That is, as $n \to \infty$ and the width of each strip tends to zero, the sums (i) will tend to a common limit which is the required area.

For a steadily falling graph like fig. 69, the rectangles $PMNR$, $SMNQ$ have respective areas $f(x_r)(x_{r+1} - x_r)$ and $f(x_{r+1})(x_{r+1} - x_r)$;

but the required area still lies between the two sums (i), and the argument is essentially unchanged.

For a curve like fig. 44 of 4.14 (a mixture of parts like figs. 68, 69) some of the strips will be like that shown enlarged in fig. 70. The area of this strip still lies between the areas of the inner and outer rectangles $R'MNR$, $SMNS'$, although their heights are no longer $f(x_r)$, $f(x_{r+1})$ but the least and greatest values of $f(x)$ in the range $x_r \leqslant x \leqslant x_{r+1}$. If $f(x)$ is continuous, we know (6.1, Property (2)) that these are attained at points $x = \xi_r$, $\eta_r$ in the range, and the areas of the rectangles are

$$f(\xi_r)\,(x_{r+1}-x_r) \quad \text{and} \quad f(\eta_r)\,(x_{r+1}-x_r).$$

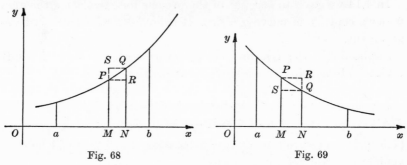

Fig. 68 Fig. 69

The total area then lies between the two sums

$$\sum_{r=0}^{n-1} f(\xi_r)\,(x_{r+1}-x_r), \quad \sum_{r=0}^{n-1} f(\eta_r)\,(x_{r+1}-x_r), \tag{ii}$$

where $x_r \leqslant \xi_r \leqslant x_{r+1}$, $x_r \leqslant \eta_r \leqslant x_{r+1}$ for each $r = 0, 1, \dots, n-1$; and we expect it to be their common limit when $n \to \infty$ and each difference $x_{r+1} - x_r \to 0$.

We pause at this stage to re-emphasise (cf. 4.16 (1)) that the discussion given so far assumes that we know what is meant by 'the area of a strip' and 'the total area under the curve'. Although in elementary work we have definitions of the term 'area' as applied to figures bounded by straight lines, we have not yet stated precisely what is to be understood by 'the area' of a figure bounded by one or more curves. The preceding considerations are therefore only *suggestive*, and are based on 'what seems reasonable'.

Although we would not so readily associate an 'area' with a curve having one or more discontinuities like that shown in fig. 71, the same argument as before can be begun. The heights of the inner and outer rectangles are now $m_r$, $M_r$, where these numbers are the lower and upper bounds (see 6.1) of $f(x)$ in $x_r \leqslant x \leqslant x_{r+1}$. (In this figure, $M_r$ is actually a value of $f(x)$, but $m_r$ will not be unless the function is so defined at the discontinuity.) The inner and outer rectangles now have areas

$$m_r(x_{r+1} - x_r), \quad M_r(x_{r+1} - x_r),$$

and we may consider the sums

$$s = \sum_{r=0}^{n-1} m_r(x_{r+1} - x_r), \quad S = \sum_{r=0}^{n-1} M_r(x_{r+1} - x_r) \tag{iii}$$

which may be called *lower* and *upper sums* for the function $f(x)$ over the range $a \leqslant x \leqslant b$.

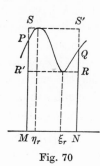

Fig. 70

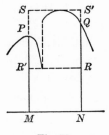

Fig. 71

## 7.12 The lower and upper sums

There is nothing vague about the sums (iii): we have obtained them by (*a*) taking a subdivision of the range $a \leqslant x \leqslant b$ by $n-1$ points $x_1 < x_2 < \ldots < x_{n-1}$; (*b*) taking the lower and upper bounds $m_r$, $M_r$ of $f(x)$ in $x_r \leqslant x \leqslant x_{r+1}$, for each of these subintervals; (*c*) adding up the products like $m_r(x_{r+1} - x_r)$ corresponding to each subinterval, to obtain the sum $s$, and similarly for $S$. We may therefore consider these sums quite independently of any notions of 'area' (although we may visualise them geometrically as sums of areas of *rectangles* associated with the curve $y = f(x)$ and the points $x = a$, $x_1, x_2, \ldots, x_{n-1}$, $b$).

We remark first that, given $f(x)$, there is still plenty of choice in the way in which we can construct $s$, $S$: the number $n-1$ of points chosen can be as large as we please and (subject only to the restriction $a < x_1 < x_2 < \ldots < x_{n-1} < b$) the points themselves can be arbitrarily placed in $a < x < b$. Thus $s$, $S$ depend on $n$, the number of subintervals; and even when $n$ is assigned, both depend on the numbers $x_1, x_2, \ldots,$ $x_{n-1}$. We express this by saying that $s$, $S$ are *functions of the subdivision* $\{x_1, x_2, \ldots, x_{n-1}\}$ of $a \leqslant x \leqslant b$.

When we vary the subdivision, $s$ and $S$ will in general vary. We may ask (in view of the intuitive considerations in 7.11) what happens to $s$ and $S$ when $n \to \infty$ and each subinterval tends to zero. This question was considered by Riemann (1826–66), and we shall state the answer without proof (because this would involve fundamental theorems on bounds of a function, and would take us too far afield). It can be proved that, *provided $f(x)$ is bounded for $a \leqslant x \leqslant b$*, there are numbers $j$, $J$ such that, when $n \to \infty$ and each interval of the subdivision tends to zero, then $s \to j$ and $S \to J$; and $j \leqslant J$.

*Definition.* If $j = J$, then $f(x)$ is said to be *integrable in $a \leqslant x \leqslant b$* (*in the sense of Riemann*).

In this case the sums like $s$ and the sums like $S$ have a common limit, viz. $j = J$. It is then natural (in view of the discussion in 7.11) to define *the area under the curve $y = f(x)$ between $x = a$, $x = b$* to be this common limit. Thus, if the lower and upper sums possess a common limit, there is defined thereby an 'area' under the curve $y = f(x)$; if the limits are different, then the term 'area' remains undefined.

It can be proved that we certainly have $j = J$ whenever (i) $f(x)$ is bounded in $a \leqslant x \leqslant b$ and continuous in $a < x < b$; or (ii) $f(x)$ is bounded and either non-decreasing or else non-increasing in $a \leqslant x \leqslant b$. That is, *every bounded continuous function and every bounded monotonic function is integrable*; and the corresponding curve has associated with it an area in the sense just defined.

### 7.13 Definite integral defined arithmetically

In 4.14 we inferred (again intuitively) that the 'area function' $A(x)$ satisfies the differential equation $dA/dx = y$ when $y = f(x)$ is continuous in $a \leqslant x \leqslant b$, and we concluded that the expression for the total area $AHKB$ (fig. 44) was

$$\left[ \int f(x)\, dx \right]_a^b, \qquad (iv)$$

which we took as the definition of $\int_a^b f(x)\, dx$. We therefore expect some close connection between the common limit of the upper and lower sums of a continuous function, and the definite integral of that function.

We remind the reader (cf. 4.16(2)) that our definition (iv) of a definite integral has a logical defect: it depends on the concept of an *indefinite integral* or *primitive function* $\phi(x) = \int f(x)\, dx$ of $f(x)$, i.e. a function $\phi(x)$ whose derivative is $f(x)$. Unless we can actually find

$\phi(x)$ explicitly, we cannot be certain that such a function even *exists*; hence any attempt at obtaining *general* properties of definite integrals (as in 4.15) would be hampered by ignorance of circumstances under which the integrals themselves have a meaning.

To give an independent account of the definite integral it is best to actually *define* the symbol $\int_a^b f(x)\,dx$ to be the common limit (when this exists) of the upper and lower sums of $f(x)$ over $a \leqslant x \leqslant b$. If $a = b$, we *define* $\int_a^a f(x)\,dx$ to be zero, and if $a > b$ we *define*

$$\int_a^b f(x)\,dx = -\int_b^a f(x)\,dx.$$

(This agrees with 4.15, properties (1), (2).)

With these definitions, the other general properties of definite integrals given in 4.15 (3)–(11) can all be established. We now prove those required for carrying out the programme outlined in 4.16 (3).

## 7.14 Properties; existence of an indefinite integral

We shall suppose that $f(x)$ is integrable in $a \leqslant x \leqslant b$ and has bounds $m, M$.

(1) $\int_a^b f(x)\,dx$ *lies between* $m(b-a)$ *and* $M(b-a)$. (Cf. 4.15 (10).)

First suppose $b > a$. Since $M_r \leqslant M$ and $m_r \geqslant m$ for each $r$, hence

$$s = \sum_{r=0}^{n-1} m_r(x_{r+1}-x_r) \quad \text{and} \quad S = \sum_{r=0}^{n-1} M_r(x_{r+1}-x_r) \quad \text{both lie between}$$

$M\Sigma(x_{r+1}-x_r) = M(b-a)$ and $m\Sigma(x_{r+1}-x_r) = m(b-a)$. Hence the common limit of $s, S$ also lies between these numbers:

$$m(b-a) \leqslant \int_a^b f(x)\,dx \leqslant M(b-a).$$

If $b < a$, apply the result just proved to $\int_b^a f(x)\,dx$, multiply the inequalities by $-1$, and use the definition of $\int_a^b f(x)\,dx$ when $b < a$. We get

$$M(b-a) \leqslant \int_a^b f(x)\,dx \leqslant m(b-a).$$

If $a = b$, the result is trivial.

COROLLARY. *If* $|f(x)| \leqslant K$ *for* $a \leqslant x \leqslant b$, *then* $\left| \int_a^b f(x)\,dx \right| \leqslant K(b-a)$.

For $K$ is the greater of $|m|, |M|$.

(2) *If* $a < c < b$, $\displaystyle\int_a^c f(x)\,dx + \int_c^b f(x)\,dx = \int_a^b f(x)\,dx$. (Cf. 4.15 (3).)

This property (at first sight rather obvious) is proved by taking $c$ as a point of the subdivision of $a \leqslant x \leqslant b$. We omit the details.

(3) *Definite integral as a function of its upper limit.* According to its definition as a limit of sums, the definite integral $\displaystyle\int_a^b f(x)\,dx$ of a given function $f(x)$ depends on $a$ and $b$, but not on the variable $x$: cf. Remark ($\beta$) of 4.15. We now allow the upper limit $b$ to vary, and consider the function

$$F(x) = \int_a^x f(t)\,dt,$$

where $a$ is still regarded as fixed.

I.  $F(x)$ *is continuous in* $a \leqslant x \leqslant b$. (Cf. 4.15 (6).)

Let $a \leqslant c < b$ and suppose $h > 0$. By (2)

$$\int_a^{c+h} f(t)\,dt = \int_a^c f(t)\,dt + \int_c^{c+h} f(t)\,dt,$$

so 
$$F(c+h) - F(c) = \int_c^{c+h} f(t)\,dt. \tag{v}$$

The bounds of $f(t)$ in $c \leqslant t \leqslant c+h$ cannot exceed the bounds $m$, $M$ of $f(t)$ in the whole interval $a \leqslant t \leqslant b$. Since $h > 0$, (1) gives

$$mh \leqslant F(c+h) - F(c) \leqslant Mh.$$

When $h \to 0+$, this shows that $F(c+h) \to F(c)$.

Similarly, if $a < c \leqslant b$ and $h < 0$, we can prove that $F(c+h) \to F(c)$ when $h \to 0-$. Hence $F(x)$ is continuous in $a \leqslant x \leqslant b$, even at the end-points.

II.  *If* $a < c < b$ *and* $f(x)$ *is continuous at* $x = c$, *then* $F(x)$ *is derivable at* $x = c$, *and* $F'(c) = f(c)$. (Cf. 4.15 (7).)

Given $\epsilon > 0$, there is a number $\eta$ such that, when $c - \eta \leqslant t \leqslant c + \eta$, $|f(t) - f(c)| < \epsilon$, i.e. 
$$f(c) - \epsilon < f(t) < f(c) + \epsilon,$$

and hence the bounds of $f(t)$ for this range lie between $f(c) \pm \epsilon$. If $0 < |h| < \eta$, then by (1)

$$\int_c^{c+h} f(t)\,dt \quad \text{lies between} \quad h\{f(c) - \epsilon\} \quad \text{and} \quad h\{f(c) + \epsilon\},$$

so that by (v)

$$\frac{F(c+h) - F(c)}{h} \quad \text{lies between} \quad f(c) \pm \epsilon,$$

i.e.
$$\left| \frac{F(c+h) - F(c)}{h} - f(c) \right| \leqslant \epsilon.$$

Hence
$$\lim_{h \to 0} \frac{F(c+h) - F(c)}{h} = f(c).$$

III. *If $\phi(x)$ possesses a derivative $f(x)$ which is bounded in $a \leqslant x \leqslant b$ and is continuous in $a < x < b$, then $\int_a^x f(t) \, dt = \phi(x) - \phi(a)$ for $a \leqslant x \leqslant b$.*

By II, $F'(x) = f(x)$ for $a < x < b$; and by hypothesis, $\phi'(x) = f(x)$ for $a < x < b$. Hence $F'(x) = \phi'(x)$ for $a < x < b$.

By I, $F(x)$ is continuous for $a \leqslant x \leqslant b$; and since by hypothesis $\phi'(x)$ exists for $a \leqslant x \leqslant b$, $\phi(x)$ must be continuous for $a \leqslant x \leqslant b$ (3.12).

Hence by 3.82, Corollary, $F(x) - \phi(x)$ is constant for $a \leqslant x \leqslant b$. Since $F(a) = 0$ by definition (7.13), we find on putting $x = a$ that this constant is $-\phi(a)$. Hence $F(x) = \phi(x) - \phi(a)$. (We now see that $F(x)$ is the 'area function' $A(x)$ of 4.14.)

**(4)** *Remarks on Theorems* II, III.

Theorem II shows that, when $f(x)$ is continuous in $a < x < b$, then in this range it is the derivative of another function, viz. $F(x) = \int_a^x f(t) \, dt$ (which certainly exists if $f(x)$ is continuous); i.e.,

$$\frac{dy}{dx} = f(x) \quad \text{is satisfied by} \quad y = F(x).$$

Theorem III shows that *any* function which satisfies $d\{\phi(x)\}/dx = f(x)$ in $a \leqslant x \leqslant b$ differs from $F(x)$ in $a \leqslant x \leqslant b$ by a constant at most, and that

$$\int_a^b f(x) \, dx = \phi(b) - \phi(a) = [\phi(x)]_a^b = \left[ \int f(x) \, dx \right]_a^b,$$

which is what we took as a definition in 4.15.

Thus, by starting from the precise idea of lower and upper sums, our theory of definite integrals does not depend on whether we can find another function $\int f(x) \, dx$ having the given $f(x)$ for derivative, but (subject to the continuity of $f(x)$) actually *defines* such a function $F(x)$. This situation has been illustrated in 4.31 where, in order to investigate $\int x^{-1} dx$ (i.e. a function satisfying $dy/dx = 1/x$) we first examined the function $\int_1^x t^{-1} dt$.

We conclude with two statements which may surprise the reader.

(i) The definite integral (as a limit of sums) may exist even when there is no function having $f(x)$ for derivative in $a \leqslant x \leqslant b$. (Of course $f(x)$ could not be continuous in such a case, by II above, but yet it may be integrable, i.e. $j = J$.)

(ii) If $g(x)$ has a derivative $g'(x)$ at each point of $a \leqslant x \leqslant b$, then $g'(x)$ is not necessarily integrable in $a \leqslant x \leqslant b$.

Examples to illustrate these statements can be constructed, but not easily.

## 7.2    Definite integral as the limit of a single summation

**7.21** Let the interval $a \leqslant x \leqslant b$ be divided by the points

$$x_0 = a < x_1 < x_2 < \ldots < x_{n-1} < b = x_n,$$

and let $\xi_r$ be any number such that $x_r \leqslant \xi_r \leqslant x_{r+1}$ for each $r = 0, 1, \ldots, n-1$. Then, if $m_r$, $M_r$ are the bounds of $f(x)$ in $x_r \leqslant x \leqslant x_{r+1}$, we have $m_r \leqslant f(\xi_r) \leqslant M_r$, so that

$$\sum_{r=0}^{n-1} m_r(x_{r+1} - x_r) \leqslant \sum_{r=0}^{n-1} f(\xi_r)(x_{r+1} - x_r) \leqslant \sum_{r=0}^{n-1} M_r(x_{r+1} - x_r),$$

i.e.
$$s \leqslant \sum_{r=0}^{n-1} f(\xi_r)(x_{r+1} - x_r) \leqslant S.$$

If $f(x)$ is integrable, then $s$ and $S$ tend to the common limit $\int_a^b f(x)\,dx$ when $n \to \infty$ and all the differences $x_{r+1} - x_r \to 0$. Hence

$$\Sigma f(\xi_r)(x_{r+1} - x_r)$$

tends to this same limit; i.e.

$$\lim \sum_{r=0}^{n-1} f(\xi_r)(x_{r+1} - x_r) = \int_a^b f(x)\,dx.$$

This result will be of practical importance to the reader. The point to grasp is that the limit of a summation of the above type (which arises in calculation of areas, volumes, arc-lengths, centres of gravity, etc. discussed later in this chapter) is a definite integral.

## 7.22    Some definite integrals calculated as limiting sums

(i) $\int_a^b x\,dx,\ b > a.$

Divide the range $a \leqslant x \leqslant b$ into $n$ intervals each of length $h$ (the calculation is simplified by choosing *equal* intervals); then

$$x_1 = a + h, \quad x_2 = a + 2h, \quad \ldots, \quad x_{n-1} = a + (n-1)h,$$

and
$$nh = b - a.$$

Again for simplicity, choose $\xi_r = x_r$ in each interval. Then the sum to be considered is

$$\sum_{r=0}^{n-1} x_r . h = h[a+(a+h)+(a+2h)+ \ldots +\{a+(n-1)\,h\}]$$

$$= h[na+h\{1+2+ \ldots +(n-1)\}]$$

$$= h[na+\tfrac{1}{2}n(n-1)\,h] \quad \text{by summing the A.P.,}$$

$$= nh\left[a+\tfrac{1}{2}nh\left(1-\frac{1}{n}\right)\right]$$

$$= (b-a)\left[a+\tfrac{1}{2}(b-a)\left(1-\frac{1}{n}\right)\right]$$

$$\to (b-a)\,[a+\tfrac{1}{2}(b-a)] \quad \text{when} \quad n \to \infty.$$

$$\therefore \quad \int_a^b x\,dx = \lim \sum_{r=0}^{n-1} x_r h = \tfrac{1}{2}(b-a)\,(b+a) = \tfrac{1}{2}(b^2-a^2).$$

(ii) $\displaystyle\int_a^b x^2\,dx$. By proceeding similarly, we consider

$$\sum_{r=0}^{n-1} x_r^2 h = h[a^2+(a+h)^2+ \ldots +\{a+(n-1)\,h\}^2]$$

$$= h[na^2+\{1+2+ \ldots +(n-1)\}\,2ah+\{1^2+2^2+ \ldots +(n-1)^2\}\,h^2]$$

$$= h[na^2+n(n-1)\,ah+\tfrac{1}{6}(n-1)\,n(2n-1)\,h^2] \quad \text{(see 12.24 (2))}$$

$$= nh\left[a^2+anh\left(1-\frac{1}{n}\right)+\tfrac{1}{6}n^2h^2\left(1-\frac{1}{n}\right)\left(2-\frac{1}{n}\right)\right]$$

$$= (b-a)\left[a^2+a(b-a)\left(1-\frac{1}{n}\right)+\tfrac{1}{6}(b-a)^2\left(1-\frac{1}{n}\right)\left(2-\frac{1}{n}\right)\right]$$

$$\to (b-a)\,[a^2+a(b-a)+\tfrac{1}{3}(b-a)^2] \quad \text{when} \quad n \to \infty.$$

$$\therefore \quad \int_a^b x^2\,dx = \lim \sum_{r=0}^{n-1} x_r^2 h = \tfrac{1}{3}(b-a)\,(b^2+ab+a^2) = \tfrac{1}{3}(b^3-a^3).$$

(iii) $\displaystyle\int_a^b e^{kx}\,dx$. This is the limit of

$$h[e^{ka}+e^{k(a+h)}+ \ldots +e^{k\{a+(n-1)h\}}]$$

$$= h\,e^{ka}\,[1+e^{kh}+ \ldots +e^{(n-1)\,kh}]$$

$$= h\,e^{ka}\,\frac{e^{nkh}-1}{e^{kh}-1} \quad \text{by summing the G.P.,}$$

$$= h\,\frac{e^{kb}-e^{ka}}{e^{kh}-1} \quad \text{since} \quad b = a+nh,$$

$$= \frac{1}{k}\,(e^{kb}-e^{ka})\,\frac{kh}{e^{kh}-1} \to \frac{1}{k}\,(e^{kb}-e^{ka})$$

when $h \to 0$, since $\lim\limits_{x\to 0}\dfrac{x}{e^x-1} = 1$ (Ex. 4 $(f)$, no. 19). Hence

$$\int_a^b e^{kx}\,dx = \frac{1}{k}\,(e^{kb}-e^{ka}).$$

(iv) $\displaystyle\int_a^b \sin x\, dx.$

We divide the range into equal intervals of length $h$, but here it is most convenient to take $\xi_r$ at the *middle* of the interval $x_r \leqslant x \leqslant x_{r+1}$, i.e.

$$\xi_r = a + (r+\tfrac{1}{2})h.$$

Then

$$\sum_{r=0}^{n-1} \sin(x_r + \tfrac{1}{2}h).h = h[\sin(a+\tfrac{1}{2}h) + \sin(a+\tfrac{3}{2}h) + \ldots + \sin\{a+(n-\tfrac{1}{2})h\}]$$

$$= \frac{\tfrac{1}{2}h}{\sin\tfrac{1}{2}h}[\cos a - \cos(a+nh)]$$

by 12.27 (1), and hence the sum considered is

$$\frac{\tfrac{1}{2}h}{\sin\tfrac{1}{2}h}(\cos a - \cos b) \to \cos a - \cos b \quad \text{when} \quad h \to 0.$$

$$\therefore \quad \int_a^b \sin x\, dx = \cos a - \cos b.$$

(v) $\displaystyle\int_a^b x^m\, dx, \ m \neq -1, \ 0 < a < b.$

Instead of dividing the interval $a \leqslant x \leqslant b$ by points whose $x$-coordinates are in A.P., as in exs. (i)–(iv), we now divide it by points

$$x_1 = ar, \quad x_2 = ar^2, \quad \ldots, \quad x_{n-1} = ar^{n-1}$$

in G.P., where $b = ar^n$ (*Wallis's method*). Taking $\xi_s = x_s$ ($s = 0, 1, \ldots, n-1$),

$$\sum_{s=0}^{n-1} x_s^m(x_{s+1} - x_s) = a^m(ar - a) + a^m r^m(ar^2 - ar)$$

$$+ a^m r^{2m}(ar^3 - ar^2) + \ldots + a^m r^{(n-1)m}(ar^n - ar^{n-1})$$

$$= a^{m+1}(r-1)[1 + r^{m+1} + r^{2(m+1)} + \ldots + r^{(n-1)(m+1)}]$$

$$= a^{m+1}(r-1)\frac{1 - r^{(m+1)n}}{1 - r^{m+1}} \quad \text{by summing the G.P.,}$$

$$= a^{m+1}(r-1)\left\{1 - \left(\frac{b}{a}\right)^{m+1}\right\}\Big/(1 - r^{m+1})$$

$$= \{b^{m+1} - a^{m+1}\}\frac{1 - r^{m+1}}{1 - r}.$$

When $n \to \infty$, then $r \to 1$ by 2.76, (i), and $(1 - r^{m+1})/(1-r) \to m+1$ (e.g. by l'Hospital's first rule, 6.92). Hence

$$\int_a^b x^m\, dx = \frac{1}{m+1}(b^{m+1} - a^{m+1}).$$

These examples show that even the simplest integrals require quite a lengthy calculation. The practical importance of 7.14 (3), Theorem III† is that it relates integration (a summation process) and the process inverse to derivation: if we know a function $\phi(x)$ such that $\phi'(x) = f(x)$ under suitable conditions, then the value of the limiting sum can be *written down* as $\phi(b) - \phi(a)$, and the tedious calculation from first principles avoided.

† The so-called 'fundamental theorem of the integral calculus'.

## 7.23 Formula for change of variable in a definite integral

In 4.22 we obtained a formula for evaluation of $\int_a^b f(x)\,dx$ by the substitution $x = g(t)$. We now prove it as a further illustration of the use of limiting sums.

Let $a = g(\alpha)$, $b = g(\beta)$. We assume $f(x)$ is continuous in $a \leqslant x \leqslant b$, that $g(t)$ is continuous in $\alpha \leqslant t \leqslant \beta$, and that $g'(t)$ is continuous in $\alpha \leqslant t \leqslant \beta$. We also assume that $g(t)$ is an increasing function in $\alpha \leqslant t \leqslant \beta$. Divide $\alpha \leqslant t \leqslant \beta$ into subintervals by $t_1, t_2, \ldots, t_{n-1}$, where

$$\alpha < t_1 < t_2 < \ldots < t_{n-1} < \beta,$$

and let the corresponding values of $x$ be $x_1, x_2, \ldots, x_{n-1}$. Since $g(t)$ increases,

$$a < x_1 < x_2 < \ldots < x_{n-1} < b.$$

By the mean value theorem,

$$x_{r+1} - x_r = (t_{r+1} - t_r)\,g'(\tau_r), \qquad\qquad \text{(i)}$$

where $t_r < \tau_r < t_{r+1}$. Let $\xi_r = g(\tau_r)$; then $x_r < \xi_r < x_{r+1}$, and

$$\sum_{r=0}^{n-1} f(\xi_r)(x_{r+1} - x_r) = \sum_{r=0}^{n-1} f\{g(\tau_r)\}\,g'(\tau_r)\,(t_{r+1} - t_r).$$

The definite integral $\int_a^b f(x)\,dx$ is the limit of the left-hand sum when $n \to \infty$ and all the differences $x_{r+1} - x_r \to 0$. In $\alpha \leqslant t \leqslant \beta$, $g'(t)$ is continuous and hence bounded (6.1(1)). Therefore, if all of $t_{r+1} - t_r$ tend to zero, so do all of $x_{r+1} - x_r$, by (i). From the hypothesis, $f\{g(t)\}\,g'(t)$ is continuous; hence

$$\int_\alpha^\beta f\{g(t)\}\,g'(t)\,dt = \lim \sum_{r=0}^{n-1} f\{g(\tau_r)\}\,g'(\tau_r)\,(t_{r+1} - t_r)$$

when $n \to \infty$ and all of $t_{r+1} - t_r \to 0$. Thus

$$\int_a^b f(x)\,dx = \int_\alpha^\beta f\{g(t)\}\,g'(t)\,dt.$$

If $g(t)$ *decreases* from $b$ to $a$ as $t$ increases from $\alpha$ to $\beta$, then to the subdivision

$$\alpha < t_1 < t_2 < \ldots < t_{n-1} < \beta$$

corresponds     $b > x_1 > x_2 > \ldots > x_{n-1} > a;$

and since $t_r < \tau_r < t_{r+1}$, we now have $x_r > \xi_r > x_{r+1}$.

$$\begin{aligned}
\int_a^b f(x)\,dx &= \lim \sum_{r=0}^{n-1} f(\xi_r)(x_r - x_{r+1}) \\
&= -\lim \Sigma f\{g(\tau_r)\}\,g'(\tau_r)\,(t_{r+1} - t_r) \\
&= -\int_\alpha^\beta f\{g(t)\}\,g'(t)\,dt \\
&= \int_\beta^\alpha f\{g(t)\}\,g'(t)\,dt \quad \text{by definition.}
\end{aligned}$$

The proof emphasises the need for $g(t)$ to be steadily increasing or steadily decreasing in $\alpha \leqslant t \leqslant \beta$, otherwise the points $x_r$ corresponding to $t_r$ would not be in order. Compare the Remark at the end of 4.22.

## Exercise 7(a)*

*Assuming the functions are continuous, prove the following properties (nos. 1–5) from 'limiting sums'.*

**1** $\displaystyle\int_a^b kf(x)\,dx = k\int_a^b f(x)\,dx$, $k$ being constant.

**2** If $f(x) \geqslant 0$ for $a \leqslant x \leqslant b$, then $\displaystyle\int_a^b f(x)\,dx \geqslant 0$. [$\Sigma m_r(x_{r+1}-x_r)$ cannot be negative.]

**3** If $f(x) = g(x) + h(x)$, then $\displaystyle\int_a^b f(x)\,dx = \int_a^b g(x)\,dx + \int_a^b h(x)\,dx$. [Use 7.21.]

**4** If $f(x) \geqslant g(x)$ for $a \leqslant x \leqslant b$, prove $\displaystyle\int_a^b f(x)\,dx \geqslant \int_a^b g(x)\,dx$. [Use nos. 2, 3.]

**5** $\displaystyle\left|\int_a^b f(x)\,dx\right| \leqslant \int_a^b |f(x)|\,dx$. [For if $\sigma = \Sigma f(\xi_r)\,(x_{r+1}-x_r)$ and

$$\bar{\sigma} = \Sigma |f(\xi_r)|\,(x_{r+1}-x_r),$$

then $|\sigma| \leqslant \bar{\sigma}$ because all terms of $\bar{\sigma}$ are positive but all those of $\sigma$ may not be.]

**6** Prove that $\displaystyle\frac{d}{dx}\int_x^b f(t)\,dt = -f(x)$ for $a < x < b$.

**7** Prove $\displaystyle\lim_{n\to\infty}\left[\frac{1}{n}\left\{f\!\left(a+\frac{h}{n}\right)+f\!\left(a+\frac{2h}{n}\right)+\ldots+f(a+h)\right\}\right] = \frac{1}{h}\int_a^{a+h} f(x)\,dx$.

(This generalises the idea of the 'average of $n$ numbers' to that of the 'mean value over an interval' of a function of a continuous variable $x$.)

**8** Prove $\displaystyle\lim_{n\to\infty} n\sum_{r=0}^{n-1}\frac{1}{n^2+r^2} = \tfrac{1}{4}\pi$. $\left[\Sigma\dfrac{n}{n^2+r^2} = \Sigma\dfrac{1/n}{1+(r/n)^2} \to \displaystyle\int_0^1\frac{dx}{1+x^2}.\right]$

## 7.3 Approximate calculation of definite integrals

**7.31** When the indefinite integral of $f(x)$ is not known, we may resort to approximative methods to calculate $\displaystyle\int_a^b f(x)\,dx$, where $a$, $b$ are given numbers. Since a definite integral measures and is represented by a plane area, we may conveniently give the following discussion in geometrical language.

The crudest method of estimating areas is by 'counting squares'; this requires the curve to be drawn on squared paper. The method is useful when the curve itself is given but its equation is not known. When the equation of the curve is known, the following methods are less tedious.

### 7.32 Trapezium rule

Divide the given area into $n$ strips by $n+1$ *equally spaced* ordinates $y_1, y_2, \ldots, y_{n+1}$ at distance $h$ apart. Let $PM$, $QN$ be two consecutive ordinates (fig. 72). The method *replaces the arc $PQ$ of the curve by the chord $PQ$* and uses the area of the trapezium $PMNQ$, viz. $\tfrac{1}{2}h(PM + QN)$, as an approximation to the area of the strip. The required area is thus approximately the sum of all the trapezia formed by joining the tops of consecutive ordinates by straight lines, viz.

$$\tfrac{1}{2}h(y_1+y_2)+\tfrac{1}{2}h(y_2+y_3)+\tfrac{1}{2}h(y_3+y_4)+\ldots+\tfrac{1}{2}h(y_n+y_{n+1})$$
$$= h\{\tfrac{1}{2}(y_1+y_{n+1})+(y_2+y_3+\ldots+y_n)\}.$$

Hence the rule is:

*Divide the given area into strips by any number of equidistant ordinates. Take the average of the first and last ordinates, and add this to the sum of the other ordinates. Multiply the result by the distance between consecutive ordinates.*

When the curve is concave down (6.71(2)) the rule clearly underestimates the area because the chords lie below their arcs. For a curve concave up, the rule overestimates.

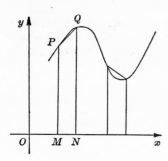

Fig. 72

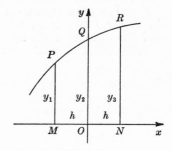

Fig. 73

## 7.33 Simpson's rule†

Instead of approximating to the curve by straight lines, we now use *parabolic arcs*.‡ First, suppose $PM$, $QO$, $RN$ are three equidistant ordinates to points $P$, $Q$, $R$ on the given curve. We may choose coordinate axes so that $Oy$ is along $OQ$. Let $PM = y_1$, $QO = y_2$, $RN = y_3$ (fig. 73).

We can choose the constants $a$, $b$, $c$ so that the curve $y = ax^2 + bx + c$ passes through the points $P(-h, y_1)$, $Q(0, y_2)$, $R(h, y_3)$; for these requirements lead to the following three equations for $a$, $b$, $c$:

$$\left.\begin{aligned} y_1 &= ah^2 - bh + c, \\ y_2 &= c, \\ y_3 &= ah^2 + bh + c. \end{aligned}\right\} \tag{i}$$

We approximate to the actual area $PMNR$ under the given curve by finding the corresponding area under the parabola $y = ax^2 + bx + c$, viz.

$$\int_{-h}^{h} (ax^2 + bx + c)\,dx = [\tfrac{1}{3}ax^3 + \tfrac{1}{2}bx^2 + cx]_{-h}^{h}$$
$$= \tfrac{2}{3}ah^3 + 2ch. \tag{ii}$$

From equations (i)

$$y_1 + y_3 = 2ah^2 + 2c = 2ah^2 + 2y_2,$$

so

$$2ah^2 = y_1 + y_3 - 2y_2.$$

† Thomas Simpson (1710–61).
‡ This title is justified by Ex. 16(e), no. 4.

Hence by (ii) the area under the parabola is

$$\tfrac{1}{3}h(y_1 + y_3 - 2y_2) + 2hy_2 = \tfrac{1}{3}h(y_1 + 4y_2 + y_3). \tag{iii}$$

This area is of course independent of any choice of position for $Oy$: we took the $y$-axis along $OQ$ merely for convenience in calculation.

If we divide the given area into any *even* number of strips, say $2n$, by ordinates $y_1, y_2, \dots, y_{2n+1}$ *equally spaced* at distance $h$ apart, we can apply (iii) to consecutive pairs of strips and hence approximate to the required area. We obtain

$$\tfrac{1}{3}h(y_1 + 4y_2 + y_3) + \tfrac{1}{3}h(y_3 + 4y_4 + y_5) + \dots + \tfrac{1}{3}h(y_{2n-1} + 4y_{2n} + y_{2n+1})$$
$$= \tfrac{1}{3}h\{(y_1 + y_{2n+1}) + 2(y_3 + y_5 + \dots + y_{2n-1}) + 4(y_2 + y_4 + \dots + y_{2n})\}.$$

Hence *Simpson's rule*:

*Divide the given area into an* EVEN *number of strips by equidistant ordinates. To the sum of the first and last add twice the sum of the remaining odd ordinates and four times the sum of all the even ordinates. Multiply the result by $\tfrac{1}{3}$ of the distance between consecutive ordinates.*

### Example

*Calculate* $\displaystyle\int_1^2 \frac{dx}{x}$ *using* (i) *the trapezium rule*; (ii) *Simpson's rule*.

The curve concerned is $y = 1/x$. We will take 10 strips, at intervals of $0\cdot1$ apart. Using tables of reciprocals, we calculate $y_1, y_2, \dots, y_{11}$ as follows.

| $x$ | First and last ordinates | Even ordinates | Odd ordinates |
|---|---|---|---|
| 1 | $y_1 = 1$ | | |
| 1·1 | | $y_2 = 0\cdot9091$ | |
| 1·2 | | | $y_3 = 0\cdot8333$ |
| 1·3 | | $y_4 = 0\cdot7692$ | |
| 1·4 | | | $y_5 = 0\cdot7143$ |
| 1·5 | | $y_6 = 0\cdot6667$ | |
| 1·6 | | | $y_7 = 0\cdot6250$ |
| 1·7 | | $y_8 = 0\cdot5882$ | |
| 1·8 | | | $y_9 = 0\cdot5556$ |
| 1·9 | | $y_{10} = 0\cdot5263$ | |
| 2 | $y_{11} = 0\cdot5$ | | |
| | 1·5 | 3·4595 | 2·7282 |

The work has been set out so as to be used more conveniently with Simpson's rule.

(i) *By the trapezium rule*,

$$\text{area} \doteqdot \tfrac{1}{10}\{\tfrac{1}{2} \times 1\cdot5 + (3\cdot4595 + 2\cdot7282)\} \doteqdot 0\cdot694.$$

(ii) *By Simpson's rule*,

$$\text{area} \doteqdot \tfrac{1}{30}\{1\cdot5 + 4 \times 3\cdot4595 + 2 \times 2\cdot7282\} \doteqdot 0\cdot693.$$

The value correct to 6 places of decimals is $0\cdot693147$.

## Exercise 7(b)*

1 From the formula $\frac{1}{4}\pi = \int_0^1 \dfrac{dx}{1+x^2}$, calculate $\frac{1}{4}\pi$ to four places of decimals (i) by the trapezium rule; (ii) by Simpson's rule, using (a) 3 ordinates; (b) 5 ordinates.

2 Verify by direct calculation that Simpson's rule is exact when $f(x)$ is a cubic polynomial.

3 (i) Prove that $\int_0^\infty \dfrac{dx}{1+x^4}$ differs from $\int_0^{10} \dfrac{dx}{1+x^4}$ by less than $\frac{1}{3} \times 10^{-3}$.

  (ii) Calculate $\int_0^2 \dfrac{dx}{1+x^4}$ by Simpson's rule with 11 ordinates.

  (iii) Calculate $\int_2^{10} \dfrac{dx}{1+x^4}$ by Simpson's rule with 9 ordinates.

  (iv) Hence find the approximate value of $\int_0^\infty \dfrac{dx}{1+x^4}$.

## 7.4 Further areas

The reader will have used the result $\int_a^b f(x)\,dx$ for finding the area under a curve in early work in calculus. We now consider some extensions.

### 7.41 Sign of an area

If $f(x) < 0$ when $a \leqslant x \leqslant b$, then the upper and lower sums of $f(x)$ (7.12) will be negative; and if they have a common limit, this will be non-positive. Hence the formula

$$A = \int_a^b f(x)\,dx$$

may give a negative value for $A$. We therefore interpret an area *below* $Ox$ as negative.

If $f(x)$ is positive in some parts of the interval and negative in others, the interval must be split up and the corresponding areas found separately.

### 7.42 Area between two curves

Suppose that $f(x) \geqslant g(x)$ throughout $a \leqslant x \leqslant b$. The areas under $y = f(x)$, $y = g(x)$ are respectively $\int_a^b f(x)\,dx$, $\int_a^b g(x)\,dx$. Hence the area between them is

$$\int_a^b \{f(x) - g(x)\}\,dx. \tag{i}$$

This formula gives the correct area whether or not the $x$-axis cuts the curves.

### 7.43 Area of certain closed curves

Given a closed curve which is met by a line parallel to $Oy$ in at most two points, then the equation of the curve provides the functions $f(x)$, $g(x)$ in 7.42, and (i) gives the area.

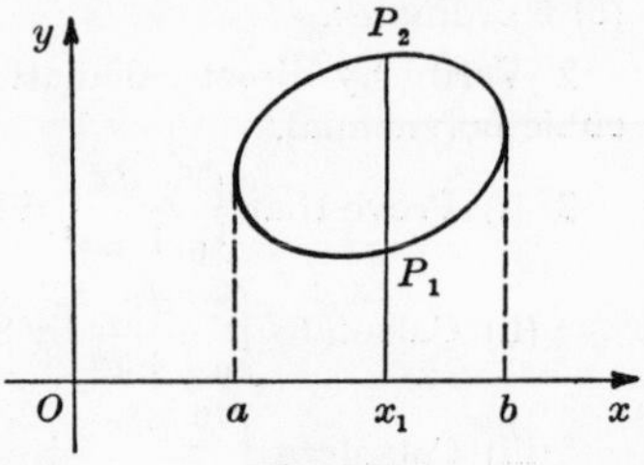

### Example

*Find the area enclosed by*

$$3x^2 - 10xy + 10y^2 + 8x - 20y + 10 = 0.$$

The two values of $y$ corresponding to a given $x$ are found by solving the equation as a quadratic in $y$

Fig. 74

$$10y^2 - 10y(x+2) + (3x^2 + 8x + 10) = 0.$$

If $y_1$, $y_2$ are the roots, where $y_1 \geqslant y_2$, then

$$y_1 + y_2 = x + 2, \quad y_1 y_2 = \tfrac{1}{10}(3x^2 + 8x + 10),$$

hence            $(y_1 - y_2)^2 = (y_1 + y_2)^2 - 4y_1 y_2 = \tfrac{1}{5}x(4-x).$

The two values of $y$ become equal when $\tfrac{1}{5}x(4-x) = 0$, i.e. when $x = 0$ or 4. These are the extreme values of $x$ for the curve. The area is

$$\int_0^4 (y_1 - y_2)\,dx = \frac{1}{\sqrt{5}} \int_0^4 \sqrt{\{x(4-x)\}}\,dx$$

$$= \frac{32}{\sqrt{5}} \int_0^{\frac{1}{2}\pi} \sin^2\theta \cos^2\theta\,d\theta = \tfrac{2}{5}\pi\sqrt{5}$$

by putting $x = 4\sin^2\theta$.

### 7.44 Generalised areas

Since    $\displaystyle\int_1^X \frac{dx}{x^2} = \left[-\frac{1}{x}\right]_1^X = 1 - \frac{1}{X} \to 1$    when    $X \to \infty$,

we may say that the infinite integral $\displaystyle\int_1^\infty \frac{dx}{x^2}$, whose value is 1, represents the 'area' under that part of the curve $y = 1/x^2$ for which $x > 1$ (cf. Ex. 4 (*d*), no. 20). This region is unbounded, and its 'area' has been defined as the limit of the area of the bounded region enclosed between $x = 1$, $x = X$ when $X \to \infty$.

Similarly, if the integral $\displaystyle\int_a^\infty f(x)\,dx$ exists, we may define its value to be the measure of the generalised area under the curve $y = f(x)$ for which $x > a$. Other types of generalised integral can also be associated with areas.

### 7.45 Area of a sector (polar coordinates)

Given a plane curve $r = f(\theta)$, where $f(\theta)$ is continuous for $\alpha \leqslant \theta \leqslant \beta$, we find a formula for the area of the sector bounded by the arc $AB$ and the radii $OA, OB$ given by $\theta = \alpha, \beta$. We first assume that $r$ steadily increases with $\theta$, and that each radius within the angle $AOB$ cuts the arc just once.

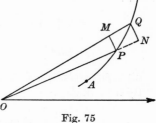

Divide the sector into $n$ elementary sectors by radii of inclinations

$$\alpha = \theta_0 < \theta_1 < \theta_2 < \ldots < \theta_{n-1} < \theta_n = \beta.$$

Fig. 75

Let $OPQ$ be a typical sector, bounded by radii $\theta = \theta_r, \theta_{r+1}$. Construct two circular arcs with centre $O$ and radii $r = f(\theta_r)$, $r + \delta r = f(\theta_{r+1})$, forming the circular sectors $OPM, ONQ$ whose areas are

$$\tfrac{1}{2}r^2(\theta_{r+1} - \theta_r), \quad \tfrac{1}{2}(r + \delta r)^2 (\theta_{r+1} - \theta_r).$$

The area of $OPQ$ lies between these. Hence that of the sector $OAB$ lies between

$$\sum_{r=0}^{n-1} \tfrac{1}{2}\{f(\theta_r)\}^2 (\theta_{r+1} - \theta_r), \quad \sum_{r=0}^{n-1} \tfrac{1}{2}\{f(\theta_{r+1})\}^2 (\theta_{r+1} - \theta_r).$$

When $n \to \infty$ and the differences $\theta_{r+1} - \theta_r \to 0$, these two sums tend to the common limit $\int_\alpha^\beta \tfrac{1}{2}\{f(\theta)\}^2 \, d\theta$. Hence

$$\text{area of sector } OAB = \frac{1}{2} \int_\alpha^\beta r^2 d\theta. \tag{ii}$$

If $r$ steadily decreases as $\theta$ increases from $\alpha$ to $\beta$, the same argument applies. In the general case we divide the arc into intervals in each of which $r$ steadily increases or steadily decreases, and apply (ii) to each part separately.

If the radius cuts the curve in more than one point (for example in the case of spirals) then as $\theta$ increases from 0 to $2\pi$, the shortest radius sweeps out a certain area; and as $\theta$ increases from $2\pi$ to $4\pi$, the area swept out includes the first one; and so on.

## Examples

(i) *Cardioid* $r = a(1 + \cos \theta)$.

The curve is symmetrical about the initial line, so the area is

$$2 \int_0^\pi \tfrac{1}{2} r^2 d\theta = \int_0^\pi a^2 (1 + \cos \theta)^2 d\theta$$

$$= a^2 \int_0^\pi (1 + 2\cos \theta + \cos^2 \theta) \, d\theta = a^2 [\theta + 2 \sin \theta + \tfrac{1}{2}(\theta + \tfrac{1}{2} \sin 2\theta)]_0^\pi$$

$$= \tfrac{3}{2} \pi a^2.$$

(ii) *Equiangular spiral* $r = a e^{k\theta}$.

By the formula, the area enclosed by the radii $r_1$, $r_2$ is

$$\frac{1}{2} \int_{\theta_1}^{\theta_2} r^2 d\theta = \tfrac{1}{2} a^2 \int_{\theta_1}^{\theta_2} e^{2k\theta} d\theta = \frac{a^2}{4k} [e^{2k\theta}]_{\theta_1}^{\theta_2}$$

$$= \frac{a^2}{4k} (e^{2k\theta_2} - e^{2k\theta_1}) = \frac{a^2}{4k} (r_2^2 - r_1^2).$$

If $\theta_2 - \theta_1 > 2\pi$, this result will include some parts of the area more than once.

## 7.46 Area of a sector (parametric formula)

Formula (ii) can be transformed as follows. Since

$$r^2 = x^2 + y^2 \quad \text{and} \quad \tan \theta = \frac{y}{x},$$

$$r^2 \frac{d\theta}{dt} = (x^2 + y^2) \frac{d}{dt} \left( \tan^{-1} \frac{y}{x} \right)$$

$$= (x^2 + y^2) \times \frac{1}{1 + (y/x)^2} \frac{x\dot{y} - y\dot{x}}{x^2}$$

$$= x\dot{y} - y\dot{x}.$$

If the given curve has parametric equations $x = \phi(t)$, $y = \psi(t)$, and the points $A$, $B$ correspond to $t = t_1, t_2$, then

$$\text{area of sector } OAB = \frac{1}{2} \int_\alpha^\beta r^2 d\theta = \frac{1}{2} \int_{t_1}^{t_2} r^2 \frac{d\theta}{dt} dt$$

$$= \frac{1}{2} \int_{t_1}^{t_2} \left( x \frac{dy}{dt} - y \frac{dx}{dt} \right) dt. \qquad \text{(iii)}$$

The change of variable is valid if $d\theta/dt$, i.e. $x\dot{y} - y\dot{x}$, retains the same sign for $t_1 < t < t_2$. If it does not, the range must be divided up and (iii) applied to the separate parts.

## Examples

(i) *Ellipse* $x = a\cos t$, $y = b\sin t$.

The curve is traced as $t$ increases from 0 to $2\pi$. Since

$$x\frac{dy}{dt} - y\frac{dx}{dt} = ab(\cos^2 t + \sin^2 t) = ab,$$

$$\text{area} = \frac{1}{2}\int_0^{2\pi} ab\,dt = \pi ab.$$

(ii) *Sketch the curve* $x = t + t^2$, $y = t^2 + t^3$, *and calculate the area of the loop.*

Since $x = t(1+t)$, $y = t^2(1+t)$, we see that $(a)$ as $t$ increases from $-\infty$ to $-1$, $x$ decreases from $+\infty$ to 0 and $y$ increases from $-\infty$ to 0; $(b)$ as $t$ increases from $-1$ to 0, $x$ varies from 0 to 0 through negative values and $y$ varies from 0 to 0 through positive values; $(c)$ as $t$ increases from 0 to $+\infty$, both $x$ and $y$ increase from 0 to $+\infty$.

The gradient of the curve is $\dot{y}/\dot{x} = (2t + 3t^2)/(1 + 2t)$; when $t = 0$ this is zero, while when $t = -1$ it is $-1$. The curve is therefore roughly as shown in fig. 76; the loop is traced when $t$ increases from $-1$ to 0.

Since $t = y/x$ in this example, we have by deriving this relation wo $t$ that $1 = (x\dot{y} - y\dot{x})/x^2$; hence $x\dot{y} - y\dot{x} = x^2$, and formula (iii) becomes

$$\frac{1}{2}\int_{-1}^0 x^2\,dt = \frac{1}{2}\int_{-1}^0 (t^2 + 2t^3 + t^4)\,dt = \tfrac{1}{60}$$

after simplification.

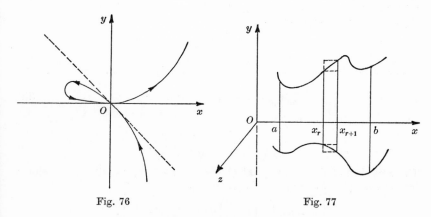

Fig. 76                    Fig. 77

## 7.5 Volume of a solid of known cross-section

Consider a surface whose sections parallel to a fixed plane are closed curves. Let $Ox$ be chosen perpendicular to this plane, and suppose that the area of the section at distance $x$ from the plane is a continuous function $\phi(x)$. We obtain a formula for the volume of the solid enclosed between the planes $x = a$, $x = b$ (fig. 77).

Divide the required volume into slices by planes through

$$a = x_0 < x_1 < x_2 < \ldots < x_{n-1} < x_n = b.$$

The slice determined by $x_r$, $x_{r+1}$ lies between inner and outer cylinders with volumes† $m_r(x_{r+1} - x_r)$ and $M_r(x_{r+1} - x_r)$, where $m_r$, $M_r$ are the least and greatest values of $\phi(x)$ for $x_r \leqslant x \leqslant x_{r+1}$. Hence the total volume lies between

$$\sum_{r=0}^{n-1} m_r(x_{r+1} - x_r) \quad \text{and} \quad \sum_{r=0}^{n-1} M_r(x_{r+1} - x_r),$$

and when $n \to \infty$ and the differences $x_{r+1} - x_r \to 0$, these sums tend to the common limit $\displaystyle\int_a^b \phi(x)\,dx$. Thus

$$V = \int_a^b \phi(x)\,dx. \tag{i}$$

In particular, if the solid is formed by revolving the plane curve $y = f(x)$ about $Ox$ though angle $2\pi$, then $\phi(x) = \pi y^2$ because the sections are now circles. The *volume of a solid of revolution* is thus

$$V = \int_a^b \pi y^2\,dx. \tag{ii}$$

**Example\***

*Find the volume of the ellipsoid*

$$\frac{x^2}{a^2} + \frac{y^2}{b^2} + \frac{z^2}{c^2} = 1.$$

The section by the plane $x = $ constant is the ellipse (see Ch. 17)

$$\frac{y^2}{b^2} + \frac{z^2}{c^2} = 1 - \frac{x^2}{a^2}, \quad x = \text{constant}, \tag{iii}$$

whose semi-axes have lengths

$$b\sqrt{\left(1 - \frac{x^2}{a^2}\right)}, \quad c\sqrt{\left(1 - \frac{x^2}{a^2}\right)}.$$

The area $\phi(x)$ of this ellipse is $\pi bc(1 - x^2/a^2)$ (by 7.46, ex. (i)). The volume between the planes $x = 0$, $x = X$ is therefore

$$V = \pi bc \int_0^X \left(1 - \frac{x^2}{a^2}\right) dx = \pi bc \left(X - \frac{X^3}{3a^2}\right).$$

The largest possible values of $x$ are given from (iii) by $x^2 = a^2$. Hence when $X = a$, we obtain half the volume of the ellipsoid, viz. $\frac{2}{3}\pi abc$. The total volume is $\frac{4}{3}\pi abc$.

*Remark.* When $b = c$, the surface is the ellipsoid of revolution obtained by rotating the ellipse $x^2/a^2 + y^2/b^2 = 1$ about its major axis $Ox$, and is called a *prolate spheroid*. When $a = c$, it is obtained by rotation about the minor axis $Oy$, and is an *oblate spheroid*. If $a = b = c$ we obtain a sphere.

† The volume of any right cylinder is defined to be 'area of cross-section × height'.

## Exercise 7(c)

*Find the area*

1 enclosed by $a^2y^2 = x^2(a^2 - x^2)$.

2 under one arch of the cycloid $x = a(\theta - \sin\theta)$, $y = a(1 - \cos\theta)$.

3 between the curves $y = x^2$, $y^2 = x^3$.

4 If $ab > h^2$, prove that the area enclosed by $ax^2 + 2hxy + by^2 = 1$ is $\pi/\sqrt{(ab - h^2)}$.

*Sketch the following curves (allowing r to take negative values) and find the area enclosed.* (Cf. Ex. 1 (e), nos. 12–15.)

5 $r = a(2 + \cos\theta)$.                   6 $r^2 = a^2\sin 2\theta$.

7 $r = a\sin 2\theta$.                        8 $r = a\cos 3\theta$.

9 Find the area of the loop of $r\cos\theta = \cos 2\theta$.

*Find the areas of the following sectors.*

10 $r = a\sec^2\frac{1}{2}\theta$, $\theta = 0$ to $\frac{1}{2}\pi$.       11 $r^2\sin 2\theta = a^2$, $\theta = \frac{1}{6}\pi$ to $\frac{1}{3}\pi$.

12 $r = e^{a\theta}$, $\theta = 0$ to $\frac{1}{2}\pi$, and included by the arc for which (i) $0 < \theta < \frac{1}{2}\pi$; (ii) $2\pi < \theta < \frac{5}{2}\pi$.

13 Sketch the curves $r = a\cos\theta$ and $r = a(1 - \cos\theta)$, and find the area common to them.

14 Sketch the curve $r = 1 + 2\cos\theta$, and prove that the area of the inner loop is $\pi - \frac{3}{2}\sqrt{3}$.

*Find the area enclosed by*

15 the hyperbola $x = a\operatorname{ch}t$, $y = b\operatorname{sh}t$ and the radii to the points where $t = 0$, $t = u$.

16 $x = a\sin^2 t$, $y = b\sin t\cos t$.       17 $x^{\frac{2}{3}} + y^{\frac{2}{3}} = a^{\frac{2}{3}}$.

18 Show that the loop of the curve $x = t/(1 + t^3)$, $y = t^2/(1 + t^3)$ is traced as $t$ varies from 0 to $\infty$. Find the area of the loop.

19 Sketch the curve $x = t(t^2 - 1)$, $y = t^4 - 1$, and find the area of its loop.

*Find the volume obtained by rotating the following curves through one revolution about Ox.*

20 $3ay^2 = x(a - x)^2$, $x = 0$ to $a$.       21 $x = a\cos^3 t$, $y = a\sin^3 t$, $t = 0$ to $\frac{1}{2}\pi$.

22 $x = a(\theta - \sin\theta)$, $y = a(1 - \cos\theta)$, $\theta = 0$ to $2\pi$.

23 $l/r = 1 + \cos\theta$, $\theta = 0$ to $\alpha$ ($0 < \alpha < \pi$).

24 The area between the parabolas $y^2 = ax$, $x^2 = ay$.

25 The inner loop of the curve in no. 14.

26 If the area under the curve $y = f(x)$ from $x = a$ to $x = b$ is rotated about $Oy$ through one revolution, prove that the volume generated is $2\pi\displaystyle\int_a^b xy\,dx$.

27 If the area between $y^2 = 4ax$ and $ay^2 = x^3$ is rotated about the line $x = -a$, prove that the volume generated is $\frac{832}{105}\sqrt{2}\,\pi a^3$.

## 7.6 Length of a curve

### 7.61 Definition, and sign conventions

(1) In attempting to make precise the intuitive idea of 'length of a curve' we meet a problem similar to that discussed for area in 7.1. We begin with straight-line approximations to the curve.

Let $AB$ be the given arc, and choose $n-1$ points $P_1, P_2, ..., P_{n-1}$ on it. Then the perimeter of the 'open polygon' $AP_1P_2...P_{n-1}B$ is

$$s_n = \sum_{r=0}^{n-1} \overline{P_r P_{r+1}},$$

where we have written $P_0$ for $A$ and $P_n$ for $B$ (fig. 78).

If $s_n$ tends to a limit $s$ when $n \to \infty$ and each chord $\overline{P_r P_{r+1}}$ tends to zero, we call $s$ the *length* of the arc $AB$.

(2) *Sign of the arc-length.* Now let $s$ denote the length of the arc $AP$, measured from the fixed point $A$ to the variable point $P$. The direction along the curve in which $s$ increases can be chosen arbitrarily in one of two ways.

If the equation of the curve is of the form $y = f(x)$, we measure $s$ to increase with $x$; if $x = g(y)$, with $y$; if $x = \phi(t)$ and $y = \psi(t)$, with $t$; if $r = f(\theta)$, with $\theta$; and if $\theta = g(r)$, with $r$. The direction of increasing $s$ may clearly be different for alternative representations of the same curve.

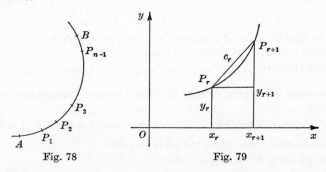

Fig. 78          Fig. 79

### 7.62 Cartesian formulae for arc-length

With the notation of 7.61 (1) let $P_r$ have coordinates $(x_r, y_r)$, and let $c_r$ be the length of the chord $P_r P_{r+1}$ (fig. 79). Then†

$$c_r = \{(x_{r+1} - x_r)^2 + (y_{r+1} - y_r)^2\}^{\frac{1}{2}}.$$

† The symbol $u^{\frac{1}{2}}$ always denotes the *positive* square root of $u$.

If the curve has equation $y = f(x)$, then $y_r = f(x_r)$ and $y_{r+1} = f(x_{r+1})$. First suppose that $x$ steadily increases as $P(x, y)$ varies along the arc from $A$ to $B$. Then by the mean value theorem

$$y_{r+1} - y_r = f'(\xi_r)\,(x_{r+1} - x_r),$$

where $x_r < \xi_r < x_{r+1}$. Hence

$$c_r = (x_{r+1} - x_r)\,[1 + \{f'(\xi_r)\}^2]^{\frac{1}{2}}$$

and

$$s_n = \sum_{r=0}^{n-1} [1 + \{f'(\xi_r)\}^2]^{\frac{1}{2}}\,(x_{r+1} - x_r).$$

If this sum tends to a limit when $n \to \infty$ and all differences $x_{r+1} - x_r$ tend to zero, then (7.21) this limit is

$$\int_a^b [1 + \{f'(x)\}^2]^{\frac{1}{2}}\,dx.$$

The limit will certainly exist if $f'(x)$ is continuous. Thus

$$s = \int_a^b \left\{ 1 + \left(\frac{dy}{dx}\right)^2 \right\}^{\frac{1}{2}} dx. \tag{i}$$

If $x$ does not steadily increase as $P$ varies from $A$ to $B$, then we first divide $AB$ into consecutive arcs, some of which have the above property and the remainder of which have $x$ decreasing steadily along them. By the sign convention (7.61 (2)) $s$ will be positive along arcs of the first sort and negative along the others. Formula (i) applies to each separate arc.

Similarly, and with like reservations, the arc-length of the curve $x = g(y)$ from $y = c$ to $y = d$ is given by

$$s = \int_c^d \left\{ 1 + \left(\frac{dx}{dy}\right)^2 \right\}^{\frac{1}{2}} dy. \tag{ii}$$

For the curve defined parametrically by $x = \phi(t)$, $y = \psi(t)$, first suppose $\phi(t)$ steadily increases from $a$ to $b$ as $t$ increases from $t_1$ to $t_2$ (so that $dx/dt$ is not negative). By changing the variable in (i) (in which $s$ is measured to increase with $x$, and hence with $t$),

$$s = \int_{t_1}^{t_2} \left\{ 1 + \left(\frac{dy}{dx}\right)^2 \right\}^{\frac{1}{2}} \frac{dx}{dt}\, dt$$

$$= \int_{t_1}^{t_2} \left\{ \left(\frac{dx}{dt}\right)^2 + \left(\frac{dy}{dt}\right)^2 \right\}^{\frac{1}{2}} dt. \tag{iii}$$

If $\phi(t)$ steadily decreases from $b$ to $a$ as $t$ increases from $t_2$ to $t_1$, then from (i),

$$s = \int_{t_1}^{t_2} \left\{ 1 + \left( \frac{dy}{dx} \right)^2 \right\}^{\frac{1}{2}} \frac{dx}{dt} \, dt = \int_{t_1}^{t_2} - \left\{ \left( \frac{dx}{dt} \right)^2 + \left( \frac{dy}{dt} \right)^2 \right\}^{\frac{1}{2}} dt,$$

with the minus because now $dx/dt$ is non-positive. This formula applies when $s$ is measured to increase with $x$, i.e. to *decrease* as $t$ increases. Hence (iii) still holds if we measure $s$ to *increase* with $t$.

### Example

*Cycloid* $x = a(\theta + \sin \theta)$, $y = a(1 - \cos \theta)$.

$$\frac{dx}{d\theta} = a(1 + \cos \theta), \quad \frac{dy}{d\theta} = a \sin \theta,$$

$$\therefore \quad \left( \frac{dx}{d\theta} \right)^2 + \left( \frac{dy}{d\theta} \right)^2 = a^2 \{ (1 + \cos \theta)^2 + \sin^2 \theta \}$$

$$= 2a^2 (1 + \cos \theta) = 4a^2 \cos^2 \tfrac{1}{2} \theta.$$

Hence in fig. 23 of 1.61, the arc $OP$ has length

$$s = \int_0^\theta 2a \cos \tfrac{1}{2} \theta \, d\theta = 4a \sin \tfrac{1}{2} \theta.$$

### 7.63 Polar formulae for arc-length

From $x = r \cos \theta$ and $y = r \sin \theta$ we find that

$$\frac{dx}{dt} = \cos \theta \frac{dr}{dt} - r \sin \theta \frac{d\theta}{dt} \quad \text{and} \quad \frac{dy}{dt} = \sin \theta \frac{dr}{dt} + r \cos \theta \frac{d\theta}{dt},$$

and hence
$$\left( \frac{dx}{dt} \right)^2 + \left( \frac{dy}{dt} \right)^2 = \left( \frac{dr}{dt} \right)^2 + r^2 \left( \frac{d\theta}{dt} \right)^2.$$

If $s$ is measured to increase with $t$, and if the arc is traced as $t$ increases from $t_1$ to $t_2$, then by (iii)

$$s = \int_{t_1}^{t_2} \left\{ \left( \frac{dr}{dt} \right)^2 + r^2 \left( \frac{d\theta}{dt} \right)^2 \right\}^{\frac{1}{2}} dt. \tag{iv}$$

When the curve has polar equation $r = f(\theta)$, the length of the arc from $\theta = \theta_1$ to $\theta = \theta_2$ is (by putting $t = \theta$)

$$s = \int_{\theta_1}^{\theta_2} \left\{ r^2 + \left( \frac{dr}{d\theta} \right)^2 \right\}^{\frac{1}{2}} d\theta, \tag{v}$$

where $s$ is measured to increase with $\theta$.

Similarly, if the curve is $\theta = g(r)$, the arc-length from $r = r_1$ to $r = r_2$ is

$$s = \int_{r_1}^{r_2} \left\{ 1 + r^2 \left( \frac{d\theta}{dr} \right)^2 \right\}^{\frac{1}{2}} dr, \tag{vi}$$

$s$ being measured to increase with $r$.

For a given arc $AB$, the cartesian and polar formulae always give the same *numerical* value for $s$, but the signs may be different. For example, this will be so for (i) and (v) when $x$ and $\theta$ do not increase together.

## 7.64 Derivative of $s$

Consider the arc of the curve $y = f(x)$ measured from $A$ (where $x = a$) to the variable point $P(x, y)$. Formula (i) gives

$$s = \int_a^x [1 + \{f'(t)\}^2]^{\frac{1}{2}}\, dt,$$

and by deriving this wo $x$,

$$\frac{ds}{dx} = [1 + \{f'(x)\}^2]^{\frac{1}{2}} = \left\{ 1 + \left(\frac{dy}{dx}\right)^2 \right\}^{\frac{1}{2}}. \tag{i}'$$

Similarly, from formulae (ii)–(vi) we find

$$\frac{ds}{dy} = \left\{ 1 + \left(\frac{dx}{dy}\right)^2 \right\}^{\frac{1}{2}}, \quad \frac{ds}{dt} = \left\{ \left(\frac{dx}{dt}\right)^2 + \left(\frac{dy}{dt}\right)^2 \right\}^{\frac{1}{2}}, \qquad \text{(ii)}', \text{(iii)}'$$

$$\frac{ds}{dt} = \left\{ \left(\frac{dr}{dt}\right)^2 + r^2\left(\frac{d\theta}{dt}\right)^2 \right\}^{\frac{1}{2}}, \quad \frac{ds}{d\theta} = \left\{ r^2 + \left(\frac{dr}{d\theta}\right)^2 \right\}^{\frac{1}{2}}, \quad \frac{ds}{dr} = \left\{ 1 + r^2\left(\frac{d\theta}{dr}\right)^2 \right\}^{\frac{1}{2}}.$$

$$\text{(iv)}', \text{(v)}', \text{(vi)}'.$$

### Exercise 7($d$)

**1** Find the length of the arc of the curve $y = c\,\mathrm{ch}\,(x/c)$ measured from $(0, c)$ to $(x, y)$.

**2** Sketch the *astroid* $x = a\cos^3 t$, $y = a\sin^3 t$, and find its total length.

**3** Sketch the curve $y = \log\sec x$ for $-\frac{1}{2}\pi < x < \frac{1}{2}\pi$. Prove

$$s = \log\tan\left(\tfrac{1}{4}\pi + \tfrac{1}{2}x\right)$$

if $s$ is measured from the origin.

**4** For the curve $x = at^2$, $y = at^3$ prove $ds/dx = \sqrt{(1 + 9x/4a)}$. Find the length of the arc from $x = 0$ to $x = c$.

**5** If $s$ is the arc-length measured from the point $u = 0$ of the *tractrix* $x = a(u - \mathrm{th}\,u)$, $y = a\,\mathrm{sech}\,u$, prove $ds/du = a\,\mathrm{th}\,u$ and $ds/dy = -a/y$. Find $y$ in terms of $s$. Prove that the length of the tangent $PT$ (5.71) at any point $P$ is $a$. (The curve is therefore the path of a particle $P$ attached to a string whose other end $T$ moves along a fixed line $Ox$; hence the name.)

**6** Find the length of the arc of the curve $x = \cos\theta\sin^2\theta$, $y = \sin\theta(1 + \cos^2\theta)$ from $\theta = 0$ to $\frac{1}{2}\pi$. [We find $(ds/d\theta)^2 = (2\cos^2\theta - \sin^2\theta)^2$, so that

$$s = \int_0^{\frac{1}{2}\pi} \sqrt{(2\cos^2\theta - \sin^2\theta)^2}\, d\theta = \int_0^{\alpha} (2\cos^2\theta - \sin^2\theta)\, d\theta + \int_{\alpha}^{\frac{1}{2}\pi} (\sin^2\theta - 2\cos^2\theta)\, d\theta,$$

where $\alpha = \tan^{-1}\sqrt{2}$.]

7 If $s$ is measured from the origin to the point $(x, y)$ of the curve

$$3y^2 = x(1-x)^2,$$

prove $3s^2 = 4x^2 + 3y^2$. Find the length of the loop of this curve.

8 Find the total length of the cardioid $r = a(1 + \cos \theta)$.

9 Prove that the length of the 'equiangular spiral' $r = a e^{k\theta}$ from $(r_1, \theta_1)$ to $(r_2, \theta_2)$ is $|r_1 - r_2| \, k^{-1} \sqrt{(1 + k^2)}$.

10 Find the length of $r = a\theta$ measured from the pole to $(r_1, \theta_1)$.

11 Sketch the curve $r = 2a \cos^3 \tfrac{1}{3}\theta$, and prove that its total length is $3\pi a$.

12 If the polar coordinates are given parametrically by

$$r = 2a \sec t, \ \theta = \tan t - t \quad (-\tfrac{1}{2}\pi < t < \tfrac{1}{2}\pi),$$

prove $s = a \tan^2 t$ if $s$ is measured from $t = 0$.

*13 A curve has equation $x = f(\theta)$, where $x$ is a cartesian coordinate, $(r, \theta)$ are polar coordinates, and $x = r \cos \theta$, $y = r \sin \theta$. Prove that the arc-length from $\theta = \theta_1$ to $\theta = \theta_2$ $(0 \leqslant \theta_1 < \theta_2 < \tfrac{1}{2}\pi)$ is

$$\int_{\theta_1}^{\theta_2} [\{f'(\theta) + f(\theta) \tan \theta\}^2 + \{f(\theta)\}^2]^{\frac{1}{2}} \sec \theta \, d\theta.$$

Check this formula by finding the length of the closed curve $x = a \cos \theta$.

*Find the curves for which*

*14 $s = \sqrt{(r^2 + 2ar)}$.  *15 $s = OT$. [Use formula (ii).]

## 7.7 Area of a surface of revolution

### 7.71 Area of a conical surface

The phrase 'area of a curved surface' needs to be defined, because so far we have discussed 'area' only for plane figures. We first consider a conical surface.

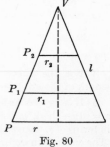

Given a right circular cone with base-radius $r$ and *slant* height $l$, take any point $P$ on the circumference of the base and join it to the vertex $V$. By cutting along this line $PV$ (called a *generator*), the surface can be *developed* (i.e. 'flattened out') into a sector of a circle. We define the area of the cone's surface to be the area of this sector, viz. $\pi r l$.

Fig. 80

Now consider the frustum of the above cone bounded by the circular sections of radii $r_1$, $r_2$; and with the notation of fig. 80, let $VP_1 = l_1$, $VP_2 = l_2$. Then the area of the frustum is $\pi r_1 l_1 - \pi r_2 l_2$; and since $r_1 : r_2 = l_1 : l_2$, this expression can be written

$$\pi r_1 (l_1 - l_2) + \pi r_2 (l_1 - l_2) = \pi (r_1 + r_2)(l_1 - l_2).$$

### 7.72 General definition

Suppose the surface of revolution is generated by rotating the continuous arc $AB$ about $Ox$, and that the arc lies entirely above $Ox$. Divide up the arc by points $P_1, P_2, \ldots, P_{n-1}$, and consider the 'open polygon' $P_1 P_2 \ldots P_{n-1}$. Let $S_n$ denote the sum of the areas of the conical surfaces generated by rotating this polygon about $Ox$. If $S_n$ tends to a limit when $n \to \infty$ and all the arcs $P_r P_{r+1}$ tend to zero, we define the *area of the surface of revolution* to be this limit.

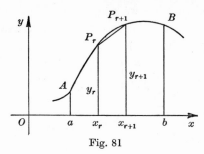

Fig. 81

Measuring arc-lengths from $A$, let $P_r$ correspond to $s = s_r$ and have coordinates $(x_r, y_r)$; let $c_r$ be the length of the chord $P_r P_{r+1}$. This chord will generate a conical frustum of area $\pi(y_r + y_{r+1}) c_r$. Hence

$$S_n = \sum_{r=0}^{n-1} \pi(y_r + y_{r+1}) c_r.$$

Since $c_r \doteqdot \operatorname{arc} P_r P_{r+1} = s_{r+1} - s_r$, we write this expression as

$$S_n = \pi \sum_{r=0}^{n-1} (y_r + y_{r+1})(s_{r+1} - s_r) - \pi \sum_{r=0}^{n-1} (y_r + y_{r+1})\{(s_{r+1} - s_r) - c_r\}.$$

Because $c_r$ is the chord of the arc $s_{r+1} - s_r$, all terms in the last sum are positive.† If $M$ denotes the greatest value of $y$ between $A$ and $B$, this sum is certainly less than

$$2\pi M \sum_{r=0}^{n-1} \{(s_{r+1} - s_r) - c_r\} = 2\pi M \left\{ \operatorname{arc} AB - \sum_{r=0}^{n-1} c_r \right\}.$$

The definition of 'length of arc $AB$' shows that the last expression tends to zero when $n \to \infty$ and all arcs $P_r P_{r+1}$ (and therefore† all chords $P_r P_{r+1}$) tend to zero.

† The intuitive property (which we assume here) that arc $AP >$ chord $AP$ can be proved by considering the difference

$$\phi(x) = \int_a^x [1 + \{f'(t)\}^2]^{\frac{1}{2}} \, dt - [(x - a)^2 + \{f(x) - f(a)\}^2]^{\frac{1}{2}}$$

and showing that $\phi'(x) > 0$ for $x > a$.

Since $\frac{1}{2}(y_r + y_{r+1})$ lies between $y_r$ and $y_{r+1}$, then by the assumed continuity of $y$ this expression is a value of $y$ taken at some point of the arc between $P_r$ and $P_{r+1}$ (2.65, Corollary). Therefore when $n \to \infty$ and all arcs $P_r P_{r+1}$ tend to zero, the first sum tends to the limit $\int_0^\sigma 2\pi y\,ds$ by 7.21 (where $s = \sigma$ at $B$), which is consequently also the limit of $S_n$. Thus

$$\text{area of the surface of revolution} = 2\pi \int_0^\sigma y\,ds.$$

By a change of variable we can adapt this formula for use with cartesian, parametric, or polar coordinates. For example, a cartesian form is (by 7.64, equation (i)$'$)

$$2\pi \int_a^b y \left\{ 1 + \left(\frac{dy}{dx}\right)^2 \right\}^{\frac{1}{2}} dx.$$

### Example

*Find the area of the surface obtained by rotating the cardioid* $r = a(1 + \cos\theta)$ *through angle* $\pi$ *about the initial line.*

By 7.64 (v)$'$,

$$\left(\frac{ds}{d\theta}\right)^2 = r^2 + \left(\frac{dr}{d\theta}\right)^2 = a^2(1 + \cos\theta)^2 + (-a\sin\theta)^2 = 2a^2(1 + \cos\theta);$$

also $y = r\sin\theta = a\sin\theta(1 + \cos\theta)$. Since rotation of the whole curve through angle $\pi$ is equivalent to rotation of the upper half through angle $2\pi$,

$$\text{surface area} = 2\pi \int_0^\pi y \frac{ds}{d\theta} d\theta$$

$$= 2\pi \int_0^\pi a\sin\theta(1 + \cos\theta) a \sqrt{2} (1 + \cos\theta)^{\frac{1}{2}} d\theta$$

$$= 2\sqrt{2}\,\pi a^2 \int_0^\pi \sin\theta(1 + \cos\theta)^{\frac{3}{2}} d\theta.$$

By putting $u = 1 + \cos\theta$ this becomes

$$2\sqrt{2}\,\pi a^2 \int_0^2 u^{\frac{3}{2}} du = \tfrac{32}{5}\pi a^2.$$

## 7.8  Centroids. The theorems of Pappus

### 7.81  Centre of mass, centroid

The *centre of mass* of particles of masses $m_1, m_2, \ldots, m_n$ at $(x_1, y_1)$, $(x_2, y_2), \ldots, (x_n, y_n)$ in a plane is defined in Statics to be the point $(\bar{x}, \bar{y})$, where

$$\bar{x} = \frac{\Sigma m_r x_r}{\Sigma m_r}, \quad \bar{y} = \frac{\Sigma m_r y_r}{\Sigma m_r}. \tag{i}$$

If the system is in space, we have a similar formula for the $z$-coordinate.

By regarding a continuous body as the limit of the sum of elementary masses $dm$, we are led to define† its centre of mass by

$$\bar{x} = \frac{\int x \, dm}{\int dm}, \quad \bar{y} = \frac{\int y \, dm}{\int dm}, \quad \bar{z} = \frac{\int z \, dm}{\int dm}, \tag{ii}$$

If the body is *uniform*, the elementary masses $dm$ are proportional to the corresponding elements of arc-length, area, or volume (according to whether the body concerned is a wire, lamina, or solid). In this case, $(\bar{x}, \bar{y}, \bar{z})$ is the same as the *centroid* of the figure, a point defined independently of statical considerations. Thus the $x$-coordinate of the centroid of an arc, an area, or a volume is respectively

$$\frac{\int x \, ds}{\int ds}, \quad \frac{\int x \, dA}{\int dA}, \quad \frac{\int x \, dV}{\int dV}. \tag{iii}$$

When the figure has axes of symmetry, the centroid must clearly lie on these.

## 7.82 Summary of well-known results

The proofs of the following results for the mass-centres of uniform bodies come directly from the definitions as an easy exercise in geometry or integration.

| UNIFORM BODY | CENTROID |
|---|---|
| Rod | Mid-point |
| Rectangle | Intersection of diagonals |
| Triangle | Intersection of medians |
| Circle | Centre |
| Solid right circular cylinder or cylindrical surface | Mid-point of axis |
| Conical surface (right circular) | $\frac{1}{3}$ way up axis from the base |
| Solid cone (right circular) | $\frac{1}{4}$ way up axis from the base |
| Hemispherical surface (or 'shell') | Mid-point of radius of symmetry |
| Solid hemisphere | $\frac{3}{8}$ way up radius of symmetry from the base |

---

† This definition involves reference to a system of coordinate axes. It is important to show that, relative to the body, we obtain the *same* point whatever axes are chosen; i.e. that the definition is actually independent of the choice of coordinate system. This will be done for a lamina in Ex. 15(e), no. 7.

## 7.83 Theorems of Pappus

These two results relate the theory of centroids to that of arc-lengths and areas, surface areas and volumes.

THEOREM I. *If an arc of a curve rotates about an axis in its plane, and the axis does not cut the arc, then the area of the surface generated is equal to the length of the arc multiplied by the length of the path of its centroid.*

*Proof.* Given an arc $AB$ of length $\sigma$, choose the axis of rotation for $Ox$ and measure arc-length from $A$ (fig. 82). If $G$ is the centroid of the arc, then by the definition its $y$-coordinate is

$$\bar{y} = \frac{\displaystyle\int_0^\sigma y\,ds}{\displaystyle\int_0^\sigma ds} = \frac{1}{\sigma}\int_0^\sigma y\,ds.$$

By 7.72 the area of the surface is $2\pi\displaystyle\int_0^\sigma y\,ds$, which is equal to $2\pi\bar{y}\sigma$; and $2\pi\bar{y}$ is the length of the (circular) path of $G$ in the rotation about $Ox$.

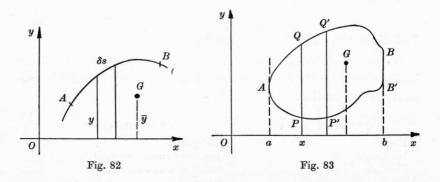

Fig. 82                              Fig. 83

THEOREM II. *If a closed curved is rotated about an axis in its own plane, and this axis does not cut the curve, then the volume of the solid generated is equal to the area of the curve multiplied by the length of the path of its centroid.*

*Proof.* Choose the axis of rotation to be $Ox$. First suppose that the closed curve is cut by a line parallel to $Oy$ in at most two points. Let $x = a$, $x = b$ correspond to the extreme ordinates (fig. 83).

Divide the area into strips parallel to $Oy$, and let the ordinate through $x$ cut the curve at $P$ and $Q$, where $P$ is $(x, y_1)$ and $Q$ is $(x, y_2)$.

The centroid of the strip $PQQ'P'$ has $y$-coordinate $\frac{1}{2}(y_1+y_2)$ and area $(y_2-y_1)\,\delta x$, approximately.† Hence for the whole area,

$$\bar{y} = \frac{\displaystyle\int_a^b \tfrac{1}{2}(y_1+y_2)(y_2-y_1)\,dx}{\displaystyle\int_a^b (y_2-y_1)\,dx}$$

$$= \frac{1}{2A}\int_a^b (y_2^2-y_1^2)\,dx,$$

by 7.43.

The volume generated is the difference of those generated by the areas under $AQQ'B$, $APP'B'$, viz. $\displaystyle\int_a^b \pi(y_2^2-y_1^2)\,dx$, and this is equal to $2\pi\bar{y}A$, which proves the theorem.

If the closed curve can be divided into a number of parts each of which is closed and has the above property, the proof applies to each part. The axis of rotation must not cut the area, but it may be part of the boundary, as in example (ii) following.

### Examples

(i) *A circle of radius $a$ is rotated about a line in its plane at distance $b\,(>a)$ from the centre. Find the surface area and volume of the anchor ring (or torus) so formed.*

Area of anchor ring is $2\pi a \cdot 2\pi b = 4\pi^2 ab$.
Volume of anchor ring is $\pi a^2 \cdot 2\pi b = 2\pi^2 a^2 b$.

(ii) *Find the centroid of (a) a semicircular arc; (b) a semicircular area, of radius $a$.*

The centroid in each case will lie on the radius of symmetry. Choose axes as shown (fig. 84).

(a) By Theorem I, the area of the surface generated is $2\pi\bar{y}\cdot\pi a$. As this surface is a sphere of radius $a$, its area is $4\pi a^2$. Hence $2\pi\bar{y}\cdot\pi a = 4\pi a^2$, and $\bar{y} = 2a/\pi$.

(b) By Theorem II, the volume of the solid generated is $2\pi\bar{y}\cdot\frac{1}{2}\pi a^2$. This solid is a sphere of radius $a$, whose volume is $\frac{4}{3}\pi a^3$. Hence

$$2\pi\bar{y}\cdot\tfrac{1}{2}\pi a^2 = \tfrac{4}{3}\pi a^3, \quad\text{and}\quad \bar{y} = \frac{4a}{3\pi}.$$

(iii) *A sector of a plane curve with polar equation $r = f(\theta)$ is rotated about the initial line. Show that the volume generated is $\frac{2}{3}\pi\int r^3\sin\theta\,d\theta$, taken between suitable limits. Calculate this volume for the cardioid $r = a(1+\cos\theta)$.*

Treating an elementary sector (fig. 85) as approximately a triangle, its centroid is $\frac{2}{3}$ the way down the median from $O$, i.e. approximately $\frac{2}{3}r\sin\theta$ from $Ox$. For a complete revolution, the length of the path of this centroid is $2\pi\cdot\frac{2}{3}r\sin\theta$.

† We are approximating 'by rectangles'. The limiting result which follows is exact, by 7.21. Examples on centroids are done in this spirit (cf. ex. (iii) following).

The area of the element rotated is $\frac{1}{2}r^2\,\delta\theta$ approximately. Hence the element of volume generated is approximately

$$2\pi \cdot \tfrac{2}{3}r\sin\theta \cdot \tfrac{1}{2}r^2\,\delta\theta = \tfrac{2}{3}\pi r^3 \sin\theta\,\delta\theta.$$

The total volume is the limit of the sum of these elements, viz. $\int \frac{2}{3}\pi r^3 \sin\theta\,d\theta$, with suitable limits for $\theta$.

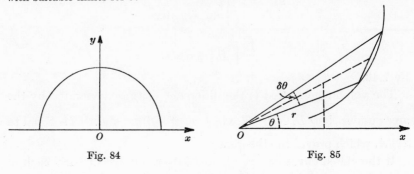

Fig. 84                    Fig. 85

For the cardioid, the volume generated (by the upper half) is

$$\tfrac{2}{3}\pi a^3 \int_0^\pi (1+\cos\theta)^3 \sin\theta\,d\theta = \tfrac{2}{3}\pi a^3 \int_0^2 u^3\,du = \tfrac{8}{3}\pi a^3$$

on putting $u = 1 + \cos\theta$.

## Exercise 7(e)

*Find the surface area generated by rotating the following curves through one revolution about Ox.*

1  $x^2 + y^2 = a^2$, $x = -a$ to $+a$.      2  $3ay^2 = x(a-x)^2$, $x = 0$ to $a$.

3  $x = a\cos^3 t$, $y = a\sin^3 t$, $t = 0$ to $\frac{1}{2}\pi$.

4  $x = a(\theta - \sin\theta)$, $y = a(1 - \cos\theta)$, $\theta = 0$ to $2\pi$.

5  $l/r = 1 + \cos\theta$, $\theta = 0$ to $\alpha$ $(0 < \alpha < \pi)$.

*Verify the results stated in 7.82 for a*

6  solid cone.              7  conical surface.

8  solid hemisphere.        9  hemispherical surface.

10  Find the coordinates of the centroid of the quadrant of the ellipse $x^2/a^2 + y^2/b^2 = 1$ for which both $x$ and $y$ are non-negative.

11  Find the position of the centroid of (i) a circular arc, (ii) a circular sector, each of angle $2\alpha$ and radius $a$.

12  Calculate the volume obtained by rotating the ellipse $x = a\cos\phi$, $y = b\sin\phi$ through one revolution about the line $x = 2a$. [Use Pappus.]

13  Find the polar coordinates of the centroid of the area of the complete cardioid $r = a(1 + \cos\theta)$, and deduce the volume formed by rotating it about the tangent $x = 2a$.

14  Find the centroid of one arch of the cycloid $x = a(\theta - \sin\theta)$, $y = a(1 - \cos\theta)$,

and deduce the area of the surface formed by rotation of the curve about the tangent $y = 2a$.

**15** Find the volume generated by revolving a loop of $r = a \sin 2\theta$ about the initial line. [See 7.83, ex. (iii).]

**16** Find the volume generated by rotating the area between the two loops of $r = 1 + 2\cos\theta$ about $Ox$ through angle $\pi$.

**17** (i) The area bounded by an arc of a curve and two straight lines through the origin is rotated through angle $2\pi$ about $Ox$. Prove that the volume generated is

$$\tfrac{2}{3}\pi \int y(x\dot{y} - y\dot{x})\, dt,$$

taken between suitable limits.

(ii) If the area bounded by the ellipse $x = a\cos t$, $y = b\sin t$ and the radii to $t = 0$, $t = \alpha$ is rotated about $Ox$, prove that the volume generated is

$$\tfrac{4}{3}\pi ab^2 \sin^2 \tfrac{1}{2}\alpha.$$

## 7.9   Moments of inertia

### 7.91   Dynamical introduction

Consider a rigid body of mass $M$ rotating with angular velocity $\omega$ about a fixed axis. A particle $P$ of mass $\delta m$ at perpendicular distance $r$ from the axis would have velocity $\omega r$ and kinetic energy $\tfrac{1}{2}\delta m(\omega r)^2$.

By regarding the continuous body as the limit of a sum of such particles, we see that its kinetic energy is

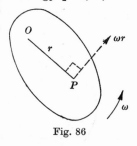

$$\lim \Sigma \tfrac{1}{2}\omega^2 r^2 \delta m = \int \tfrac{1}{2}\omega^2 r^2\, dm = \tfrac{1}{2}\omega^2 \int r^2\, dm,$$

since $\omega$ is the same for all particles of the body.

The expression

$$I = \int r^2\, dm$$

Fig. 86

is called the *moment of inertia* (M.I.) of the body about the axis of rotation. It is usually given in the standard form $Mk^2$, where $k$ is called the *radius of gyration* about the axis.

### 7.92   Examples

We suppose in each case that the body is of mass $M$ and uniform density $\rho$.

(i) *Rod of length $2a$; axis along its perpendicular bisector.*

An element $PQ$, where (fig. 87) $OP = x$ and $PQ = \delta x$, has mass $\rho\,\delta x$. Hence the M.I. of rod is

$$\int_{-a}^{a} \rho x^2\, dx = \tfrac{2}{3}\rho a^3 = \tfrac{1}{3}Ma^2,$$

since $M = 2a\rho$.

(ii) *Rectangular lamina; axis bisecting sides of length* $2a$.

Let $AB = 2a$. Divide the rectangle into strips of mass $\delta m$ parallel to $AB$. Since by ex. (i)

$$\text{M.I. of each strip} = \tfrac{1}{3}\delta m\, a^2,$$

$$\therefore \quad \text{M.I. of rectangle} = \Sigma \tfrac{1}{3}a^2\, \delta m = \tfrac{1}{3}a^2 \Sigma\, \delta m = \tfrac{1}{3}Ma^2.$$

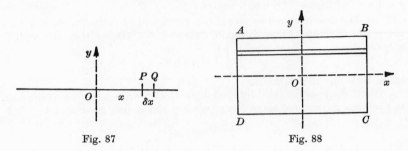

Fig. 87                                    Fig. 88

(iii) *Circular disc of radius $a$; perpendicular axis through the centre.*

Divide the disc into concentric rings, and consider the ring bounded by circles of radii $x$, $x + \delta x$. Approximately, the mass is $2\pi x\, \delta x \rho$ and the M.I. is $(2\pi x\, \delta x \rho)\, x^2$.

$$\therefore \quad \text{M.I. of disc} = \lim \Sigma 2\pi \rho x^3\, \delta x$$

$$= \int_0^a 2\pi \rho x^3\, dx = \tfrac{1}{2}\pi \rho a^4 = \tfrac{1}{2}Ma^2,$$

since $M = \pi a^2 \rho$.

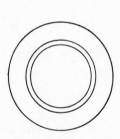
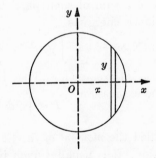

Fig. 89                                    Fig. 90

(iv) *Solid sphere of radius $a$; axis along a diameter $Ox$.*

Divide the sphere into circular discs perpendicular to $Ox$. The disc of radius $y$ at distance $x$ from $O$ and of thickness $\delta x$ has mass $\pi y^2\, \delta x \rho$. Hence by ex. (iii), its M.I. about $Ox$ is $\tfrac{1}{2}(\pi y^2\, \delta x \rho)\, y^2 = \tfrac{1}{2}\pi \rho(a^2 - x^2)^2\, \delta x$, since $x^2 + y^2 = a^2$.

$$\therefore \quad \text{M.I. of sphere} = \lim \Sigma \tfrac{1}{2}\pi \rho(a^2 - x^2)^2\, \delta x$$

$$= \int_{-a}^a \tfrac{1}{2}\pi \rho(a^2 - x^2)^2\, dx$$

$$= \tfrac{8}{15}\pi \rho a^5 = \tfrac{2}{5}Ma^2,$$

since $M = \tfrac{4}{3}\pi a^3 \rho$.

(v) *Spherical shell of radius a; axis along a diameter.*

Take rectangular axes $Ox$, $Oy$, $Oz$ at the centre (fig. 91). By the symmetry,

$$\int x^2\,dm = \int y^2\,dm = \int z^2\,dm.$$

$$\text{M.I. about } Oz = \int ON^2\,dm = \int (x^2+y^2)\,dm$$

$$= \frac{2}{3}\int (x^2+y^2+z^2)\,dm$$

$$= \frac{2}{3}\int a^2\,dm = \tfrac{2}{3}Ma^2,$$

since $\qquad x^2+y^2+z^2 = ON^2+z^2 = OP^2 = a^2.$

The result can also be found by direct integration: see Ex. 7 $(f)$, no. 10.

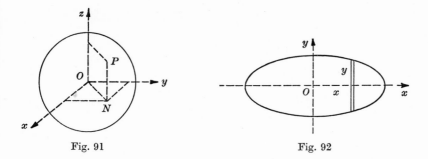

Fig. 91          Fig. 92

(vi) *Ellipse $x^2/a^2 + y^2/b^2 = 1$; about major axis $Ox$.*

Divide the area into strips parallel to $Oy$. By ex. (i) the M.I. of a typical strip (fig. 92) is $\frac{1}{3}(2y\,\delta x\rho)\,y^2$.

$$\therefore \quad \text{M.I. of ellipse} = \int_{-a}^{a} \tfrac{2}{3}\rho y^3\,dx = 2\int_{0}^{a} \tfrac{2}{3}\rho y^3\,dx$$

by the symmetry about $Oy$.

On the ellipse, $x = a\cos\phi$ and $y = b\sin\phi$ (see 17.31); hence

$$\text{M.I.} = 2\int_{\frac{1}{2}\pi}^{0} \tfrac{2}{3}\rho b^3 \sin^3\phi(-a\sin\phi)\,d\phi$$

$$= \tfrac{4}{3}\rho ab^3 \int_{0}^{\frac{1}{2}\pi} \sin^4\phi\,d\phi = \tfrac{4}{3}\rho ab^3 \frac{3.1}{4.2}\frac{\pi}{2}$$

$$= \tfrac{1}{4}\pi\rho ab^3 = \tfrac{1}{4}Mb^2,$$

since by 7.46, ex. (i), $M = \pi ab\rho$.

Similarly, M.I. about the minor axis is $\frac{1}{4}Ma^2$.

When the M.I. of a body about a certain axis is known, that about other axes can often be written down without further integration by use of the following theorems.

## 7.93 Theorem of parallel axes

*The* M.I. *of a body about any axis is equal to the* M.I. *about a parallel axis through the mass-centre, plus* $Mh^2$, *where* $h$ *is the distance between the two axes.*

*Proof.* Choose the mass-centre $G$ of the body for origin of co-ordinates, the $z$-axis along the line through $G$ parallel to the given axis, the $x$-axis along the perpendicular $GA$ from $G$ to this axis, and $Gy$ perpendicular to $Gx$, $Gz$ (cf. 21.11).

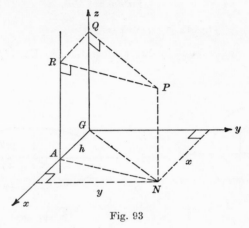

Fig. 93

Let the particle $P$ of the body have coordinates $(x, y, z)$ and mass $\delta m$. Then, with the notation of the figure, the M.I. of the body about $Gz$ is

$$I_G = \int PQ^2 dm = \int (x^2 + y^2)\, dm.$$

Since $GA = h$, the M.I. about the given axis is

$$I = \int PR^2 dm = \int \{(x-h)^2 + y^2\}\, dm$$

$$= \int \{(x^2 + y^2) + h^2 - 2hx\}\, dm$$

$$= \int (x^2 + y^2)\, dm + h^2 \int dm - 2h \int x\, dm$$

$$= I_G + h^2 M$$

as $\int x\, dm = 0$; this is because the mass-centre $G$ is the origin $(0, 0, 0)$, so that in particular $\bar{x} = 0$ from (ii) of 7.81.

### Example

*Rod of length 2a about a perpendicular axis through one end.*
Here $h = a$ and $I_G = \frac{1}{3}Ma^2$ by 7.92, ex. (i). So

$$\text{M.I. about an end} = \tfrac{1}{3}Ma^2 + Ma^2 = \tfrac{4}{3}Ma^2,$$

which is easily verified by direct integration.

The result also holds for the M.I. of a *rectangular lamina* about a side of length $2b$, by 7·92, ex. (ii).

The above theorem applies to any body, but the following is true only for a *lamina*.

### 7.94 Theorem of perpendicular axes for a lamina

*If the* M.I. *of a lamina about two perpendicular axes* $Ox$, $Oy$ *in its plane are* $I_x$, $I_y$, *then the* M.I. *about the axis* $Oz$ *perpendicular to its plane is* $I_x + I_y$.

*Proof.* With the notation of fig. 94,

$$I_z = \int r^2 dm = \int (x^2 + y^2)\, dm$$

$$= \int x^2 dm + \int y^2 dm$$

$$= I_y + I_x.$$

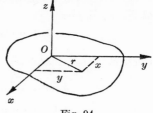

Fig. 94

### Examples

(i) *Rectangular lamina of sides* $2a$, $2b$; *perpendicular axis through its centre.*

By 7.92, ex. (ii) the M.I. about axes through the centre and parallel to the sides are $\frac{1}{3}Ma^2$, $\frac{1}{3}Mb^2$.

$$\therefore \quad \text{M.I. about perpendicular axis} = \tfrac{1}{3}M(a^2 + b^2).$$

(ii) *Cuboid of sides* $2a$, $2b$, $2c$; *axis through the centre and parallel to sides* $2c$.

By dividing into rectangular laminae perpendicular to the edges $2c$ and summing, we find by ex. (i) that M.I. $= \frac{1}{3}M(a^2 + b^2)$.

(iii) *Circular disc of radius* $a$; *axis along a diameter.*

By symmetry the M.I. about any two diameters are the same, say $I$. Since the M.I. about a perpendicular axis through the centre is $\frac{1}{2}Ma^2$ (7.92, ex. (iii)), hence

$$I + I = \tfrac{1}{2}Ma^2, \quad \text{i.e.} \quad I = \tfrac{1}{4}Ma^2.$$

(This is the particular case when $b = a$ in 7.92, ex. (vi), and could be obtained similarly by direct integration.)

*(iv) Ellipsoid $x^2/a^2 + y^2/b^2 + z^2/c^2 = 1$; about axis $Ox$.*

Divide the solid into laminae by planes parallel to $yOz$. As in the example in 7.5, the section at distance $x$ is an ellipse of area $\pi bc(1 - x^2/a^2)$. Hence the mass of a lamina of thickness $\delta x$ is approximately $\pi bc(1-x^2/a^2)\,\delta x\rho$, and by using ex. (vi) of 7.92 its M.I. about $Ox$ is

$$\frac{1}{4}\left\{\pi bc\left(1-\frac{x^2}{a^2}\right)\delta x\rho\right\}\left\{b^2\left(1-\frac{x^2}{a^2}\right)+c^2\left(1-\frac{x^2}{a^2}\right)\right\}$$

$$= \tfrac{1}{4}\pi\rho bc(b^2+c^2)\left(1-\frac{x^2}{a^2}\right)^2\delta x.$$

∴ M.I. of ellipsoid about $Ox$ is (from the symmetry about plane $yOz$)

$$2\int_0^a \tfrac{1}{4}\pi\rho bc(b^2+c^2)\left(1-\frac{x^2}{a^2}\right)^2 dx = \tfrac{1}{2}\pi\rho bc(b^2+c^2)\times\tfrac{8}{15}a = \tfrac{1}{5}M(b^2+c^2),$$

since $M = \tfrac{4}{3}\pi abc\rho$ by 7.5, example.

## 7.95 Routh's rule

This summarises many of the preceding results, and is a useful aid to memory.

*If $Gx$, $Gy$, $Gz$ are perpendicular axes of symmetry*† *which cut the body at $X$, $Y$, $Z$, then*

$$\text{M.I. } about\ Gx = mass \times \frac{GY^2+GZ^2}{3\ or\ 4\ or\ 5},$$

*where the denominator is*

3 *for a rod, rectangular lamina, or cuboid;*

4 *for a circular or elliptic disc;*

5 *for a solid sphere or ellipsoid.*

For a lamina, the intercept perpendicular to its plane is zero; and for a rod, both intercepts perpendicular to it are zero.

The reader should verify the correctness of this rule in all cases.

### Exercise 7(f)

*Assume in this exercise that each body has mass $M$, and is uniform unless otherwise stated. The results of 7·92, exs. (i), (iii) may be quoted.*

*Find by direct integration the M.I. of the following bodies about the axes stated.*

1 Circular wire of radius $a$; perpendicular axis through the centre.

2 Rod of length $2a$; axis meeting it at distance $c$ from the centre and inclined to the rod at angle $\theta$.

3 Lamina bounded by concentric circles of radii $a$, $b$; perpendicular axis through centre.

4 Solid right circular cylinder of radius $a$; axis of symmetry.

5 Isosceles triangle of height $h$; axis through the vertex and parallel to the base.

† If a body has an axis of symmetry, this must pass through the centroid $G$.

6 Solid right circular cone of base-radius $r$; axis of symmetry.

7 Segment of parabola $y^2 = 4ax$ cut off by line $x = b$; (i) about $Ox$; (ii) about $x = b$.

8 Solid formed by revolving the segment in no. 7 about $Ox$; about $Ox$.

9 Surface of a right circular cone of base-radius $r$; axis of symmetry.

10 Spherical shell of radius $a$; about a diameter.

*Use the theorems of* 7.93, 7.94 *to write down the* M.I. *of the following.*

11 Circular wire of radius $a$; about (i) a diameter; (ii) an axis perpendicular to its plane and through a point on the circumference.

12 Rectangular lamina of sides $2a$, $2b$; perpendicular axis through a corner.

13 Square of side $2a$; about *any* line through the centre and in its plane.

14 Solid right circular cylinder of radius $a$; axis along a generator.

15 Solid anchor ring formed by rotating a circle of radius $a$ about a line in its plane at distance $b$ from the centre $(b > a)$; about axis of rotation.

16 Isosceles triangle of height $h$; axis through the centroid parallel to the base. [Use no. 5.]

17 If $I_A$ is the M.I. of a body about an axis through $A$, show that the formula $I_B = I_A + Mh^2$ will give correctly the M.I. about a parallel axis through $B$, at distance $h$ from the first, only if *either* (i) $A$ is $G$, *or* (ii) $GAB$ is a right-angle.

*Calculate the* M.I. *of the following* (nos. 18, 19).

18 Solid right circular cylinder of radius $a$ and height $h$; about (i) a diameter of an end; (ii) an axis through the centre perpendicular to the axis of symmetry; (iii) a tangent to an end.

19 Solid hemisphere of radius $a$; about (i) a diameter of the base; (ii) a tangent parallel to the base.

*20 The density of a rod $AB$ of length $2a$ varies as the square of the distance from $A$. Calculate the M.I. about a perpendicular axis through $A$.

*21 The density at any point of a circular lamina of radius $a$ varies as its distance from the centre. Find the M.I. about (i) a perpendicular axis through the centre; (ii) a diameter; (iii) a tangent.

*22 Find the M.I. of the area enclosed by $r^2 = a^2 \cos 2\theta$ about a perpendicular axis through the pole. [Divide into sectors, regarding each as a rod whose density is proportional to the distance from the pole.]

## Miscellaneous Exercise 7(g)

1 Prove that the parabola $y^2 = x$ divides the circle $x^2 + y^2 = 2$ into two parts whose areas are in the ratio $(3\pi + 2) : (9\pi - 2)$.

2 Sketch the curve $y^2 = a^2 x/(2a - x)$, and prove that the area enclosed by the curve and the line $x = a$ is $(\pi - 2) a^2$. Also prove that the volume obtained by rotating this area about $Ox$ through two right angles is $\pi a^3 (\log 4 - 1)$.

3 Sketch the curve $x = a \cos^2 t \sin t$, $y = a \cos t \sin^2 t$, and find the area of a loop.

**4** If $r = f(\theta)$, $r = g(\theta)$ are closed curves surrounding the pole, and the second lies entirely within the first, prove that the area of the annular region enclosed between them is

$$\frac{1}{2} \int_0^{2\pi} [\{f(\theta)\}^2 - \{g(\theta)\}^2] \, d\theta.$$

$P$ is a variable point of the ellipse $l/r = 1 - e\cos\theta$, and $Q$ is on $OP$ produced such that $PQ = c$. Show that the area between the ellipse and the locus of $Q$ is $\pi c(2b + c)$, where $b$ is the minor semi-axis of the ellipse.

**5** Sketch the curve $r = a \th \frac{1}{2}\theta$. Find the length $s$ of the arc and the area $A$ of the sector measured from $\theta = 0$ to $\theta = \alpha$, and prove $2A = a(s - r)$.

**6** The length of the tangent to a curve is constant and equal to $c$; prove

$$x = c(\log\tan\tfrac{1}{2}\psi + \cos\psi), \quad y = c\sin\psi$$

if $x = 0$ and $y = c$ when $\psi = \frac{1}{2}\pi$. Also find $s$ in terms of $\psi$, measuring it from the point where $\psi = \frac{1}{2}\pi$.

**7** If the curve $3ay^2 = x(a - x)^2$ is represented parametrically by

$$x = 3at^2, \quad y = a(t - 3t^3),$$

show how the curve is traced as $t$ increases from $-\infty$ to $+\infty$. Prove that the arc-length from the origin to the point of the loop at which the tangent makes angle $\lambda$ with $Oy$ is $\frac{1}{9}a(3\tan\frac{1}{2}\lambda + \tan^3\frac{1}{2}\lambda)$.

**8** $A$, $B$ are the points on $y = c \ch(x/c)$ corresponding to $x = a$, $x = b$ $(a < b)$. The area under the arc $AB$ is rotated through $360°$ about $Ox$. If $V$ is the volume of the solid generated, and $S$ is its surface area, prove $V = \frac{1}{2}cS$, and find $V$.

**9** The circle $x^2 + y^2 = a^2$ cuts $Ox$, $Oy$ at $P$, $Q$. Find the centroid of the quadrant $OPQ$. If the semicircle with $OQ$ for diameter is cut away, find the centroid of the remaining area.

**10** Find the centroid of the area enclosed by one loop of $r^2 = a^2\cos 2\theta$.

***11** Prove that the centroid of the half of the ellipsoid

$$\frac{x^2}{a^2} + \frac{y^2}{b^2} + \frac{z^2}{c^2} = 1$$

for which $x \geqslant 0$ is the point $(\frac{3}{8}a, 0, 0)$.

**12** A semicircle $ABC$ of radius $a$ rotates about a line in its plane parallel to and at distance $b$ from the bounding diameter $AC$, so that the area is on the side of $AC$ remote from the line. Find the surface area and volume generated.

**13** Find the area of the sector bounded by $\theta = 0$, $\theta = \frac{1}{2}\pi$, and the curve $r = a\sin^2\theta$. Also find the volume obtained by rotating this sector about the line $\theta = \frac{1}{2}\pi$.

**14** A circle of radius $a$ is divided into two segments by a line distant $\lambda a$ from the centre, and the major segment is rotated through angle $2\pi$ about this line. Find the volume generated. [Use Ex. 7 (e), no. 17 (i).]

**15** Sketch the curve $y^2 = x^3/(a - x)$ $(a > 0)$. Find the area between the curve and its asymptote, and the volume obtained by rotating the area about this asymptote. Hence obtain the coordinates of the centroid of the area.

**16** A surface is generated by rotating $y = ax^2 + 2bx + c$ about $Ox$. Show that the volume bounded by this and the planes $x = \pm h$ is $\frac{1}{3}h(A_1 + 4A_2 + A_3)$, where

$A_1$, $A_2$, $A_3$ are the areas of the cross-sections by the planes $x = -h$, $x = 0$, $x = +h$ respectively. Prove also that the $x$-coordinate of the centroid of this volume is $h(A_3 - A_1)/(A_1 + 4A_2 + A_3)$.

17 For the area enclosed by $Ox$ and the curve $y = a\sin x$ between $(0, 0)$ and $(\pi, 0)$, find the centroid and the M.I. about $Ox$.

*Find the M.I. about the axis stated of the following (assumed uniform).*

18 Equilateral triangle of side $2a$; perpendicular axis through the centroid.

19 Solid right circular cone of height $h$, base-radius $r$; axis through the vertex and parallel to the base.

20 Rectangular lamina of sides $2a$, $2b$; about a diagonal. [Use Ex. 7 $(f)$, no. 2.]

21 The density of a solid sphere of radius $a$ varies as the square of the distance from the centre. Find the M.I. about a diameter. [Method of 7.92, ex. (v).]

# 8

# FURTHER GEOMETRICAL APPLICATIONS
# OF THE CALCULUS

## 8.1  Relations involving arc-length

## 8.11  Sign conventions

In this chapter we give some applications of the calculus to the geometry of plane curves, a subject known as *plane differential geometry*. As the emphasis is now geometrical, *we shall assume* (without further comment) *the continuity and derivability up to any required order of all functions which occur in the general discussion*.

We begin by extending the work already done about arc-length in 7.6. It is desirable to introduce more sign conventions (we defined the sign of $s$ in 7.61 (2)).

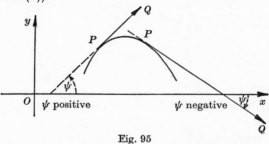

Fig. 95

(*a*) *The positive tangent.* A line drawn from $P$ along the tangent at $P$ in the direction of increasing $s$ is the *positive tangent* at $P$.

Since the direction in which $s$ increases may be different for two representations of the same curve (7.61 (2)), the positive tangents at $P$ may be opposite for such representations.

(*b*) *The angle $\psi$.* The angle made by the positive tangent with the positive direction of $Ox$, measured in the sense from positive $Ox$ to positive $Oy$, is denoted by $\psi$. It is determined to within an integral multiple of $2\pi$, and is usually chosen so that it varies continuously along the curve.

Reversal of the sense of the positive tangent is equivalent to replacing $\psi$ by $\pi + \psi$. For either sense, we still have the gradient $dy/dx$ equal to $\tan\psi$, since $\tan(\pi + \psi) = \tan\psi$.

## 8.12 Differential relations (cartesian coordinates)

From formula (i)' of 7.64 we have

$$\frac{ds}{dx} = \left\{ 1 + \left(\frac{dy}{dx}\right)^2 \right\}^{\frac{1}{2}}, \tag{i}$$

so that

$$\left(\frac{ds}{dx}\right)^2 = 1 + \left(\frac{dy}{dx}\right)^2,$$

and hence

$$ds^2 = dx^2 + dy^2 \tag{ii}$$

on converting to differentials. This result also embodies formulae (ii)', (iii)' of 7.64.

Since $dy/dx = \tan \psi$, (i) also shows that

$$\frac{ds}{dx} = \{1 + \tan^2 \psi\}^{\frac{1}{2}} = \pm \sec \psi.$$

By 7.61 (2), $s$ is measured to increase with $x$. It follows that (a) $ds/dx \geqslant 0$; and (b) the positive tangent is in the direction of $x$ increasing, so that $\psi$ *is a positive or negative* ACUTE *angle*, and $\sec \psi \geqslant 0$. Hence the positive sign must be chosen: $ds/dx = \sec \psi$. We now have

$$dx : dy : ds = \cos \psi : \sin \psi : 1. \tag{iii}$$

If the direction of the positive tangent is reversed (e.g. owing to use of a different representation $x = g(y)$ of the curve), then $\cos \psi$, $\sin \psi$ and $ds$ all change sign, but equations (iii) are unaltered. They are therefore valid for all representations of the curve.

In fig. 36 of 3.11, $PR = dx$ and $RT = dy$, so by (ii)

$$ds^2 = PR^2 + RT^2 = PT^2,$$

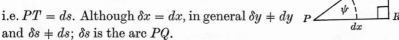

i.e. $PT = ds$. Although $\delta x = dx$, in general $\delta y \neq dy$ and $\delta s \neq ds$; $\delta s$ is the arc $PQ$.

Fig. 96

THEOREM. *When* $P \to Q$, $(chord\ PQ)/(arc\ PQ) \to 1$.

For

$$\frac{\text{chord } PQ}{\text{arc } PQ} = \frac{\{(\delta x)^2 + (\delta y)^2\}^{\frac{1}{2}}}{\delta s}$$

$$= \left\{ \left(\frac{\delta x}{\delta s}\right)^2 + \left(\frac{\delta y}{\delta s}\right)^2 \right\}^{\frac{1}{2}}$$

$$\to \left\{ \left(\frac{dx}{ds}\right)^2 + \left(\frac{dy}{ds}\right)^2 \right\}^{\frac{1}{2}} = 1$$

when $\delta s \to 0$, i.e. when $P \to Q$ along the curve.

*Remarks*

($\alpha$) The result can be expressed as follows:

*when $\delta s$ is small,*    $(\delta x)^2 + (\delta y)^2 \doteqdot (\delta s)^2$.

Compare the exact relation (ii), from which this approximation arises.

($\beta$) When the curve is a circle of radius $a$, let arc $PQ$ subtend angle $2\theta$ at the centre. Then arc $PQ = 2a\theta$, chord $PQ = 2a\sin\theta$, and the theorem gives

$$\lim_{\theta \to 0} \frac{\sin\theta}{\theta} = 1.$$

Compare 2.12, where this result was obtained from assumptions about 'area'.

## 8.13 Intrinsic equation

A relation between $s$ (measured from a fixed point $A$ of the curve to a variable point $P$) and $\psi$ (the angle between the positive tangent at $P$ and a fixed line) is called the *intrinsic equation* of the curve. Arbitrary constants can be added to $s$ and $\psi$, since the fixed point and line are arbitrary.

### Examples

(i) *Find the intrinsic equation of the cycloid* $x = a(\theta + \sin\theta)$, $y = a(1 - \cos\theta)$.

$$\tan\psi = \frac{dy}{dx} = \frac{\sin\theta}{1 + \cos\theta} = \tan\tfrac{1}{2}\theta.$$

Hence $\psi = \tfrac{1}{2}\theta + n\pi$; and $n = 0$ since $\psi = 0$ when $\theta = 0$ (see 1.61, fig. 23). By the example in 7.62, we have

$$s = 4a\sin\psi.$$

Sometimes the intrinsic equation arises naturally, as in ex. (ii).

(ii) *The catenary.* The curve assumed by a uniform thin flexible chain hanging freely from fixed ends is called a *catenary*.

Let $A$ be the point at which the tangent is horizontal, and $P$ be any point $(s, \psi)$ where $s$ is measured from $A$ and $\psi$ from the horizontal. Let $T$, $T_0$ denote the tensions at $P$, $A$, and suppose $w$ is the weight per unit length of the chain ($w$ is constant because the chain is 'uniform').

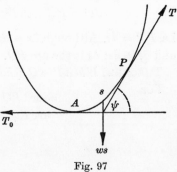

Fig. 97

The part $AP$ of the chain is in equilibrium under the three forces $T$, $T_0$, and weight $ws$. Resolving vertically and horizontally,

$$T\sin\psi = ws, \quad T\cos\psi = T_0.$$

By division, $\qquad\qquad\qquad \tan\psi = ws/T_0;$

i.e. $\qquad\qquad\qquad\qquad s = c\tan\psi,$

where $c = T_0/w$.

(iii) *Given the intrinsic equation, to find parametric equations.*

We illustrate with the catenary $s = c\tan\psi$.

Since $dy/ds = \sin\psi$ by 8.12, equations (iii),

$$\frac{dy}{d\psi} = \frac{dy}{ds}\frac{ds}{d\psi} = \sin\psi\,.\,c\sec^2\psi.$$

$$\therefore\quad y = \int c\sec\psi\tan\psi\,d\psi = c\sec\psi + \text{constant}.$$

Suppose axes chosen so that $y = c$ when $\psi = 0$; then $y = c\sec\psi$.

Also, since $dx/ds = \cos\psi$,

$$\frac{dx}{d\psi} = \frac{dx}{ds}\frac{ds}{d\psi} = \cos\psi\,.\,c\sec^2\psi = c\sec\psi.$$

$$\therefore\quad x = \int c\sec\psi\,d\psi = c\log\left(\sec\psi + \tan\psi\right)$$

if axes are chosen so that $x = 0$ when $\psi = 0$.

*Alternatively*, we may obtain $x$ as follows, working with $s$ instead of $\psi$:

$$\frac{dx}{ds} = \cos\psi = \frac{1}{\sqrt{(1+\tan^2\psi)}} = \frac{c}{\sqrt{(s^2+c^2)}},$$

$$\therefore\quad x = \int\frac{c\,ds}{\sqrt{(s^2+c^2)}} = c\,\text{sh}^{-1}\frac{s}{c} + \text{constant}$$

$$= c\,\text{sh}^{-1}(\tan\psi) + \text{constant}$$

$$= c\log\left(\tan\psi + \sec\psi\right) + \text{constant}$$

on using the logarithmic expression for $\text{sh}^{-1}u$ in 4.45(1). If $x = 0$ when $\psi = 0$, the constant is found to be zero.

*Remarks*

($\alpha$) $y^2 = c^2\sec^2\psi = c^2 + c^2\tan^2\psi = c^2 + s^2$.

($\beta$) $\dfrac{dy}{dx} = \tan\psi = \dfrac{s}{c} = \dfrac{1}{c}\sqrt{(y^2 - c^2)}$, by ($\alpha$).

Integrating by separation of the variables, we find

$$\text{ch}^{-1}\frac{y}{c} = \frac{x}{c} + A.$$

If axes are chosen (as before) so that $y = c$ when $x = 0$, then $A = 0$ and

$$y = c\,\text{ch}\frac{x}{c},$$

which is the *cartesian equation of the catenary*. It could also be obtained by elimination of $\psi$ from the parametric equations above.

### Exercise 8(a)

**1** Find the intrinsic equation of the curve

$$x = a(2\cos\theta + \cos 2\theta), \quad y = a(2\sin\theta + \sin 2\theta).$$

**2** Obtain parametric equations of the cycloid $s = 4a\sin\psi$, taking the origin at $\psi = 0$, $s = 0$.

**3** Show that the cartesian equation of $s = \log\tan\frac{1}{2}\psi$ can be written

$$x = \log\sin y.$$

**4** Find the cartesian equation of $y = \log\operatorname{ch} s$ by first proving $\sin\psi = \operatorname{th} s$ and $\tan\psi = \operatorname{sh} s$.

**5** Prove that the curve $s = c\psi$ is a circle of radius $c$.

**6** Find parametric equations for the curve $s = a\psi^2$.

## 8.14 Differential relations (polar coordinates)

From $x = r\cos\theta$, $y = r\sin\theta$ we have

$$dx = \cos\theta\,dr - r\sin\theta\,d\theta, \quad dy = \sin\theta\,dr + r\cos\theta\,d\theta,$$

and hence from formula (ii) of 8.12 we deduce

$$ds^2 = dr^2 + r^2 d\theta^2. \tag{iv}$$

Equivalently, this follows from (v)′ of 7.64, and embodies (iv)′ and (vi)′ there also.

We now introduce an angle $\phi$ whose role in polar coordinates is similar to that of $\psi$ in cartesians.

*Definition.* $\phi$ is the angle from the radius vector to the positive tangent at $P$.

Fig. 98 is the simple 'standard' case; fig. 100 illustrates a case when $r$ is negative.

We may choose $\phi$ to be in the range $0 \leqslant \phi < 2\pi$. In all cases we have

$$\theta + \phi = \psi$$

to within an integral multiple of $2\pi$. Hence

$$\cos\phi = \cos(\psi - \theta)$$

$$= \cos\psi\cos\theta + \sin\psi\sin\theta$$

$$= \frac{dx}{ds}\frac{x}{r} + \frac{dy}{ds}\frac{y}{r} \quad \text{by 8.12, (iii)}$$

$$= \frac{x\,dx + y\,dy}{r\,ds}.$$

Since $r^2 = x^2 + y^2$, therefore $r\,dr = x\,dx + y\,dy$, and so

$$\cos\phi = \frac{dr}{ds}.$$

Similarly, $\quad \sin\phi = \sin(\psi - \theta) = \sin\psi\cos\theta - \cos\psi\sin\theta$

$$= \frac{dy}{ds}\frac{x}{r} - \frac{dx}{ds}\frac{y}{r} = \frac{x\,dy - y\,dx}{r\,ds};$$

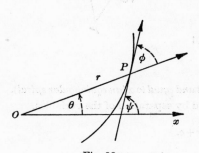       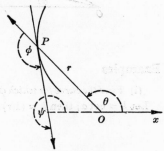

Fig. 98                     Fig. 99

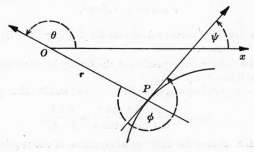

Fig. 100

and from $y/x = \tan\theta$ we have $(x\,dy - y\,dx)/x^2 = \sec^2\theta\,d\theta = (r^2/x^2)\,d\theta$, so

$$\sin\phi = r\frac{d\theta}{ds}.$$

(Since $s$ is measured to increase with $\theta$, this last result shows that $\sin\phi$ and $r$ have the same sign: $0 \leqslant \phi \leqslant \pi$ when $r > 0$, and $\pi \leqslant \phi \leqslant 2\pi$ when $r < 0$; cf. fig. 100.)

By division we find $\qquad \cot\phi = \dfrac{1}{r}\dfrac{dr}{d\theta}.$

The results can be summarised as

$$\cos\phi : \sin\phi : 1 = dr : r\,d\theta : ds, \qquad\qquad \text{(v)}$$

and may be remembered by thinking of the triangle $P'Q'R'$ suggested by the approximately triangular part $PQR$ of the main figure, in which $PR$ is an arc of a circle of centre $O$ and radius $r$.

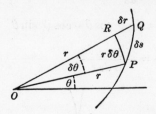

Fig. 101

## Examples

(i) *Find the curve for which $\phi$ is constant and equal to $\alpha$ (an equiangular spiral).*

Let $k = \cot\alpha$; then $k = (1/r)\,(dr/d\theta)$, and by separation of the variables,

$$\theta = \frac{1}{k}\log r + c.$$

If $r = a$ when $\theta = 0$, then $c = -(1/k)\log a$, and $\theta = (1/k)\log(r/a)$, i.e.

$$r = a\,e^{k\theta} = a\,e^{\theta\cot\alpha}.$$

(ii) *Find the family of curves which cuts the family $r = a\theta$ (spirals of Archimedes) at constant angle $\alpha$.*

First form the differential equation of the family by eliminating $a$. We get $r\,d\theta/dr = \theta$, and hence $\tan\phi = \theta$.

If $\phi'$ is the corresponding angle for the required family, then $\phi' = \phi \pm \alpha$, and so (if $\alpha \neq \tfrac{1}{2}\pi$)

$$\tan\phi' = \frac{\tan\phi \pm \tan\alpha}{1 \mp \tan\phi\,\tan\alpha} = \frac{\theta + k}{1 - k\theta},$$

where $k = \pm\tan\alpha$. Hence the differential equation of the required family is

$$r\frac{d\theta}{dr} = \frac{\theta + k}{1 - k\theta},$$

and by separation of the variables we find

$$r = c(\theta + k)^{k^2 + 1}e^{-k\theta},$$

where $c$ is an arbitrary constant.

If $\alpha = \tfrac{1}{2}\pi$, then on the orthogonal curve we have

$$\cot\phi' = -\tan\phi = -\theta,$$

i.e.

$$\frac{1}{r}\frac{dr}{d\theta} = -\theta,$$

from which

$$r^2 = c\,e^{-\theta^2}.$$

(iii) *Orthogonal trajectories in polar coordinates.*

Suppose the differential equation of the given family is $f(r, \theta, dr/d\theta) = 0$. This determines $dr/d\theta$ for the member(s) of the family through the point $(r, \theta)$, say $dr/d\theta = g(r, \theta)$. Then $\cot\phi = (1/r)\,g(r, \theta)$.

For the orthogonal curve(s) through $(r, \theta)$, $\phi' = \phi \pm \frac{1}{2}\pi$, so that

$$\cot \phi' = -\tan \phi = -\frac{r}{g(r, \theta)}.$$

Hence on the orthogonal curve(s) we have

$$\frac{1}{r}\frac{dr}{d\theta} = -\frac{r}{g(r, \theta)},$$

i.e.

$$-r^2 \Big/ \frac{dr}{d\theta} = g(r, \theta),$$

and therefore

$$f\left(r, \theta, -r^2 \Big/ \frac{dr}{d\theta}\right) = 0.$$

This is the differential equation of the orthogonal family.

(iv) *Reflector property of the ellipse.*

If $S, S'$ are the foci and $P$ is any point on the ellipse, it is known (17.22) that

$$SP + S'P = 2a,$$

where $2a$ is the major axis. Writing $SP = r$, $S'P = r'$ (i.e. using two systems of polar coordinates with poles $S, S'$ respectively), the relation is

$$r + r' = 2a.$$

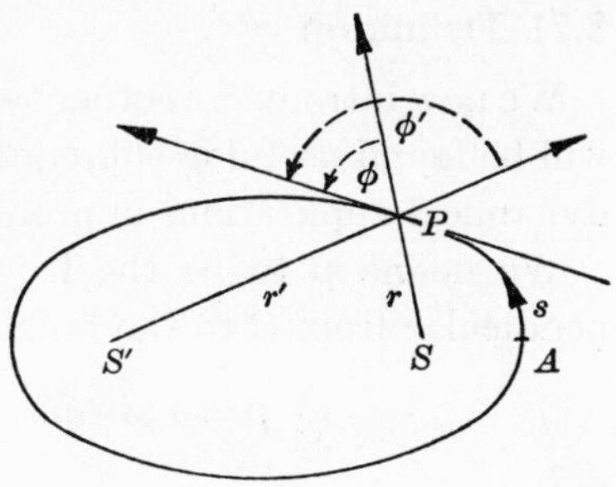

Deriving wo $s$,

$$\frac{dr}{ds} + \frac{dr'}{ds} = 0,$$

$$\therefore \quad \cos \phi + \cos \phi' = 0,$$

i.e.     $\cos \phi = -\cos \phi' = \cos(\pi - \phi')$   and so   $\phi = \pi - \phi'$.

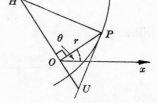

Fig. 102

Hence *the angles between the focal radii and the tangent at $P$ are equal;* i.e. *the tangent and normal at $P$ are the angle-bisectors of $SPS'$.* Cf. Ex. 17 (b), no. 14.

## Exercise 8(b)

**1** Find $\phi$ for the curve $r^n = a^n \cos n\theta$.

**2** Prove that the tangent to the cardioid $r = a(1 - \cos \theta)$ makes angle $\frac{3}{2}\theta$ with $Ox$.

**3** Prove that the curves

$$r = a \cos \theta, \ r = a(1 - \cos \theta)$$

intersect at the point $(\frac{1}{2}a, \frac{1}{3}\pi)$, and find their angle of intersection.

**4** Find the curves for which $\phi = n\theta$.

*If the line through $O$ perpendicular to $OP$ meets the tangent and normal at $U, H$, then $OU, OH, PU, PH$ are called the* polar subtangent, subnormal, tangent, normal *respectively. If $r'$ denotes $dr/d\theta$, prove*

Fig. 103

**5** $OH = r'$.        **6** $OU = r^2/r'$.        **7** $PH = \sqrt{(r^2 + r'^2)}$.

**8** $PU = r\sqrt{(r^2 + r'^2)}/r'$.

*Find the curves for which*

9  $OH = a.$     10  $OU = a.$     11  $OH = OU.$     12  $PH = a.$

*Find the orthogonal trajectories of the following families, a being the parameter.*

13  $r = a(1 + \cos\theta).$     14  $r^2 = a^2\cos 2\theta.$     15  $r^3 = a^3\cos 3\theta.$

16  $r = a(1 + 2\cos\theta).$     17  $x^2 + y^2 = a(x - y)$ [first convert to polars].

18  $x^2 + y^2 = ax.$     19  $x(x^2 + y^2) + a(x^2 - y^2) = 0.$

*20  Loci $f(r, \theta) = 0, f(r', \theta') = 0$ for which $rr' = k^2$ and $\theta = \theta'$ are called *inverse curves*. Prove that these curves cut the radius vector at supplementary angles. [Derive $rr' = k^2$ logarithmically wo $\theta$.]

## 8.2  (p, r) equation

### 8.21  Definition

We now introduce another type of equation for a plane curve which will be found useful in subsequent work on curvature, and which has dynamical applications in problems on central orbits.

We define $p$ to be the length of the perpendicular from $O$ to the tangent at $P$. Then

$$p = r\sin\phi.$$

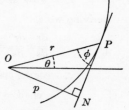

Since $r\sin\phi = r^2 d\theta/ds$ and $s$ is measured to increase with $\theta$, it follows that when polar coordinates are used, $p$ is never negative.

The relation between the $p$ and $r$ corresponding to a general point $P$ of the curve is called the $(p, r)$ *equation* or *pedal equation* of the curve.

Fig. 104

### 8.22  (p, r) equation from polar equation

Write $u = 1/r$; then

$$\frac{du}{d\theta} = -\frac{1}{r^2}\frac{dr}{d\theta} = -\frac{1}{r}\cot\phi.$$

Since $p = r\sin\phi$,

$$\frac{1}{p^2} = \frac{\operatorname{cosec}^2\phi}{r^2} = \frac{1 + \cot^2\phi}{r^2} = \frac{1}{r^2}\left\{1 + \frac{1}{r^2}\left(\frac{dr}{d\theta}\right)^2\right\}, \tag{i}$$

or

$$\frac{1}{p^2} = u^2 + \left(\frac{du}{d\theta}\right)^2. \tag{ii}$$

The $(p, r)$ equation is therefore obtained by eliminating $\theta$ from (i) and the polar equation of the curve.

## Examples

(i) *Equiangular spiral.* Since $\phi = \alpha$ by definition (see 8.14, ex. (i)), the $(p, r)$ equation is $p = r \sin \alpha$.

(ii) *Lemniscate* $r^2 = a^2 \cos 2\theta$.

Derive wo $\theta$: $\qquad\qquad 2r \dfrac{dr}{d\theta} = -2a^2 \sin 2\theta.$

$$\therefore \quad \left(\frac{dr}{d\theta}\right)^2 = \frac{a^4 \sin^2 2\theta}{r^2} = \frac{a^4(1 - \cos^2 2\theta)}{r^2} = \frac{a^4 - r^4}{r^2}.$$

Hence by equation (i),

$$\frac{1}{p^2} = \frac{1}{r^2}\left\{1 + \frac{1}{r^2}\left(\frac{a^4 - r^4}{r^2}\right)\right\} = \frac{a^4}{r^6},$$

so that $r^6 = a^4 p^2$ and† $\pm r^3 = a^2 p$.

(iii) *Cardioid* $r = a(1 + \cos \theta)$.

Since $r = 2a \cos^2 \tfrac{1}{2}\theta$, by deriving logarithmically wo $\theta$ we have

$$\frac{1}{r}\frac{dr}{d\theta} = -\tan \tfrac{1}{2}\theta.$$

$$\therefore \quad \cot \phi = -\tan \tfrac{1}{2}\theta = \cot(\tfrac{1}{2}\theta + \tfrac{1}{2}\pi),$$

and $\qquad\qquad\qquad \phi = \tfrac{1}{2}\theta + \tfrac{1}{2}\pi.$

As $p = r \sin \phi = r \cos \tfrac{1}{2}\theta$, therefore $p^2 = r^2 \cos^2 \tfrac{1}{2}\theta = r^3/2a$, and‡ $2ap^2 = r^3$.

(iv) *Rectangular hyperbola* $x^2 - y^2 = a^2$.

First transform to polars: $r^2 \cos 2\theta = a^2$, i.e. $a^2 u^2 = \cos 2\theta$. Derive wo $\theta$:

$$a^2 u \frac{du}{d\theta} = -\sin 2\theta.$$

Hence by equation (ii),

$$\frac{1}{p^2} = u^2 + \left(\frac{\sin 2\theta}{a^2 u}\right)^2 = u^2 + \frac{\sin^2 2\theta}{a^4 u^2}$$

$$= u^2 + \frac{1 - a^4 u^4}{a^4 u^2} = \frac{1}{a^4 u^2}.$$

$$\therefore \quad p^2 = a^4 u^2 = a^4/r^2 \quad \text{and†} \quad \pm pr = a^2.$$

## 8.23 Polar equation from $(p, r)$ equation

From $p = r \sin \phi$ and $\cot \phi = (1/r)\,dr/d\theta$, elimination of $\phi$ gives a relation of the form $dr/d\theta = f(r)$ which, on integrating, leads to the required polar equation.

Alternatively, equation (i) of 8.22, where $p$ is a known function of $r$ from the given $(p, r)$ equation, is the relation $dr/d\theta = f(r)$.

---

† $\pm$ since $r$ may be negative, but $p$ is always positive.
‡ $+$ since $r \geqslant 0$ from the equation of the curve.

## Example

*Find the polar equation of* $2ap = r^2$.

Since

$$\sin^2\phi = \frac{p^2}{r^2} = \frac{r^2}{4a^2},$$

$$\csc^2\phi = \frac{4a^2}{r^2} \quad \text{and} \quad \cot^2\phi = \frac{4a^2}{r^2} - 1.$$

Therefore from

$$\cot\phi = \frac{1}{r}\frac{dr}{d\theta},$$

$$\frac{1}{r}\frac{dr}{d\theta} = \pm\frac{1}{r}\sqrt{(4a^2 - r^2)},$$

$$\theta + c = \pm \int \frac{dr}{\sqrt{(4a^2 - r^2)}} = \pm \sin^{-1}\left(\frac{r}{2a}\right),$$

and

$$r = \pm 2a\sin(\theta + c).$$

The constant $c$ arises because the choice of initial line is arbitrary—the $(p, r)$ equation makes no reference to $\theta$.

When the polar equation is known, the cartesian equation can be found from it.

### Exercise 8(c)

*Find the* $(p, r)$ *equation of*

**1** $r\theta = a$.      **2** $r^n \sin n\theta = a^n$.      **3** $lu = 1 + e\cos\theta$.

**4** $xy = 2a^2$ (*O* being the pole).      **5** $x^2/a^2 + y^2/b^2 = 1$ (*O* being the pole).

**6** Prove that the tangent at the point $t$ to the curve $x = a\cos^3 t$, $y = a\sin^3 t$ is $x\sin t + y\cos t = a\sin t\cos t$. Hence find the $(p, r)$ equation wo $O$ as pole.

*Find the polar equation of*

**7** $p^2 = ar$.      **8** $2ap^2 = r^3$.      **9** $r^2 = ap$.

**\*10** Use $p = r\sin\phi$ and $\sin\phi\,dr = r\cos\phi\,d\theta$ to prove that $dp = r\cos\phi\,(d\theta + d\phi)$. Deduce that $dp/d\psi = r\cos\phi$.

*Definition.* The relation between $p$ and $\psi$ for a curve is called its $(p, \psi)$ *equation* or *tangential polar equation.*

**\*11** Prove $p^2 + (dp/d\psi)^2 = r^2$, and deduce that $p + d^2p/d\psi^2 = r\,dr/dp$.

**\*12** If $p = f(\psi)$, prove

$$x = \cos\psi f'(\psi) + \sin\psi f(\psi) \quad \text{and} \quad y = \sin\psi f'(\psi) - \cos\psi f(\psi).$$

[Use no. 9 and $x = r\cos\theta = r\cos(\psi - \phi) = \ldots$.] (The results are parametric equations of a curve whose $(p, \psi)$ equation is given. Elimination of $\psi$ gives the cartesian equation.)

## 8.3   Curvature

### 8.31   Definitions

The concept of *curvature* arises from the need to measure the rate of bending of a curve, i.e. the rate at which the tangent turns as the point of contact varies along the curve.

(*a*) The *mean curvature* of the arc $PQ$ is (with the notation of fig. 105) $(\psi_Q - \psi_P)/(s_Q - s_P)$.

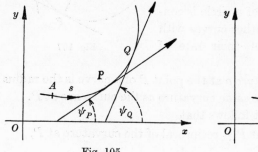

Fig. 105

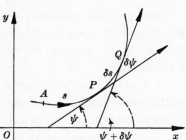

Fig. 106

(*b*) The *curvature at P* is

$$\lim_{Q \to P} \frac{\psi_Q - \psi_P}{s_Q - s_P}.$$

With the notation of fig. 106, the curvature at $P$ is

$$\lim_{\delta s \to 0} \frac{\delta \psi}{\delta s} = \frac{d\psi}{ds}.$$

Writing $\kappa$ for the curvature, we have

$$\kappa = \frac{d\psi}{ds}.$$

*Remarks*

($\alpha$) $\kappa$ does not depend on the point $A$ from which $s$ is measured, nor on the line $Ox$ from which $\psi$ is measured; but its *sign* depends on the sense in which $s$ is measured.

($\beta$) We shall have $\kappa = 0$ at points where $\psi$ is stationary. These usually correspond to points of inflexion of the curve (3.71).

($\gamma$) Since $\psi$ = constant for a straight line, we have the natural result $\kappa \equiv 0$ (that the line does not bend anywhere).

($\delta$)  *For a circle of radius $R$, the curvature is constant and equal to $1/R$.*
Let arc $PQ$, of length $\delta s$, subtend angle $\delta\psi$ at the centre. Then $\delta\psi$
is also the angle between the tangents at $P$ and $Q$, and $\delta s = R\,\delta\psi$.

$$\therefore \quad \frac{\delta\psi}{\delta s} = \frac{1}{R} \quad \text{and} \quad \kappa = \lim_{\delta s \to 0} \frac{\delta\psi}{\delta s} = \frac{1}{R}.$$

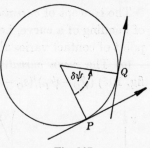

This is also a natural result: since the
circle is symmetrical, it bends everywhere
at the same rate.

Because the curvature of a circle is constant, we may compare other curves with
the circle in respect of their rate of
bending.

Fig. 107

(c)  The *radius of curvature $\rho$* at the point $P$ of a curve is the radius
of the circle which has the same curvature as the curve has at $P$.

By Remark ($\delta$) above it follows that

radius of curvature at $P$ = reciprocal of the curvature at $P$,

i.e. $$\rho = \frac{1}{\kappa} = \frac{ds}{d\psi}.$$

*Remarks*

($\epsilon$)  $\rho$ may be positive or negative: it has the same sign as $\kappa$.

($\zeta$)  When $\kappa = 0$ at $P$, there is no radius of curvature at $P$ since no
circle can have zero curvature. But as a circle of *large* radius has a
*small* curvature, we may conveniently think of points where $\kappa = 0$
as giving an 'infinite radius of curvature'.

## 8.32  Formulae for $\kappa$ or $\rho$

The definition $\kappa = d\psi/ds$ can be applied directly only when the
intrinsic equation of the curve is known. We now seek other more
convenient formulae.

**(1)  *Curve $y = f(x)$.***

From
$$\frac{dy}{dx} = \tan\psi, \quad \psi = \tan^{-1}\left(\frac{dy}{dx}\right),$$

and
$$\kappa = \frac{d\psi}{ds} = \frac{d\psi}{dx}\frac{dx}{ds} = \frac{d}{dx}\left\{\tan^{-1}\left(\frac{dy}{dx}\right)\right\} \times \cos\psi$$

$$= \frac{\dfrac{d^2y}{dx^2}}{1 + \left(\dfrac{dy}{dx}\right)^2} \times \frac{1}{\left\{1 + \left(\dfrac{dy}{dx}\right)^2\right\}^{\frac{1}{2}}},$$

the positive sign being chosen since $\cos \psi = dx/ds \geqslant 0$ because $s$ is measured to increase with $x$. Thus

$$\kappa = \frac{\dfrac{d^2y}{dx^2}}{\left\{1 + \left(\dfrac{dy}{dx}\right)^2\right\}^{\frac{3}{2}}}.$$

*Remarks*

($\alpha$) $\kappa$ has the same sign as $d^2y/dx^2$; it is positive if the curve is concave upwards (6.71 (2)) at $P$, negative if concave down.

($\beta$) At a point of inflexion $\kappa$ vanishes and changes sign; for $d^2y/dx^2$ has this property (see 3.71).

($\gamma$) At a point where $\psi = 0$, i.e. $dy/dx = 0$, we have $\kappa = d^2y/dx^2$. When discussing the curvature of a curve at a *particular* point $P$ we may always choose axes with $P$ as origin, $Px$ along the tangent at $P$, and $Py$ along the normal. With this choice, the curvature at $P$ is $f''(0)$.

(**2**) *Curve* $x = g(y)$.

We find similarly that

$$\kappa = -\frac{\dfrac{d^2x}{dy^2}}{\left\{1 + \left(\dfrac{dx}{dy}\right)^2\right\}^{\frac{3}{2}}},$$

where the minus sign is correct if $s$ is chosen to increase with $y$.

(**3**) *Curve* $x = x(t),\ y = y(t)$.

$$\tan \psi = \frac{dy}{dx} = \frac{\dot{y}}{\dot{x}}, \qquad \therefore \ \psi = \tan^{-1} \frac{\dot{y}}{\dot{x}}.$$

$$\kappa = \frac{d\psi}{ds} = \frac{d\psi}{dt} \bigg/ \frac{ds}{dt} = \frac{d}{dt}\left\{\tan^{-1}\frac{\dot{y}}{\dot{x}}\right\} \div \frac{ds}{dt}$$

$$= \left\{\frac{1}{1 + \left(\dfrac{\dot{y}}{\dot{x}}\right)^2} \frac{\dot{x}\ddot{y} - \ddot{x}\dot{y}}{\dot{x}^2}\right\} \bigg/ (\dot{x}^2 + \dot{y}^2)^{\frac{1}{2}}$$

$$= \frac{\dot{x}\ddot{y} - \ddot{x}\dot{y}}{(\dot{x}^2 + \dot{y}^2)^{\frac{3}{2}}}.$$

(**4**) *Curve* $r = f(\theta)$.

Since

$$\psi = \theta + \phi \quad \text{and} \quad \tan\phi = r \bigg/ \frac{dr}{d\theta},$$

$$\kappa = \frac{d\psi}{ds} = \frac{d\theta}{ds} + \frac{d\phi}{ds} = \frac{d\theta}{ds}\left(1 + \frac{d\phi}{d\theta}\right)$$

$$= \frac{d\theta}{ds}\left[1 + \frac{d}{d\theta}\left\{\tan^{-1}\left(r\bigg/\frac{dr}{d\theta}\right)\right\}\right].$$

Now

$$\frac{d}{d\theta}\left\{\tan^{-1}\left(r\Big/\frac{dr}{d\theta}\right)\right\} = \frac{1}{1+r^2\Big/\left(\frac{dr}{d\theta}\right)^2} \frac{\left(\frac{dr}{d\theta}\right)^2 - r\frac{d^2r}{d\theta^2}}{\left(\frac{dr}{d\theta}\right)^2}$$

$$= \frac{\left(\frac{dr}{d\theta}\right)^2 - r\frac{d^2r}{d\theta^2}}{\left(\frac{dr}{d\theta}\right)^2 + r^2}.$$

Also

$$\frac{ds}{d\theta} = \left\{r^2 + \left(\frac{dr}{d\theta}\right)^2\right\}^{\frac{1}{2}}.$$

$$\therefore \quad \kappa = \frac{1}{\left\{r^2 + \left(\frac{dr}{d\theta}\right)^2\right\}^{\frac{1}{2}}}\left[1 + \frac{\left(\frac{dr}{d\theta}\right)^2 - r\frac{d^2r}{d\theta^2}}{\left(\frac{dr}{d\theta}\right)^2 + r^2}\right]$$

$$= \frac{r^2 + 2\left(\frac{dr}{d\theta}\right)^2 - r\frac{d^2r}{d\theta^2}}{\left\{r^2 + \left(\frac{dr}{d\theta}\right)^2\right\}^{\frac{3}{2}}}.$$

See also Ex. 8 (d), no. 14, and Ex. 8 (g), no. 20.

Owing to the complexity of this formula and the simplicity of the one now to be obtained for $(p, r)$ equations, it is best to start by transforming the given polar equation into $(p, r)$ coordinates (8.22). We may finally have to eliminate $p$ from the expression for $\kappa$.

(5) $(p, r)$ equation.

$$\kappa = \frac{d\psi}{ds} = \frac{d\theta}{ds} + \frac{d\phi}{ds},$$

also

$$\sin\phi = r\frac{d\theta}{ds}, \quad \cos\phi = \frac{dr}{ds}.$$

$$\therefore \quad \kappa = \frac{1}{r}\sin\phi + \frac{d\phi}{dr}\cos\phi$$

$$= \frac{1}{r}\frac{d}{dr}(r\sin\phi)$$

$$= \frac{1}{r}\frac{dp}{dr}.$$

**Examples**

(i) *Find $\rho$ at the vertex of the cycloid $s = 4a\sin\psi$.*

$$\rho = \frac{ds}{d\psi} = 4a\cos\psi.$$

At the vertex, $\psi = 0$, and hence $\rho = 4a$.

(ii) *Find $\rho$ at the point $(3, 4)$ of the curve $xy = 12$.*

$$y = \frac{12}{x}, \quad y' = -\frac{12}{x^2} \quad \text{and} \quad y'' = \frac{24}{x^3}.$$

At $(3, 4)$,  $y' = -\frac{4}{3}$  and  $y'' = \frac{8}{9}$,  so  $\rho = (1 + \frac{16}{9})^{\frac{3}{2}}/\frac{8}{9} = \frac{125}{24}$.

(iii) *Find $\rho$ at the point $\phi$ of the ellipse $x = a\cos\phi, y = b\sin\phi$.*

$$\frac{dx}{d\phi} = -a\sin\phi, \quad \frac{dy}{d\phi} = b\cos\phi,$$

$$\frac{d^2x}{d\phi^2} = -a\cos\phi, \quad \frac{d^2y}{d\phi^2} = -b\sin\phi.$$

$$\therefore \quad \frac{dx}{d\phi}\frac{d^2y}{d\phi^2} - \frac{dy}{d\phi}\frac{d^2x}{d\phi^2} = ab \quad \text{and} \quad \left(\frac{dx}{d\phi}\right)^2 + \left(\frac{dy}{d\phi}\right)^2 = a^2\sin^2\phi + b^2\cos^2\phi,$$

and

$$\rho = \frac{1}{ab}\{a^2\sin^2\phi + b^2\cos^2\phi\}^{\frac{3}{2}}.$$

This result can be written $\rho = OD^2/p$, where $p$ is the length of the perpendicular from the centre $O$ to the tangent at $P$, and $OD$ is a semi-diameter conjugate to $OP$: see 17.64, ex. (i).

(iv) *Find $\rho$ for the cardioid $r = a(1 + \cos\theta)$.*

First we find the $(p, r)$ equation as in 8.22, ex. (iii), viz.

$$2ap^2 = r^3.$$

From this,

$$4ap = 3r^2\frac{dr}{dp}.$$

$$\rho = r\frac{dr}{dp} = r\frac{4ap}{3r^2} = \frac{4a}{3r}\sqrt{\frac{r^3}{2a}} = \frac{2}{3}\sqrt{(2ar)}$$

$$= \frac{2}{3}\sqrt{(2a \cdot 2a\cos^2\tfrac{1}{2}\theta)} \quad \text{since} \quad r = 2a\cos^2\tfrac{1}{2}\theta,$$

$$= \frac{4}{3}a\cos\tfrac{1}{2}\theta \quad (-\pi < \theta \leqslant \pi).$$

*\*(v) Find the radii of curvature of $x^3 + y^3 = x(x - y)$ at (a) $(1, 0)$; (b) $(0, 0)$.*

(a) We can find $dy/dx$, and then $d^2y/dx^2$, by deriving the given equation wo $x$ as in 3.41:

$$3x^2 + 3y^2y' = 2x - y - xy',$$

$$\therefore \quad (3y^2 + x)y' = 2x - y - 3x^2. \tag{$\alpha$}$$

At $(1, 0)$, this shows $y' = -1$.

Deriving $(\alpha)$ wo $x$:

$$(3y^2 + x)y'' + 6yy'^2 + y' = 2 - y' - 6x,$$

and hence when $x = 1$, $y = 0$ and $y' = -1$, then $y'' = -2$.

Therefore at $(1, 0)$,

$$\rho = \frac{\{1 + 1\}^{\frac{3}{2}}}{-2} = -\sqrt{2}.$$

(b) Equation $(\alpha)$ does not determine $y'$ when $x = 0$ and $y = 0$ because both sides vanish. We may avoid this difficulty by making successive approximations to the equation of the curve near the origin as follows.

If $x$ and $y$ are small, then the terms $x^3$, $y^3$ are small compared with $x(x-y)$, so that the given equation is approximately $x(x-y) = 0$. Hence near $O$ the curve has two branches approximating to $x = 0$, $x-y = 0$.

The given equation can be written $x = (x^3+y^3)/(x-y)$, so that on the branch for which $x \doteqdot 0$ we have $x \doteqdot (y^3)/(-y) = -y^2$, which gives a closer approximation to this branch. To get a still better one, consider

$$x+y^2 = \frac{x^3+y^3}{x-y}+y^2 = \frac{x^3+xy^2}{x-y}$$

(again from the given equation); since $x \doteqdot -y^2$, this gives

$$x+y^2 \doteqdot \frac{-y^4}{-y} = y^3.$$

Thus $x = -y^2+y^3$ is a third approximation to this branch.

Also, the equation can be written $x-y = (x^3+y^3)/x$, so on the branch for which $y \doteqdot x$ we have $x-y \doteqdot (2x^3)/x = 2x^2$. Hence $y = x-2x^2$ is an approximation to this branch.

On the first branch we have at $O$ that

$$\frac{dx}{dy} = 0, \quad \frac{d^2x}{dy^2} = -2, \quad \text{so} \quad \rho = -\frac{\{1+0^2\}^{\frac{3}{2}}}{-2} = \tfrac{1}{2}.$$

At $O$ on the second branch, $dy/dx = 1$ and $d^2y/dx^2 = -4$; hence

$$\rho = \frac{\{1+1^2\}^{\frac{3}{2}}}{-4} = -\tfrac{1}{2}\sqrt{2}.$$

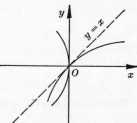

Fig. 108

## Exercise 8(d)

1　Find the radius of curvature of the catenary $s = c\tan\psi$ at the vertex.

2　Find the radius of curvature of the tractrix $e^{-s/a} = \cos\psi$ in terms of $\psi$.

3　If $\rho^2 = c^2-s^2$, find the intrinsic equation of the curve.

4　Find the curvature of the catenary $y = c\,\mathrm{ch}\,(x/c)$.

5　Find the radius of curvature of the cycloid $x = a(\theta+\sin\theta)$, $y = a(1-\cos\theta)$.

6　Find $\rho$ for $x = at^2$, $y = at^3$.

7　A curve which touches $Ox$ at $O$ has $\rho = c\sec\psi$. Find parametric equations and prove that the cartesian equation is $y = c\log\sec(x/c)$.

*8　Prove that

$$\kappa = -\frac{d^2x}{ds^2}\bigg/\frac{dy}{ds} = +\frac{d^2y}{ds^2}\bigg/\frac{dx}{ds} \quad \text{and deduce} \quad \kappa^2 = \left(\frac{d^2x}{ds^2}\right)^2+\left(\frac{d^2y}{ds^2}\right)^2.$$

*9　Prove that

$$\frac{d\kappa}{ds} = \frac{dx}{ds}\frac{d^3y}{ds^3}-\frac{dy}{ds}\frac{d^3x}{ds^3}.$$

10　If $p^2 = ar$, find $\rho$ in terms of $r$.

11　Find $\rho$ in terms of $r$ for the equiangular spiral $r = a\,e^{\theta\cot\alpha}$.

12　Find $\rho$ in terms of $r$ for the lemniscate $r^2 = a^2\cos 2\theta$.

13　If $s$ is measured from the point $\theta = 0$ of the cardioid $r = a(1+\cos\theta)$, prove that $9\rho^2+s^2 = 16a^2$ for $0 < \theta < \pi$.

*14 Denoting $du/d\theta$ by $u'$, derive each of the equations $x = r\cos\theta$, $y = r\sin\theta$ twice wo $\theta$ to prove

    (a) $s'\cos\psi = -r\sin\theta + r'\cos\theta$,

    (b) $s'\sin\psi = r\cos\theta + r'\sin\theta$,

    (c) $s''\cos\psi - \kappa s'^2\sin\psi = -r\cos\theta - 2r'\sin\theta + r''\cos\theta$,

    (d) $s''\sin\psi + \kappa s'^2\cos\psi = -r\sin\theta + 2r'\cos\theta + r''\sin\theta$.

By subtracting the product of (a) and (c) from the product of (b) and (d), prove

$$\kappa s'^3 = r^2 - rr'' + 2r'^2,$$

and deduce the formula for $\kappa$ given in 8.32 (4).

*Find the radius of curvature of*

*15   $2x^3 + y^3 = y^2 - x^2$ at $(-\frac{1}{2}, 1)$.     *16   $xy(x^2 + y^2) = x^3 - y^3$ at $(0, 0)$.

*Find the radii of curvature at the origin for the branches of*

*17   $2x^3 + y^3 = y^2 - x^2$.          *18   $x(y - 2x) = y^3$.

*19   If $p = f(\psi)$, prove $\rho = p + d^2p/d\psi^2$. [Use Ex. 8 (c), no. 11.]

20   What are the curves which satisfy

$$\left\{1 + \left(\frac{dy}{dx}\right)^2\right\}^{\frac{3}{2}} = c\,\frac{d^2y}{dx^2}?$$

21   Find the radius of curvature of $r = a\cos\theta$.

*22   Find the cartesian equation of the curves for which $k$ times the radius of curvature is equal to the cube of the normal (see 5.71).

## 8.4   Circle and centre of curvature

### 8.41   Osculating circle

Suppose a curve $y = f(x)$ and a point $P$ on it are given, and also a family of curves $y = g(x, \alpha_1, \alpha_2, \ldots)$ depending on a number of parameters. If the parameters are chosen so that the corresponding member of the family has the highest possible order of contact (see 6.72) with $y = f(x)$ at $P$, then the two curves are said to *osculate* there.

(1) *Osculating line*. If $y = g(x) = ax + b$, then the two constants $a$, $b$ can be chosen to satisfy the two conditions for first-order contact with $y = f(x)$ at $x = x_0$, viz.

$$f(x_0) = ax_0 + b, \quad f'(x_0) = a.$$

These give $b = f(x_0) - x_0 f'(x_0)$, so that the osculating line is

$$y - f(x_0) = f'(x_0)\,(x - x_0),$$

i.e. the tangent to $y = f(x)$ at $x = x_0$.

This line will not in general have contact of order higher than the first; but if $f''(x_0) = 0$, then the order is at least two.

(2) *Osculating circle.* The three constants $\alpha$, $\beta$, $R$ in the equation

$$(x-\alpha)^2 + (y-\beta)^2 = R^2$$

can in general be chosen to satisfy the three conditions for second-order contact with $y = f(x)$ at $x = x_0$, viz.

$$f(x_0) = y_0, \quad f'(x_0) = y_0', \quad f''(x_0) = y_0'',$$

where $y'$, $y''$ are given by

$$(x-\alpha) + (y-\beta)\,y' = 0,$$

$$1 + y'^2 + (y-\beta)\,y'' = 0.$$

Hence (dropping the suffix 0 in $x_0$ for convenience) we require

$$(x-\alpha)^2 + \{f(x)-\beta\}^2 = R^2,$$

$$(x-\alpha) + f'(x)\{f(x)-\beta\} = 0,$$

$$1 + \{f'(x)\}^2 + f''(x)\{f(x)-\beta\} = 0.$$

If $f''(x) \neq 0$ these give in turn, starting from the last:

$$f(x) - \beta = -\frac{1+f'^2}{f''},$$

$$x - \alpha = \frac{f'(1+f'^2)}{f''},$$

$$R^2 = \frac{(1+f'^2)^3}{f''^2},$$

which are sufficient to determine $\alpha$, $\beta$, $R^2$ uniquely.

We observe that $R$ is numerically equal to the radius of curvature $\rho$ at $P(x,y)$. The circle just determined is therefore called the *circle of curvature* at $P$, and its centre $C(\alpha, \beta)$ is the *centre of curvature* at $P$. Thus the centre of curvature at $P(x,y)$ is

$$\alpha = x - \frac{f'(1+f'^2)}{f''}, \quad \beta = f + \frac{1+f'^2}{f''}. \tag{i}$$

Since the circle touches the curve at $P$, therefore $C$ *lies on the normal at $P$*. (This is also easily verified from equations (i).) Intuitively we should expect the circle of closest contact to 'bulge' the same way as the curve does at $P$, as in fig. 109 rather than 110; i.e. that $C$ lies on the 'inward normal' at $P$. This is so since in determining the circle we chose $y'' = f''$ at $P$, so that the senses of the concavities at $P$ are the same (6.71 (2)).

As $\rho = (1 + f'^2)^{\frac{3}{2}}/f''$, hence by (i)

$$\alpha = x - \frac{\rho f'}{(1 + f'^2)^{\frac{1}{2}}}, \quad \beta = y + \frac{\rho}{(1 + f'^2)^{\frac{1}{2}}}.$$

Since $f' = \tan \psi$ and $\psi$ is acute (positive or negative), by 8.12, we have

$$\alpha = x - \rho \sin \psi, \quad \beta = y + \rho \cos \psi. \tag{ii}$$

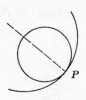 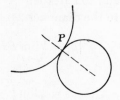

Fig. 109             Fig. 110

If $f''(x) = 0$ (which is so in particular at a point of inflexion), the equations for $\alpha$, $\beta$, $R$ cannot be satisfied: there is no osculating circle. However, the tangent then has at least second-order contact by (1), and we may conveniently say that 'the "circle" of curvature at $P$ is the tangent'. There is no centre of curvature.

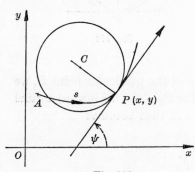 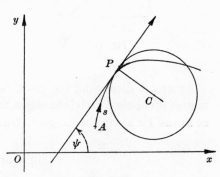

Fig. 111                 Fig. 112

In general *the circle of curvature will cross the curve at $P$*; for $m = 2$ in the theorem of 6.72. The relation of the curve to a circle of curvature is illustrated in figs. 111, 112.

Formulae (ii) are still valid in fig. 112, since $\rho$ is negative there. Fig. 111, the simplest case, can be taken as the standard, and used to write down the expressions for $\alpha$, $\beta$ quickly (see fig. 113).†

† An easier method is given in 8.53.

### 8.42 Newton's formula for ρ

(1) We should expect from intuitive considerations that the circle having closest contact with a given curve $y = f(x)$ at a point $P$ could be obtained by considering a circle touching the curve at $P$ and passing through a neighbouring point $Q$, and then taking the limit by letting $Q$ tend to $P$ along the curve. Assuming this for the moment, let $PR$ be the diameter through $P$ of the circle $PQ$, and draw $QN$ perpendicular to $PR$; then $PN \cdot NR = QN^2$. In the limit, the diameter $PR$ tends to the diameter $2|\rho|$ of the osculating circle at $P$; so

$$|\rho| = \tfrac{1}{2} \lim_{Q \to P} PR = \tfrac{1}{2} \lim (NR + PN)$$

$$= \tfrac{1}{2} \lim \left( \frac{QN^2}{PN} + PN \right) = \tfrac{1}{2} \lim \frac{QN^2}{PN}$$

since $PN \to 0$.

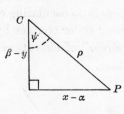

Fig. 113

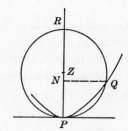

Fig. 114

As we are discussing the given curve at the particular point $P$, we may first choose $P$ to be the origin, taking $Py$ along the inward normal at $P$ and $Px$ along the tangent. Our result then becomes

$$|\rho| = \lim_{x \to 0} \frac{x^2}{2y}.$$

This is *Newton's formula*, and gives the value of $|\rho|$ at the origin.

Analytically, the result can be obtained from Maclaurin's theorem:

$$y = f(x) = f(0) + xf'(0) + \tfrac{1}{2}x^2 f''(0) + \tfrac{1}{6}x^3 f'''(\theta x).$$

Choosing axes as before, we have $f(0) = 0$, $f'(0) = 0$, and $\kappa = f''(0)$ (see 8.32, Remark $(\gamma)$). Hence

$$y = \tfrac{1}{2}\kappa x^2 + \tfrac{1}{6}x^3 f'''(\theta x),$$

and
$$\kappa = \lim_{x \to 0} \left\{ \frac{2y}{x^2} - \tfrac{1}{3}xf'''(\theta x) \right\}$$

$$= \lim_{x \to 0} \frac{2y}{x^2}$$

if $xf'''(\theta x) \to 0$ when $x \to 0$; this certainly happens if $f'''(x)$ is bounded for all $x$ sufficiently small.

## Examples

(i) *Find* $|\rho|$ *at the origin for the parabola* $x^2 = 4ay$.
We have
$$|\rho| = \lim_{x \to 0} \frac{x^2}{2y} = \lim_{y \to 0} \frac{4ay}{2y} = 2a.$$

(ii) *Find* $|\rho|$ *at the pole for* $r = a \sin 2\theta$.
Since $x = r \cos \theta = a \sin 2\theta \cos \theta$, therefore on the branches touching the initial line $Ox$ at $O$ (see Ex. 1 $(e)$, no. 13) we have $\theta \to 0$ when $x \to 0$. Hence

$$|\rho| = \lim_{x \to 0} \frac{x^2}{2y} = \lim_{\theta \to 0} \frac{r^2 \cos^2 \theta}{2r \sin \theta} = \lim_{\theta \to 0} \frac{a \sin 2\theta \cos^2 \theta}{2 \sin \theta}$$

$$= \lim_{\theta \to 0} (a \cos^3 \theta) = a.$$

The symmetry of the curve shows that $|\rho|$ has this value for the branches touching $Oy$ at $O$.

(iii) *Find* $|\rho|$ *at the point* $(a, 0)$ *of the ellipse*

$$\frac{x^2}{a^2} + \frac{y^2}{b^2} = 1.$$

Change the origin to $A(a, 0)$ by writing $x - a$ instead of $x$ (see 15.73 (2)). The equation becomes
$$\frac{(x-a)^2}{a^2} + \frac{y^2}{b^2} = 1$$

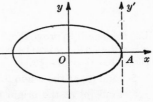

Fig. 115

referred to axes $Ax$, $Ay'$. The tangent at the new origin is not $Ax$ but $Ay'$; hence we must interchange $x$ and $y$ before applying Newton's formula. We obtain

$$|\rho| = \lim_{x \to 0} \frac{y^2}{2x} = \lim_{x \to 0} \frac{1}{2x} \left\{ b^2 - \frac{b^2}{a^2}(x-a)^2 \right\} = \lim_{x \to 0} \left( \frac{b^2}{a} - \frac{b^2 x}{2a^2} \right) = \frac{b^2}{a}.$$

(2) To justify the assumption made at the beginning of this section, let $Z$ be the centre of the circle $PQR$. Assuming the continuity and derivability of $f(x)$, the distance $r$ from $Z$ to points of the arc $PQ$ is also continuous and derivable, and is such that $ZP = ZQ$. Hence by Rolle's theorem there must be a point $P_1$ of the arc between $P$ and $Q$ such that $ZP_1$ is a stationary value of $r$.

Taking $Z$ as pole, we now show that *when the radius vector of a curve is stationary, it is normal to the curve.* For $dr/d\theta = 0$ gives $\cot \phi = 0$, i.e. $\phi = \tfrac{1}{2}\pi$. Hence $ZP_1$ is normal to the arc at $P_1$, and so $Z$ is the meet of normals at $P$ and $P_1$.

We show finally that *the intersection Z of normals at neighbouring points $P, P_1$ tends to the centre of curvature C at P when $P_1 \to P$ along the curve.* Choose axes as before; $C$ is then the point $(0, \rho)$, i.e. $(0, 1/f''(0))$. The equation of the normal at $P_1(\xi, f(\xi))$ is       $f'(\xi)\{y - f(\xi)\} + (x - \xi) = 0.$

This cuts $x = 0$, the normal at $P$, where $y = f(\xi) + \xi/f'(\xi)$, and this is therefore the ordinate of $Z$. When $P_1 \to P$, $\xi \to 0$ and $f(\xi) \to f(0) = 0$, while

$$\frac{\xi}{f'(\xi)} = \frac{\xi - 0}{f'(\xi) - f'(0)} \to \frac{1}{f''(0)},$$

the ordinate of $C$. Hence $Z \to C$.

Our original assumption is now justified since, when $Q \to P$ along the curve, the point $P_1$ (which lies between $Q$ and $P$) also tends to $P$.

## Exercise 8(e)

**1** Find the centre of curvature at $(3, 4)$ for the curve $xy = 12$.

**2** Find the centre of curvature at the point $(a \cos \phi, b \sin \phi)$ of the ellipse

$$\frac{x^2}{a^2} + \frac{y^2}{b^2} = 1.$$

**3** Prove that the centre of curvature $(\alpha, \beta)$ is given parametrically by

$$\alpha = x - \dot{y} \frac{\dot{x}^2 + \dot{y}^2}{\dot{x}\ddot{y} - \ddot{x}\dot{y}}, \quad \beta = y + \dot{x} \frac{\dot{x}^2 + \dot{y}^2}{\dot{x}\ddot{y} - \ddot{x}\dot{y}}.$$

**4** Show that the coordinates $(\alpha, \beta)$ of the centre of curvature at $(x, y)$ can be written

$$\alpha = x - \frac{dy}{d\psi}, \quad \beta = y + \frac{dx}{d\psi}.$$

Use this result to prove that the centre of curvature at the point $\theta$ of the cycloid $x = a(\theta - \sin \theta)$, $y = a(1 - \cos \theta)$ is $\alpha = a(\theta + \sin \theta)$, $\beta = -a(1 - \cos \theta)$. [First show that $\psi = \frac{1}{2}\pi - \frac{1}{2}\theta$.]

**5** Find $|\rho|$ at the origin for $x^2 + 3y^2 = 2y$.

**6** Find $|\rho|$ at the origin for $x^2/a^2 + (y - b)^2/b^2 = 1$. Interpret the result in terms of the ellipse $x^2/a^2 + y^2/b^2 = 1$.

**7** Find $|\rho|$ at the pole for $r = a \sin 3\theta$.

**\*8** Find $|\rho|$ at the origin for $y = 2x^2 - 3xy + 4y^2$.

## 8.5   Envelope of a family of curves

In this section the technique of partial derivation is required as far as 9.42.

## 8.51 Definition and determination of the envelope

(1) Consider the curve

$$f(x, y, \alpha) = 0 \qquad\qquad C_\alpha$$

depending on a parameter $\alpha$. As $\alpha$ varies, we obtain a *family of curves* (cf. 5.72).

If, for each $\alpha$, the curve $C_\alpha$ touches a definite curve $\mathscr{E}$, then $\mathscr{E}$ is called the *envelope* of the family.

Let $C_\alpha$ touch $\mathscr{E}$ at $P_\alpha$. The coordinates of $P_\alpha$ are certain functions (at present not determined) of $\alpha$, say

$$x = x(\alpha), \quad y = y(\alpha).$$

Then $\mathscr{E}$ has parametric equations

$$x = x(t), \quad y = y(t);$$

it touches $C_\alpha$ at the point where $t = \alpha$.

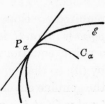

Fig. 116

Assuming that the family possesses an envelope, we now find equations to determine it.

(2) *The coordinates $(x, y)$ of any point on $\mathscr{E}$ satisfy*

$$f(x, y, \alpha) = 0 \quad and \quad \frac{\partial}{\partial \alpha} f(x, y, \alpha) = 0.$$

*Proof.* For each $\alpha$, the point $\big(x(\alpha), y(\alpha)\big)$ lies on $C_\alpha$; hence

$$f\big(x(\alpha), y(\alpha), \alpha\big) = 0 \tag{i}$$

for all $\alpha$. Deriving this wo $\alpha$, we have (by 9.41, equation (iv), extended to a function of three variables)

$$\frac{\partial f}{\partial x} x'(\alpha) + \frac{\partial f}{\partial y} y'(\alpha) + \frac{\partial f}{\partial \alpha} = 0. \tag{ii}$$

Since $\mathscr{E}$ *touches* $C_\alpha$ at $t = \alpha$ (and this implies that $C_\alpha$ possesses a definite tangent at $t = \alpha$, i.e. that not both of $\partial f/\partial x$, $\partial f/\partial y$ vanish there: cf. 9.42), therefore

$$\left( \frac{dy}{dx} \text{ on } \mathscr{E} \text{ at } t = \alpha \right) = \left( \frac{dy}{dx} \text{ on } C_\alpha \text{ at } \big(x(\alpha), y(\alpha)\big) \right),$$

i.e. by 9.42, $\qquad \dfrac{y'(\alpha)}{x'(\alpha)} = -\left[ \dfrac{\partial f}{\partial x} \Big/ \dfrac{\partial f}{\partial y} \right]_{\substack{x=x(\alpha) \\ y=y(\alpha)}}.$

Thus at $\big(x(\alpha), y(\alpha)\big)$ we have

$$\frac{\partial f}{\partial x} x'(\alpha) + \frac{\partial f}{\partial y} y'(\alpha) = 0. \tag{iii}$$

By subtracting (iii) from (ii), we see that at $\big(x(\alpha), y(\alpha)\big)$, $\partial f/\partial \alpha = 0$. Consequently $x(\alpha)$ and $y(\alpha)$ satisfy the equations

$$f(x, y, \alpha) = 0, \quad \frac{\partial}{\partial \alpha} f(x, y, \alpha) = 0. \tag{iv}$$

The cartesian equation of the envelope is found by eliminating $\alpha$ from equations (iv). Frequently, however, the parametric equations obtained by solving (iv) for $x$ and $y$ in terms of $\alpha$ are more convenient.

*Remark.* The solution of (iv) may include loci other than the required envelope, viz. *loci of singular points of the family*. A *singular point* $(x, y)$ on $C_\alpha$ is one at which both $\partial f/\partial x = 0$ and $\partial f/\partial y = 0$. At such a point, equations (i) and (iii) are satisfied, and hence also $\partial f/\partial \alpha = 0$ by (ii), so that equations (iv) still hold. To ensure that we have found the genuine envelope we must verify that our solution of (iv) actually *touches* $C_\alpha$ at $t = \alpha$.

## 8.52 Examples

(i) If the family of lines $x - \alpha y + \alpha^2 = 0$ has an envelope, it is given by

$$x - \alpha y + \alpha^2 = 0, \quad -y + 2\alpha = 0,$$

from which $y = 2\alpha$ and $x = \alpha^2$.

The parabola $x = \alpha^2$, $y = 2\alpha$ is actually the envelope because its gradient at the point $\alpha$ is

$$\frac{dy}{d\alpha} \bigg/ \frac{dx}{d\alpha} = \frac{1}{\alpha},$$

which is also the gradient of the line $x - \alpha y + \alpha^2 = 0$ (fig. 117). Also see ex. (v).

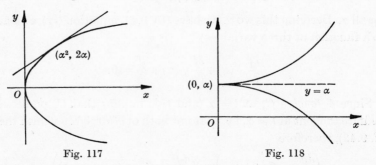

Fig. 117        Fig. 118

(ii) The family $(y - \alpha)^2 - x^3 = 0$ has no envelope. For the equations to determine it are

$$(y - \alpha)^2 = x^3, \quad -2(y - \alpha) = 0,$$

and the only points satisfying these lie on $x = 0$, i.e. the $y$-axis, which does not touch any member of the family (the reader should verify this by finding the gradient of the given curve at $(0, \alpha)$).

The $y$-axis is a locus of singular points of the family, for such points are given by

$$0 = \frac{\partial f}{\partial x} = -3x^2 \quad \text{and} \quad 0 = \frac{\partial f}{\partial y} = 2(y - \alpha),$$

and are thus the cusps $(0, \alpha)$. See fig. 118.

(iii) If the family $3(y - \alpha)^2 = 2(x - \alpha)^3$ has an envelope, it is given by

$$3(y - \alpha)^2 = 2(x - \alpha)^3, \quad 6(y - \alpha) = 6(x - \alpha)^2,$$

from which $\quad x = \alpha + \tfrac{2}{3}, \quad y = \alpha + \tfrac{4}{9}, \quad \text{or} \quad x = \alpha, \quad y = \alpha.$

The first solution corresponds to the line $x - y = \frac{2}{9}$, which touches the curve at $(\alpha + \frac{2}{3}, \alpha + \frac{4}{9})$; this should be verified by the reader.

The second solution corresponds to the line $y = x$; this does not touch the curve at $(\alpha, \alpha)$, and in fact is the locus of the singular points (cusps) of the family (fig. 119).

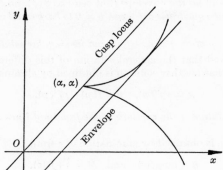

Fig. 119

(iv) *Find the envelope of a family of coaxal ellipses for which the sum of the semi-axes is constant.*

Choosing the common axes as $Ox$, $Oy$, the family is

$$\frac{x^2}{\alpha^2} + \frac{y^2}{\beta^2} = 1,$$

where
$$\alpha + \beta = c.$$

Regarding $\beta$ as a function of $\alpha$ given by $\alpha + \beta = c$, the envelope is determined by the above equations together with

$$-2\frac{x^2}{\alpha^3} - 2\frac{y^2}{\beta^3}\frac{d\beta}{d\alpha} = 0, \quad \frac{d\beta}{d\alpha} = -1.$$

From the last two, $x^2/\alpha^3 = y^2/\beta^3$. Hence by using properties of equal ratios, we have

$$\frac{\alpha}{x^{\frac{2}{3}}} = \frac{\beta}{y^{\frac{2}{3}}} = \frac{\alpha + \beta}{x^{\frac{2}{3}} + y^{\frac{2}{3}}} = \frac{c}{x^{\frac{2}{3}} + y^{\frac{2}{3}}},$$

from which we find $\alpha$, $\beta$ in terms of $x$ and $y$. Substituting into the equation of the ellipse,

$$\frac{x^2}{x^{\frac{4}{3}}c^2}(x^{\frac{2}{3}} + y^{\frac{2}{3}})^2 + \frac{y^2}{y^{\frac{4}{3}}c^2}(x^{\frac{2}{3}} + y^{\frac{2}{3}})^2 = 1,$$

from which
$$x^{\frac{2}{3}} + y^{\frac{2}{3}} = c^{\frac{2}{3}}.$$

Since an ellipse has no singular points, we may conclude that this locus is in fact the envelope of the family.

*Alternatively* we may begin by eliminating $\beta$, and then proceed in the usual way. The method given preserves the symmetry in $\alpha$, $\beta$.

(v) *Family which is quadratic in the parameter.*

If
$$\alpha^2 f(x, y) + \alpha g(x, y) + h(x, y) = 0$$

has an envelope, it is given by this equation and

$$2\alpha f(x, y) + g(x, y) = 0.$$

Eliminating $\alpha$,
$$\left(-\frac{g}{2f}\right)^2 f + \left(-\frac{g}{2f}\right)g + h = 0,$$

i.e.
$$g^2 = 4fh.$$

In general this will be the envelope (but see Ex. 8 $(f)$, no. 13). *The equation is the condition for the quadratic* $\alpha^2 f + \alpha g + h = 0$ *to have equal roots.*

*(vi) Clairaut's equation* $y = px + f(p)$ *and its singular solution.*

In 5.27 we showed that the general solution of this differential equation is $y = cx + f(c)$, and that another solution is obtained by eliminating $p$ from

$$x = -f'(p), \quad y = f(p) - pf'(p).$$

*This singular solution is the envelope of the family of lines represented by the general solution.*

For the envelope is obtained by elimination of $c$ from

$$y = cx + f(c) \quad \text{and} \quad 0 = x + f'(c),$$

i.e. from
$$x = -f'(c), \quad y = f(c) - cf'(c).$$

The curve thus determined is a genuine envelope since its gradient is†

$$\frac{dy}{dx} = \frac{dy}{dc}\bigg/\frac{dx}{dc} = \frac{f'(c) - f'(c) - cf''(c)}{-f''(c)} = c,$$

which is the gradient of the line $y = cx + f(c)$.

(vii) *Limiting intersections of neighbouring curves of a family.*

Consider the neighbouring curves $C_\alpha$, $C_{\alpha+\delta\alpha}$:

$$f(x, y, \alpha) = 0, \quad f(x, y, \alpha + \delta\alpha) = 0.$$

If these intersect, their common point(s) satisfy

$$f(x, y, \alpha) = 0 \quad \text{and} \quad f(x, y, \alpha + \delta\alpha) - f(x, y, \alpha) = 0,$$

i.e.
$$f(x, y, \alpha) = 0 \quad \text{and} \quad f_\alpha(x, y, \xi) = 0$$

for some number $\xi$ between $\alpha$ and $\alpha + \delta\alpha$.

When $\delta\alpha \to 0$, then by our general hypothesis of continuity in this chapter, the curve $C_{\alpha+\delta\alpha}$ approaches the curve $C_\alpha$, and the coordinates of the limit of their intersection(s) satisfy

$$f(x, y, \alpha) = 0 \quad \text{and} \quad \frac{\partial}{\partial\alpha} f(x, y, \alpha) = 0.$$

Hence *if neighbouring curves intersect, the limit of these intersections lies on the envelope or on a locus of singular points of the family.*

*Remark.* The envelope of a family used to be defined as the locus of the limiting intersections of neighbouring members. This definition is not consistent with the concept of the envelope as a curve which touches all members of the family, because an envelope (in our sense) may exist even when neighbouring curves do not intersect. For example, no two curves of the family $y = (x - \alpha)^3$

---

† We are assuming throughout (cf. 5.27) that $f(x)$ is not a linear function.

intersect, but $y = 0$ is the envelope (which also happens to be a locus of inflexions) as is easily verified. The old definition would not admit that a curve is the envelope of its own circles of curvature: see 8.54, Corollary 2.

## 8.53 The evolute of a curve

*Definition.* The locus of the centre of curvature $C$ as the corresponding point $P$ varies along the given curve is called the *evolute* of this curve.

If the curve is $y = f(x)$, then equations (i) of 8.41 give a parametric representation of the evolute, the parameter being $x$. The following theorem, which relates 'curvature' and 'envelopes', also provides a neat way of finding the coordinates of $C$ and the value of $|\rho|$, instead of using the formulae in 8.32 (3) and Ex. 8 (*e*), no. 3.

*The evolute of a curve is the envelope of the normals to the curve.*

*Proof.* If $P(x, y)$ is any point on the curve and $C(\alpha, \beta)$ the corresponding centre of curvature, then (8.41, equations (ii))

$$\alpha = x - \rho \sin \psi, \quad \beta = y + \rho \cos \psi.$$

$$\therefore \quad d\alpha = dx - \rho \cos \psi \, d\psi - \sin \psi \, d\rho$$

$$= dx - \frac{ds}{d\psi} \frac{dx}{ds} d\psi - \sin \psi \, d\rho$$

$$= -\sin \psi \, d\rho,$$

and similarly $\quad d\beta = \cos \psi \, d\rho.$

$$\therefore \quad \frac{d\beta}{d\alpha} = -\cot \psi = \text{gradient of the normal at } P \text{ to the curve.}$$

But $d\beta/d\alpha$ is the gradient of the evolute at $C$. Hence the normal at $P$ to the curve is parallel to the tangent at $C$ to the evolute. Since $C$ lies on the normal at $P$, *this normal touches the evolute at $C$, the corresponding centre of curvature.*

Since $CP^2 = \rho^2$, the 'distance formula' of coordinate geometry gives the value of $|\rho|$ when the coordinates of $C$ have been found.

## Examples

(i) *Find the centre of curvature at the point $t$ of the parabola $x = at^2$, $y = 2at$.*

The normal at $(at^2, 2at)$ is (see 16.31)

$$y + (x - 2a) t - at^3 = 0.$$

Its envelope is given by this equation and

$$(x - 2a) - 3at^2 = 0.$$

Hence $\quad x = 2a + 3at^2 \quad \text{and} \quad y = -2at^3.$

These expressions are the coordinates of the centre of curvature at $(at^2, 2at)$, and are also the parametric equations of the evolute. The cartesian equation, found by eliminating $t$, is $4(x - 2a)^3 = 27ay^2$.

Also,
$$CP^2 = (2a + 2at^2)^2 + (-2at^3 - 2at)^2$$
$$= 4a^2(1 + t^2)^3$$

on factorising, so that $|\rho| = 2a(1 + t^2)^{\frac{3}{2}}$.

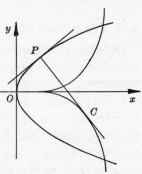

Fig. 120

(ii) *Find the centre of curvature at the point* $\phi$ *of the ellipse* $x = a\cos\phi$, $y = b\sin\phi$.

The normal at $(a\cos\phi, b\sin\phi)$ is (see 17.51)

$$y - b\sin\phi = \frac{a}{b}\tan\phi(x - a\cos\phi),$$

i.e.

$$y - \frac{a}{b}x\tan\phi = b\sin\phi - \frac{a^2}{b}\sin\phi = \frac{b^2 - a^2}{b}\sin\phi.$$

Its envelope is given by this and

$$-\frac{a}{b}x\sec^2\phi = \frac{b^2 - a^2}{b}\cos\phi,$$

from which

$$x = \frac{a^2 - b^2}{a}\cos^3\phi.$$

We find that

$$y = -\frac{a^2 - b^2}{b}\sin^3\phi$$

*either* by direct substitution in the equation of the normal, *or* by re-writing this equation with the $x$-term independent of $\phi$ as

$$y\cot\phi - \frac{a}{b}x = \frac{b^2 - a^2}{b}\cos\phi$$

and then deriving partially wo $\phi$ to get its envelope afresh.

The required centre of curvature is thus

$$\left(\frac{a^2 - b^2}{a}\cos^3\phi, \; -\frac{a^2 - b^2}{b}\sin^3\phi\right).$$

The evolute has cartesian equation $(ax)^{\frac{2}{3}} + (by)^{\frac{2}{3}} = (a^2 - b^2)^{\frac{2}{3}}$.

## 8.54 Arc of the evolute

Provided that $\rho$ steadily increases or steadily decreases along the given curve, *the arc-length of the evolute is equal to the difference between the radii of curvature corresponding to its extremities.*

*Proof.* Let $A$ be the fixed point from which $s$ is measured on the curve $AP$; let $C_0$, $C$ be the centres of curvature at $A$, $P$, and let $\sigma$ be the arc $C_0C$ of the evolute, measured from $C_0$ *in the sense for which $s$ increases.*

If $C$ is $(\alpha, \beta)$, then from the proof in 8.53,

$$d\alpha = -\sin\psi\,d\rho, \quad d\beta = \cos\psi\,d\rho.$$

Therefore by 8.12, equation (ii) applied to the evolute,

$$d\sigma^2 = d\alpha^2 + d\beta^2$$
$$= \sin^2 \psi \, d\rho^2 + \cos^2 \psi \, d\rho^2$$
$$= d\rho^2.$$

If $\rho$ steadily increases with $s$, i.e. with $\sigma$, then $d\sigma = +d\rho$ and $\sigma = \rho + c$. When $s = 0$, $\sigma = 0$ and $\rho = \rho_0$, so $c = -\rho_0$ and $\sigma = \rho - \rho_0$.

If $\rho$ steadily decreases as $s$ increases, then $d\sigma = -d\rho$ and $\sigma = \rho_0 - \rho$.

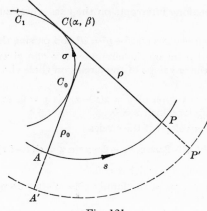

Fig. 121

**COROLLARY 1.** *The circle is the only plane curve with constant $\rho$.*

If $\rho = $ constant, then $d\sigma = 0$ and so $\sigma = $ constant. Hence the centre of curvature is the same for every point of the curve, and every such point lies at (constant) distance $\rho$ from this fixed point; i.e. the curve is a circle.

**COROLLARY 2.** *The circles of curvature at neighbouring points of a curve do not intersect.*

The difference between the radii of curvature at neighbouring points is equal to the arc of the evolute between the corresponding centres of curvature, and this arc is greater than the distance between these centres. Hence one circle completely encloses the other (illustrate this with a sketch), and so they do not intersect.

Yet these circles have an envelope, viz. the curve itself.

**COROLLARY 3.** *Involutes of a given curve.*

Given the evolute $C_0 C$, we can generate the original curve mechanically as follows. Let a string be wound along the evolute from a fixed point $C_1$ to $C_0$, and leave the evolute tangentially at $C_0$, continuing as far as $A$. Unwind the string, keeping it taut; then the end $A$ will trace the original curve.

For, when the end has any position $P$, the length of the unrolled part of the string is $PC = AC_0 + \text{arc } C_0 C = \rho_0 + \sigma = \rho$, so that $P$ lies on the original curve, which is called an *involute* of $C_0 C C_1$.

Given $C_0 C C_1$, there are infinitely many involutes: any point $A'$ on $C_0 A$ traces a curve $A'P'$ to which $P'C$ is normal. Since tangents at such corresponding points $P$, $P'$ are parallel, the involutes are all parallel curves.

## Exercise 8($f$)

*Find the envelope of the following families of curves (parameter $\alpha$).*

**1**   $y = \dfrac{\alpha}{x} + \alpha^2.$        **2**   $y = \alpha x + \tfrac{1}{3}\alpha^3.$

**3**   $y = x\tan\alpha - \dfrac{gx^2}{2v^2}\sec^2\alpha;$ interpret dynamically.

**4**   $x\sec\alpha + y\cosec\alpha = a.$        **5**   $\dfrac{x^2}{\alpha} + \dfrac{y^2}{\beta} = 1,$ where $\alpha + \beta = c.$

**6** Straight lines making intercepts on the axes whose sum is constant and equal to $a$.

**7** Circles with their centres on $x^2 + y^2 = a^2$ and passing through $(0, c)$.

\*8 Through a fixed point on the circumference of a given circle chords are drawn. Show that the envelope of the circles on these chords as diameters is a cardioid.

**9** Show that the envelope of $y = \alpha(x-\alpha)^2$ is $y = 0$, and that this is the member of the family for which $\alpha = 0$.

**10** Prove that the envelope of the circles

$$x^2 + y^2 - 2cx\alpha\cos\alpha - 2cy\alpha\sin\alpha = c^2(1 - 2\alpha)$$

is the circle of the family for which $\alpha = 0$.

**11** Verify that the curves of the family $y^2 = (x-\alpha)^3$ do not intersect. What is their envelope?

**12** Show that the curves $x^{\frac{2}{3}} + y^{\frac{2}{3}} = \alpha^{\frac{2}{3}}$ do not intersect. What is their envelope?

\*13 Verify that the process for finding envelopes applied to the family $\alpha^2 f + 2\alpha g + g = 0$, where $f$, $g$ are functions of $x$ and $y$, leads to $g(f-g) = 0$; and that $g = 0$, $f - g = 0$ are the members of the family corresponding to $\alpha = 0$, $\alpha = -1$ respectively. Assuming that $g = 0$, $f - g = 0$ do not touch, verify that no other members of the family touch either $g = 0$ or $f - g = 0$, so that the family has no envelope.

*Find the centre of curvature at the point $t$, and the evolute of*

**14**   $x = at^2,\ y = at^3.$        **15**   $x = ct,\ y = c/t.$

**16**   $x = a\operatorname{ch}t,\ y = b\operatorname{sh}t.$        **17**   $x = a(t - \operatorname{th}t),\ y = a\operatorname{sech}t.$

\*18   $x = a\cos^3 t,\ y = a\sin^3 t.$

## Miscellaneous Exercise 8($g$)

**1** Prove that $\cot\phi = (1/r)\,dr/d\theta$. If $P$ is any given point on the cardioid $r = a(1 - \cos\theta)$, find two other points $Q$, $R$ on the curve such that the tangents at $P$, $Q$, $R$ are all parallel, and show that the sum of the ordinates of $P$, $Q$, $R$ is zero.

**2** Find the values of $\theta$ at those points of the curve $r = a\sin 2\theta$ at which the tangent is parallel to the initial line. Sketch the curve.

**3** Prove that the curves $r^2\cos(2\theta - \alpha) = a^2\sin 2\alpha$, $r^2 = 2a^2\sin(2\theta + \alpha)$ cut orthogonally.

**4**  The tangent and normal at $P$ to the curve $r = f(\theta)$ meet the line through $O$ perpendicular to $OP$ at $A$, $B$. Prove $OB:OA = (dr/d\theta)^2 : r^2$.

A curve is such that (i) $OB/OA = \theta^2$; (ii) $r \to \infty$ when $\theta \to \infty$; and (iii) $r = a$ when $\theta = 0$. Find its equation, and show that triangle $APB$ has area

$$\tfrac{1}{2}a^2 e^{\theta^2}(\theta + \theta^{-1}).$$

**5**  If $\cot\phi$ varies as the ordinate of the point of contact of the tangent, prove that the curve concerned is a conic having a focus at the pole and eccentricity equal to the value of $\cot\phi$ when $\theta = \tfrac{1}{2}\pi$.

**6**  If $s = f(\psi)$, show how to find $x$, $y$ in terms of $\psi$. A curve is given by

$$5s = 4a(5 + \tan^2 \tfrac{1}{2}\psi)\,\sqrt{(\tan \tfrac{1}{2}\psi)};$$

if axes are chosen so that $x$, $y$, $s$, $\psi$ vanish simultaneously, find $x$ and $y$ in terms of $\psi$, and verify that $5x^2 + 9y^2 = 5s^2$.

**7**  Show that the tangent at the point $\theta$ of the curve

$$x = a(4\cos\theta - \cos 4\theta), \quad y = a(4\sin\theta - \sin 4\theta)$$

is

$$x\sin\tfrac{5}{2}\theta - y\cos\tfrac{5}{2}\theta = 5a\sin\tfrac{3}{2}\theta.$$

With $O$ as pole, prove that the $(p, r)$ equation is $r^2 = 9a^2 + \tfrac{16}{25}p^2$, and deduce the radius of curvature at the point $\theta$. Find the arc-length from $\theta = 0$ to $2\pi$.

**8**  A circle of radius $a$ rolls without slipping on the outside of a fixed circle of radius $2a$ and centre $O$; $P$ is a point fixed on the circumference of the rolling circle, and $A$ is the point of the fixed circle with which $P$ initially coincided. With $O$ for origin and $OA$ for $x$-axis, show that $P$ has coordinates

$$x = 3a\cos\theta - a\cos 3\theta, \quad y = 3a\sin\theta - a\sin 3\theta,$$

where $\theta$ is the angle turned through by the line of centres. Find the $(p, r)$ equation, and prove $\rho = 3a\sin\theta$. Sketch the curve and find its area.

**9**  If $r = a(1 - \cos\theta)$, prove $p = 2a\sin^3\tfrac{1}{3}\psi$.

**10**  If $p = a\cos 3\psi$, obtain $x$ and $y$ in terms of $\psi$, and prove that the $(p, r)$ equation is $r^2 + 8p^2 = 9a^2$.

**11**  At the point $P(ct, c/t)$ of $xy = c^2$, prove $\kappa = 2c^2/r^3$, where $r = OP$.

**12**  Find the radius of curvature of $r = a\theta\sec\tfrac{1}{2}\theta$ at the pole by Newton's formula.

**13**  If $x = c\log(\sec\theta + \tan\theta)$ and $y = c\sec\theta$, prove $\rho = c\sec^2\theta$.

**14**  If $r = a\cos n\theta$, find $\rho$ in terms of $r$. Show that $\rho = a/(1 + n^2)$ when $r = a$.

**15**  Prove that $\rho = \operatorname{cosec}\psi\,dy/d\psi$. For the curve $y = a\sec^n\psi$, prove $\rho$ is $n$ times the length of the normal.

**16**  Find the $(p, r)$ equation of the curve $r^n = a^n\sin n\theta$, and prove that $\rho r^{n-1} = a^n/(n+1)$. Prove also that the length intercepted on the radius vector by the circle of curvature is $2r/(n+1)$.

**17**  Find the equation of the circle of curvature of $axy = x^3 - 2a^3$ at $(-a, 3a)$.

**18**  Prove that the centre of curvature $C(\alpha, \beta)$ at $P(x, y)$ is given by

$$\alpha = x - \frac{dy}{d\psi}, \quad \beta = y + \frac{dx}{d\psi}.$$

If $P$ is the point $(x, y)$ on the curve $x = a\cos^3(\tfrac{1}{3}t)$, $y = a\sin^3(\tfrac{1}{3}t)$, and $Q$ divides $CP$ internally in the ratio $2 : 1$, find the locus of $Q$.

**19** Find the equation of the normal at the point $\theta$ of the cycloid $x = a(\theta + \sin\theta)$, $y = a(1 + \cos\theta)$, and hence the centre of curvature. Show that its locus is another cycloid having the same dimensions as the given one.

**20** Prove that $1/p^2 = u^2 + u'^2$, where $u = 1/r$ and $u' = du/d\theta$. Hence prove

$$\rho = \frac{(u^2 + u'^2)^{\frac{3}{2}}}{u^3(u + u'')}.$$

**21** Show that a curve whose radius of curvature is equal to the length of the normal is a circle or a catenary. $[\rho = \pm PG$ with notation of 5.71.]

**22** Find the two systems of curves for which the radius of curvature is half the normal.

**23** What are the curves for which $\rho = TG$?

**24** Sketch the curve $(x-y)(2x+y) = x^2(x^2+y)$ near the origin, and find the radii of curvature of the branches at $O$.

**25** Find the envelope of circles whose centres lie on $xy = c^2$ and which pass through $O$.

# 9

## FUNCTIONS OF SEVERAL VARIABLES

### 9.1 Introduction

### 9.11 Functions, limits, continuity

In 1.13 we discussed the concept of 'function' for a single independent variable, and mentioned that it extends to more than one. We now begin the differential calculus of functions of several variables.

Suppose $x, y$ are two variables whose values (perhaps within certain ranges) can be assigned arbitrarily, so that $x, y$ are not related in any way. If one or more values of $u$ are determined when values are given to $x$ and $y$, we say that $u$ is a *function of the pair of variables* $(x, y)$, and write $u = f(x, y)$. We call $x, y$ the *independent variables* and $u$ the *dependent variable*.

The notation $u = f(x, y)$ is still used even when $x$ and $y$ are related, although we are not then dealing with a genuine function of *two* independent variables.

In this chapter we give our attention to functions of two variables. This case is typical, and the results can be extended at once to functions of three or more.

As in Ch. 2, we can define the *limit* of a function of $(x, y)$ when these variables tend to given values, and the meaning of *continuity in* $(x, y)$ at a given point. Thus, if a positive number $\epsilon$ (however small) is given, and a positive number $\eta$ can be found such that, whenever

$$|x-a| < \eta \quad \text{and} \quad |y-b| < \eta, \quad \text{but } not\ both \text{ are zero,}$$

we have
$$|f(x, y) - l| < \epsilon,$$

then we say that $f(x, y)$ *tends to the limit* $l$ when $x$ tends to $a$ and $y$ tends to $b$, and we write
$$\lim_{(x, y) \to (a, b)} f(x, y) = l.$$

If also $f(a, b)$ exists and is equal to $l$, then $f(x, y)$ is *continuous in* $(x, y)$ at 'the point' $(a, b)$.

We do not develop these matters here because we shall be concerned only with certain special limits (of incrementary ratios) associated with the process of derivation. However, the reader is warned that the situation is more complicated than would be expected at first sight: continuity in the *pair* of variables $(x, y)$ requires more of a function than its continuity (as defined in 2.61) in $x$ alone and in $y$ alone.

## 9.12 Economy in functional notation

It will be convenient to economise in symbols. Instead of writing $y = f(x)$, $u = \phi(x, y)$, ..., we now write $y = y(x)$, $u = u(x, y)$, .... That is, *we use the same letter for the dependent variable and the functional symbol*. No confusion can arise unless we require to make a substitution or change of variables, and then we can revert to the former functional notation whenever necessary.

Thus, if $u = u(x, y)$ and $x = x(\xi, \eta)$, $y = y(\xi, \eta)$, then by substitution for $x$, $y$ we can construct from these three functions a *new* function $u = f(\xi, \eta)$, where $f(\xi, \eta) = u(x(\xi, \eta), y(\xi, \eta))$. It would be wrong to call this new function $u(\xi, \eta)$ because this symbol already has a meaning, viz. the value of $u(x, y)$ when we replace $x$ by $\xi$ and $y$ by $\eta$. For example, if

$$u = x^2 + y^2, \quad x = \xi^2 - \eta^2 \quad \text{and} \quad y = 2\xi\eta,$$

then we find $u = (\xi^2 + \eta^2)^2$.

## 9.2 Partial derivatives

### 9.21 Definitions

We cannot usefully speak of *the* derivative of a function $u(x, y)$ because $x$ and $y$ may vary separately or together in any way. However, we can define the derivative of $u$ wo $x$, or of $u$ wo $y$, as follows.

Suppose $y$ is kept constant and $x$ is allowed to vary alone. Then a change $\delta x$ in $x$ causes a change $\delta u$ in $u$ given by

$$\delta u = u(x + \delta x, y) - u(x, y),$$

so that
$$\frac{\delta u}{\delta x} = \frac{u(x + \delta x, y) - u(x, y)}{\delta x}.$$

If $\delta u / \delta x$ tends to a limit when $\delta x \to 0$, this limit is called *the partial derivative of $u$ wo $x$*, and is written $\partial u / \partial x$.

Similarly, if $x$ is kept constant and $y$ varies alone, and if

$$\lim_{\delta y \to 0} \frac{u(x, y + \delta y) - u(x, y)}{\delta y}$$

exists, then this limit is the partial derivative of $u$ wo $y$, written $\partial u / \partial y$.

The reader should notice that, according to these definitions, *no new process is involved*: partial derivatives are calculated in the same way as 'ordinary' derivatives, by first treating all but one of the independent variables as if they were constants and then finding the derivative of the function with respect to the remaining independent variable. In particular, all the rules of derivation (product, quotient, function of a function) in Ch. 3 continue to apply.

## Examples

(i) If
$$u = x^3 + 3x^2y^2 + 5xy^4 - y^3,$$

then (by treating $y$ as if it were constant)

$$\frac{\partial u}{\partial x} = 3x^2 + 6xy^2 + 5y^4,$$

and (treating $x$ as if it were constant)

$$\frac{\partial u}{\partial y} = 6x^2y + 20xy^3 - 3y^2.$$

(ii) If $u = \tan^{-1}(y/x)$, put $z = y/x$ so that $u = \tan^{-1}z$. The rule for 'function of a function' is

$$\frac{\partial u}{\partial x} = \frac{du}{dz}\frac{\partial z}{\partial x},$$

so that
$$\frac{\partial u}{\partial x} = \frac{1}{1+z^2}\left(-\frac{y}{x^2}\right) = -\frac{y}{x^2+y^2}.$$

Similarly,
$$\frac{\partial u}{\partial y} = \frac{1}{1+z^2}\left(\frac{1}{x}\right) = \frac{x}{x^2+y^2}.$$

In the statement of the rule for 'function of a function', we do not write $\partial u/\partial z$ because $u$ is a function of the *single* variable $z$; but we write $\partial z/\partial x$ because $z$ is a function of both $x$ and $y$.

## 9.22 Other notations for the partial derivatives

Instead of $\partial u/\partial x$, the functional notations $u_x(x,y)$, $u_x'(x,y)$, or just $u_x$ are often used. The former symbols are appropriate for indicating particular values of the partial derivative; thus $u_x(a,b)$ denotes the value of $u_x(x,y)$ when $x = a$ and $y = b$.

Similarly, $u_y(x,y)$, $u_y'(x,y)$ and $u_y$ mean the same as $\partial u/\partial y$.

## 9.23 Geometrical meaning of $\partial u/\partial x$, $\partial u/\partial y$

Just as a relation $y = f(x)$ between two variables $x$, $y$ is represented by a plane curve (1.6), so a relation $z = u(x,y)$ between three variables $x$, $y$, $z$ is represented geometrically by a locus in space called a *surface* (fig. 122).

Let $P(a,b)$ be a point on the surface $z = u(x,y)$. The plane through $P$ and parallel to the coordinate plane $xOz$ cuts the surface in a curve whose gradient $\tan\psi_x$ at $P$ is the value of $\partial u/\partial x$ when $x = a$ and $y = b$, viz. $u_x(a,b)$. Similarly $\partial u/\partial y$ at $P$ measures the gradient of the curve of section of the surface by the plane $x = a$.

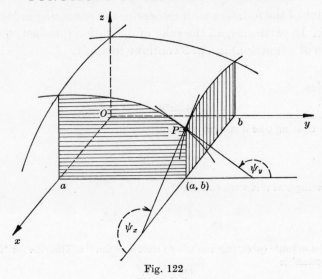

Fig. 122

## 9.24 Partial derivatives of second and higher orders

**(1)** The functions $\partial u/\partial x$, $\partial u/\partial y$ of $(x, y)$ may themselves possess partial derivatives wo $x$ and wo $y$, viz.

$$\frac{\partial}{\partial x}\left(\frac{\partial u}{\partial x}\right), \quad \frac{\partial}{\partial y}\left(\frac{\partial u}{\partial x}\right), \quad \frac{\partial}{\partial x}\left(\frac{\partial u}{\partial y}\right), \quad \frac{\partial}{\partial y}\left(\frac{\partial u}{\partial y}\right),$$

written
$$\frac{\partial^2 u}{\partial x^2}, \quad \frac{\partial^2 u}{\partial y \, \partial x}, \quad \frac{\partial^2 u}{\partial x \, \partial y}, \quad \frac{\partial^2 u}{\partial y^2},$$

or
$$u_{xx}, \quad u_{yx}, \quad u_{xy}, \quad u_{yy}.$$

Similarly the partial derivatives of these give eight third-order partial derivatives of $u$, and so on.

### Examples

(i) In 9.21, ex. (i) we have

$$\frac{\partial^2 u}{\partial x^2} = 6x + 6y^2, \quad \frac{\partial^2 u}{\partial y \, \partial x} = 12xy + 20y^3,$$

$$\frac{\partial^2 u}{\partial x \, \partial y} = 12xy + 20y^3, \quad \frac{\partial^2 u}{\partial y^2} = 6x^2 + 60xy^2 - 6y.$$

*Remark.* We notice that the expressions for the mixed derivatives $\partial^2 u/\partial y \, \partial x$, $\partial^2 u/\partial x \, \partial y$ are the same in this example. It is true that, for most functions which we meet, the mixed derivatives are equal; for some exceptions see Ex. 9 (*a*), nos. 29, 30. The question is considered in (2) below. For all 'well-behaved' functions there are only *three* distinct second derivatives; and similarly only *four* distinct third derivatives, and so on.

(ii) *If* $u = x^n f(y/x)$, *prove that*

$$(a) \quad x\frac{\partial u}{\partial x} + y\frac{\partial u}{\partial y} = nu, \quad (b) \quad x^2\frac{\partial^2 u}{\partial x^2} + 2xy\frac{\partial^2 u}{\partial x\,\partial y} + y^2\frac{\partial^2 u}{\partial y^2} = n(n-1)\,u,$$

*assuming the mixed derivatives are equal.*

(*a*) By the product rule,

$$\frac{\partial u}{\partial x} = nx^{n-1}f\left(\frac{y}{x}\right) + x^n\frac{\partial}{\partial x}f\left(\frac{y}{x}\right);$$

and on putting $z = y/x$ and using 'function of a function',

$$\frac{\partial}{\partial x}f\left(\frac{y}{x}\right) = \frac{df}{dz}\frac{\partial z}{\partial x} = f'(z)\left(-\frac{y}{x^2}\right) = -\frac{y}{x^2}f'\left(\frac{y}{x}\right).$$

Hence

$$\frac{\partial u}{\partial x} = nx^{n-1}f\left(\frac{y}{x}\right) - x^{n-2}yf'\left(\frac{y}{x}\right).$$

Similarly

$$\frac{\partial u}{\partial y} = x^n f'\left(\frac{y}{x}\right)\frac{1}{x} = x^{n-1}f'\left(\frac{y}{x}\right).$$

$$\therefore \quad x\frac{\partial u}{\partial x} + y\frac{\partial u}{\partial y} = nx^n f\left(\frac{y}{x}\right) - x^{n-1}yf'\left(\frac{y}{x}\right) + x^{n-1}yf'\left(\frac{y}{x}\right)$$

$$= nu.$$

*Alternatively*, to calculate $\partial u/\partial x$ we may first take logarithms:

$$\log u = n\log x + \log f\left(\frac{y}{x}\right);$$

and then derive wo $x$:

$$\frac{1}{u}\frac{\partial u}{\partial x} = \frac{n}{x} + \frac{1}{f}\frac{\partial f}{\partial x}.$$

We find $\partial f/\partial x$ as before.

(*b*) Instead of finding the second derivatives and verifying result (*b*), it is easier to proceed as follows. Derive result (*a*) wo $x$, and also wo $y$:

$$x\frac{\partial^2 u}{\partial x^2} + \frac{\partial u}{\partial x} + y\frac{\partial^2 u}{\partial x\,\partial y} = n\frac{\partial u}{\partial x},$$

$$x\frac{\partial^2 u}{\partial y\,\partial x} + y\frac{\partial^2 u}{\partial y^2} + \frac{\partial u}{\partial y} = n\frac{\partial u}{\partial y}.$$

Hence $\quad x\dfrac{\partial^2 u}{\partial x^2} + y\dfrac{\partial^2 u}{\partial x\,\partial y} = (n-1)\dfrac{\partial u}{\partial x}, \quad x\dfrac{\partial^2 u}{\partial x\,\partial y} + y\dfrac{\partial^2 u}{\partial y^2} = (n-1)\dfrac{\partial u}{\partial y}.$

Multiply the first of these equations by $x$, the second by $y$, and add:

$$x^2\frac{\partial^2 u}{\partial x^2} + 2xy\frac{\partial^2 u}{\partial x\,\partial y} + y^2\frac{\partial^2 u}{\partial y^2} = (n-1)\left(x\frac{\partial u}{\partial x} + y\frac{\partial u}{\partial y}\right)$$

$$= (n-1)\,nu$$

by result (*a*).

(2) *Equality of the mixed derivatives.* Consider the second derivatives of $u(x, y)$ at $(a, b)$. By definition,

$$u_y(x, b) = \lim_{k \to 0} \frac{u(x, b+k) - u(x, b)}{k},$$

and

$$u_{xy}(a, b) = \lim_{h \to 0} \frac{u_y(a+h, b) - u_y(a, b)}{h}$$

$$= \lim_{h \to 0} \frac{1}{h} \left\{ \lim_{k \to 0} \frac{u(a+h, b+k) - u(a+h, b)}{k} - \lim_{k \to 0} \frac{u(a, b+k) - u(a, b)}{k} \right\}$$

$$= \lim_{h \to 0} \lim_{k \to 0} \frac{1}{hk} \{ u(a+h, b+k) - u(a+h, b) - u(a, b+k) + u(a, b) \}$$

$$= \lim_{h \to 0} \lim_{k \to 0} \phi(h, k),$$

where

$$\phi(h, k) = \frac{u(a+h, b+k) - u(a+h, b) - u(a, b+k) + u(a, b)}{hk}.$$

Similarly

$$u_{yx}(a, b) = \lim_{k \to 0} \lim_{h \to 0} \phi(h, k).$$

Thus, in calculating the two mixed derivatives, we are considering the limit of $\phi(h, k)$ in two ways: (i) when $k \to 0$, and then $h \to 0$ in the result; (ii) when $h \to 0$ first, then $k \to 0$. It is easy to show that these two limits may be different; for example, if $\phi(h, k)$ turned out to be $(h+k)/(h-k)$, then

$$\lim_{h \to 0} \lim_{k \to 0} \frac{h+k}{h-k} = \lim_{h \to 0} \frac{h}{h} = 1, \quad \lim_{k \to 0} \lim_{h \to 0} \frac{h+k}{h-k} = \lim_{k \to 0} \frac{k}{-k} = -1.$$

In general, the result of a double limiting process depends on the order in which the two stages of the process are carried out. For the case of the mixed derivatives, we can show that *the results will certainly be the same whenever $u_{xy}$ and $u_{yx}$ are continuous at $(a, b)$.* This condition is satisfied by all 'ordinary' functions.

Write

$$f(x) = \frac{u(x, b+k) - u(x, b)}{hk}.$$

Then

$$\phi(h, k) = f(a+h) - f(a)$$

$$= hf'(a+\theta h) \quad (0 < \theta < 1),$$

by the mean value theorem (6.33 (1)) applied to $f(x)$ in $a \leqslant x \leqslant a+h$; here $\theta$ depends on $b$, $k$ as well as $a$, $h$, since the former appear as parameters in $f(x)$.

Since

$$f'(x) = \frac{u_x(x, b+k) - u_x(x, b)}{hk},$$

we have

$$\phi(h, k) = \frac{u_x(a+\theta h, b+k) - u_x(a+\theta h, b)}{k}$$

$$= u_{yx}(a+\theta h, b+\theta' k) \quad (0 < \theta' < 1),$$

by the mean value theorem applied to the function $u_x(a+\theta h, y)$ of $y$ in $b \leqslant y \leqslant b+k$.

Similarly, writing

$$g(y) = \frac{u(a+h, y) - u(a, y)}{hk},$$

we have $\quad \phi(h,k) = g(b+k) - g(b) = k g'(b+\theta_1 k) \quad (0 < \theta_1 < 1)$

$$= \frac{u_y(a+h, b+\theta_1 k) - u_y(a, b+\theta_1 k)}{h}$$

$$= u_{xy}(a + \theta_1' h, b + \theta_1 k) \quad (0 < \theta_1' < 1).$$

Hence $\quad u_{yx}(a+\theta h, b + \theta' k) = u_{xy}(a + \theta_1' h, b + \theta_1 k).$

If $u_{yx}$, $u_{xy}$ are both *continuous* at $(a, b)$, these expressions tend to $u_{yx}(a, b)$, $u_{xy}(a, b)$ respectively when $h$ and $k$ tend to zero; therefore $u_{yx}(a, b) = u_{xy}(a, b)$.

## 9.25 Partial differential equations

(1) *Their construction.* A relation between partial derivatives of a function, perhaps also including the independent variables and the function itself, is called a *partial differential equation* (cf. 5.12). The results of 9.24, ex. (ii) are examples; the first, which does not involve the function $f$ explicitly, can be regarded as the result of *eliminating the arbitrary†* *function f* from the equation $u = x^n f(y/x)$. Nos. 10, 13, 14, 17 in Ex. 9 (a) can be interpreted similarly.

Partial differential equations also arise from elimination of parameters from a given function (cf. 5.11).

### Examples

(i) *Eliminate a and p from* $y = a e^{pt} \sin px$.

Partial derivatives are appropriate here because $y$ is a function of *two* variables $t$, $x$. We have

$$\frac{\partial y}{\partial t} = ap\, e^{pt} \sin px, \quad \frac{\partial^2 y}{\partial t^2} = ap^2\, e^{pt} \sin px,$$

$$\frac{\partial y}{\partial x} = ap\, e^{pt} \cos px, \quad \frac{\partial^2 y}{\partial x^2} = -ap^2\, e^{pt} \sin px.$$

Hence $$\frac{\partial^2 y}{\partial t^2} + \frac{\partial^2 y}{\partial x^2} = 0.$$

(ii) *Eliminate a, b, c from* $u = a(x+y) + b(x-y) + abz + c$.

$$\frac{\partial u}{\partial x} = a+b, \quad \frac{\partial u}{\partial y} = a-b, \quad \frac{\partial u}{\partial z} = ab.$$

Since $$(a+b)^2 - (a-b)^2 = 4ab,$$

$$\therefore \quad \left(\frac{\partial u}{\partial x}\right)^2 - \left(\frac{\partial u}{\partial y}\right)^2 = 4\frac{\partial u}{\partial z}.$$

† The function $f(z)$ is supposed to possess a derivative.

**(2)** *Solutions having a given form.*

## Examples

(iii) *If*
$$\frac{\partial^2 u}{\partial r^2} + \frac{1}{r}\frac{\partial u}{\partial r} + \frac{1}{r^2}\frac{\partial^2 u}{\partial \theta^2} = 0,$$

*find the most general solution $u$ if* (a) *$u$ is a function of $r$ only;* (b) *$u = r^n f(\theta)$ where $n$ is constant.*

(a) If $u = u(r)$, the equation becomes
$$\frac{d^2 u}{dr^2} + \frac{1}{r}\frac{du}{dr} = 0.$$

Writing $v = du/dr$, we find $(1/v)\,dv/dr = -1/r$, $\log v = -\log r + C$, and $v = B/r$. From $du/dr = B/r$, $u = A + B\log r$, where $A$, $B$ are arbitrary constants.

(b) If $u = r^n f(\theta)$, the equation can be written
$$\left\{ n(n-1)\,r^{n-2} + \frac{1}{r}\,nr^{n-1} \right\} f(\theta) + \frac{1}{r^2}\{r^n f''(\theta)\} = 0,$$

i.e.
$$n^2 r^{n-2} f(\theta) + r^{n-2} f''(\theta) = 0.$$

Excluding the trivial case $r = 0$ (which corresponds to the solution $u = 0$),
$$f''(\theta) + n^2 f(\theta) = 0,$$

so
$$f(\theta) = A\cos n\theta + B\sin n\theta$$

and
$$u = r^n(A\cos n\theta + B\sin n\theta).$$

*\*(iv) Find a general solution of the form $u = R\Theta$ for the differential equation in ex. (iii), where $R$ is a function of $r$ only and $\Theta$ is a function of $\theta$ only.*†

When $u = R\Theta$, the equation becomes
$$\left( \frac{d^2 R}{dr^2} + \frac{1}{r}\frac{dR}{dr} \right)\Theta + \frac{R}{r^2}\frac{d^2\Theta}{d\theta^2} = 0,$$

i.e.
$$\frac{r^2}{R}\left( \frac{d^2 R}{dr^2} + \frac{1}{r}\frac{dR}{dr} \right) = -\frac{1}{\Theta}\frac{d^2\Theta}{d\theta^2}.$$

The left-hand side is independent of $\theta$; the same must therefore be true of the right-hand side. Since the latter is also independent of $r$, it must be constant. Hence each side is constant, say equal to $n^2$.

From the right,
$$\frac{d^2\Theta}{d\theta^2} + n^2\Theta = 0$$

and (if $n \neq 0$)
$$\Theta = A\cos n\theta + B\sin n\theta.$$

From the left,
$$r^2\frac{d^2 R}{dr^2} + r\frac{dR}{dr} - n^2 R = 0,$$

an equation of Euler's type (5.61). The substitution $r = e^t$ reduces it to
$$\frac{d^2 R}{dt^2} - n^2 R = 0,$$

so (if $n \neq 0$)
$$R = C e^{nt} + D e^{-nt} = Cr^n + Dr^{-n}.$$

---

† A solution of this form is said to be *separable*.

A general solution having the form specified is therefore

$$u = (A\cos n\theta + B\sin n\theta)(Cr^n + Dr^{-n}) \quad (n \neq 0).$$

When $n = 0$ the corresponding solution is easily found to be

$$u = (A\theta + B)(C\log r + D).$$

## Exercise 9(a)

*Calculate $\partial u/\partial x$, $\partial u/\partial y$ for the following functions $u(x, y)$.*

**1** $3x^2 - 2xy + 5y^2$.  **2** $x/y$.  **3** $x^2 y^3$.  **4** $\sin^{-1}(y/x)$.  **5** $\log(x^2 + y^2)$.

*Calculate $u_{xx}$, $u_{yx}$, $u_{xy}$, $u_{yy}$ and verify that $u_{yx} = u_{xy}$ for the following.*

**6** $x^3 + 3x^2y + y^3$.  **7** $x\sin y + y\sin x$.  **8** $e^{xy}$.  **9** $\operatorname{ch} x \operatorname{ch} y$.

*In the following, use the rule for 'function of a function'; when necessary, assume that the mixed derivatives are equal.*

**10** If $u = f(y/x)$, prove $xu_x + yu_y = 0$.

**11** If $u = \log(x^2 + y^2)$, prove $u_{xx} + u_{yy} = 0$.

**12** If $u = \tan^{-1}\{(x+y)/z\}$, prove $xu_x + yu_y + zu_z = 0$.

**13** If $u = f(x+ct) + g(x-ct)$, prove $\partial^2 u/\partial t^2 = c^2 \partial^2 u/\partial x^2$.

**14** If $u = (1/r)\{f(ct+r) + g(ct-r)\}$, prove

$$\frac{\partial^2 u}{\partial t^2} = \frac{c^2}{r^2}\frac{\partial}{\partial r}\left(r^2\frac{\partial u}{\partial r}\right). \quad [\text{Put } v = ru.]$$

**15** If $u = f(z)$, where $z$ is a function of $x$ and $y$, prove

$$\frac{\partial^2 u}{\partial x^2} = f''(z)\left(\frac{\partial z}{\partial x}\right)^2 + f'(z)\frac{\partial^2 z}{\partial x^2},$$

and find $\partial^2 u/\partial x\,\partial y$ similarly.

**16** If $u = x^2\tan^{-1}(y/x) - y^2\tan^{-1}(x/y)$, calculate $x\,\partial u/\partial x + y\,\partial u/\partial y$. [Use 9.24, ex. (ii)(a).]

**17** If $u = f(x^2 + y^2)$, prove

$$\text{(i)} \quad y\frac{\partial u}{\partial x} - x\frac{\partial u}{\partial y} = 0;$$

$$\text{(ii)} \quad y^2\frac{\partial^2 u}{\partial x^2} - 2xy\frac{\partial^2 u}{\partial x\,\partial y} + x^2\frac{\partial^2 u}{\partial y^2} = x\frac{\partial u}{\partial x} + y\frac{\partial u}{\partial y}.$$

[Method of 9.24, ex. (ii)(b).]

**18** If $u = (y/x)f(x+y)$ and $f'(x+y)$ denotes $f'(t)$ when $t = x+y$, prove that

$$xu_x + yu_y = \frac{y}{x}(x+y)f'(x+y), \quad x^2u_{xx} + 2xyu_{xy} + y^2u_{yy} = \frac{y}{x}(x+y)^2f''(x+y),$$

and calculate $\qquad x^3u_{xxx} + 3x^2yu_{xxy} + 3xy^2u_{xyy} + y^3u_{yyy}$.

**19** If $u = (x^2 - y^2)f(t)$ where $t = xy$, prove

(i) $\dfrac{\partial^2 u}{\partial x^2} + \dfrac{\partial^2 u}{\partial y^2} = (x^4 - y^4)f''(xy);$  (ii) $\dfrac{\partial^2 u}{\partial x\,\partial y} = (x^2 - y^2)\{3f'(xy) + xyf''(xy)\}.$

**20** Eliminate $a$ and $b$ from $z = ax + by + ab$.

**21** Eliminate $a$ and $p$ from $y = ae^{-p^2t}\cos px$.

**22** Eliminate $a$ and $p$ from $u = a\,e^{-px}\sin(2p^2ct - px)$.

**23** If $z^2 = \dfrac{(x+b)^2}{a^2-y^2}$, prove $yz\left(\dfrac{\partial z}{\partial x}\right)^2 = \dfrac{\partial z}{\partial y}$.

**24** If $u$, $v$ are functions of $x$ and $y$, eliminate the function $f$ from $u = f(v)$.

**25** Find $Y$, a function of $y$ only, if $Y\cos px$ satisfies $\partial u/\partial y = c^2\,\partial^2 u/\partial x^2$.

**26** Find $Y$, a function of $y$ only, if $Y\sin px$ satisfies $\partial^2 u/\partial x^2 + \partial^2 u/\partial y^2 = 0$. Find a solution which vanishes when $y = -1$, and equals $\sin x$ when $y = +1$.

**27** If $\dfrac{\cos r}{r}\phi(t)$ satisfies $\dfrac{\partial^2 u}{\partial t^2} = c^2\left(\dfrac{\partial^2 u}{\partial r^2} + \dfrac{2}{r}\dfrac{\partial u}{\partial r}\right)$, find $\phi(t)$.

**\*28** If $r = \sqrt{(x^2+y^2)}$ and $u = f(r)$, prove that

$$\frac{\partial r}{\partial x} = \frac{x}{r} \quad\text{and}\quad \frac{\partial r}{\partial y} = \frac{y}{r},$$

and hence that

$$\frac{\partial^2 u}{\partial x^2} + \frac{\partial^2 u}{\partial y^2} = f''(r) + \frac{1}{r}f'(r).$$

Find $u$ in terms of $r$ if

$$\frac{\partial^2 u}{\partial x^2} + \frac{\partial^2 u}{\partial y^2} = 0.$$

**\*29** If $f(x,y) = xy(x^2-y^2)/(x^2+y^2)$ when $x$, $y$ are not both zero, and $f(0,0) = 0$,

(i) prove that when $x$, $y$ are not both zero,

$$\frac{\partial f}{\partial x} = y\left\{\frac{x^2-y^2}{x^2+y^2} + \frac{4x^2y^2}{(x^2+y^2)^2}\right\}, \quad \frac{\partial f}{\partial y} = x\left\{\frac{x^2-y^2}{x^2+y^2} - \frac{4x^2y^2}{(x^2+y^2)^2}\right\};$$

(ii) calculate

$$\lim_{h\to 0}\frac{f(h,0)-f(0,0)}{h},$$

i.e. $f_x(0,0)$; and similarly find $f_y(0,0)$;

(iii) hence calculate

$$\lim_{h\to 0}\frac{f_y(h,0)-f_y(0,0)}{h} = f_{xy}(0,0) \quad\text{and}\quad \lim_{k\to 0}\frac{f_x(0,k)-f_x(0,0)}{k} = f_{yx}(0,0).$$

(iv) What follows from the results of (iii)?

**\*30** If $f(x,y) = x^2\tan^{-1}(y/x) - y^2\tan^{-1}(x/y)$ when $x \neq 0$, $y \neq 0$, and

$$f(x,0) = f(0,y) = f(0,0) = 0,$$

prove that $f_{xy}(0,0) = +1$, $f_{yx}(0,0) = -1$.

## 9.3   The total variation of $u(x, y)$.  Small changes

### 9.31   Total variation

Partial derivatives are defined by changing $x$ or $y$ alone. When $x$ and $y$ vary together, a change $\delta x$ in $x$ and a change $\delta y$ in $y$ cause a change $\delta u$ in $u$, given by

$$\delta u = u(x + \delta x, y + \delta y) - u(x, y).$$

For example, taking $u = x^2y^3$ and writing $h = \delta x$, $k = \delta y$, we have

$$\delta u = (x+h)^2(y+k)^3 - x^2y^3$$
$$= (x^2 + 2xh + h^2)(y^3 + 3ky^2 + 3k^2y + k^3) - x^2y^3$$
$$= 3x^2y^2k + 3x^2yk^2 + x^2k^3 + 2xy^3h + 6xy^2hk + 6xyhk^2 + 2xhk^3$$
$$+ y^3h^2 + 3y^2h^2k + 3yh^2k^2 + h^2k^3$$
$$= 2xy^3\,\delta x + 3x^2y^2\,\delta y + \delta x\{6xy^2k + 6xyk^2 + 2xk^3 + y^3h + 3y^2hk$$
$$+ 3yhk^2 + hk^3\} + \delta y\{3x^2yk + x^2k^2\}.$$

Hence, in this example, $\delta u$ is of the form

$$A\,\delta x + B\,\delta y + \epsilon_1\,\delta x + \epsilon_2\,\delta y,$$

where $A$ and $B$ are functions of $x$ and $y$, but not of $\delta x$ or $\delta y$; and $\epsilon_1$, $\epsilon_2$ are functions of $x$, $y$, $\delta x$, $\delta y$ which tend to zero when $\delta x$ and $\delta y$ both tend to zero.

## 9.32 Definition of 'differentiable function of $(x, y)$'

The function $u = u(x, y)$ is said to be a *differentiable function of* $(x, y)$ if arbitrary changes $\delta x$, $\delta y$ in $x$, $y$ cause a change $\delta u$ in $u$ which can be expressed in the form

$$\delta u = A\,\delta x + B\,\delta y + \epsilon_1\,\delta x + \epsilon_2\,\delta y, \tag{i}$$

where $A$, $B$ are independent of $\delta x$ and $\delta y$ but are in general functions of $x$ and $y$, and $\epsilon_1$, $\epsilon_2$ are functions of $\delta x$, $\delta y$ (and possibly also $x$, $y$) which tend to zero when both $\delta x$ and $\delta y$ tend to zero in any manner.

Notice that the changes $\delta x$, $\delta y$ are *arbitrary*, so that one of them can be taken equal to zero.

The work in 9.31 shows that $x^2y^3$ is a differentiable function of $(x, y)$ for all values of $x$ and $y$. On the other hand, the function defined by

$$u(x, y) = \frac{xy(x+y)}{x^2+y^2}$$

when $x$, $y$ are not both zero, and $u(0, 0) = 0$, is not differentiable at $(0, 0)$. For, writing $h = \delta x$ and $k = \delta y$,

$$\delta u = u(h, k) - u(0, 0) = \frac{hk(h+k)}{h^2+k^2},$$

and if $u(x, y)$ were differentiable at $(0, 0)$ we should have

$$\frac{hk(h+k)}{h^2+k^2} = Ah + Bk + \epsilon_1 h + \epsilon_2 k.$$

Take $k = 0$, and let $h \to 0$ after dividing by $h$: then we find $0 = A$. Taking $h = 0$ and letting $k \to 0$ similarly gives $0 = B$. Finally, letting $h = k$, the relation becomes

$$h = (A+B)h + (\epsilon_1 + \epsilon_2)h;$$

and on dividing by $h$ ($\neq 0$) and then letting $h \to 0$, we get $1 = A + B$. The three relations for $A$, $B$ are clearly inconsistent.

*Identification of A, B in* (i).

In equation (i) put $\delta y = 0$; then

$$\delta u = A\,\delta x + \epsilon_1\,\delta x,$$

where $\epsilon_1$ is a function of $x$, $y$, $\delta x$ which tends to zero when $\delta x \to 0$. We have

$$\frac{\delta u}{\delta x} = A + \epsilon_1,$$

and by letting $\delta x \to 0$, the right-hand side tends to $A$. The left-hand side tends to $\partial u/\partial x$ by definition of this symbol (9.21). Hence $A = \partial u/\partial x$. Similarly, $B = \partial u/\partial y$, and so (i) becomes

$$\delta u = \frac{\partial u}{\partial x}\delta x + \frac{\partial u}{\partial y}\delta y + \epsilon_1\,\delta x + \epsilon_2\,\delta y. \tag{ii}$$

This work also shows that if a function is differentiable, then it possesses partial derivatives; but *the converse may be false,* as in the example above: $u(x, y)$ is not differentiable at $(0, 0)$, but

$$u_x(0, 0) = \lim_{h \to 0} \frac{u(h, 0) - u(0, 0)}{h} = \lim_{h \to 0} \frac{0}{h} = 0,$$

and similarly $u_y(0, 0)$ exists and has the value 0. See also Ex. 9(b), nos. 8, 9. Thus (cf. the end of 3.93) 'derivability' and 'differentiability' are not equivalent concepts.

## 9.33 Small changes

Equation (ii) leads to the *approximation*

$$\delta u \doteqdot \frac{\partial u}{\partial x}\delta x + \frac{\partial u}{\partial y}\delta y \tag{iii}$$

when $\delta x$, $\delta y$ are small. Comparing this with the corresponding result (iii) in 3.91 for functions of a single variable, we see that

$\dfrac{\partial u}{\partial x}\delta x$ is the approximate change in $u$ for a change $\delta x$ in $x$,

*keeping y constant;*

$\dfrac{\partial u}{\partial y}\delta y$ is the approximate change in $u$ for a change $\delta y$ in $y$,

*keeping x constant.*

(The constancy of $y$, $x$ respectively is indicated by the partial derivative notation.) Also (cf. 3.91) these approximations are correct to first order in $\delta x$, $\delta y$ respectively. We thus obtain

*The principle of superposition of small changes.*

*The change in a function when both variables alter* (i.e. the total variation) *is approximately equal to the sum of the changes arising when each varies separately, this approximation being correct to the first order in these changes.*

### Example

*Estimate the possible error in $S$ calculated from $S = \frac{1}{2}bc\sin A$ when there are errors $\delta b$, $\delta c$, $\delta A$ in $b$, $c$, $A$.* (Cf. 3.91, ex. (iv).)

Since
$$\frac{\partial S}{\partial b} = \tfrac{1}{2}c\sin A, \quad \frac{\partial S}{\partial c} = \tfrac{1}{2}b\sin A, \quad \frac{\partial S}{\partial A} = \tfrac{1}{2}bc\cos A,$$

$$\therefore \quad \delta S \doteqdot \tfrac{1}{2}c\sin A\,\delta b + \tfrac{1}{2}b\sin A\,\delta c + \tfrac{1}{2}bc\cos A\,\delta A.$$

Division by $S$ gives the *relative error*
$$\frac{\delta S}{S} \doteqdot \frac{\delta b}{b} + \frac{\delta c}{c} + \cot A\,\delta A,$$

from which the *percentage error* can be found. The last approximation could be obtained directly by applying the principle to
$$u = \log S = \log\tfrac{1}{2} + \log b + \log c + \log\sin A$$

and using the fact that
$$\delta u \doteqdot \frac{\partial(\log S)}{\partial S}\,\delta S = \frac{\delta S}{S}.$$

As a numerical illustration, suppose $S$ is calculated when $b = 5$, $c = 2$ and $A = 30°$, where $b$ and $c$ may each be in error by $0\cdot05$ and $A$ by $10'$. Then $S = \frac{1}{2} \times 5 \times 2\sin 30 = 2\cdot5$, $\delta b = \delta c = 0\cdot05$, $\delta A = \pi/(180 \times 6)$ (working in *radians*), and the numerically largest possible error in $S$ is given by
$$\frac{\delta S}{2\cdot5} \doteqdot \left(\frac{0\cdot05}{5} + \frac{0\cdot05}{2} + \frac{\pi\sqrt{3}}{1080}\right),$$

from which $\delta S \doteqdot 0\cdot1$. The relative error is $0\cdot04$.

### Exercise 9(*b*)

**1** If $pv = RT$ where $R$ is constant, find the approximate change in $v$ caused by small changes $\delta p$, $\delta T$ in $p$, $T$.

**2** If $y = uv/w$, prove
$$\frac{\delta y}{y} \doteqdot \frac{\delta u}{u} + \frac{\delta v}{v} - \frac{\delta w}{w}.$$

**3** If $T = 2\pi\sqrt{(l/g)}$, find the approximate error in $T$ due to small errors $\delta l$, $\delta g$ in $l$, $g$.

**4** If $f(x,y) = xe^{xy}$, and the values of $x$ and $y$ are slightly changed from 1 and 0 to $1 + \delta x$ and $\delta y$ respectively so that $\delta f$, the change in $f$, is very nearly $3\,\delta x$, show that $\delta y$ is approximately $2\,\delta x$.

**5** If $z = \sin\theta\sin\phi/\sin\psi$ and $z$ is calculated for the values $\theta = 30°$, $\phi = 45°$, $\psi = 60°$, find approximately the change in $z$ if each of $\theta$, $\psi$ is increased by the same small angle $\alpha°$ and $\phi$ is decreased by $\frac{1}{2}\alpha°$.

**6** In triangle $ABC$, angle $A$ is known accurately, but the measurements of sides $b$, $c$ may be in error by $\delta b$, $\delta c$ respectively. Find approximately the error obtained by calculating the side $a$ from $b$, $c$, $A$. What shape should the triangle have in order to make as small as possible the effect of the error $\delta b$?

**7** The points $A$ and $B$, at distance $a$ apart on a horizontal plane, are in line with the base $C$ of a vertical tower and on the same side of $C$. The elevations of the top of the tower from $A$, $B$ are observed to be $\alpha°$, $\beta°$ $(\alpha < \beta)$. Show that $BC = a\sin\alpha\cos\beta\,\mathrm{cosec}\,(\beta-\alpha)$.

If the observations of the angles are uncertain by 4 minutes, show that the maximum possible percentage error in the calculated value of $BC$ is approximately

$$\frac{\pi\sin(\alpha+\beta)}{27\sin\alpha\cos\beta\tan(\beta-\alpha)}.$$

**\*8** If $u(x,y) = (x^3-y^3)/(x^2+y^2)$ when $x$, $y$ are not both zero, and $u(0,0) = 0$, prove $u$ is not differentiable at $(0,0)$, but that $\partial u/\partial x$, $\partial u/\partial y$ exist at $(0,0)$ and have the values $+1$, $-1$ respectively.

**\*9** If $u(x,y) = \sqrt{|xy|}$, show $u_x(0,0) = 0 = u_y(0,0)$, but that $u$ is not differentiable at $(0,0)$.

## 9.4 Extensions of 'function of a function'

### 9.41 Function of two functions of $t$

If $u = u(x,y)$, where $x = x(t)$ and $y = y(t)$, then

$$u = u\big(x(t), y(t)\big)$$

is a new function $u = f(t)$ of one independent variable $t$, and its derivative $du/dt = f'(t)$ may be obtained directly. However, this substitution for $x$ and $y$ is not always convenient, and we proceed indirectly as follows.

Assuming that all the functions concerned are differentiable, a change $\delta t$ in $t$ causes a change $\delta x$ in $x$ and a change $\delta y$ in $y$, which in turn cause a change $\delta u$ in $u$, where† by 9.32, equation (ii),

$$\delta u = \frac{\partial u}{\partial x}\delta x + \frac{\partial u}{\partial y}\delta y + \epsilon_1\delta x + \epsilon_2\delta y.$$

Since $x(t)$, $y(t)$ are differentiable, they are certainly continuous functions of $t$ (3.12 and 3.93); hence when $\delta t \to 0$, also $\delta x \to 0$ and $\delta y \to 0$. Therefore $\epsilon_1 \to 0$ and $\epsilon_2 \to 0$ when $\delta t \to 0$. Now

$$\frac{\delta u}{\delta t} = \frac{\partial u}{\partial x}\frac{\delta x}{\delta t} + \frac{\partial u}{\partial y}\frac{\delta y}{\delta t} + \epsilon_1\frac{\delta x}{\delta t} + \epsilon_2\frac{\delta y}{\delta t},$$

† $\epsilon_1$ and $\epsilon_2$ are not defined when $\delta x$, $\delta y$ are both zero. To cover the case when $\delta x = 0 = \delta y$ for some values of $\delta t \neq 0$ (e.g. $x = y = \phi(t)$ near $t = 0$, with $\phi(t)$ as on p. 64), we may 'complete the definition' by taking $\epsilon_1 = 0$, $\epsilon_2 = 0$ when $\delta x = 0 = \delta y$. Cf. p. 63, footnote.

and when $\delta t \to 0$, the right-hand side tends to

$$\frac{\partial u}{\partial x}\frac{dx}{dt} + \frac{\partial u}{\partial y}\frac{dy}{dt}.$$

Hence $\delta u/\delta t$ tends to this limit, i.e.

$$\frac{du}{dt} = \frac{\partial u}{\partial x}\frac{dx}{dt} + \frac{\partial u}{\partial y}\frac{dy}{dt}. \qquad \text{(iv)}$$

The formula (iv) may be false if $u(x, y)$ is not differentiable; this is shown by the example already considered in 9.32, where $u_x(0, 0) = 0 = u_y(0, 0)$. Putting $x = t$, $y = t$, we have $f(t) = u(t, t) = t$ if $t \neq 0$, and $f(0) = u(0, 0) = 0$; hence $f'(t) = 1$ for all $t$, even when $t = 0$. But when $t = 0$,

$$\frac{\partial u}{\partial x}\frac{dx}{dt} + \frac{\partial u}{\partial y}\frac{dy}{dt} = 0.1 + 0.1 = 0.$$

The formula fails because this function $u(x, y)$ is not differentiable at $(0, 0)$.

## 9.42 Total derivative; application to implicit functions

When $t = x$ in 9.41, we have $u = u(x, y)$ where $y = y(x)$; (iv) becomes

$$\frac{du}{dx} = \frac{\partial u}{\partial x} + \frac{\partial u}{\partial y}\frac{dy}{dx}, \qquad \text{(v)}$$

which is sometimes called the formula for the *total derivative* of $u$ wo $x$.

It is at this stage that a special notation $\partial u/\partial x$ for partial derivatives becomes essential: $du/dx$ means the derivative of the function $u(x, y)$ in which we have substituted $y = y(x)$ *before* derivation; $\partial u/\partial x$ means the derivative of $u(x, y)$ wo $x$, *carried out as if $y$ were constant*, where the substitution $y = y(x)$ is made *after* derivation. Thus $\partial u/\partial x$ symbolises a 'formal' derivation wo $x$. Similar details of interpretation apply to (iv); the reader should thoroughly understand their significance before proceeding further, and should do Ex. 9 (c), no. 1.

In particular, suppose $y$ is defined implicitly as a function† $y(x)$ of $x$ by the equation $u(x, y) = 0$. Then $u(x, y(x)) = 0$ for all $x$, and so $du/dx = 0$; (v) becomes

$$0 = \frac{\partial u}{\partial x} + \frac{\partial u}{\partial y}\frac{dy}{dx},$$

$$\therefore \quad \frac{dy}{dx} = -\frac{\partial u}{\partial x}\bigg/\frac{\partial u}{\partial y}. \qquad \text{(vi)}$$

† Assumed differentiable.

## Example

*Find $dy/dx$ and $d^2y/dx^2$ if $x^3 + y^3 = 3xy$.*

We will find $dy/dx$ (a) by the elementary method of 3.41; (b) by formula (vi).

(a) Derive both sides wo $x$, treating $y$ as a function of $x$ (defined implicitly):

$$3x^2 + 3y^2 \frac{dy}{dx} = 3 \left( y + x \frac{dy}{dx} \right),$$

$$\therefore \quad (y^2 - x) \frac{dy}{dx} = y - x^2,$$

and

$$\frac{dy}{dx} = \frac{y - x^2}{y^2 - x}.$$

(b) Taking $u = x^3 + y^3 - 3xy$, $\partial u/\partial x = 3x^2 - 3y$ and $\partial u/\partial y = 3y^2 - 3x$. Therefore by (v),

$$\frac{dy}{dx} = -\frac{x^2 - y}{y^2 - x} = \frac{y - x^2}{y^2 - x}.$$

Methods (a), (b) are really the same *process*, as the above working shows; but we now have a general notation available to formulate the process (a).

To find $d^2y/dx^2$ we may either proceed as in 3.53, ex. (iii), or apply formula (v) to the function $u = dy/dx$:

$$\frac{d^2y}{dx^2} = \frac{\partial}{\partial x} \left( \frac{y - x^2}{y^2 - x} \right) + \frac{\partial}{\partial y} \left( \frac{y - x^2}{y^2 - x} \right) \cdot \frac{dy}{dx}$$

$$= \frac{-2x(y^2 - x) + (y - x^2)}{(y^2 - x)^2} + \frac{(y^2 - x) - 2y(y - x^2)}{(y^2 - x)^2} \cdot \frac{y - x^2}{y^2 - x}$$

$$= \frac{(x^2 + y - 2xy^2)(y^2 - x) + (2x^2y - y^2 - x)(y - x^2)}{(y^2 - x)^3}.$$

The numerator reduces to

$$2xy(3xy - x^3 - y^3 - 1),$$

which by the given equation is equal to $-2xy$. Hence

$$\frac{d^2y}{dx^2} = \frac{2xy}{(x - y^2)^3}.$$

[*The work on envelopes in 8.5 could be read at this stage.*]

## 9.43 Function of two functions of ($\xi, \eta$)

If $u = u(x, y)$ and $x = x(\xi, \eta)$, $y = y(\xi, \eta)$, then direct substitution shows that $u$ is a function $f(\xi, \eta)$ of $\xi$ and $\eta$, where

$$f(\xi, \eta) = u(x(\xi, \eta), y(\xi, \eta)).$$

To find $\partial u/\partial \xi$ and $\partial u/\partial \eta$ (or, more precisely, $f_\xi(\xi, \eta)$ and $f_\eta(\xi, \eta)$), we observe that provided all three functions are differentiable, formula (iv) still applies, except that the notation in it is now modified into *partial* derivatives:

$$\frac{\partial u}{\partial \xi} = \frac{\partial u}{\partial x}\frac{\partial x}{\partial \xi} + \frac{\partial u}{\partial y}\frac{\partial y}{\partial \xi} \left.\begin{array}{c}\\\\\\\\\end{array}\right\}.$$

$$\frac{\partial u}{\partial \eta} = \frac{\partial u}{\partial x}\frac{\partial x}{\partial \eta} + \frac{\partial u}{\partial y}\frac{\partial y}{\partial \eta}$$

(vii)

For, the proof of either of results (vii) is the same as that of (iv): to get (say) $\partial u/\partial \xi$ we are keeping $\eta$ constant, and are thus dealing with a function of effectively only one variable $\xi$.

## Examples

(i) *If $u$ is a function of $x$ and $y$, and $x = e^{2\xi}\sin\eta$, $y = e^{2\xi}\cos\eta$, prove*

(a) $\dfrac{\partial u}{\partial \xi} = 2x\dfrac{\partial u}{\partial x} + 2y\dfrac{\partial u}{\partial y}$,    (b) $\dfrac{\partial u}{\partial x} = \tfrac12 e^{-2\xi}\left(\sin\eta\dfrac{\partial u}{\partial \xi} + 2\cos\eta\dfrac{\partial u}{\partial \eta}\right)$.

By (vii),

$$\frac{\partial u}{\partial \xi} = \frac{\partial u}{\partial x}(2e^{2\xi}\sin\eta) + \frac{\partial u}{\partial y}(2e^{2\xi}\cos\eta)$$

$$= 2x\frac{\partial u}{\partial x} + 2y\frac{\partial u}{\partial y},$$

which proves (a). Similarly

$$\frac{\partial u}{\partial \eta} = \frac{\partial u}{\partial x}(e^{2\xi}\cos\eta) + \frac{\partial u}{\partial y}(-e^{2\xi}\sin\eta)$$

$$= y\frac{\partial u}{\partial x} - x\frac{\partial u}{\partial y}.$$

To solve these for $\partial u/\partial x$, multiply the first by $x$, the second by $2y$, and add:

$$x\frac{\partial u}{\partial \xi} + 2y\frac{\partial u}{\partial \eta} = 2(x^2 + y^2)\frac{\partial u}{\partial x},$$

i.e.    $e^{2\xi}\sin\eta\dfrac{\partial u}{\partial \xi} + 2e^{2\xi}\cos\eta\dfrac{\partial u}{\partial \eta} = 2e^{4\xi}\dfrac{\partial u}{\partial x},$

so that    $\dfrac{\partial u}{\partial x} = \tfrac12 e^{-2\xi}\left(\sin\eta\dfrac{\partial u}{\partial \xi} + 2\cos\eta\dfrac{\partial u}{\partial \eta}\right).$

*Remarks*

($\alpha$) After observing what has to be proved in (b), the reader may be tempted to begin directly with the formula

$$\frac{\partial u}{\partial x} = \frac{\partial u}{\partial \xi}\frac{\partial \xi}{\partial x} + \frac{\partial u}{\partial \eta}\frac{\partial \eta}{\partial x}.$$

Further progress could then be made only by expressing $\xi$ and $\eta$ as functions of $x$ and $y$. We should obtain

$$\xi = \tfrac14\log(x^2 + y^2), \quad \eta = \tan^{-1}(x/y),$$

and the work could be completed by direct calculations. We emphasise that in general $\partial\xi/\partial x \neq 1/(\partial x/\partial\xi)$, etc., so that the preceding steps cannot be short-cut; here, for example,

$$\frac{\partial\xi}{\partial x} = \frac{\frac{1}{2}x}{x^2+y^2} = \frac{e^{2\xi}\sin\eta}{2e^{4\xi}} = \tfrac{1}{2}e^{-2\xi}\sin\eta,$$

while

$$\frac{1}{\partial x/\partial\xi} = \frac{1}{2e^{2\xi}\sin\eta} = \tfrac{1}{2}e^{-2\xi}\operatorname{cosec}\eta.$$

When deciding on the formula with which to start, we must take account of the *given* functional relationships.

($\beta$) When we have four variables $x$, $y$, $\xi$, $\eta$ connected by two relations, the symbol $\partial x/\partial\xi$ standing alone is ambiguous. It implies that $x$ has been expressed as a function of $\xi$ and another variable, which could be either $\eta$ or $y$. In the present case the relations are $x = e^{2\xi}\sin\eta$, $y = e^{2\xi}\cos\eta$, and when $x$ is a function of $(\xi,\eta)$ the first gives $\partial x/\partial\xi = 2e^{2\xi}\sin\eta$; the second is not used. If instead we express $x$ as a function of $(\xi, y)$ by eliminating $\eta$, we find that $x = \sqrt{(e^{4\xi}-y^2)}$, from which

$$\frac{\partial x}{\partial\xi} = \frac{2e^{4\xi}}{\sqrt{(e^{4\xi}-y^2)}} = \frac{2e^{4\xi}}{x} = 2e^{2\xi}\operatorname{cosec}\eta.$$

To distinguish the two meanings we write $(\partial x/\partial\xi)_\eta$, $(\partial x/\partial\xi)_y$ respectively, with a similar notation for other partial derivatives.

(ii) *If $u = f(x-y, y-z, z-x)$, prove*

$$\frac{\partial u}{\partial x} + \frac{\partial u}{\partial y} + \frac{\partial u}{\partial z} = 0.$$

Put $\xi = x-y$, $\eta = y-z$, $\zeta = z-x$, so that $u = f(\xi,\eta,\zeta)$. By formula (vii) extended to a function of three functions $\xi$, $\eta$, $\zeta$ of $(x,y,z)$,

$$\frac{\partial u}{\partial x} = \frac{\partial u}{\partial\xi}\frac{\partial\xi}{\partial x} + \frac{\partial u}{\partial\eta}\frac{\partial\eta}{\partial x} + \frac{\partial u}{\partial\zeta}\frac{\partial\zeta}{\partial x}$$

$$= \frac{\partial u}{\partial\xi}.1 + \frac{\partial u}{\partial\eta}.0 + \frac{\partial u}{\partial\zeta}(-1)$$

$$= \frac{\partial u}{\partial\xi} - \frac{\partial u}{\partial\zeta}.$$

Similarly

$$\frac{\partial u}{\partial y} = -\frac{\partial u}{\partial\xi} + \frac{\partial u}{\partial\eta}, \quad \frac{\partial u}{\partial z} = -\frac{\partial u}{\partial\eta} + \frac{\partial u}{\partial\zeta}.$$

The result follows by adding.

### Exercise 9(c)

1 Find $du/dt$ if $u = x^2+y^2$ and $x = 3t^2$, $y = 2t^3$ (*a*) by direct substitution; (*b*) by using formula (iv) in 9.41.

*Find $dy/dx$ for the functions defined implicitly by the following equations.*

2 $(x+y)^3 = 3xy$.　　　　　　　　3 $x^5+y^5 = 5ax^2y^2$ (*a* being constant).

4 $x\cos y = y\sin x$.　　　　　　　5 $f(xy) = 0$.

*6 If the curves $f(x,y) = 0$, $g(x,y) = 0$ intersect, show that they do so at angle $\tan^{-1}\{(f_xg_y-f_yg_x)/(f_xg_x+f_yg_y)\}$.

*7 If $x^4+y^4 = 4xy$, prove

$$\frac{d^2y}{dx^2} = \frac{2xy(x^2y^2+3)}{(x-y^3)^3}.$$

*8 If $f(x, y) = 0$, prove that

$$\frac{d^2y}{dx^2} = -\frac{f_y^2 f_{xx} - 2f_x f_y f_{xy} + f_x^2 f_{yy}}{f_y^3}.$$

Show that the curvature of $f(x, y) = 0$ at $(x, y)$ is

$$\kappa = -\frac{f_y^2 f_{xx} - 2f_x f_y f_{xy} + f_x^2 f_{yy}}{(f_x^2 + f_y^2)^{\frac{3}{2}}}.$$

9 If $u$ is a function of $(x, y)$ and $x = \frac{1}{2}(\xi^2 + \eta^2)$, $y = \xi\eta$, find $\partial u/\partial\xi$, $\partial u/\partial\eta$.

10 If $u = u(x, y)$ and $x = r\cos\theta$, $y = r\sin\theta$, prove

$$r\frac{\partial u}{\partial r} = x\frac{\partial u}{\partial x} + y\frac{\partial u}{\partial y},$$

and express $\partial u/\partial\theta$ similarly.

11 If $u = u(x, y)$ and $\xi = e^{xy}$, $\eta = x^2 + y^2$, express $\partial u/\partial x$, $\partial u/\partial y$ in terms of $\partial u/\partial\xi$, $\partial u/\partial\eta$, $x$ and $y$.

12 If $u = u(x, y)$ and $x = \xi + \eta$, $y = \xi\eta$, prove

(i) $(\xi - \eta)\dfrac{\partial u}{\partial x} = \xi\dfrac{\partial u}{\partial\xi} - \eta\dfrac{\partial u}{\partial\eta};$   (ii) $(\xi - \eta)\dfrac{\partial u}{\partial y} = \dfrac{\partial u}{\partial\eta} - \dfrac{\partial u}{\partial\xi}.$

13 If $u = u(x, y)$ and $x = \xi^2 - \eta^2$, $y = 2\xi\eta$, prove

(i) $2(\xi^2 + \eta^2)\dfrac{\partial u}{\partial x} = \xi\dfrac{\partial u}{\partial\xi} - \eta\dfrac{\partial u}{\partial\eta};$

(ii) $\dfrac{\partial^2 u}{\partial\xi^2} = 2\dfrac{\partial u}{\partial x} + 4\xi^2\dfrac{\partial^2 u}{\partial x^2} + 8\xi\eta\dfrac{\partial^2 u}{\partial x\,\partial y} + 4\eta^2\dfrac{\partial^2 u}{\partial y^2},$

and hence that   (iii) $\dfrac{\partial^2 u}{\partial x^2} + \dfrac{\partial^2 u}{\partial y^2} = \dfrac{1}{4(\xi^2 + \eta^2)}\left(\dfrac{\partial^2 u}{\partial\xi^2} + \dfrac{\partial^2 u}{\partial\eta^2}\right).$

14 If $u$, $v$ are functions of $(x, y)$ which satisfy

$$\frac{\partial u}{\partial x} = \frac{\partial v}{\partial y}, \quad \frac{\partial u}{\partial y} = -\frac{\partial v}{\partial x},$$

and if $x = r\cos\theta$, $y = r\sin\theta$, prove that

$$\frac{\partial u}{\partial r} = \frac{1}{r}\frac{\partial v}{\partial\theta}, \quad \frac{1}{r}\frac{\partial u}{\partial\theta} = -\frac{\partial v}{\partial r} \quad \text{and} \quad \frac{\partial^2 u}{\partial r^2} + \frac{1}{r}\frac{\partial u}{\partial r} + \frac{1}{r^2}\frac{\partial^2 u}{\partial\theta^2} = 0.$$

15 If $u = x^n f(y/x, z/x)$, prove that

$$x\frac{\partial u}{\partial x} + y\frac{\partial u}{\partial y} + z\frac{\partial u}{\partial z} = nu.$$

16 If $x = r\cos\theta$, $y = r\sin\theta$, prove

$$\left(\frac{\partial r}{\partial x}\right)_y = \left(\frac{\partial x}{\partial r}\right)_\theta \quad \text{and find} \quad \left(\frac{\partial\theta}{\partial x}\right)_y, \quad \left(\frac{\partial x}{\partial\theta}\right)_r.$$

Interpret these four partial derivatives geometrically, and indicate why

$$\frac{\partial r}{\partial x} \neq 1\left/\frac{\partial x}{\partial r}\right., \quad \frac{\partial\theta}{\partial x} \neq 1\left/\frac{\partial x}{\partial\theta}\right..$$

17 In no. 16 find $(\partial x/\partial r)_y$, $(\partial y/\partial\theta)_x$.

## 9.44 Further examples

(1) *Euler's theorem on homogeneous functions*

We defined 'homogeneous function of degree $n$ in $(x, y)$' in 1.52(4); the definition extends obviously to three or more independent variables.

*If $u(x, y)$ is homogeneous of degree $n$, then*

$$x\frac{\partial u}{\partial x} + y\frac{\partial u}{\partial y} = nu$$

*for all $x$, $y$ for which the function $u$ is differentiable.*

The result has already been verified by direct calculation in 9.24, ex. (ii)(a), since every such function can be expressed in the form $x^n f(y/x)$: see Ex. 1(d), no. 7. We now give a proof based on formula (vii).†

Put $\xi = tx$, $\eta = ty$ in the definition

$$u(tx, ty) = t^n u(x, y)$$

of 'homogeneous function'; then

$$u(\xi, \eta) = t^n u(x, y).$$

Derive both sides wo $t$:

$$\frac{\partial u}{\partial \xi}\frac{\partial \xi}{\partial t} + \frac{\partial u}{\partial \eta}\frac{\partial \eta}{\partial t} = nt^{n-1}u(x, y),$$

i.e.
$$x\frac{\partial u}{\partial \xi} + y\frac{\partial u}{\partial \eta} = nt^{n-1}u(x, y). \tag{a}$$

Since
$$\frac{\partial}{\partial \xi}u(\xi, \eta) = \left[\frac{\partial}{\partial x}u(x, y)\right]_{\substack{x=\xi \\ y=\eta}}, \quad \text{etc.},$$

equation (a) becomes

$$x\left[\frac{\partial u}{\partial x}\right]_{\xi, \eta} + y\left[\frac{\partial u}{\partial y}\right]_{\xi, \eta} = nt^{n-1}u(x, y).$$

Putting $t = 1$, this becomes

$$x\frac{\partial u}{\partial x} + y\frac{\partial u}{\partial y} = nu(x, y),$$

since $\xi = x$ and $\eta = y$ when $t = 1$.

## Example

*Obtain the result of 9.24, ex. (ii)(b) similarly.*

Derive equation (a) above wo $t$, applying formula (vii) to each of the functions $\partial u/\partial \xi$, $\partial u/\partial \eta$:

$$x\left(\frac{\partial^2 u}{\partial \xi^2}\frac{\partial \xi}{\partial t} + \frac{\partial^2 u}{\partial \eta \partial \xi}\frac{\partial \eta}{\partial t}\right) + y\left(\frac{\partial^2 u}{\partial \xi \partial \eta}\frac{\partial \xi}{\partial t} + \frac{\partial^2 u}{\partial \eta^2}\frac{\partial \eta}{\partial t}\right) = n(n-1)t^{n-2}u(x, y),$$

i.e.
$$x^2\frac{\partial^2 u}{\partial \xi^2} + 2xy\frac{\partial^2 u}{\partial \xi \partial \eta} + y^2\frac{\partial^2 u}{\partial \eta^2} = n(n-1)t^{n-2}u(x, y).$$

Now put $t = 1$, and the result ('Euler's theorem of second order') follows.

† See also Ex. 9(c), no. 15, and Ex. 9(f), no. 23.

(2) *Laplace's equation* $\quad \dfrac{\partial^2 u}{\partial x^2} + \dfrac{\partial^2 u}{\partial y^2} = 0.$

Interpreting $(x, y)$ as cartesian coordinates, we investigate what becomes of this partial differential equation when we change to polar coordinates by the substitutions $x = r\cos\theta$, $y = r\sin\theta$. The new independent variables are $r$ and $\theta$, and we have to express partial derivatives of $u$ wo $x$, $y$ in terms of those wo $r$, $\theta$. We have by (vii):

$$\frac{\partial u}{\partial r} = \frac{\partial u}{\partial x}\frac{\partial x}{\partial r} + \frac{\partial u}{\partial y}\frac{\partial y}{\partial r} = \frac{\partial u}{\partial x}\cos\theta + \frac{\partial u}{\partial y}\sin\theta,$$

$$\frac{\partial u}{\partial \theta} = \frac{\partial u}{\partial x}\frac{\partial x}{\partial \theta} + \frac{\partial u}{\partial y}\frac{\partial y}{\partial \theta} = -\frac{\partial u}{\partial x}r\sin\theta + \frac{\partial u}{\partial y}r\cos\theta.$$

Solving for $\partial u/\partial x$ and $\partial u/\partial y$, we find that

$$\frac{\partial u}{\partial x} = \cos\theta\frac{\partial u}{\partial r} - \frac{\sin\theta}{r}\frac{\partial u}{\partial\theta}, \quad \frac{\partial u}{\partial y} = \sin\theta\frac{\partial u}{\partial r} + \frac{\cos\theta}{r}\frac{\partial u}{\partial\theta}. \tag{b}$$

The first of equations (b) shows that the operators $\partial/\partial x$ and

$$\cos\theta\frac{\partial}{\partial r} - \frac{\sin\theta}{r}\frac{\partial}{\partial\theta}$$

*have the same effect* on $u$, where on the left $u$ is a function of $(x, y)$ and on the right it has been expressed in terms of $r$ and $\theta$. Using these *equivalent operators* (cf. 5.62) to find $\partial^2 u/\partial x^2$, we have

$$\frac{\partial^2 u}{\partial x^2} = \left(\cos\theta\frac{\partial}{\partial r} - \frac{\sin\theta}{r}\frac{\partial}{\partial\theta}\right)\left(\cos\theta\frac{\partial u}{\partial r} - \frac{\sin\theta}{r}\frac{\partial u}{\partial\theta}\right)$$

$$= \cos\theta\left(\cos\theta\frac{\partial^2 u}{\partial r^2} + \frac{\sin\theta}{r^2}\frac{\partial u}{\partial\theta} - \frac{\sin\theta}{r}\frac{\partial^2 u}{\partial r\,\partial\theta}\right)$$

$$- \frac{\sin\theta}{r}\left(-\sin\theta\frac{\partial u}{\partial r} + \cos\theta\frac{\partial^2 u}{\partial\theta\,\partial r} - \frac{\cos\theta}{r}\frac{\partial u}{\partial\theta} - \frac{\sin\theta}{r}\frac{\partial^2 u}{\partial\theta^2}\right)$$

$$= \cos^2\theta\frac{\partial^2 u}{\partial r^2} - \frac{2\sin\theta\cos\theta}{r}\frac{\partial^2 u}{\partial r\,\partial\theta} + \frac{\sin^2\theta}{r^2}\frac{\partial^2 u}{\partial\theta^2} + \frac{\sin^2\theta}{r}\frac{\partial u}{\partial r} + \frac{2\sin\theta\cos\theta}{r^2}\frac{\partial u}{\partial\theta}.$$

Similarly, or see Remark ($\alpha$) below,

$$\frac{\partial^2 u}{\partial y^2} = \sin^2\theta\frac{\partial^2 u}{\partial r^2} + \frac{2\sin\theta\cos\theta}{r}\frac{\partial^2 u}{\partial r\,\partial\theta} + \frac{\cos^2\theta}{r^2}\frac{\partial^2 u}{\partial\theta^2} + \frac{\cos^2\theta}{r}\frac{\partial u}{\partial r} - \frac{2\sin\theta\cos\theta}{r^2}\frac{\partial u}{\partial\theta}.$$

Adding, $\qquad \dfrac{\partial^2 u}{\partial x^2} + \dfrac{\partial^2 u}{\partial y^2} = \dfrac{\partial^2 u}{\partial r^2} + \dfrac{1}{r^2}\dfrac{\partial^2 u}{\partial\theta^2} + \dfrac{1}{r}\dfrac{\partial u}{\partial r}.$ \hfill (c)

*Remarks*

($\alpha$) Since the expression for $\partial u/\partial y$ is obtainable from that for $\partial u/\partial x$ by putting $\theta - \frac{1}{2}\pi$ for $\theta$, therefore $\partial^2 u/\partial y^2$ can be found from $\partial^2 u/\partial x^2$ by the same substitution.

($\beta$) The result can be written

$$\frac{1}{r}\frac{\partial}{\partial r}\left(r\frac{\partial u}{\partial r}\right) + \frac{1}{r^2}\frac{\partial^2 u}{\partial\theta^2}.$$

($\gamma$) The expression on the left-hand side of (c) is often written $\nabla^2 u$, where $\nabla$ is called *del* or *nabla*; $\nabla^2$ is the *Laplace operator*.

(3) *The wave equation* $\dfrac{\partial^2 y}{\partial x^2} = \dfrac{1}{c^2}\dfrac{\partial^2 y}{\partial t^2}$.

To find the general solution of this, we change the independent variables from $x$, $t$ to $u$, $v$ by the substitutions $u = x - ct$, $v = x + ct$ which are suggested by Ex. 9 (a), no. 13. We have

$$\frac{\partial y}{\partial x} = \frac{\partial y}{\partial u}\frac{\partial u}{\partial x} + \frac{\partial y}{\partial v}\frac{\partial v}{\partial x} = \frac{\partial y}{\partial u} + \frac{\partial y}{\partial v};$$

and by using equivalent operators,

$$\frac{\partial^2 y}{\partial x^2} = \left(\frac{\partial}{\partial u} + \frac{\partial}{\partial v}\right)\left(\frac{\partial y}{\partial u} + \frac{\partial y}{\partial v}\right) = \frac{\partial^2 y}{\partial u^2} + 2\frac{\partial^2 y}{\partial u\,\partial v} + \frac{\partial^2 y}{\partial v^2}.$$

Similarly,

$$\frac{\partial y}{\partial t} = -c\frac{\partial y}{\partial u} + c\frac{\partial y}{\partial v} \quad\text{and}\quad \frac{\partial^2 y}{\partial t^2} = c^2\left(\frac{\partial^2 y}{\partial u^2} - 2\frac{\partial^2 y}{\partial u\,\partial v} + \frac{\partial^2 y}{\partial v^2}\right).$$

Hence the given equation becomes

$$4\frac{\partial^2 y}{\partial u\,\partial v} = 0.$$

Since this can be written $\quad\dfrac{\partial}{\partial u}\left(\dfrac{\partial y}{\partial v}\right) = 0,$

we see that $\partial y/\partial v$ is independent of $u$ (3.82), and is therefore a function of $v$ only, say $\partial y/\partial v = F(v)$. Integrating wo $v$,

$$y = \int F(v)\,dv + g(u),$$

where $g(u)$ is an 'arbitrary' function of $u$. Writing $f(v) = \int F(v)\,dv$, the required general solution is $\quad y = f(v) + g(u) = f(x + ct) + g(x - ct),$

where $f$, $g$ are 'arbitrary' functions.

## Example

*Find the solution for which* $y = e^x - 1$ *and* $\partial y/\partial t = e^x$ *when* $t = 0$.

Since $\qquad\dfrac{\partial y}{\partial t} = cf'(x + ct) - cg'(x - ct),$

we see that when $t = 0$,

$$y = f(x) + g(x) \quad\text{and}\quad \frac{\partial y}{\partial t} = cf'(x) - cg'(x).$$

Hence we require

$$f(x) + g(x) = e^x - 1 \quad\text{and}\quad f'(x) - g'(x) = \frac{1}{c}e^x,$$

i.e. $\qquad f(x) + g(x) = e^x - 1 \quad\text{and}\quad f(x) - g(x) = \frac{1}{c}e^x + a$

for arbitrary $a$. Therefore, on solving for $f(x)$, $g(x)$, we find

$$f(x) = \frac{1}{2}\left(1 + \frac{1}{c}\right)e^x + \tfrac{1}{2}(a - 1), \quad g(x) = \frac{1}{2}\left(1 - \frac{1}{c}\right)e^x - \tfrac{1}{2}(a + 1),$$

and the required solution is

$$y = \frac{1}{2}\left(1 + \frac{1}{c}\right)e^{x+ct} + \frac{1}{2}\left(1 - \frac{1}{c}\right)e^{x-ct} - 1.$$

## Exercise 9(d)*

**1** *Write down* the expression for

$$x\frac{\partial u}{\partial x} + y\frac{\partial u}{\partial y} + z\frac{\partial u}{\partial z}$$

when $u$ is

(i) $\tan^{-1}\dfrac{y}{\sqrt{(x^2+y^2+z^2)}}$;  (ii) $\dfrac{\sqrt{(x+y-z)}}{x^2+y^2+z^2}$;  (iii) $f\left(\dfrac{x}{z}, \dfrac{z}{y}\right)$.

**2** For any differentiable functions $\phi(t)$, $\psi(t)$, show that $u = x\phi(y/x) + \psi(y/x)$ satisfies

$$x^2\frac{\partial^2 u}{\partial x^2} + 2xy\frac{\partial^2 u}{\partial x\,\partial y} + y^2\frac{\partial^2 u}{\partial y^2} = 0.$$

**3** If $f(x,y)$ is homogeneous of degree $n$ in $(x,y)$, prove that $f_x(x,y)$, $f_y(x,y)$ are homogeneous of degree $n-1$. [Derive $f(tx, ty) = t^n f(x, y)$ partially wo $x$.]

**4** If $x = r\cos\theta$, $y = r\sin\theta$ and $u = r^n\cos\theta$, find the possible values of the constant $n$ if $\partial^2 u/\partial x^2 + \partial^2 u/\partial y^2 = 0$. [Use 9.44 (2).]

**5** Find the value of the constant $\lambda$ if $u = x^3 + \lambda xy^2$ satisfies

$$\frac{\partial^2 u}{\partial x^2} + \frac{\partial^2 u}{\partial y^2} = 0.$$

With this value of $\lambda$, show that if $z = r^n u$ where $r^2 = x^2 + y^2$, then

$$\frac{\partial^2 z}{\partial x^2} + \frac{\partial^2 z}{\partial y^2} = n(n+6)\, r^{n-2}u.$$

[Express $z$ in polar coordinates.]

**6** If $z$ is a function of $x$ and $y$, and the variables are changed to $u$, $v$ by

$$u = \alpha x^2 + \beta y^2, \quad v = \alpha x^2 - \beta y^2$$

(where $\alpha$, $\beta$ are constants), prove that

(i) $x\dfrac{\partial z}{\partial x} + y\dfrac{\partial z}{\partial y} = 2\left(u\dfrac{\partial z}{\partial u} + v\dfrac{\partial z}{\partial v}\right)$;

(ii) $x\dfrac{\partial z}{\partial x} - y\dfrac{\partial z}{\partial y} = 2\left(v\dfrac{\partial z}{\partial u} + u\dfrac{\partial z}{\partial v}\right)$;

(iii) $x^2\dfrac{\partial^2 z}{\partial x^2} - 2xy\dfrac{\partial^2 z}{\partial x\,\partial y} + y^2\dfrac{\partial^2 z}{\partial y^2} = 4\left(v^2\dfrac{\partial^2 z}{\partial u^2} + 2uv\dfrac{\partial^2 z}{\partial u\,\partial v} + u^2\dfrac{\partial^2 z}{\partial v^2}\right) + 2\left(u\dfrac{\partial z}{\partial u} + v\dfrac{\partial z}{\partial v}\right)$.

[For (iii) use equivalent operators obtained from (ii).]

**7** If $x = e^u + e^{-v}$, $y = e^v + e^{-u}$, prove

$$\frac{\partial^2 z}{\partial u^2} - 2\frac{\partial^2 z}{\partial u\,\partial v} + \frac{\partial^2 z}{\partial v^2} = x\frac{\partial z}{\partial x} + y\frac{\partial z}{\partial y} + x^2\frac{\partial^2 z}{\partial x^2} - 2xy\frac{\partial^2 z}{\partial x\,\partial y} + y^2\frac{\partial^2 z}{\partial y^2}.$$

**8** If $x = u + v$ and $y = uv$, prove

$$\frac{\partial^2 z}{\partial u^2} - 2\frac{\partial^2 z}{\partial u\,\partial v} + \frac{\partial^2 z}{\partial v^2} = (x^2 - 4y)\frac{\partial^2 z}{\partial y^2} - 2\frac{\partial z}{\partial y}.$$

9   Obtain the solution of $\partial^2 z/\partial x\,\partial y = \sin x \sin y$ for which $\partial z/\partial y = -2\sin y$ when $x = 0$, and $z = 0$ when $y$ is an odd multiple of $\frac{1}{2}\pi$.

10   Find the general solution of $\partial^2 z/\partial x^2 - \partial^2 z/\partial y^2 = \sin(x+y)\cos(x-y)$. [Use 9.44 (3).]

## 9.5   Differentials

### 9.51   Definition

Suppose that $u(x, y)$ is a differentiable function of the independent variables $x$, $y$. Then the approximation (iii) in 9.33 and the procedure in 3.92 (1) suggest that we define $du$, *the differential of u*, by the exact relation

$$du = \frac{\partial u}{\partial x}\,\delta x + \frac{\partial u}{\partial y}\,\delta y. \qquad \text{(viii)}$$

Thus $du$ is that part of the total variation $\delta u$ which is linear in $\delta x$ and in $\delta y$ (the *'principal part'* of $\delta u$). It is defined whenever both $\partial u/\partial x$, $\partial u/\partial y$ exist.

This definition of the differential of the dependent variable $u$ also defines the differentials $dx$, $dy$ of the independent variables $x$, $y$. For $dx$ *is du* when $u$ is the function $x$, in which case $\partial u/\partial x = 1$ and $\partial u/\partial y = 0$, and (viii) gives

$$dx = 1\,\delta x + 0\,\delta y = \delta x.$$

Hence $dx = \delta x$: the differential of $x$ is identical with the arbitrary increment $\delta x$. Similarly (viii) shows that $dy = \delta y$. Consequently (viii) can be written

$$du = \frac{\partial u}{\partial x}\,dx + \frac{\partial u}{\partial y}\,dy. \qquad \text{(ix)}$$

*Remarks*

($\alpha$) Although $dx = \delta x$ and $dy = \delta y$, in general $du \neq \delta u$ since by (ii) of 9.32 and (viii) we have
$$\delta u = du + \epsilon_1\,\delta x + \epsilon_2\,\delta y,$$

and in general $\epsilon_1$, $\epsilon_2$ are not both zero.

($\beta$) We shall see (9.53, Remark) that formula (ix) is more general than (viii).

($\gamma$) The partial derivatives $\partial u/\partial x$, $\partial u/\partial y$ appear in (ix) as the coefficients of the differentials $dx$, $dy$. For this reason these partial derivatives were (and sometimes still are) called *partial differential coefficients*.

($\delta$) To *differentiate* a function $u$ is to write down the expression (ix) for $du$. Contrast this with the process of *deriving u wo x* (or wo $y$), in which we merely write down the expression for $\partial u/\partial x$ (or $\partial u/\partial y$).

### 9.52   Principle of equating differential coefficients

*If*             $A\,dx + B\,dy = C\,dx + D\,dy,$

*where $A$, $B$, $C$, $D$ are functions of the* INDEPENDENT *variables $x$, $y$ (or are possibly constants), then $A = C$ and $B = D$.*

*Proof.* Since $dx$, $dy$ are arbitrary, being identical with the increments $\delta x$, $\delta y$ respectively, we may take $dy = 0$, $dx \neq 0$ in the given relation and obtain $A\,dx = C\,dx$, whence $A = C$. Similarly, taking $dx = 0$ and $dy \neq 0$ shows $B = D$.

It is essential here for $x$ and $y$ to be the independent variables, otherwise $dx$, $dy$ would *not* be arbitrary.

## 9.53 Invariance of the expression for the differential

(1) If we interpret the derivatives $du/dt$, $dx/dt$, $dy/dt$ in formula (iv) of 9.41 as quotients of differentials, we may multiply both sides of (iv) by $dt$ and obtain (ix). Now (ix) was proved on the assumption that in $u(x, y)$, $x$ and $y$ are the *independent* variables; whereas under the circumstances in which (iv) holds both $x$ and $y$ are functions of a variable $t$. It thus appears that (ix) holds more generally when $x$, $y$ are themselves functions of an independent variable $t$.

(2) Next, suppose as in 9.43 that $u = u(x, y)$ has been expressed as a function $u = f(\xi, \eta)$ of the independent variables $\xi, \eta$ by means of the substitutions $x = x(\xi, \eta)$, $y = y(\xi, \eta)$. By the definition of 'differential' applied to the function $u = f(\xi, \eta)$,

$$du = \frac{\partial u}{\partial \xi} d\xi + \frac{\partial u}{\partial \eta} d\eta$$

$$= \left( \frac{\partial u}{\partial x} \frac{\partial x}{\partial \xi} + \frac{\partial u}{\partial y} \frac{\partial y}{\partial \xi} \right) d\xi + \left( \frac{\partial u}{\partial x} \frac{\partial x}{\partial \eta} + \frac{\partial u}{\partial y} \frac{\partial y}{\partial \eta} \right) d\eta \quad \text{by (vii),}$$

$$= \frac{\partial u}{\partial x} \left( \frac{\partial x}{\partial \xi} d\xi + \frac{\partial x}{\partial \eta} d\eta \right) + \frac{\partial u}{\partial y} \left( \frac{\partial y}{\partial \xi} d\xi + \frac{\partial y}{\partial \eta} d\eta \right) \quad \text{on rearranging,}$$

$$= \frac{\partial u}{\partial x} dx + \frac{\partial u}{\partial y} dy$$

since by the definition of 'differential' applied to the functions $x = x(\xi, \eta)$, $y = y(\xi, \eta)$,

$$dx = \frac{\partial x}{\partial \xi} d\xi + \frac{\partial x}{\partial \eta} d\eta \quad \text{and} \quad dy = \frac{\partial y}{\partial \xi} d\xi + \frac{\partial y}{\partial \eta} d\eta.$$

Thus *du is still given in terms of dx and dy by the formula* (ix), JUST AS IF $(x, y)$ WERE THE INDEPENDENT VARIABLES.

*Remark.* In contrast to this result, formula (viii) is false when $x$, $y$ are not the independent variables. For, since we are assuming $x(\xi, \eta)$ and $y(\xi, \eta)$ to be differentiable functions of $(\xi, \eta)$ as in 9.43,

$$\delta x = \frac{\partial x}{\partial \xi} \delta\xi + \frac{\partial x}{\partial \eta} \delta\eta + \epsilon_1 \delta\xi + \epsilon_2 \delta\eta$$

$$= dx + \epsilon_1 \delta\xi + \epsilon_2 \delta\eta$$

by definition of $dx$; and similarly

$$\delta y = dy + \epsilon_3 \delta\xi + \epsilon_4 \delta\eta.$$

The right-hand side of (viii) is therefore

$$\frac{\partial u}{\partial x} (dx + \epsilon_1 \delta\xi + \epsilon_2 \delta\eta) + \frac{\partial u}{\partial y} (dy + \epsilon_3 \delta\xi + \epsilon_4 \delta\eta),$$

which in general is not equal to

$$\frac{\partial u}{\partial x} dx + \frac{\partial u}{\partial y} dy, \quad \text{i.e.} \quad du,$$

the left-hand side. Thus (ix) is more general than (viii) because it holds whether or not the variables are independent.

(3) The above invariance property holds for the differential $du$ of any differentiable function $u$ of $n$ variables $x, y, z, \ldots$ which are themselves differentiable functions of $m$ *independent* variables $\xi, \eta, \zeta, \ldots$. The proof would proceed similarly from the corresponding extensions of formulae (vii) and of the definition of 'differential'. The technical convenience of differentials depends on this property; see 9.6.

## 9.6   Further implicit functions

### 9.61   Differentiation of equations

Let $f(r, s, t)$ be a differentiable function of the independent variables $r, s, t$. Suppose that functions $u = u(x, y)$, $v = v(x, y)$, $w = w(x, y)$ of independent variables $x, y$ are such that
$$f(u, v, w) = 0$$

for all $x$ and $y$ for which the functions are defined. We then say that $u, v, w$ satisfy a *differentiable equation* $f(u, v, w) = 0$. For example, if
$$f(r, s, t) = r^2 + s^2 + t^2 - 1,$$
then          $u = \sin x \cos y, \quad v = \sin x \sin y, \quad w = \cos x$

satisfy $f(u, v, w) = 0$.

On substituting for $u, v, w$ in $f$, we obtain a function of the *independent* variables $(x, y)$ which is zero for all $x, y$ concerned. Hence
$$\frac{\partial f}{\partial x} = 0 \quad \text{and} \quad \frac{\partial f}{\partial y} = 0,$$

and so        $df = \dfrac{\partial f}{\partial x} dx + \dfrac{\partial f}{\partial y} dy = 0\, dx + 0\, dy = 0.$

By the invariance property we always have
$$df = \frac{\partial f}{\partial u} du + \frac{\partial f}{\partial v} dv + \frac{\partial f}{\partial w} dw.$$

Hence        $\dfrac{\partial f}{\partial u} du + \dfrac{\partial f}{\partial v} dv + \dfrac{\partial f}{\partial w} dw = 0.$

Therefore, *given a differentiable equation involving differentiable functions, we may differentiate both sides* AS IF *the functions in it were independent variables.*

### 9.62   Derivatives of functions defined implicitly

*In this section we assume that all functions involved are differentiable.* We consider general functions rather than specific examples because we thereby exhibit more clearly the process used, unencumbered by details of calculation. The following cases are typical.

(a) In 9.42 we regarded the equation $f(x, y) = 0$ as determining $y$ as an implicit function of $x$, say $y = y(x)$; then $f(x, y(x)) = 0$ for all $x$ for which the functions are defined. The same equation $f(x, y) = 0$ could also determine $x$ as an implicit function of $y$.

(b) Similarly, $f(x, y, u) = 0$ may define $u$ as a function of $(x, y)$, say $u = u(x, y)$; and then $f(x, y, u(x, y)) = 0$ for all $x, y$ concerned. The given equation might also define $y$ as a function of $(x, u)$, or $x$ as a function of $(y, u)$.

(c) The pair $f(x, u, v) = 0$, $g(x, u, v) = 0$ may be thought of as simultaneous equations capable of being solved for $u$ and $v$ in terms of $x$,† i.e. as defining implicit functions $u = u(x)$, $v = v(x)$ for which

$$f(x, u(x), v(x)) = 0 \quad \text{and} \quad g(x, u(x), v(x)) = 0$$

for all $x$ concerned.

(d) Likewise the pair of equations $f(x, y, u, v) = 0$, $g(x, y, u, v) = 0$ could define $u = u(x, y)$, $v = v(x, y)$.†

*Assuming throughout that these implicit functions actually exist and are differentiable,‡ we employ 9.61 and 9.52 to obtain expressions for their derivatives. The results need not be memorised because they can be found in any particular case by the same methods. The first step is always to differentiate the given equation(s).*

(a) From $f(x, y) = 0$ we have

$$\frac{\partial f}{\partial x} dx + \frac{\partial f}{\partial y} dy = 0,$$

so that

$$\frac{dy}{dx} = -\frac{\partial f}{\partial x} \bigg/ \frac{\partial f}{\partial y}$$

as in (vi) of 9.42.

(b) From $f(x, y, u) = 0$,

$$\frac{\partial f}{\partial x} dx + \frac{\partial f}{\partial y} dy + \frac{\partial f}{\partial u} du = 0.$$

Solving for $du$,

$$du = -\frac{f_x}{f_u} dx - \frac{f_y}{f_u} dy.$$

But

$$du = \frac{\partial u}{\partial x} dx + \frac{\partial u}{\partial y} dy,$$

and since $(x, y)$ are the independent variables, 9.52 gives (on equating coefficients of $dx$ and of $dy$)

$$\frac{\partial u}{\partial x} = -\frac{f_x}{f_u} \quad \text{and} \quad \frac{\partial u}{\partial y} = -\frac{f_y}{f_u}.$$

(c) We have

$$\frac{\partial f}{\partial x} dx + \frac{\partial f}{\partial u} du + \frac{\partial f}{\partial v} dv = 0 \quad \text{and} \quad \frac{\partial g}{\partial x} dx + \frac{\partial g}{\partial u} du + \frac{\partial g}{\partial v} dv = 0.$$

Solving for $du$, $dv$ in terms of $dx$ gives expressions for $du/dx$, $dv/dx$.

(d) Differentiating,

$$\frac{\partial f}{\partial x} dx + \frac{\partial f}{\partial y} dy + \frac{\partial f}{\partial u} du + \frac{\partial f}{\partial v} dv = 0,$$

with a similar equation for $g$. We can solve these for $du$, $dv$ in terms of $dx$ and $dy$ in the form

$$du = P\,dx + Q\,dy, \quad dv = R\,dx + S\,dy.$$

Since $(x, y)$ are the independent variables, we deduce

$$\frac{\partial u}{\partial x} = P, \quad \frac{\partial u}{\partial y} = Q, \quad \frac{\partial v}{\partial x} = R, \quad \frac{\partial v}{\partial y} = S.$$

† Other choices are possible, of course.
‡ General conditions sufficient for this can be formulated.

*Remarks*

($\alpha$) The equations in differentials are always *linear*, and therefore easily solvable, whereas the given equations will not in general be linear in the corresponding variables.

($\beta$) We have already remarked that the choice of independent variable(s) can be made in more than one way. For example, in ($d$) we can select two independent variables from among $x$, $y$, $u$, $v$ in $^4C_2 = 6$ ways, and for each choice there are 4 equations giving the partial derivatives of the remaining variables wo these, making a total of 24 such equations. The *two* equations obtained by differentiating $f = 0$ and $g = 0$ are symmetrical in the differentials, and convey all this information. Also see 9.43, Remark ($\beta$).

($\gamma$) The reader who nevertheless wishes to avoid differentials can obtain the results described above by using the obvious extensions of formulae (v) or (vii) to functions of more than two variables, in the manner that (vi) was obtained in 9.42. Thus in ($b$), by deriving $f(x, y, u(x, y)) = 0$ wo $x$ (using the extension of (vii) to a function of *three* functions of $(x, y)$) we have

$$\frac{\partial f}{\partial x} + \frac{\partial f}{\partial u} \frac{\partial u}{\partial x} = 0, \quad \text{so that} \quad \frac{\partial u}{\partial x} = -\frac{f_x}{f_u}.$$

## Examples

(i) *If* $u = \sin x \, \text{sh} \, y$ *and* $\log(x+y) + 2y - 3\log z = 4$, *calculate* $\partial u / \partial x$ *when* $x$ *and* $z$ *are the independent variables.*

Differentiating both equations,

$$du = \cos x \, \text{sh} \, y \, dx + \sin x \, \text{ch} \, y \, dy$$

and

$$\frac{1}{x+y}(dx + dy) + 2 \, dy - \frac{3}{z} \, dz = 0.$$

From the latter we find

$$dy\left(\frac{1}{x+y} + 2\right) = \frac{3}{z} \, dz - \frac{1}{x+y} \, dx,$$

and on using this to eliminate $dy$ from the former,

$$du = \cos x \, \text{sh} \, y \, dx + \sin x \, \text{ch} \, y \left(\frac{3}{z} \, dz - \frac{1}{x+y} \, dx\right) \bigg/ \left(\frac{1}{x+y} + 2\right).$$

Since $(x, z)$ are the independent variables, $\partial u / \partial x$ will be the coefficient of $dx$ in the expression for $du$, viz.

$$\cos x \, \text{sh} \, y - \frac{\sin x \, \text{ch} \, y}{1 + 2x + 2y}.$$

(ii) *If* $f(x, y, z) = 0$, *prove*

$$\frac{\partial x}{\partial y} = -\frac{\partial z}{\partial y}\bigg/\frac{\partial z}{\partial x} \quad \text{and} \quad \frac{\partial x}{\partial z} = 1\bigg/\frac{\partial z}{\partial x}.$$

*First method.* By differentiating the given equation,

$$\frac{\partial f}{\partial x} \, dx + \frac{\partial f}{\partial y} \, dy + \frac{\partial f}{\partial z} \, dz = 0. \tag{A}$$

The expressions $\partial z/\partial x$, $\partial z/\partial y$ in the required results indicate that $z$ is an implicit function of the independent variables $x$, $y$. To find them, solve (A) for $dz$:

$$dz = -\frac{f_x}{f_z}\,dx - \frac{f_y}{f_z}\,dy.$$

Hence

$$\frac{\partial z}{\partial x} = -\frac{f_x}{f_z} \quad \text{and} \quad \frac{\partial z}{\partial y} = -\frac{f_y}{f_z}.$$

Similarly, the appearance of $\partial x/\partial y$, $\partial x/\partial z$ indicates that we regard the same equation $f(x, y, z) = 0$ as defining $x$ as a function of $(y, z)$. Solving (A) for $dx$,

$$dx = -\frac{f_y}{f_x}\,dy - \frac{f_z}{f_x}\,dz,$$

from which

$$\frac{\partial x}{\partial y} = -\frac{f_y}{f_x} \quad \text{and} \quad \frac{\partial x}{\partial z} = -\frac{f_z}{f_x}.$$

The two results now follow.

The preceding is hardly more than a verification of results already stated. A more subtle use of differentials leads directly to the results themselves.

*Second method.* Regarding $z$ as an implicit function of $(x, y)$,

$$dz = \frac{\partial z}{\partial x}\,dx + \frac{\partial z}{\partial y}\,dy.$$

Next, regarding $x$ as an implicit function of $(y, z)$,

$$dx = \frac{\partial x}{\partial y}\,dy + \frac{\partial x}{\partial z}\,dz.$$

Eliminate $dx$ from these by substituting the second in the first:

$$dz = \left(\frac{\partial z}{\partial x}\frac{\partial x}{\partial y} + \frac{\partial z}{\partial y}\right)dy + \frac{\partial z}{\partial x}\frac{\partial x}{\partial z}\,dz.$$

Equating coefficients of $dy$, $dz$ gives the respective results. Observe that the function $f$ does not enter the calculations. For Ex. 9 (e), no. 16 the first method would be even more prolix.

## Exercise 9(e)*

*Assume that all the functions concerned are differentiable.*

**1**   $f(x, y, z) = $ constant and $xyz = $ constant; prove that

$$\frac{dy}{dx} = -\frac{y(xf_x - zf_z)}{x(yf_y - zf_z)}.$$

**2**   If $f(x, y, z) = $ constant and $x^2 + y^2 + z^2 = $ constant, find $dy/dx$.

**3**   $u = f(x, y)$ and $\phi(x, y) = 0$; prove that $du/dx = (f_x\phi_y - \phi_x f_y)/\phi_y$.

**4**   If $y = f(x, z)$ and $z = g(x, y)$, find $dy/dx$.

**5**   A curve in the $(x, y)$-plane is given by the equations $f(x, y, \alpha) = 0$ and $f_\alpha(x, y, \alpha) = 0$. Prove that its gradient is $-f_x/f_y$.

**6**   Elimination of $t$ from the equations $y = f(x, t)$, $z = g(x, t)$ leads to the relation $z = \phi(x, y)$. Prove that $\phi_x = (f_t\,g_x - f_x\,g_t)/f_t$ and $\phi_y = g_t/f_t$.

**7**   If $x = e^{2\xi}\sin\eta$ and $y = e^{2\xi}\cos\eta$, prove that $(\partial x/\partial\xi)_y = 2e^{2\xi}\operatorname{cosec}\eta$. [Notation of 9.43, Remark $(\beta)$. Cf. the direct method shown there.]

**8** If $u = xy$ and $ax + by + cz = 1$, where $a$, $b$, $c$ are constants, find $\partial u/\partial x$ under all possible meanings.

**9** If $u = x^2 + y^2 + z^2$ and $z = xyt$, find $\partial u/\partial x$ under all possible meanings.

**10** If $u = x^3 + y^3 + z^3$ and $x^2 + y^2 + z^2 = $ constant, find $(\partial u/\partial x)_y$.

**11** If $u = x^3 y^2 z$ where $x^2 + y^2 + z^2 = 3xyz$, calculate $(\partial u/\partial x)_y$ and $(\partial u/\partial x)_z$ at $(1, 1, 1)$.

**12** If $f(x, y, z) = 0$, explain in detail the meaning of the symbols $\partial y/\partial x$, $\partial x/\partial y$. Prove that

$$\text{(i)} \quad \frac{\partial x}{\partial y} \frac{\partial y}{\partial x} = 1; \quad \text{(ii)} \quad \frac{\partial x}{\partial y} \frac{\partial y}{\partial z} = -\frac{\partial x}{\partial z}.$$

**13** Given that $p$, $v$, $\theta$ are connected by a single equation, prove that

$$\left(\frac{\partial p}{\partial \theta}\right)_v \left(\frac{\partial \theta}{\partial v}\right)_p \left(\frac{\partial v}{\partial p}\right)_\theta = -1.$$

**14** The variables $p$, $v$, $\theta$, $\phi$ are related by $pv = R\theta$, $\phi = C_v \log p + C_p \log v$, where $C_p$, $C_v$, $R$ are constants and $C_p - C_v = R$. Prove that

$$\frac{\partial \theta}{\partial p} \frac{\partial \phi}{\partial v} - \frac{\partial \theta}{\partial v} \frac{\partial \phi}{\partial p} = 1 = \frac{\partial p}{\partial \theta} \frac{\partial v}{\partial \phi} - \frac{\partial p}{\partial \phi} \frac{\partial v}{\partial \theta}.$$

[The context makes clear that in the left-hand side $\theta$, $\phi$ are functions of $(p, v)$, while in the right-hand side $p$, $v$ are functions of $(\theta, \phi)$.]

**15** The equation $dE = \theta \, d\phi - p \, dv$ arises in Thermodynamics. If $E$ is assumed to be a differentiable function of $(\phi, v)$, show that

$$\theta = \left(\frac{\partial E}{\partial \phi}\right)_v \quad \text{and} \quad p = -\left(\frac{\partial E}{\partial v}\right)_\phi,$$

and deduce that the four variables $p$, $v$, $\theta$, $\phi$ are related by two equations. Prove that

$$\text{(i)} \quad \left(\frac{\partial \theta}{\partial v}\right)_\phi = -\left(\frac{\partial p}{\partial \phi}\right)_v; \quad \text{(ii)} \quad \left(\frac{\partial p}{\partial \theta}\right)_v = \left(\frac{\partial \phi}{\partial v}\right)_\theta.$$

[For (ii) put $\psi = \theta\phi - E$, show $d\psi = \phi \, d\theta + p \, dv$, and hence get

$$\phi = \left(\frac{\partial \psi}{\partial \theta}\right)_v, \quad p = \left(\frac{\partial \psi}{\partial v}\right)_\theta; \quad \text{then} \quad \frac{\partial p}{\partial \theta} = \frac{\partial^2 \psi}{\partial \theta \, \partial v} = \frac{\partial \phi}{\partial v}.]$$

**16** If $x$, $y$, $u$, $v$ are connected by two relations, show that

$$\left(\frac{\partial x}{\partial u}\right)_v \left(\frac{\partial u}{\partial x}\right)_y + \left(\frac{\partial x}{\partial v}\right)_u \left(\frac{\partial v}{\partial x}\right)_y = 1 \quad \text{and} \quad \left(\frac{\partial y}{\partial u}\right)_v \left(\frac{\partial u}{\partial x}\right)_y + \left(\frac{\partial y}{\partial v}\right)_u \left(\frac{\partial v}{\partial x}\right)_y = 0.$$

[Write down $du$, $dv$ and $dx$, $dy$. Substitute the first set in the second, and then equate coefficients of $dx$ and of $dy$.]

## Miscellaneous Exercise 9(f)

**1** If $u = y \log(y + r) - r$ where $r^2 = x^2 + y^2$, prove

$$\frac{\partial^2 u}{\partial x^2} + \frac{\partial^2 u}{\partial y^2} = \frac{1}{y + r}.$$

**2** If $x = r \cos \theta$, $y = r \sin \theta$, prove

$$\frac{\partial^2 \theta}{\partial x \, \partial y} = -\frac{\cos 2\theta}{r^2}.$$

3 If $z = f(x, y) + g(t)$ where $t = xy$, prove

$$V = x\frac{\partial z}{\partial x} - y\frac{\partial z}{\partial y}$$

is independent of $g$. Calculate $V$ when $f(x, y) = xy\, e^{x-y}$.

4 If $u = x^2 - y^2$, find a function $v(x, y)$ for which

$$\frac{\partial v}{\partial x} = -\frac{\partial u}{\partial y} \quad \text{and} \quad \frac{\partial v}{\partial y} = \frac{\partial u}{\partial x}$$

for all $x$ and $y$.

5 If $z = f(x^2 + y^2)$, prove

(i) $y\dfrac{\partial z}{\partial x} = x\dfrac{\partial z}{\partial y}$; (ii) $(x^2 - y^2)\dfrac{\partial^2 z}{\partial x\,\partial y} = xy\left(\dfrac{\partial^2 z}{\partial x^2} - \dfrac{\partial^2 z}{\partial y^2}\right)$.

6 If $u = f(x^2 + y^2 + z^2)$, prove

$$\frac{\partial^2 u}{\partial x^2} + \frac{\partial^2 u}{\partial y^2} + \frac{\partial^2 u}{\partial z^2} = 4(x^2 + y^2 + z^2)f''(x^2 + y^2 + z^2) + 6f'(x^2 + y^2 + z^2).$$

7 If $u = x^n f(y/x)$, prove

$$x\frac{\partial u}{\partial x} + y\frac{\partial u}{\partial y} = nu.$$

If $v = u\, e^{ax+by}$ with $u$ as before, prove

$$x\frac{\partial v}{\partial x} + y\frac{\partial v}{\partial y} = (ax + by + n)\, v.$$

8 If $z = f(x+y)\, g(x-y)$, prove

$$z\left(\frac{\partial^2 z}{\partial x^2} - \frac{\partial^2 z}{\partial y^2}\right) = \left(\frac{\partial z}{\partial x}\right)^2 - \left(\frac{\partial z}{\partial y}\right)^2.$$

9 If $z = xf(x+y) + yg(x+y)$, prove

$$\frac{\partial^2 z}{\partial x^2} - 2\frac{\partial^2 z}{\partial x\,\partial y} + \frac{\partial^2 z}{\partial y^2} = 0.$$

10 Find the constant $n$ if $u = t^n e^{-r^2/4t}$ satisfies

$$\frac{1}{r^2}\frac{\partial}{\partial r}\left(r^2\frac{\partial u}{\partial r}\right) = \frac{\partial u}{\partial t}.$$

11 If $u = f(z)$ where $z^2 = t^2 - x^2$, prove $z\,\partial u/\partial t = tf'(z)$ and show that if $u$ satisfies

$$\frac{\partial^2 u}{\partial x^2} - u = \frac{\partial^2 u}{\partial t^2},$$

then $f(z)$ satisfies

$$\frac{d^2 f}{dz^2} + \frac{1}{z}\frac{df}{dz} + f = 0.$$

12 If $u = f(z)$ and $z^2 = x^2/4t$, calculate $\partial u/\partial t$, $\partial^2 u/\partial x^2$. If $u$ satisfies $\partial u/\partial t = \partial^2 u/\partial x^2$, prove $f(z) = A\int e^{-z^2}\,dz + B$, where $A$, $B$ are arbitrary constants.

13 Find the gradient of the curve $x^2\tan^{-1}(y/x) - y^2\tan^{-1}(x/y) = c$ at the general point $(r\cos\theta,\, r\sin\theta)$.

14 Show that the value of $d^2y/dx^2$ at the point $(a, a)$ of the curve

$$x^3 + y^3 = axy + a^3 \quad \text{is} \quad -7/a.$$

**15** (i) Find the equations of the tangent and normal at $(x_1, y_1)$ on the curve $f(x, y) = 0$.

(ii) Calculate the lengths of the subtangent and subnormal.

**16** Prove that the curves $x^2 + y^2 = e^{2x}$, $y = x \tan y$ cut orthogonally.

**17** If $u = u(x, y)$ and $x = a + ht$, $y = b + kt$, where $a, b, h, k$ are constants, prove

$$\frac{d^2u}{dt^2} = h^2 \frac{\partial^2 u}{\partial x^2} + 2hk \frac{\partial^2 u}{\partial x \partial y} + k^2 \frac{\partial^2 u}{\partial y^2}.$$

**18** If $x = r \cos\theta$, $y = r \sin\theta$, express $(\partial u/\partial x)^2 + (\partial u/\partial y)^2$ in terms of $r$, $\theta$ and partial derivatives of $u$ wo $r$, $\theta$.

**19** If $u = u(x, y)$ and $x = \xi \cos\alpha - \eta \sin\alpha$, $y = \xi \sin\alpha + \eta \cos\alpha$ (where $\alpha$ is constant), prove

(i) $\left(\dfrac{\partial u}{\partial x}\right)^2 + \left(\dfrac{\partial u}{\partial y}\right)^2 = \left(\dfrac{\partial u}{\partial \xi}\right)^2 + \left(\dfrac{\partial u}{\partial \eta}\right)^2$;

(ii) $\dfrac{\partial^2 u}{\partial x^2} + \dfrac{\partial^2 u}{\partial y^2} = \dfrac{\partial^2 u}{\partial \xi^2} + \dfrac{\partial^2 u}{\partial \eta^2}.$

**20** If $u, v$ are expressed as functions of $(x, y)$ by the formulae

$$u + \lambda v = f(x + \lambda y), \quad u - \lambda v = f(x - \lambda y),$$

where $\lambda$ is a non-zero constant, prove

$$\frac{\partial u}{\partial x} = \frac{\partial v}{\partial y}, \quad \frac{\partial u}{\partial y} = \lambda^2 \frac{\partial v}{\partial x} \quad \text{and} \quad \frac{\partial^2 u}{\partial y^2} = \lambda^2 \frac{\partial^2 u}{\partial x^2}.$$

**21** If $x = (1/u) \cos\theta$, $y = (1/u) \sin\theta$, calculate $(\partial x/\partial u)_\theta$ and $(\partial u/\partial x)_y$, and show that their product is $\cos^2\theta$.

If $V(x, y)$ is transformed into a function of $(u, \theta)$ by these relations, prove

(i) $\dfrac{\partial V}{\partial x} = -u^2 \cos\theta \dfrac{\partial V}{\partial u} - u \sin\theta \dfrac{\partial V}{\partial \theta}$;

(ii) $\left(\dfrac{\partial V}{\partial x}\right)^2 + \left(\dfrac{\partial V}{\partial y}\right)^2 = u^4 \left(\dfrac{\partial V}{\partial u}\right)^2 + u^2 \left(\dfrac{\partial V}{\partial \theta}\right)^2.$

**22** If $x = e^{-r} \sin\theta$, $y = e^{-r} \cos\theta$, change the variables from $(x, y)$ to $(r, \theta)$ in

$$y^2 \frac{\partial^2 u}{\partial x^2} - 2xy \frac{\partial^2 u}{\partial x \partial y} + x^2 \frac{\partial^2 u}{\partial y^2}.$$

**23** Writing $G(x, y) = x^n H(x, y)$, and changing the variables from $(x, y)$ to $(u, v)$ where $u = y/x$, $v = xy$, transform the equations

$$x \frac{\partial G}{\partial x} + y \frac{\partial G}{\partial y} = nG, \quad x \frac{\partial F}{\partial x} - y \frac{\partial F}{\partial y} = 0,$$

where $F = F(x, y)$. Hence prove $G = x^n \phi(y/x)$ and $F = \psi(xy)$, where $\phi(t)$ and $\psi(t)$ are arbitrary differentiable functions of $t$.

**24** If $u, v$ are functions of $(x, y)$ which are connected by a relation $\phi(u, v) = 0$ for all $x, y$, prove

$$\frac{\partial u}{\partial x} \frac{\partial v}{\partial y} - \frac{\partial v}{\partial x} \frac{\partial u}{\partial y} = 0.$$

*25 (i) If $u, v, w$ are functions of $(x, y, z)$ which satisfy the relation $\phi(u, v, w) = 0$ for all $x, y, z$, prove that the determinant

$$\begin{vmatrix} \dfrac{\partial u}{\partial x} & \dfrac{\partial u}{\partial y} & \dfrac{\partial u}{\partial z} \\[2mm] \dfrac{\partial v}{\partial x} & \dfrac{\partial v}{\partial y} & \dfrac{\partial v}{\partial z} \\[2mm] \dfrac{\partial w}{\partial x} & \dfrac{\partial w}{\partial y} & \dfrac{\partial w}{\partial z} \end{vmatrix}$$

is zero for all $x, y, z$.

(ii) If $u = ax^2 + 2hxy + by^2 + 2gxz + 2fyz + cz^2$ has linear factors

$$v = lx + my + nz, \quad w = l'x + m'y + n'z,$$

prove

$$\begin{vmatrix} a & h & g \\ h & b & f \\ g & f & c \end{vmatrix} = 0.$$

[The functional relation is $u - vw = 0$.]

# ADDENDA TO VOLUME I

## Notes to 3.64

[1] We employ the main result and Corollary 1 in 3.83 ahead, which is independent of the present section. This rearrangement is necessary because the proof given in the first edition of this book contained a subtle fallacy.

It was asserted that ' since by hypothesis (ii) we have $f'(x_1) < 0$ for any $x_1$ less than and sufficiently near $a$, hence by 3.61 $f(x)$ is decreasing at $x_1$, *that is* $f(x_1) > f(a)$'. The correct inference is that $f(x_1) > f(x)$ *for all $x$ greater than and sufficiently near $x_1$,* i.e. throughout some interval $x_1 < x \leqslant x_1 + \eta(x_1)$; but $a$ may not be ' sufficiently near', since possibly $x_1 + \eta(x_1) < a$ for *all* $x_1 < a$, and so nothing follows about $f(a)$. Similarly, '$f'(x_2) < 0$ for any $x_2$ greater than and sufficiently near $a$' does not imply $f(x_2) > f(a)$.

[2] Hypothesis (i) in the text can be replaced by the following weaker condition:

(i)' *$f'(x)$ exists in the neighbourhood of $x = a$, and $f(x)$ is continuous at* $x = a$.

The proof in the text holds almost without change. In this form the test applies to the function $f(x) = x^{\frac{2}{3}}$ at $x = 0$: cf. 3.63, Remark $(\beta)$. (If it is also known that $f'(a)$ *exists*, then by 3.63 we must have $f'(a) = 0$.)

## Notes to 4.84

[3] If $m = -1$ (so that $n \neq 1$),

$$u_{-1, n} = \int \frac{\cos^n x}{\sin x} dx = \int \frac{(1 - \sin^2 x) \cos^{n-2} x}{\sin x} dx$$

$$= \int \frac{\cos^{n-2} x}{\sin x} dx - \int \sin x \cos^{n-2} x \, dx$$

$$= u_{-1, n-2} + \frac{1}{n-1} \cos^{n-1} x.$$

Thus (ii) holds for $m = -1$ also. Hence (vi) holds for $m = -1$, provided that $n \neq -1$ as before.

[4] Similarly,

$$u_{m, -1} = \int \frac{\sin^m x}{\cos x} dx = \int \frac{\sin^{m-2} x}{\cos x} dx - \int \cos x \sin^{m-2} x \, dx$$

$$= u_{m-2, -1} - \frac{1}{m-1} \sin^{m-1} x \quad \text{since} \quad m \neq 1.$$

Thus (iv) holds even when $n = -1$, and hence so does (v) provided that $m \neq -1$.

# ANSWERS TO VOLUME I

## Exercise 1(*a*), p. 10

1 (i) $-3 < x < \frac{3}{4}$; (ii) $2 < x < 2\frac{1}{2}$, $x > 3$.

13 $w = x = y = \frac{1}{6}$, $z = \frac{1}{2}$. 14 (i) $2^5/(3^7\sqrt{3})$; (ii) $(4/3^6)\sqrt[3]{\frac{2}{3}}$.

## Exercise 1(*b*), p. 14

1 $x < \frac{1}{2}$, $x > 3$. 2 $-1 < x < \frac{2}{3}$. 3 All $x$.

4 $x \neq -\frac{1}{2}$. 5 $x < -1$, $x > 1$. 6 $1 < x < 2$, $x > 3$.

7 $1 < x < 3$, $x < 1$. 8 $-1 < x < 1$, $x > 3$.

9 $x < -1$, $1 < x < 2$, $x > 4$. 10 $4 < x < 3+\sqrt{3}$, $3-\sqrt{3} < x < 2$.

11 $4\frac{1}{3}$, $-\frac{1}{8}$. 14 No.

15 $\lambda = 2$, $-\frac{3}{2}$; $p = -2$, $q = 1$, $A = 3$, $B = 2$, $C = 2$, $D = -1$.

## Exercise 1(*c*), p. 19

16 Graph like fig. 9, p. 17. 17 Graph like fig. 10, p. 18.

18 Graph like fig. 39, p. 79. 19 $y \geqslant -\frac{1}{8}$; minimum at $(-3, -\frac{1}{8})$.

## Exercise 1(*d*), p. 29

1 (i) $O$, $2\pi$; (ii) $E$, $\pi$; (iii) $E$; (iv) $O$; (v) neither; (vi) $E$;

(vii) $E$, $2\pi$; (viii) $E$; (ix) $E$; (x) $O$; (xi) neither; (xii) $O$.

5 (i) $y^2 - 2xy + x^2 - x = 0$; (ii) $y^4 - 2xy^2 + x^2 - x = 0$;

(iii) $xy^2 - 2xy + x - 4y = 0$; (iv) $3xy^2 + 6xy - y^2 + 2y - 1 = 0$.

6 (i) 5; (iii) 1; (iv) $-\frac{3}{2}$; (v) $-2$; (vi) 0.

8 Homogeneous of degree (i) $m+n$, (ii) $m-n$; (iii) not homogeneous unless $m = n$, and then of degree $m$.

11 (ii), (iii) $\sin(1/x)$ oscillates between $\pm 1$ increasingly often as $x$ approaches 0.

## Exercise 1(*e*), p. 34

1 (i) Raise $Ox$ $c$ units; (ii) alter $y$-scale in ratio $c:1$; (iii) move $Oy$ back $c$ units.

2 Not $(-a, 0)$. 3 $x = at/(1+t^3)$, $y = at^2/(1+t^3)$.

5 $r^2 \cos 2\theta = a^2$. 6 $r^2 \sin 2\theta = 2c^2$.

7 $r^2 = \frac{1}{4}a^2 \sin 4\theta$. 8 $(x^2 + y^2)^3 = 4a^2 x^2 y^2$.

17 (i) Produce $r$ by an amount $c$; (ii) produce $r$ in the ratio $c:1$; (iii) rotate initial line clockwise through angle $c$.

18 Cardioid.

## Miscellaneous Exercise 1(*f*), p. 35

5  $d = \frac{1}{3}(a+b+c)$.

6  By Cauchy, $(\Sigma abc)^2 \leqslant (\Sigma a^2 b^2)(\Sigma c^2)$; and $\Sigma a^2 b^2 < (\Sigma a^2)(\Sigma b^2)$ unless all $a$'s or all $b$'s are zero.

7  $a > 0$, $b^2 - 4ac < 0$; *or* $a = b = 0$, $c > 0$; $\lambda = -4$, $\mu = 8$, $\rho = 16$. Expression $= (x^2 - 4x + 12)(x-2)^2$.

8  All values.                9  Outside $-1$, $3$.         10  Between $2$, $5$.

12  $-\sqrt{2} < x < 1$, $x < -2$, $x > \sqrt{2}$.        13  $-\frac{7}{6} < x < -1$, $x < -2$.

16  $(bb' - 2ac' - 2a'c)^2 = (b^2 - 4ac)(b'^2 - 4a'c')$, which can be reduced to $(ca' - c'a)^2 = (ab' - a'b)(bc' - b'c)$.

19  $X = ax + hy + g$, $Y = y - (gh - af)/(ab - h^2)$; $a \neq 0$, $ab = h^2$, $af = gh$, $ac = g^2$.

21  $x = a(2\cos\theta + \cos 2\theta)$, $y = a(2\sin\theta - \sin 2\theta)$.

## Exercise 2(*a*), p. 50

1  $2a$.           2  (i) $-1$; (ii) $0$; (iii) $0$; (iv) $-3$.         3  $5a^4$.

4  $1$.           5  $3$.           6  $\frac{2}{5}$.           7  $\frac{1}{2}$.

8  $1$.           9  $1$.           10  $0$.           11  $\frac{1}{2}$.

12  (i) $0$; (ii) osc. inf. $\pm\infty$.           13  (i) $0$; (ii) $1$.

14  (i) osc. fin. $\pm 1$; (ii) $1$.           15  (i) $0$; (ii) osc. inf. $\pm\infty$.

16  $2(p^2 - q^2)/r^2$.    17  $\sin x$.           18  $-\sin x$.

## Exercise 2(*b*), p. 58

1  $\to\infty$.        2  $\to 1$.        3  osc. fin. $0$, $2$.    4  osc. inf. $0$, $\infty$.

5  $\to\infty$.        6  osc. inf. $\pm\infty$.    7  $\to\infty$.        8  osc. fin. $\pm 1$.

9  $\to 0$.        10  $\to 0$.        11  $\to 0$.        12  osc. fin. $\pm 1$.

13  $\frac{1}{2}$.        14  $1$.        15  $\pi$.        16  $\frac{3}{8}\theta^2$.

17  $1$.        18  $a$ if $a \geqslant b$, $b$ if $b > a$.        19  $0$.

20  If $a \geqslant 1$, $\to\infty$; if $a \leqslant -1$, osc. inf. $\pm\infty$.

22  $g(n)$ has limit $\geqslant l$.

## Miscellaneous Exercise 2(*c*), p. 58

2  $1$.           3  $\lim\limits_{n\to\infty} u_n = -\sqrt{a}$.    5  $3 + 4^{2-n}$.        8  $u_n \to 0$ for all $a$.

9  If $|a| < 1$, $u_n \to 0$; if $|a| > 1$, $|u_n| \to\infty$.

10  If $|a| < 1$, $u_n \to 0$; if $|a| > 1$, $|u_n| \to\infty$.

11  (i) $u \to -m$; (ii) $u \to 0$, $-\infty$, or $+\infty$, depending along which part of the curve $O$ is approached.

12  $y = 1 - \lim\limits_{m\to\infty} [\lim\limits_{n\to\infty} \{\cos(m!\,\pi x)\}^{2n}]$.

## Exercise 3(a), p. 68

1 (i) $-1/x^2$; (ii) $1/(2\sqrt{x})$; (iii) $\frac{3}{2}\sqrt{x}$; (iv) $\sec^2 x$.

2 (i) $30x^2(x^3-3)^9$; (ii) $x/\sqrt{(a^2+x^2)}$; (iii) $8x/(1-x^2)^5$.

3 (i) $n\sin^{n-1}x\cos x$; (ii) $m\cos mx$; (iii) $mn\sin^{n-1}mx\cos mx$;
 (iv) $n\sin^{n-1}x\sec^{n+1}x$.

5 (i) $1/(2y+5)$; (ii) $\pm(4x+17)^{-\frac{1}{2}}$.     8 $f'(x)\,\phi'(t)\,F'(u)$.

17 $5/\sqrt{(1-25x^2)}$.     18 $2/(1+x^2)$.     19 $\sin^{-1}x$.

20 (i) $1/\sqrt{(a^2-x^2)}$; (ii) $-1/\sqrt{(a^2-x^2)}$.

21 (i) $-1/\sqrt{(a^2-x^2)}$; (ii) $1/\sqrt{(a^2-x^2)}$.

22 $a/(a^2+x^2)$.     27 $b^2x/a^2y$.     28 $-(by/ax)^{\frac{1}{2}}$.

29 $(ay-x^2)/(y^2-ax)$.     30 $xy^2(2y+3xy')$.     31 $-1/t^2$.

32 $-(b/a)\cot\phi$.     33 $t(2-t^3)/(1-2t^3)$.

34 $x-ty+at^2=0$; $t=\cot\psi$; $tx+y=2at+at^3$.

35 $x\sec^2\theta+y\operatorname{cosec}^2\theta=a$.

## Exercise 3(b), p. 73

1 $2(x^3-3x^2-12x+4)/(x^2+4)^3$.     2 $6x(x^2+2x+4)/(x+1)^2(x-2)^2$.

3 $3x(1-x^2)^{-\frac{5}{2}}$.     4 $\sin x(9\sin^2 x-7)$.

5 $\{(2-x^2)\sin x-2x\cos x\}/x^3$.

6 $-m(1-x^2)^{-\frac{3}{2}}\{x\sin(m\sin^{-1}x)+m(1-x^2)^{\frac{1}{2}}\cos(m\sin^{-1}x)\}$.

7 $-m^2y$.     10 $y_0'=1$, $y_0''=0$, $y_0'''=2$.

11 (i) $m(m-1)\dots(m-n+1)x^{m-n}$ if $m\geqslant n$, 0 if $m<n$;
 (ii) $m(m-1)\dots(m-n+1)x^{m-n}$; (iii) $(-1)^n n!x^{-n-1}$.

12 $\sin(x+\frac{1}{2}n\pi)$.     13 $\cos(x+\frac{1}{2}n\pi)$.

14 $m(m-1)\dots(m-n+1)a^n(ax+b)^{m-n}$, unless $m$ is a positive integer less than $n$, when the result is 0.

15 $a^n\sin(ax+b+\frac{1}{2}n\pi)$.     17 $vu^{(4)}+4v'u'''+6v''u''+4v'''u'+v^{(4)}u$.

18 $2a/y$, $-4a^2/y^3$.     19 $(ay-x^2)/(y^2-ax)$, $-2a^3xy/(y^2-ax)^3$.

20 $x(2ay^2-x^3)/y(y^3-2ax^2)$, $2ax^4y(25a^2-8xy)/(y^3-2ax^2)^3$.

21 $1/t$, $-1/(2at^3)$.     22 $-(b/a)\cot\phi$, $-(b/a^2)\operatorname{cosec}^3\phi$.

23 $(b/a)\tan\theta$, $(b\sec^3\theta)/(a^2\theta)$.     24 $\dot y/\dot x$, $(\dot x\ddot y-\ddot x\dot y)/\dot x^3$.

25 $n\cos n\theta\sec\theta$, $n(\sin\theta\cos n\theta-n\cos\theta\sin n\theta)\sec^3\theta$.     26 $(\alpha^2-\beta\tau)/\tau^6$.

## Exercise 3(c), p. 84

1 $x=1$, min.; $x=-2$, max.; $x=-\frac{1}{2}$, inflex.     2 $x=0$, inflex.

3 $x=-1$, min.; $x=-2$, inflex.     4 $x=-\frac{1}{2}$, max.; $x=1\frac{3}{5}$, min.

5 $x=\alpha+2n\pi$, max.; $x=\alpha+(2n-1)\pi$, min., where $n$ is an integer and $\alpha$ is given by $\cos\alpha:\sin\alpha:1=a:b:\sqrt{(a^2+b^2)}$.

6 $x = n\pi$, min.; $x = (n+\frac{1}{2})\pi$, max.; $x = \frac{1}{4}(2n+1)\pi$, inflex.

7 $x = \frac{1}{6}\pi + 2n\pi$, max.; $x = \frac{5}{6}\pi + 2n\pi$, min.; $x = 2n\pi - \frac{1}{2}\pi$, inflex.

8 $x = -5$, max.; inflexions at $x = -2$, $-2 \pm \frac{3}{2}\sqrt{2}$.

9 $x = \frac{1}{2}$, $\frac{1}{3}(6+\sqrt{3})$, min.; $x = \frac{1}{3}(6-\sqrt{3})$, max.; $x = 2$, inflex.

10 The value $-1$ of $f(x)$ is not a minimum value.

11 To a given value of $x$ for which $|x| \leqslant 1$ there correspond infinitely many values of $\theta$.

13 (i) $2$ cu.ft.; (ii) $8/\pi$ cu.ft. \qquad 14 $\frac{1}{3}AB$.

15 $1:\sqrt{2}$. \qquad 17 Equality.

19 $\sin x \geqslant 2(\pi-x)/\pi$ when $\frac{1}{2}\pi \leqslant x \leqslant \pi$.

21 $x^r - 1 < r(x-1)$ if $x > 0$, $x \neq 1$, $0 < r < 1$.

## Exercise 3 (d), p. 87

1 $1\cdot 0093$. \qquad 2 $4\pi r^2 \delta r$. \qquad 5 (i) $-\dfrac{RT}{p^2}\delta p$; (ii) $\dfrac{R}{p}\delta T - \dfrac{RT}{p^2}\delta p$.

7 (i) $h$; (ii) $3ah + h^2$; (iii) $h/a^2(a+h)$; when $a = 0$.

## Miscellaneous Exercise 3 (e), p. 90

1 $-\sin(\sin x)\cos x$. \qquad\qquad 2 $1 - x^2 - 3x(1-x^2)^{\frac{1}{2}}\sin^{-1}x$.

3 $n\sec^2 x/(1 + n^2\tan^2 x)$. \qquad\qquad 4 $-(b^2-a^2)^{\frac{1}{2}}/(b + a\cos x)$.

5 $-2/\sqrt{(1-x^2)}$. \qquad\qquad 6 $1/(x+1)\sqrt{x}$.

7 $-1/\sqrt{(1-x^2)}$; because $\sin^{-1}\sqrt{(1-x^2)} = \cos^{-1}x$ when $0 < x < 1$.

9 $\frac{1}{4}(-1)^n n!\{3(x-2)^{-n-1} + (x+2)^{-n-1}\}$.

10 $(-1)^n n!\{8(x-2)^{-n-1} - (x-1)^{-n-1}\}$ if $n > 1$; $1 + (x-1)^{-2} - 8(x-2)^{-2}$ if $n = 1$.

11 $2^{n-1}\cos(2x + \frac{1}{2}n\pi) - \frac{1}{2}.5^n\cos(5x + \frac{1}{2}n\pi)$.

12, 13 See 10.43. \qquad 21 $d^2y/dt^2 + y = 0$. \qquad 25 $a = -\frac{3}{4}, b = \frac{1}{16}$.

26 $[2(x+y)\cos\{(x+y)^2\}]/[1 - 2(x+y)\cos\{(x+y)^2\}]$.

27 $x = -1$, min.; $x = -5$, max. \qquad 28 None. \qquad 29 None.

30 $x = 2n\pi$, min.; $x = (2n+1)\pi$, max., provided $ad > bc$; reverse if $ad < bc$; none if $ad = bc$.

31 $x = a^{\frac{2}{3}}b^{\frac{2}{3}}/(a^{\frac{2}{3}} + b^{\frac{2}{3}})^{\frac{1}{2}}$, max.; $x = $ negative of this, min.

32 $(27 + 3\sqrt{78})^{\frac{1}{3}}$, max.; $(27 - 3\sqrt{78})^{\frac{1}{3}}$, min.

35 $x = \frac{1}{4}\pi$, $\frac{2}{3}\pi$, $\frac{3}{4}\pi$. (i) Incr. for $0 < x < \frac{1}{4}\pi$, $\frac{2}{3}\pi < x < \frac{3}{4}\pi$; (ii) decr. for $\frac{1}{4}\pi < x < \frac{2}{3}\pi$, $\frac{3}{4}\pi < x < \pi$. Greatest value at $x = \frac{1}{4}\pi$.

39 $2ab$. \qquad\qquad 40 $a + b$.

## Exercise 4 (a), p. 99

1 $\frac{1}{3}\{(x+1)^{\frac{3}{2}} + (x-1)^{\frac{3}{2}}\}$. \quad 2 $\frac{1}{2}x + \frac{1}{4}\sin 2x$. \quad 3 $\frac{1}{2}x - \frac{1}{4}\sin 2x$.

4 $\tan x - x$, \qquad 5 $\frac{1}{12}(\sin 3x + 9\sin x)$. \quad 6 $x - \tan^{-1}x$.

7 $\frac{1}{3}x^3 - x + \tan^{-1}x$.    8 $\frac{2}{3}(1+x)^{\frac{3}{2}} - 2(1+x)^{\frac{1}{2}}$.    9 $-1/x$.

10 $\frac{1}{8}(2\cos 2x - \cos 4x)$.    11 $\dfrac{(ax+b)^{n+1}}{a(n+1)}$.    12 $\dfrac{1}{a}\sin(ax+b)$.

13 $-(1/a)\cos(ax+b)$.    14 $(1/a)\tan^{-1}(x/a)$.    15 $\sin^{-1}(x/a)$.

16 $\frac{1}{3}\tan^{-1}x - \frac{1}{6}\tan^{-1}\frac{1}{2}x$.    17 $-1/(x+3)$.

18 $\tan^{-1}(x+3)$.    19 $\sin^{-1}(x-1)$.

## Exercise 4(b), p. 104

1 $\frac{1}{8}(x-5)^8 + \frac{6}{7}(x-5)^7$.    2 $\dfrac{3}{2(x+3)^2} - \dfrac{1}{x+3}$.    3 $\frac{2}{3}(x+3)^{\frac{3}{2}} - 6(x+3)^{\frac{1}{2}}$.

4 $-\frac{1}{12}(1-x^2)^6$.    5 $-\dfrac{1}{16(4x^2+1)^2}$.    6 $\frac{1}{2}\surd(1+x^4)$.

7 $\frac{1}{4}\sin^4 x$.    8 $\frac{1}{4}\tan^4 x$.    9 $\frac{1}{2}\sin^{-1}\left(\dfrac{2x}{3}\right)$.

10 $\frac{1}{6}\tan^{-1}\left(\dfrac{2x}{3}\right)$.    11 $\dfrac{x}{9\surd(9-x^2)}$.    12 $\dfrac{x}{9\surd(x^2+9)}$.

13 $-\dfrac{x}{9\surd(x^2-9)}$.    14 $-\dfrac{2x^2+1}{4(x^2+1)^2}$.    15 $-\dfrac{(1+x^2)^{\frac{3}{2}}}{5x^5}$.

16 $\sec^{-1}x$.    17 $\dfrac{1}{x}\surd(x^2-1)$.    18 $\frac{1}{2}(\tan^{-1}x)^2$.

19 $\cos^{-1}(1-x)$.    20 $\frac{1}{5}\sin^5 x - \frac{1}{7}\sin^7 x$.    21 $\sin^{-1}x - \surd(1-x^2)$.

22 $\sec^{-1}x - \dfrac{1}{x}\surd(x^2-1)$.    23 $\frac{2}{3}(\surd 11 - \surd 2)$.    24 $\frac{1}{9}(3\surd 3 - 1)$.

25 $\frac{2}{15}$.    26 $\frac{1}{4}$.    27 $\frac{1}{15}\tan^{-1}\frac{5}{3}$.

28 $\frac{32}{9}$.    29 2.    31 $\frac{1}{4}\pi$.

## Exercise 4(c), p. 109

1 $\sin x - x\cos x$.    2 $(x/a)\sin ax - (1/a^2)\cos ax$.

3 $(2-x^2)\cos x + 2x\sin x$.    4 $\frac{1}{8}(2x^2 + 2x\sin 2x + \cos 2x)$.

5 $\frac{1}{7}(x+1)(x-5)^7 - \frac{1}{56}(x-5)^8$.

6 $-\dfrac{x}{n+1}(1-x)^{n+1} - \dfrac{1}{(n+1)(n+2)}(1-x)^{n+2}$.

7 $x\cos^{-1}x - \surd(1-x^2)$.    8 $-\dfrac{1}{2x} - \dfrac{1}{2x^2}(1+x^2)\tan^{-1}x$.

9 $\frac{1}{32}\{\sin 4x - 4x\cos 4x + 4\sin 2x - 8x\cos 2x\}$.

10 $\frac{1}{3}\cos^3 x - \cos x$.    11 $\frac{1}{2}x^2\sec^{-1}x - \frac{1}{2}\surd(x^2-1)$.

12 $\dfrac{n}{a^2}x^{n-1}\sin ax - \dfrac{x^n}{a}\cos ax - \dfrac{n(n-1)}{a^2}s_{n-2} = s_n$;

$(5x^4 - 60x^2 + 120)\sin x - (x^5 - 20x^3 + 120x)\cos x$.

13 $\frac{1}{25}\pi$.    14 $8\pi - 16$.    15 $\frac{2}{3}$.

16 $\dfrac{2}{(n+1)(n+2)(n+3)}$.               17 $\frac{1}{4}\pi - \frac{1}{2}$.

18 $x - \sqrt{(1-x^2)}\sin^{-1}x$.                19 $2\sqrt{(1-x^2)}\sin^{-1}x + x(\sin^{-1}x)^2 - 2x$.

20 $\frac{1}{4}(2x^2-1)\sin^{-1}x + \frac{1}{4}x\sqrt{(1-x^2)}$.     21 $(1-x^2)y + 3x^2 + Ax + B$.

## Exercise 4(d), p. 114

1 $1/x$.         2 $2/x$.         3 $-1/x$.         4 $\cot x$.

5 $2\cot x$.     6 $-\sec x\,\mathrm{cosec}\,x$.  7 $2/(1-x^2)$.      8 $1/(x\log x)$.

9 $(\log x - 1)/(\log x)^2$.               10 $(2/x)\log x$.      12 $(1/x)\log_{10}e$.

13 $\mathrm{cosec}\,x$.   14 $\sec x$.         15 $2\sec x$.          16 $0$.

17 $1/e$.                   21 $\frac{1}{3}\log x$.                22 $\log(1+x)$.

23 $-\log(1-x)$.           24 $\frac{1}{2}\log(1+x^2)$.          25 $\frac{1}{4}\log(1+x^4)$.

26 $\log(3x^2 - 7x + 5)$.      27 $-\frac{1}{6}\log(3x^2 - 12x + 7)$.

28 $\log\sin x$.           29 $-\frac{1}{3}\log\cos 3x$.          30 $-\frac{1}{5}\log(3+5\cos^2 x)$.

31 $\log(\cos x + \sin x)$.    32 $\log\log x$.             33 $2\log\tan\frac{1}{2}x$; $\log\tan\frac{1}{2}x$.

34 $\dfrac{x^{m+1}}{m+1}\left(\log x - \dfrac{1}{m+1}\right)$; $x(\log x - 1)$.          35 $\frac{1}{2}(\log x)^2$.

36 $x\tan^{-1}x - \frac{1}{2}\log(1+x^2)$.          37 $\frac{1}{3}x^3(\log x)^2 - \frac{2}{9}x^3\log x + \frac{2}{27}x^3$.

38 $x\tan x + \log\cos x - \frac{1}{2}x^2$.          39 $\frac{1}{4}\log 2$.

40 $\frac{1}{3}\log\frac{5}{2}$.             41 $\log 2$.              42 $1 - \log 2$.

43 $\frac{1}{2}$.               44 $\frac{1}{4}$.               45 $\frac{2}{27}(2e^3+1)$.

46 $x$, $\log|a\cos x + b\sin x|$; $\frac{3}{50}\pi + \frac{4}{25}\log\frac{4}{3}$.

47 $\lambda = 3$, $\mu = -2$; $3x - 2\log(\sin x + 2\cos x)$.

## Exercise 4(e), p. 119

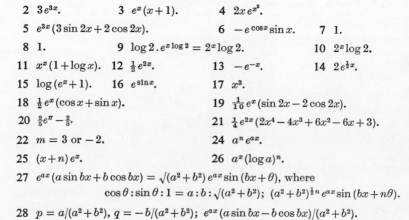

2 $3e^{3x}$.             3 $e^x(x+1)$.          4 $2xe^{x^2}$.

5 $e^{3x}(3\sin 2x + 2\cos 2x)$.             6 $-e^{\cos x}\sin x$.    7 $1$.

8 $1$.           9 $\log 2.\,e^{x\log 2} = 2^x\log 2$.          10 $2^x\log 2$.

11 $x^x(1+\log x)$.   12 $\frac{1}{2}e^{2x}$.     13 $-e^{-x}$.         14 $2e^{\frac{1}{2}x}$.

15 $\log(e^x + 1)$.   16 $e^{\sin x}$.          17 $x^3$.

18 $\frac{1}{2}e^x(\cos x + \sin x)$.             19 $\frac{1}{10}e^x(\sin 2x - 2\cos 2x)$.

20 $\frac{2}{5}e^\pi - \frac{2}{5}$.                       21 $\frac{1}{4}e^{2x}(2x^4 - 4x^3 + 6x^2 - 6x + 3)$.

22 $m = 3$ or $-2$.                          24 $a^n e^{ax}$.

25 $(x+n)e^x$.                            26 $a^x(\log a)^n$.

27 $e^{ax}(a\sin bx + b\cos bx) = \sqrt{(a^2+b^2)}\,e^{ax}\sin(bx+\theta)$, where
   $\cos\theta : \sin\theta : 1 = a : b : \sqrt{(a^2+b^2)}$; $(a^2+b^2)^{\frac{1}{2}n}e^{ax}\sin(bx+n\theta)$.

28 $p = a/(a^2+b^2)$, $q = -b/(a^2+b^2)$; $e^{ax}(a\sin bx - b\cos bx)/(a^2+b^2)$.

29 $e^{ax}(b\sin bx + a\cos bx)/(a^2+b^2)$.     30 $x=0$, max.; $x=\pm 1/\sqrt{2}$, inflex.

31 $x=0$, min.; $x=2$, max.; $x=2\pm\sqrt{2}$, inflex.     32 $n^n e^{-n}$.

36 (i) $3e^{2t}+A+Bt+Ct^2$;   (ii) $3x^2+A+B\log x+C(\log x)^2$.

## Exercise 4($f$), p. 124

1 $(1-x^2)^{-\frac{3}{2}}$.

2 $\dfrac{(x+1)^4(2x+3)^2}{(3x-5)^3}\left\{\dfrac{4}{x+1}+\dfrac{4}{2x+3}-\dfrac{9}{3x-5}\right\}$.

3 $(x-2)^{\frac{1}{2}}(3x+2)^{\frac{1}{5}}(2x+5)^2\left\{\dfrac{1}{2(x-2)}+\dfrac{3}{5(3x+2)}+\dfrac{4}{2x+5}\right\}$.

4 $x^5 e^{3x}\sin 2x(5/x+3+2\cot 2x)$.     5 $e^{x+e^x}$.

6 $e^x\sin x(\log x)^2\{1+\cot x+2/(x\log x)\}$.

7 $xe^x\tan x(1/x+1+2\operatorname{cosec}2x)$.

8 $(1+x)^{1/x}\left\{\dfrac{1}{x(1+x)}-\dfrac{1}{x^2}\log(1+x)\right\}$.     9 $-\dfrac{\log 2}{x(\log x)^2}$.

10 $(\log x)^x\left\{\log\log x+\dfrac{1}{\log x}\right\}$.     18 $-\dfrac{1}{x^2}$.     21 $e^{-2}$.

22 1.     23 $e^3$.     26 $\log 2$.     27 $\log\frac{3}{2}$.

28 $\log p$.     29 $\log(p/q)$.

## Exercise 4($g$), p. 128

8 (i) $\frac{1}{2}\operatorname{ch}(x+y)-\frac{1}{2}\operatorname{ch}(x-y)$;   (ii) $\frac{1}{2}\operatorname{sh}(x+y)+\frac{1}{2}\operatorname{sh}(x-y)$;
(iii) $\frac{1}{2}\operatorname{ch}(x+y)+\frac{1}{2}\operatorname{ch}(x-y)$.

16 $\log\frac{13}{9}$ or $-\log 3$.     22 $-\operatorname{th}x\operatorname{sech}x$.     23 $-\operatorname{coth}x\operatorname{cosech}x$.

24 $-\operatorname{cosech}^2 x$.     25 $\operatorname{coth}x$.     26 $(1/x)\operatorname{ch}(\log x)$.

27 $\operatorname{cosech}x$.     28 $-\operatorname{sech}2x$.     29 $\frac{1}{2}\operatorname{sech}x$.

30 $2\operatorname{sh}x\cos x$.     31 $x^{\operatorname{ch}x}\{\operatorname{sh}x\log x+(1/x)\operatorname{ch}x\}$.

32 $(\operatorname{ch}x)^x(\log\operatorname{ch}x+x\operatorname{th}x)$.     35 $\frac{1}{2}\operatorname{ch}2x$.

36 $\frac{1}{3}\operatorname{th}3x$.     37 $2\log\operatorname{ch}\frac{1}{2}x$.     38 $e^x\operatorname{th}x$.

39 If $a\neq b$, $\dfrac{1}{2}\left\{\dfrac{1}{a+b}e^{(a+b)x}+\dfrac{1}{a-b}e^{(a-b)x}\right\}$; if $a=b$, $\dfrac{1}{2}\left\{\dfrac{1}{2a}e^{2ax}+x\right\}$.

40 $\frac{1}{4}\operatorname{sh}2x-\frac{1}{2}x$.     41 $\frac{1}{2}x+\frac{1}{4}\operatorname{sh}2x$.     42 $x-\operatorname{th}x$.

43 $\frac{1}{8}\operatorname{ch}4x-\frac{1}{4}\operatorname{ch}2x$.     44 $\frac{1}{3}\operatorname{ch}^3 x-\operatorname{ch}x$.

45 $\frac{1}{4}(\frac{1}{6}\operatorname{sh}6x+\frac{1}{4}\operatorname{sh}4x+\frac{1}{2}\operatorname{sh}2x+x)$.     46 $\frac{1}{4}x^2+\frac{1}{2}\log x$.

47 $x\operatorname{sh}x-\operatorname{ch}x$.     48 $x\operatorname{th}x-\log\operatorname{ch}x$.     49 $(x^2+2)\operatorname{ch}x-2x\operatorname{sh}x$.

52 $\log\operatorname{th}(\frac{1}{2}x)$.     55 1; 1.

## Exercise 4($h$), p. 132

2 $\dfrac{2}{\sqrt{(25+4x^2)}}$.     3 $\dfrac{1}{\sqrt{(x^2-x)}}$, $x>1$.     4 $\dfrac{2}{1-4x^2}$, $|x|<\frac{1}{2}$.

**5** $\frac{1}{2}\sec x$, $|x| < \frac{1}{2}\pi$.    **6** $2/(1-x^2)$, $x \neq 0$, $\pm 1$.    **7** $|\sec x|$.

**8** $\sec x$.    **9** $\sec x$ if $\tan x > 0$, $-\sec x$ if $\tan x < 0$.

**10** $\mathrm{sh}^{-1}x$.    **11** $\mathrm{ch}^{-1}x$.

**16** $c' = c - \log a$;   $\log\{x + \sqrt{(x^2 - a^2)}\} + c'$;   $\dfrac{1}{2a}\log\dfrac{a+x}{a-x} + c'$.

**17** $\mathrm{sh}^{-1}\frac{1}{5}x$.    **18** $\mathrm{ch}^{-1}\frac{1}{6}x$ $(x > 6)$.    **19** $\frac{1}{2}\mathrm{sh}^{-1}(\frac{2}{5}x)$.

**20** $\sin^{-1}\frac{1}{6}x$.    **21** $\frac{1}{2}\mathrm{ch}^{-1}\frac{2}{3}x$ $(x > \frac{3}{2})$.    **22** $\mathrm{sh}^{-1}(x+1)$.

**23** $\mathrm{ch}^{-1}(x+1)$ $(x > 0)$.    **25** $\frac{1}{2}x\sqrt{(x^2-a^2)} - \frac{1}{2}a^2\mathrm{ch}^{-1}(x/a)$.

**26** $\frac{1}{2}x\sqrt{(9x^2 - 4)} - \frac{2}{3}\mathrm{ch}^{-1}(\frac{3}{2}x)$.    **27** $\sqrt{40} + \frac{2}{3}\log(3 + \sqrt{10})$.

**28** $\frac{1}{2}x\sqrt{(x^2-1)} + \frac{1}{2}\mathrm{ch}^{-1}x$.    **29** $\log\dfrac{\sqrt{10}+3}{\sqrt{2}+1} - \frac{1}{3}\sqrt{10} + \sqrt{2}$.

**31** (i), (ii) $\mathrm{ch}^{-1}(\frac{13}{5}) - \mathrm{ch}^{-1}(\frac{12}{5})$.    **32** $x\mathrm{ch}^{-1}x - \sqrt{(x^2-1)}$.

**33** $x\mathrm{th}^{-1}x + \frac{1}{2}\log(1-x^2)$.    **34** $\frac{1}{4}(2x^2+1)\mathrm{sh}^{-1}x - \frac{1}{4}x\sqrt{(x^2+1)}$.

**35** $\frac{1}{3}x^3\mathrm{th}^{-1}x + \frac{1}{6}x^2 + \frac{1}{6}\log(1-x^2)$.

**36** $c_n = \dfrac{x^n}{a}\mathrm{sh}\,ax - \dfrac{n}{a^2}x^{n-1}\mathrm{ch}\,ax + \dfrac{n(n-1)}{a^2}c_{n-2}$;

$$s_n = \dfrac{x^n}{a}\mathrm{ch}\,ax - \dfrac{n}{a^2}x^{n-1}\mathrm{sh}\,ax + \dfrac{n(n-1)}{a^2}s_{n-2}.$$

**37** $2\,\mathrm{th}^{-1}(\cot\frac{1}{2}x)$, $= \log\cot(\frac{1}{2}x - \frac{1}{4}\pi)$.

**38** $0 < |x| < 1$; $x \neq 0$; $|x| > 1$.

## Exercise 4(*i*), p. 141

**1** $\frac{5}{8}\log(x-2) + \frac{5}{8}\log(x+2) - \frac{1}{4}\log x$.    **2** $2\log\dfrac{x-1}{x+2} - \dfrac{1}{x-1}$.

**3** $28\log\dfrac{x-3}{x-2} + \dfrac{26}{x-2} + \dfrac{15}{2(x-2)^2}$.    **4** $\log\dfrac{x^2}{x^2+2} + \dfrac{1}{\sqrt{2}}\tan^{-1}\dfrac{x}{\sqrt{2}}$.

**5** $\frac{3}{5}\tan^{-1}\dfrac{x}{3} - \frac{2}{5}\tan^{-1}\dfrac{x}{2}$.

**6** $\frac{1}{2}x^2 + x + \frac{2}{3}\log(x-1) - \frac{1}{3}\log(x^2+x+1)$.

**7** $\log(x-1) - \frac{1}{4}\log(1+x^2) - \frac{1}{2}\tan^{-1}x + \dfrac{1}{x^2+1}$.

**8** $\frac{1}{2}\log\dfrac{x-1}{x+1}$.    **9** $x + \log\dfrac{x-1}{x+1}$.    **10** $\log(x-1) - \dfrac{2}{x-1}$.

**11** $\log(x-2) - \dfrac{4}{x-2} - \dfrac{5}{2(x-2)^2}$.    **12** $3\log(x+1) - 2\log x$.

**13** $\log(x-1) - \frac{1}{2}\log(4x-1)$.    **14** $\log(x^2-x-6)$.

**15** $x^2 - x + 4\log x - 3\log(x+1)$.

**16** $7\log(x+1) + \log(x+2) - 8\log(2x+3)$.

**17** $2\log x - 2\log(x-3) - \dfrac{7}{x-3}$.    **18** $\log(x-1) - \log(x+2) + \dfrac{1}{x+2}$.

19 $\log x + \dfrac{2}{x} - \log(x+1) - \dfrac{4}{x+1}$.

20 $-2\log(x+1) - \dfrac{1}{x+1} - \dfrac{1}{2(x+1)^2} + 2\log(x-2)$.

21 $3\log x - \tfrac{3}{2}\log(x^2+1) + 2\tan^{-1}x$.  22 $x + 4\log(x-2) - \dfrac{1}{x-2} - \tfrac{1}{2}\tan^{-1}\dfrac{x}{2}$.

23 $-3\log(1-2x) + \tfrac{3}{2}\log(x^2+3) + \dfrac{1}{x^2+3}$.

24 $\sqrt{2}\tan^{-1}(x\sqrt{2}) - \sqrt{\tfrac{3}{2}}\tan^{-1}(x\sqrt{\tfrac{3}{2}})$.  25 $\tfrac{1}{6}\log(x^2-3) - \tfrac{1}{3}\log(2x+3)$.

26 $\log(x-2) - \tfrac{1}{2}\log(x^2+4x-1)$.

27 $\tfrac{1}{2}\log(x^2+2) + \dfrac{1}{\sqrt{2}}\tan^{-1}\dfrac{x}{\sqrt{2}} + \tfrac{1}{2}\log(x^2-2)$.

## Exercise 4(j), p. 145

1 $\tfrac{1}{12}\tan^{-1}\dfrac{3x+2}{4}$.

2 $\tfrac{1}{24}\log\dfrac{3x-2}{3x+6}$.

3 $\tan^{-1}(x+1)$.

4 $\tfrac{1}{6}\log\dfrac{x+1}{x-5}$.

5 $\tfrac{3}{2}\log(x^2+2x-10) - \dfrac{2}{\sqrt{11}}\log\dfrac{x+1-\sqrt{11}}{x+1+\sqrt{11}}$.

6 $\tfrac{1}{3}\log(9x^2-18x+25) + \tfrac{11}{12}\tan^{-1}\tfrac{3}{4}(x-1)$.

7 $\log(1+x) + 2\tan^{-1}x$.

8 $\tfrac{1}{2}\log\dfrac{x^2+2x+2}{x^2-1} + 3\tan^{-1}(x+1)$.

9 $\log(x^2-x+1) - \tfrac{1}{2}\log(x^2+x+1) + 3\sqrt{3}\tan^{-1}(x+\tfrac{1}{2}) + \tfrac{2}{3}\sqrt{3}\tan^{-1}(x-\tfrac{1}{2})$.

10 $\dfrac{1}{250}\left\{\dfrac{5x}{x^2+25} + \tan^{-1}\dfrac{x}{5}\right\}$.

11 $\dfrac{1}{128}\left\{\tfrac{3}{2}\tan^{-1}\dfrac{x}{2} + \dfrac{3x^3+20x}{(x^2+4)^2}\right\}$.

12 $\dfrac{x-3}{8(x^2+2x+5)} + \tfrac{1}{16}\tan^{-1}\dfrac{x+1}{2}$.

13 $\tfrac{1}{8}\sqrt{2}\left\{\log\dfrac{1-x\sqrt{2}+x^2}{1+x\sqrt{2}+x^2} + 2\tan^{-1}\dfrac{x\sqrt{2}}{1-x^2}\right\}$.

14 $\tfrac{1}{8}\sqrt{2}\left\{\log\dfrac{1+x\sqrt{2}+x^2}{1-x\sqrt{2}+x^2} + 2\tan^{-1}\dfrac{x\sqrt{2}}{1-x^2}\right\}$.

15 $\tfrac{1}{4}\log\dfrac{1+x+x^2}{1-x+x^2} + \dfrac{2}{\sqrt{3}}\tan^{-1}\dfrac{x\sqrt{3}}{1-x^2}$.

## Exercise 4(k), p. 150

1 $\log x - 2\sqrt{x}$.  2 $\tfrac{4}{3}(x+1)^{\frac{3}{2}} - 2\sqrt{(x+1)}$.  3 $2\log(1+\sqrt{x})$.

4 $\tfrac{3}{2}x^{\frac{2}{3}} - 3x^{\frac{1}{3}} + 3\log(1+x^{\frac{1}{3}})$.

5 $\tfrac{1}{2}\log\dfrac{\sqrt{(x+4)}-2}{\sqrt{(x+4)}+2}$.

6 $\tfrac{6}{7}x^{\frac{7}{6}} + \tfrac{6}{5}x^{\frac{5}{6}} + 3x^{\frac{2}{3}} + 2x^{\frac{1}{2}} + 6x^{\frac{1}{3}} + 6x^{\frac{1}{6}} + 9\log(x^{\frac{1}{6}}+1) + 3\log(x^{\frac{1}{6}}-1)$.

7 $2\{\sqrt{(2x+1)} + \sqrt{x} - \tan^{-1}\sqrt{(2x+1)} - \tan^{-1}\sqrt{x}\}$.

8   $\frac{2}{3}(1+x)^{\frac{3}{2}} + \frac{3}{4}(1+x)^{\frac{4}{3}} + 1 + x + \frac{6}{5}(1+x)^{\frac{5}{6}} + \frac{3}{2}(1+x)^{\frac{2}{3}}.$

9   $\text{sh}^{-1}\,\frac{1}{2}(x+3).$        10   $\frac{1}{\sqrt{2}}\,\text{ch}^{-1}\{\frac{1}{3}\sqrt{2}(x+1)\}.$    11   $\sin^{-1}\dfrac{x-3}{\sqrt{2}}.$

12   $\sin^{-1}\dfrac{2x+1}{5}.$        13   $\sqrt{(x^2+1)} + 2\,\text{sh}^{-1}x.$

14   $\sqrt{(x^2+6x+13)} - 3\,\text{sh}^{-1}\dfrac{x+3}{2}.$

15   $\frac{3}{2}\sqrt{(2x^2+4x-7)} - 4\sqrt{2}\,\text{ch}^{-1}\{\frac{1}{3}\sqrt{2}(x+1)\}.$

16   $6\sin^{-1}\dfrac{2x+1}{5} - 2\sqrt{(6-x-x^2)}.$     17   $\sqrt{(x^2-1)} + \text{ch}^{-1}x.$

18   $\frac{1}{2}x\sqrt{(x^2+1)} - \frac{1}{2}\text{sh}^{-1}x.$       19   $\frac{1}{3}(x^2+1)^{\frac{3}{2}} - \sqrt{(x^2+1)} + \text{sh}^{-1}x.$

20   $-\frac{1}{2}\text{sh}^{-1}\dfrac{2}{x}.$       21   $-2\sqrt{\left(2+\dfrac{1}{x}\right)}.$      22   $\cos^{-1}\dfrac{x-1}{2x}.$

23   $3\,\text{sh}^{-1}\dfrac{x}{3} - \dfrac{1}{\sqrt{10}}\text{sh}^{-1}\dfrac{x+9}{3x-3}.$     24   $\dfrac{1}{\sqrt{2}}\sin^{-1}\dfrac{x\sqrt{2}}{x+1}.$

25   $\frac{1}{2}x\sqrt{(a^2-x^2)} + \frac{1}{2}a^2\sin^{-1}\dfrac{x}{a}.$      26   $\frac{1}{2}x\sqrt{(x^2-a^2)} - \frac{1}{2}a^2\,\text{ch}^{-1}\dfrac{x}{a}.$

27   $-\dfrac{25\sqrt{3}}{72}\,\text{ch}^{-1}\dfrac{6x-5}{5} + \frac{1}{12}(6x-5)\sqrt{(3x^2-5x)}.$

28   $\frac{1}{4}(2x-5)\sqrt{(x^2-5x+6)} - \frac{1}{8}\text{ch}^{-1}(2x-5).$

29   $\frac{1}{2}(2x^2-6x+1)^{\frac{3}{2}} + \frac{11}{8}(2x-3)\sqrt{(2x^2-6x+1)} - \dfrac{77\sqrt{2}}{16}\text{ch}^{-1}\dfrac{2x-3}{\sqrt{7}}.$

30   $\dfrac{1}{2\sqrt{5}}\log\dfrac{\sqrt{(x^2+9)} - \sqrt{5}}{\sqrt{(x^2+9)} + \sqrt{5}}.$      31   $-\dfrac{1}{2\sqrt{5}}\tan^{-1}\left\{\dfrac{2}{x}\sqrt{\left(\dfrac{x^2+9}{5}\right)}\right\}.$

32   $2(\beta-\alpha)^2\int\sin^2\theta\,\cos^2\theta\,d\theta,\ \int 2\,d\theta,\ 2(\beta-\alpha)\int\sin^2\theta\,d\theta.$

33   $2(\alpha-\beta)^2\int\text{sh}^2u\,\text{ch}^2u\,du,\ \int 2\,du,\ 2(\alpha-\beta)\int\text{sh}^2u\,du.$

## Exercise 4(*l*), p. 156

1   $-\cot\frac{1}{2}x.$        2   $\dfrac{-2}{1+\tan\frac{1}{2}x}.$       3   $\frac{1}{2}\tan^{-1}(2\tan\frac{1}{2}x).$

4   $\dfrac{1}{\sqrt{2}}\log\tan\left(\frac{1}{8}\pi+\frac{1}{2}x\right)$   *or*   $\dfrac{1}{\sqrt{2}}\log\dfrac{\tan\frac{1}{2}x - 1 + \sqrt{2}}{\tan\frac{1}{2}x - 1 - \sqrt{2}}.$

5   $\log(1+\tan\frac{1}{2}x).$        6   $\tan^{-1}\{\frac{1}{2}(\tan\frac{1}{2}x+1)\}.$

7   $-\log(\sin x+\cos x).$       8   $\frac{1}{2}x + \frac{1}{2}\log(\sin x+\cos x).$

9   $2x + \log(2\cos x+\sin x+3) + \tan^{-1}\{\frac{1}{2}(\tan\frac{1}{2}x+1)\}.$

10   $-\sqrt{2}\,\text{sh}^{-1}\{(1/\sqrt{3})\tan(\frac{1}{4}\pi-\frac{1}{2}x)\}.$      11   $x\tan\frac{1}{2}x.$

12   $\frac{1}{7}\sin^7x - \frac{1}{9}\sin^9x.$     13   $-\frac{1}{3}\cos^3x + \frac{2}{5}\cos^5x - \frac{1}{7}\cos^7x.$

14   $\tan x - 2\cot x - \frac{1}{3}\cot^3x.$       15   $-\text{cosec}\,x - \sin x.$

16   $\frac{1}{2}\tan^2x + \log\cos x.$     17   $\frac{1}{8}\tan^2\frac{1}{2}x - \frac{1}{8}\cot^2\frac{1}{2}x + \frac{1}{2}\log\tan\frac{1}{2}x.$

18 $\sqrt{2}\tan^{-1}(\sqrt{2}\tan x)-x.$

19 $\frac{1}{4}\log\cos 2x.$

20 $\frac{1}{20}(5\cos 2x-\cos 10x).$

21 $\frac{1}{96}(3\cos 8x-4\cos 6x-12\cos 2x).$

22 $-\frac{1}{420}(15\cos 7x+42\cos 5x+35\cos 3x).$

23 $\frac{1}{16}(5\sin x-\frac{1}{3}\sin 3x-\frac{3}{5}\sin 5x-\frac{1}{7}\sin 7x).$

24 $R=\sqrt{(a^2+b^2)}$, $\alpha$ is given by $\cos\alpha:\sin\alpha:1=a:b:R.$ Put $u=x-\alpha.$

25 (iii) $(1/b)\cot\frac{1}{2}x.$

26 If $b^2<4ac$, $\dfrac{2}{\sqrt{(4ac-b^2)}}\tan^{-1}\dfrac{2c\tan x+b}{\sqrt{(4ac-b^2)}}$; if $b^2=4ac$, $\dfrac{-1}{c\tan x+\frac{1}{2}b}$;

if $b^2>4ac$, $\dfrac{1}{\sqrt{(b^2-4ac)}}\log\dfrac{2c\tan x+b-\sqrt{(b^2-4ac)}}{2c\tan x+b+\sqrt{(b^2-4ac)}}.$

27 If $a\neq b$, $\dfrac{1}{2(a^2-b^2)}\cdot\dfrac{x}{a^2\cos^2 x+b^2\sin^2 x}-\dfrac{1}{2ab(a^2-b^2)}\tan^{-1}\left(\dfrac{b}{a}\tan x\right)$;

if $a=b$, $\dfrac{1}{8a^4}(\sin 2x-2x\cos 2x).$

28 (ii) $\frac{1}{2}x+\dfrac{1}{4m}\sin 2mx,\quad \frac{1}{2}x-\dfrac{1}{4m}\sin 2mx,\quad -\dfrac{1}{4m}\cos 2mx.$

## Exercise 4(m), p. 163

1 $\frac{1}{6}\cos^5 x\sin x+\frac{5}{24}\cos^3 x\sin x+\frac{5}{16}\cos x\sin x+\frac{5}{16}x.$

2 $-\frac{1}{5}\sin^4 x\cos x-\frac{4}{15}\sin^2 x\cos x-\frac{8}{15}\cos x.$

3 $\frac{1}{7}\cos^3 x\sin^4 x+\frac{3}{35}\cos x\sin^4 x+\frac{1}{35}\cos^3 x-\frac{3}{35}\cos x.$

4 $\frac{1}{10}\cos^5 x\sin^5 x+\frac{1}{16}\cos^3 x\sin^5 x+\frac{1}{32}\cos x\sin^5 x$
$\qquad\qquad -\frac{1}{128}\sin^3 x\cos x-\frac{3}{256}\sin x\cos x+\frac{3}{256}x.$

5 $\frac{1}{5}\sin x\sec^5 x+\frac{4}{15}\sin x\sec^3 x+\frac{8}{15}\tan x.$

6 $\frac{1}{6}\sin^3 x\sec^6 x-\frac{1}{8}\sin x\sec^4 x+\frac{1}{16}\sin x\sec^2 x+\frac{1}{16}\log\tan(\frac{1}{2}x+\frac{1}{4}\pi).$

7 $\frac{8}{15}.$      8 $\frac{5}{32}\pi.$      9 $\frac{3}{8}\pi.$      10 $\frac{8}{315}.$

11 $\frac{63}{8}\pi.$      12 $\frac{64}{21}.$      13 $\frac{5}{256}\pi.$      14 $\frac{7}{256}\pi.$

15 $m!\,n!\,a^{m+n+1}/(m+n+1)!.$      16 $\frac{35}{64}\pi.$      17 $\frac{1}{12}.$

18 $0.$      20 $u_n=(1/a)\{x^n e^{ax}-nu_{n-1}\}.$

21 $u_n=(1/n)\operatorname{ch}^{n-1}x\operatorname{sh}x+(1-1/n)u_{n-2}.$

22 $u_n=(1/n)\operatorname{sh}^{n-1}x\operatorname{ch}x-(1-1/n)u_{n-2}.$

23 $s_n=(1/a^2)\{x^{n-1}(n\sin ax-ax\cos ax)-n(n-1)s_{n-2}\}.$

24 $c_n=(1/a^2)\{x^{n-1}(ax\operatorname{sh}ax-n\operatorname{ch}ax)+n(n-1)c_{n-2}\}.$

25 $s_n=(1/a^2)\{x^{n-1}(ax\operatorname{ch}ax-n\operatorname{sh}ax)+n(n-1)s_{n-2}\}.$

26 $u_n=\{e^{ax}\sin^{n-1}bx(a\sin bx-nb\cos bx)+n(n-1)b^2 u_{n-2}\}/(a^2+n^2b^2).$

27 If $m\neq-1$, $u_{m,n}=\{x^{m+1}(\log x)^n-nu_{m,n-1}\}/(m+1)$;
if $n\neq-1$, $u_{-1,n}=(\log x)^{n+1}/(n+1)$; $u_{-1,-1}=\log\log x.$

28 $\frac{5}{12}$.

29 $u_{m,n} = \dfrac{\sin^{m-1} x(n \sin x \cos nx - m \cos x \sin nx) + m(m-1) u_{m-2,n}}{m^2 - n^2}$;   $\frac{8}{315}$.

30 $-\frac{1}{2}\pi$.                                  34   Put $y = \pi - x$.

## Exercise 4(n), p. 171

[∗ means 'the integral does not exist']

1   1.          2   ∗.          3   $\frac{1}{4}\pi$.          4   ∗.

5   1.          6   $\frac{1}{2}\log 3$.          7   $-1$.          8   $\frac{1}{3}\pi$.

9   $\frac{1}{2}\pi$.          10   ∗.          11   $\frac{3}{2}$.          12   ∗.

13   ∗.          14   $\frac{1}{2}\pi$.          15   ∗.          16   $\frac{1}{4}e^2$.

17   2.          20   $\dfrac{\pi}{2(a+b)}$.          21   $-\dfrac{\pi}{2ab}$.          22   ∗.

29   $\frac{1}{2}\pi$.

32   $m$ odd, $\dfrac{(m-1)(m-3)\dots 2}{m(m-2)\dots 1}$;   $m$ even, $\dfrac{(m-1)(m-3)\dots 1}{m(m-2)\dots 2}\dfrac{\pi}{2}$.

33   $\dfrac{(m-1)(m-3)\dots(2n-m-3)(2n-m-5)\dots}{(2n-2)(2n-4)\dots 2}$, $m$ odd; same with factor $\frac{1}{2}\pi$
if $m$ is even.

35   $\frac{1}{2}\pi(a+b)$.

## Miscellaneous Exercise 4(o), p. 173

1   $x + \frac{3}{2}\log(2x+1) - \log(x^2+2) - \dfrac{1}{\sqrt{2}}\tan^{-1}\dfrac{x}{\sqrt{2}}$.      2   $1 + 3\log 2 - \frac{4}{3}\log 5$.

3   $\frac{1}{2}\log(x^2+4) - \log(x-1) - \dfrac{1}{x-1} + \frac{1}{2}\tan^{-1}\frac{1}{2}x$.      4   $\frac{1}{8}\pi - \frac{1}{4}$.

5   $\frac{5}{4} - \frac{3}{8}\pi$.      6   $\frac{1}{4}\pi - \frac{1}{3}$.      7   $\frac{1}{6}\log\dfrac{(x-1)^2}{x^2+x+1} - \dfrac{1}{\sqrt{3}}\tan^{-1}\dfrac{2x+1}{\sqrt{3}}$.

8   $\frac{1}{6}\log\dfrac{x-1}{x+1} + \dfrac{\sqrt{2}}{3}\tan^{-1}\dfrac{x}{\sqrt{2}}$.      9   $\frac{1}{2}\log\dfrac{x^2+x+1}{x^2+1} + \dfrac{1}{\sqrt{3}}\tan^{-1}\dfrac{2x+1}{\sqrt{3}}$.

10   $\dfrac{1}{2a^3}\tan^{-1}\dfrac{x}{a} + \dfrac{1}{4a^3}\log\dfrac{a+x}{a-x}$.

11   $\log x - \frac{1}{2}\log(1+x) - \frac{1}{4}\log(1+x^2) - \frac{1}{2}\tan^{-1}x$.

12   $\frac{1}{3}\log(x+2) - \frac{1}{12}\log(1-x) - \frac{1}{4}\log(x+3)$.

13   $2\log(x+1) - \log(2x-1) - \log(x-1)$.

14   $\frac{1}{4}\log\frac{32}{17}$ [put $x = \sqrt{\tan\theta}$].      15   $\dfrac{1}{\sqrt{2}}\tan^{-1}\left(\dfrac{x\sqrt{2}}{1-x^2}\right)$ $\left[\text{put } u = \dfrac{x\sqrt{2}}{1-x^2}\right]$.

16   $\frac{5}{3} - \log 4$.      17   $\dfrac{1}{3a}\{(x+a)^{\frac{3}{2}} + (x-a)^{\frac{3}{2}}\}$.

18   $\frac{1}{8}\tan^{-1}x + \dfrac{2x^4+x^3-x}{8(1+x^2)^2}$ [put $x = \tan\theta$].      19   $a\sin^{-1}(x/a) - \sqrt{(a^2-x^2)}$.

**20** $\frac{1}{4}a^2(\pi-2)$ [put $x^2 = a^2\cos 2\theta$].      **21** $\frac{1}{2}\pi\{1-\sqrt{(b^2-a^2)}/|b|\}$.

**22** $\sqrt{(ax+x^2)}+a\log\{\sqrt{x}+\sqrt{(a+x)}\}$ [put $x=t^2$].    **23** $\dfrac{2x^2-1}{3x^3}\sqrt{(1+x^2)}$.

**24** $\frac{1}{6}\log(t^2-t+1)-\frac{1}{3}\log(t+1)+\dfrac{1}{\sqrt{3}}\tan^{-1}\dfrac{2t-1}{\sqrt{3}}$, where $t^3 = \dfrac{1}{x^3}-1$.

**25** $\operatorname{sh}^{-1}(x+1/x)$ [put $t=x+1/x$].

**26** $(1/a^4)\{x(a^2-x^2)^{-\frac{1}{2}}+\frac{1}{3}x^3(a^2-x^2)^{-\frac{3}{2}}\}$.

**27** $-\dfrac{1}{6x^2}-\frac{1}{3}\log x-\dfrac{(1-x^2)^{\frac{3}{2}}}{3x^3}\sin^{-1}x$.

**28** $\frac{1}{2}\theta^2-\theta\tan\theta-\log\cos\theta$, where $x=\cot\theta$.     **29** $\frac{1}{2}\log\frac{5}{3}$.

**30** $-\operatorname{ch}^{-1}\{(1+x)/(x\sqrt{2})\}$.     **31** $3\pi$.     **32** $\log(1+\sqrt{2})$.

**33** $(1/\sqrt{2})\log(1+\sqrt{2})$.     **34** $\frac{1}{2}$.    **35** $\frac{1}{5}e^{-x}(2\sin 2x-\cos 2x)$.

**36** $\frac{3}{10}(1+e^{-\pi/3})\doteqdot 0\cdot405$.     **37** $\frac{1}{4}(x^4-1)\log(1+x^2)-\frac{1}{8}x^4+\frac{1}{4}x^2$.

**38** $\log x(\log\log x-1)$.    **39** $2-\frac{1}{2}\pi$.     **40** $1/\sqrt{2}+\frac{1}{2}\log(\tan\frac{3}{8}\pi)$.

**41** $\frac{1}{4}x-\frac{1}{16}\sin 4x$.     **42** $\frac{1}{2}\log 2$.

**43** (i) $(1-\log\sin x)\cos x+\log\tan\frac{1}{2}x$;   (ii) $\log 2-1$.     **44** $1-\frac{2}{3}c^2$.

**45** $\frac{1}{4}(\pi-1)$.     **46** $\frac{1}{4}\log 3$.     **47** $(1/\sqrt{2})\log\tan(\frac{1}{2}x-\frac{1}{8}\pi)$.

**48** $-\log(1-\tan\frac{1}{2}x)$.     **49** $\frac{1}{5}\log 6$.     **50** $\pi/\sqrt{(ab)}$.

**51** If $\alpha \neq \frac{1}{2}\pi$, $\frac{1}{2}\sec\alpha\log|\tan(\alpha-\frac{1}{4}\pi)|$; if $\alpha=\frac{1}{2}\pi$, integral does not exist.

**52** $1$.               **53** $\frac{8}{45}$.

**54** $\frac{1}{16}\{3\sqrt{2}\log(1+\sqrt{2})-2\}$ [put $\cos 2\phi = \tan^2 x$].

**55** $(1/\sqrt{6})\{\tan^{-1}\sqrt{6}-\tan^{-1}\frac{1}{2}\sqrt{6}\} = (1/\sqrt{6})\tan^{-1}(\frac{1}{8}\sqrt{6})$.

**56** $\frac{4}{3}$.        **57** $\dfrac{a^2-1}{a^3}\log(1+a)+\dfrac{2-a}{2a^2}$ if $a\neq 0$; $\frac{2}{3}$ if $a=0$.

**58** $\frac{5}{12}-\frac{1}{2}\log 2$.    **59** $\frac{1}{2}\pi a^3$.     **60** $\frac{5}{16}\pi a^3$ [put $x=a\sin^2\theta$].

**61** $x(\log x-1) = x\log(x/e)$. Put $x=1-e\cos\theta$.

**65** $\frac{7}{8}\pi a^5$.          **66** $\frac{8}{315}$.

**67** We have $(m+n)J(m,n) = mI(m-1,n-1)-\cos\frac{1}{2}n\pi$, hence

     $(m+n)(m+n-2)I(m,n) = (2m+n-2)\sin\frac{1}{2}n\pi - m(m-1)I(m-2,n-2)$.

**68** $\frac{23}{15}$.        **69** $n!\,q^n/\{(nq+p+1)(\overline{n-1}q+p+1)\dots(q+p+1)(p+1)\}$.

**70** $u_n = \dfrac{x}{n+1}(x^2+a^2)^{\frac{1}{2}n}+\dfrac{na^2}{n+1}u_{n-1}$.

**71** $n$ even, $u_n = \dfrac{1}{n-1}-\dfrac{1}{n-3}+\dfrac{1}{n-5}-\dots+(-1)^{\frac{1}{2}n-1}+(-1)^{\frac{1}{2}n}\frac{1}{4}\pi$;

   $n$ odd, $u_n = \dfrac{1}{n-1}-\dfrac{1}{n-3}+\dfrac{1}{n-5}-\dots+(-1)^{\frac{1}{2}(n+1)}\frac{1}{2}+(-1)^{\frac{1}{2}(n+3)}\frac{1}{2}\log 2$.

**73** $c = \{a\operatorname{sh} ax\cos bx+b\operatorname{ch} ax\sin bx\}/(a^2+b^2)$,

   $s = \{a\operatorname{ch} ax\sin bx-b\operatorname{sh} ax\cos bx\}/(a^2+b^2)$.

74  $\{a \operatorname{sh} ax \sin bx - b \operatorname{ch} ax \cos bx\}/(a^2 + b^2)$;
    $\{a \operatorname{ch} ax \cos bx + \operatorname{sh} ax \sin bx\}/(a^2 + b^2)$.

77  Put $y = \frac{1}{4}\pi - x$ in the first integral.          78  $n\int_0^\pi f(\cos^2 x)\, dx.$

79  (ii) $\frac{1}{8}\pi \log 2.$

## Exercise 5(a), p. 178

1  $x \log x \, dy/dx = y.$

2  Tangent to a circle is perpendicular to the radius.

4  $y'' = p^2 y.$          6  $y = 3xy' + y^4 y'^3.$

7  Differential equation of all parabolas with $Ox$ for axis of symmetry.

11  $y''' - 6y'' + 11y' - 6y = 0.$

## Exercise 5(b), p. 181

1  $y = \frac{5}{3}x^3 + c.$        2  $y^4(c - 12x) = 1.$        3  $\tan y = x + c.$

4  $y = \pm \frac{2}{5}x^{\frac{5}{2}} + c.$        5  $y^{\frac{3}{2}} = \frac{2}{3}x + c.$        6  $y = \operatorname{sh}(x + c).$

7  $y = \frac{1}{2}x^2 - 1/(2x^2).$        8  $\cos y = c \cos x.$        9  $y^2 = 1 + 2x - x^2.$

10  $xy = x + c.$        11  $\log y = x(\log x - 1).$        13  $x + y = \tan(x + c).$

14  $(x^2 + 2y + c)(y + c e^{2x}) = 0.$        15  $y^3 = xy + c.$

16  $xy = \log x + c.$        17  $v \to k.$

## Exercise 5(c), p. 185

1  $2xy = x^2 + c.$        2  $x^2 - 2xy - y^2 = 9.$        3  $y^2 = x^2 \log(cx^2).$

4  $(x - y)^2 = cxy^2.$        5  $2\sin(y/x) = x.$

6  $y = x \operatorname{sh}(\log x + c) = \frac{1}{2}(c_1 x^2 - 1/c_1).$        7  $(x - y)^3(x^3 - y^3) = c.$

8  $(x - y - 2)^3(x + y) = c.$        9  $\log(x + y) = x - y + c.$

10  $4(x - 2) = (x - y - 3)\log\{c(x - 2)\}.$

11  $2(x + 1)^3 + 3(x + 1)^2(y - 2) + (y - 2)^3 = c.$

## Exercise 5(d), p. 189

1  $y = e^x(x + c).$        2  $y = \frac{2}{3}\sin^2 x + c \operatorname{cosec} x.$

3  $y = 2\sin x \log(\sin x) + c \sin x.$          4  $y = 1 + e^{1 - x^2}.$

5  $y \log x = (\log x)^2 + c.$    7  $y^2(1 + c e^{2x}) = 1.$    8  $y^2 = -x - \frac{1}{2}.$

9  $y^2 = cx^6 - 2x^3.$        10  $xy^2 = c - \cos x.$        11  $y^2(c - 2\tan x) = \cos^2 x.$

12  $y = A x e^{-1/x}.$        13  $y^2(c - 5y) = x.$        14  $x = c e^y - \frac{1}{2}e^{-y}.$

15  $y = x + a + b e^{-x}.$        16  $y = cx + 1/c,\ y^2 = 4x.$

17  $y = cx \pm \sqrt{(c^2 + 1)},\ x^2 + y^2 = 1.$

18  $y = cx + c^3,\ 4x^3 + 27y^2 = 0;\ c = -1, -2, 3.$

19  $y = cx + \sin c$; s.s. given parametrically by $x = -\cos c,\ y = \sin c - c \cos c.$

21  $(y - px)^2 = -4pk^2.$

## Exercise 5(e), p. 192

1   $y = \dfrac{1}{2x} + ax + b.$     2   $y^2 = c^2 + \left(\dfrac{a}{c}x + b\right)^2.$

3   $y = a\,e^{nx} + b\,e^{-nx} = a_1\,\text{ch}\,(nx + b_1).$      4   $y = a + b\,e^{3x}.$

5   $y = \tfrac{1}{2}x^4 + 1.$     6   $a - x = y^3 + by, y = c.$   7   $y = b \pm \text{ch}\,(x + a).$

8   $y = 1 + x^2.$     9   $y = x^2.$      10   $u = \left(\dfrac{1}{c} - \dfrac{\mu}{h^2}\right)\cos\theta + \dfrac{\mu}{h^2}.$

11   If $\mu < h^2, u = \dfrac{1}{c}\cos\left\{\theta\Big/\left(1 - \dfrac{\mu}{h^2}\right)\right\}$; if $\mu = h^2, u = \dfrac{1}{c}$;

    if $\mu > h^2, u = \dfrac{1}{c}\,\text{ch}\left\{\theta\Big/\left(\dfrac{\mu}{h^2} - 1\right)\right\}.$

12   $u^2 = \dfrac{1}{c^2}\cos^2\theta + \dfrac{\mu}{h^2}c^2\sin^2\theta.$

## Exercise 5(f), p. 198

1   $y = A\,e^x + B\,e^{3x}.$     2   $y = A\,e^{4x} + B\,e^{-3x}.$     3   $y = (Ax + B)\,e^{-bx}.$

4   $y = A\,e^{2x} + B\,e^{-2x}.$     5   $y = A\cos 2x + B\sin 2x.$

6   $y = (A\cos 3x + B\sin 3x)\,e^{-2x}.$

7   $y = e^{\frac{1}{2}x}\{A\cos(\tfrac{1}{2}x\sqrt{3}) + B\sin(\tfrac{1}{2}x\sqrt{3})\}.$      8   $y = A + B\,e^{-3x}.$

9   $A = 2\tfrac{1}{2}.$     10   $A = 2, B = \tfrac{4}{3}.$

11   $A = -2, B = -10, C = -27.$     12   $A = -7.$

13   $A = \tfrac{22}{13}, B = -\tfrac{19}{13}.$    14   $A = 1, B = -\tfrac{1}{2}.$

15   $A = \tfrac{1}{5}, B = -\tfrac{3}{25}, C = \tfrac{6}{125}.$     16   $A = 1.$

17   $A = 1.$     18   $A = \tfrac{1}{3}, B = \tfrac{1}{2}.$     19   $A = \tfrac{1}{2}, B = \tfrac{3}{4}.$

20   $A = \tfrac{1}{2}, B = -1.$     21   $A = \tfrac{1}{6}.$     22   $A = \tfrac{1}{2}, B = \tfrac{3}{2}, C = \tfrac{7}{4}.$

23   $A = -\tfrac{1}{3}, B = -1, C = -2.$     24   $A = \tfrac{1}{12}.$

## Exercise 5(g), p. 201

1   $y = A\,e^x + B\,e^{2x} + 3.$

2   $y = e^x\{A\cos(x\sqrt{2}) + B\sin(x\sqrt{2})\} + 2x + 1.$

3   $y = A\,e^{2x} + B\,e^{-x} + \tfrac{1}{6}(2x^2 - 2x + 3).$

4   $y = A\,e^x + B\,e^{\frac{3}{2}x} + 4e^{2x};\ y = 4(e^{2x} + e^x - 2e^{\frac{3}{2}x}).$

5   $y = (A + \tfrac{2}{7}x)\,e^{4x} + B\,e^{-3x}.$

6   $y = e^x(A\cos x + B\sin x) + \tfrac{1}{85}(6\cos 3x - 7\sin 3x).$

7   $y = e^{7x}(A\cos x + B\sin x) + \tfrac{2}{371}(7\cos x - 2\sin x).$

8   $y = A\cos 4x + (B + \tfrac{1}{8}x)\sin 4x;\ y = (A + \tfrac{1}{8}x)\sin 4x.$

9   $y = e^{-4x}(3x^2 + Ax + B).$      10   $y = A + B\,e^{-6x} + \tfrac{1}{9}(3x^2 - x).$

11   $y = \tfrac{1}{24}\cos 4x + A\cos 2x + (B + \tfrac{1}{8}x)\sin 2x.$

12  $y = e^{\frac{3}{2}x}\{A\cos\left(\frac{3}{2}\sqrt{7}x\right) + B\sin\left(\frac{3}{2}\sqrt{7}x\right)\} + \frac{1}{32}e^{2x} - \frac{1}{56}e^{-2x}.$

13  $y = A\,e^{2x} + (B + \frac{1}{2}x)\,e^{4x} - \frac{1}{40}(\cos 2x - 3\sin 2x).$

14  $y = e^{-x}(A - \frac{1}{2}\sin x - \frac{1}{2}\cos x) + B\,e^{-2x}.$

15  $y = e^{x}(A + 18x - 3x^2 + \frac{1}{3}x^3) + B\,e^{\frac{3}{2}x}.$

16  $y = e^{-3x}(\frac{1}{6}x^3 + \frac{1}{2}x^2 + Ax + B).$

17  $y = e^{-\frac{1}{2}x}\{A\cos\left(\frac{1}{2}\sqrt{3}x\right) + B\sin\left(\frac{1}{2}\sqrt{3}x\right)\} + e^{x}\{\frac{1}{3}(x-1) + \frac{1}{13}(2\cos x + 3\sin x)\}.$

18  $y = e^{2x}\{Ax + B + (3 - 2x^2)\sin 2x - 4x\cos 2x\}.$

## Exercise 5(h), p. 208

1  (ii) $x = a(\cos pt - \cos nt)/(n^2 - p^2).$    2  (ii) $x = (at/2n)\sin nt.$

8  $x = e^{-Rt/2L}(A\cos nt + B\sin nt),$ where $n^2 = 1/LC - R^2/4L^2.$

9  $b = EC\{(CLp^2 - 1)^2 + C^2R^2p^2\}^{-\frac{1}{2}},\ \tan\alpha = CRp/(CLp^2 - 1).$

13  $y = \dfrac{\phi(-a^2)(\lambda\cos ax + \mu\sin ax) + a\psi(-a^2)(\lambda\sin ax - \mu\cos ax)}{\{\phi(-a^2)\}^2 + a^2\{\psi(-a^2)\}^2}.$

## Exercise 5(i), p. 211

1  $x = A\,e^{t} + B\,e^{-t},\ y = \sin t - A\,e^{t} + B\,e^{-t}.$

2  $x = A + B\,e^{-2t} + \frac{2}{3}e^{t},\ y = A - B\,e^{-2t} + \frac{1}{3}e^{t}.$

3  $x = A\,e^{-3t} + B\,e^{-t},\ y = A\,e^{-3t} - B\,e^{-t}.$

4  $x = \frac{3}{5} + \frac{1}{6}e^{t} - \frac{3}{4}e^{-t} - \frac{1}{60}e^{-5t},\ y = \frac{2}{5} + \frac{1}{3}e^{t} - \frac{3}{4}e^{-t} + \frac{1}{60}e^{-5t}.$

5  $x = A\,e^{3t} + (B - t)\,e^{2t} - \frac{3}{2}e^{t},\ y = 2e^{t} - 2A\,e^{3t} - (B + 2 - t)\,e^{2t}.$

6  $x = \frac{6}{7}(e^{-2t} - e^{-3t/5}) + \sin t,\ y = 2\cos t - \frac{2}{7}(6e^{-2t} + e^{-3t/5}).$

7  $u = e^{t} + a,\ v = \frac{1}{3}e^{t} + b\,e^{-2t}.$

9  $x = (A - \frac{1}{2}t^2)\cos t + (B + \frac{1}{2}t)\sin t,\ y = (3A - \frac{3}{2} - \frac{3}{2}t^2)\sin t + (\frac{11}{2}t - 3B)\cos t.$

10  $u = A\cos\omega t + B\sin\omega t,\ v = E/H - B\cos\omega t + A\sin\omega t.$

## Exercise 5(j), p. 216

1  $y = x^2\{A + B\log x + \frac{1}{6}(\log x)^3\}.$    2  $y = \frac{1}{16}x^2(A + 4\log x) + B/x^2.$

3  $y = x^{\frac{1}{2}}(\log x - 6) + \log x + 6.$    4  $y = Ax^4 + B/x - \frac{1}{4}x^3 - \frac{1}{2}\log x + \frac{3}{8}.$

5  $y = (1 + 2x)^2\{A\log(1 + 2x) + B\} + \frac{1}{2}\log(1 + 2x) + \frac{1}{2}.$

6  $x = At^2 + B/t^2 + C\cos(2\log t) + D\sin(2\log t),$
$\quad y = At^2 + B/t^2 - C\cos(2\log t) - D\sin(2\log t).$

7  $y = A\sqrt{(1 - x^2)} + Bx - \frac{1}{2}\sqrt{(1 - x^2)}\sin^{-1}x.$

8  $y = \frac{1}{4}(\mathrm{sh}^{-1}x)^4 + A\,\mathrm{sh}^{-1}x + B.$

9  $y = A\dfrac{1 - x^2}{1 + x^2} + B\dfrac{x}{1 + x^2}.$    10  $y = \dfrac{1}{x^3}(A\,e^{-x} + B\,e^{-3x}).$

11  $n = -2,\ a = 4,\ b = 0;\ y = \dfrac{1}{x^2}\{A + B\,e^{-4x} + \frac{1}{17}(4\sin x - \cos x)\}.$

12 $y = x(A e^{-2/x} + B)$.

13 $y = e^x \{A/(x+1) + B\}$.

14 $y = e^{2x} + (Ax^3 + B) e^x$.

15 $y = \frac{1}{6}x^3 \log x + Ax^{-3} + Bx^3$.

16 $y = e^x (\frac{1}{2}x - 1 + 1/x) + A e^{-x} (x + 2 + 2/x) + B/x$.

19 $xy(x^2 + c) = 3x^2 + c$.

## Exercise 5(k), p. 219

1 $y^2 = kx + c$.

2 $y = A e^{x/k}$.

3 $y = k \operatorname{ch} \{(x-c)/k\}$.

4 $xy = c^2$.

5 $x - \sqrt{(x^2 + y^2)} = c$.

6 $x + \sqrt{(x^2 + y^2)} = c$.

7 $x^2 - y^2 = c$.

8 $x = y(c - k \log y)$.

9 $(x-a)^2 + y^2 = k^2$.

10 $ky = \operatorname{ch}(kx + c)$.

11 $x^2 - y^2 = b$.

12 $2x^2 + y^2 = b$.

13 $2x^2 + 3y^2 = b$.

14 $y^{\frac{2}{3}} = x^{\frac{2}{3}} + b^{\frac{2}{3}}$.

15 $x^3 - 3xy^2 = b$.

16 $x + 1 = b e^{x + \frac{1}{2}y^2}$.

17 (i) Differential equation is $y/y' - yy' = 2x$.

18 Given $x$ and $y$, equation is a quadratic for $dy/dx$, giving the two directions through $(x, y)$. Condition for equal roots.

19 $\frac{1}{4}\pi$, $\tan^{-1}\frac{1}{8}$.

20 $y = -1$.

## Miscellaneous Exercise 5(l), p. 220

1 $x^2(1 + y'^2) = y^2(\operatorname{sh}^{-1} y')^2$.

2 $(1 - x^2) y'' = xy'$.

3 Since $A \cos^{-1} x + B \sin^{-1} x = A \cos^{-1} x + B(\frac{1}{2}\pi - \cos^{-1} x) = A' \cos^{-1} x + B'$.

4 $x^4 y'' + y = 0$.

5 $y'' = 0$.

6 $(xy' - y)^2 = \pm 2xy(1 + y'^2)$.

7 $(xy' - y)^2 + 4y' = 0$.

8 $y = (x + c)/(1 - cx)$.

9 $y = cx e^x$.

10 $y = c + \tan^{-1}(x + y)$.

11 $\log(x^2 + y^2) = 2 \tan^{-1}(y/x) + c$.

12 $\tan^{-1}(y/x) + 2 \log(x^2 + y^2) = c$.

13 $2(x - y)^2 - 2(3x + y) + \log(2x - 2y - 1) = c$.

14 $x^2 + y^2 = cy$.

15 $7x^2 - 5y^2 - 6xy + 4x - 2y = c$.

16 $\tan(y/x) = \log(c/x)$.

17 $\log(1 + y) = \frac{1}{2}x^2 + c$.

18 $xy = \frac{1}{4}x^4 + c$.

19 $y = x + c\sqrt{(1 + x^2)}$.

20 $1/y^2 = \frac{1}{2} + x^2 + c e^{2x^2}$.

21 $1/x = 2 - y^2 + c e^{-\frac{1}{2}y^2}$.

22 $1/y = 1 + cx + \log x$.

23 $y = (e^x + c) \sin x$.

24 G.S. $y = cx + \frac{1}{3}c^3$, S.S. $4x^3 + 9y^2 = 0$.

25 G.S. $y = cx + e^c$, S.S. $y = x \log(-x/e)$.

26 $y = \log(x + a) + b$.

27 $y = a \operatorname{sh}^{-1} x + b$.

28 $y^2 = x^2 + ax + b$.

29 $y^2 = a \log x + b$.

30 $y = A e^{2x} + B e^{-x}$.

31 $y = (Ax + B) e^x$.

32 $y = Ax + B + C e^{3x}$.

33 $y = (Ax^2 + Bx + C) e^{-x}$.

34 $y = A e^{4x} + B e^{-x} - \frac{1}{5}(4 \cos 2x + 3 \sin 2x)$.

35 $y = A e^{2x} + B e^{3x} + e^x (2x^2 + 6x + 7)$.

**36** $y = A\,e^x + B\,e^{3x} + \frac{1}{3}x^3 + \frac{4}{3}x^2 + \frac{26}{9}x + \frac{80}{27}$.

**37** $y = e^{5x}(A\cos 2x + B\sin 2x) - \frac{1}{4}x\,e^{5x}\cos 2x$.   **38** $y = Ax^5 + Bx^{-4} - \frac{1}{2}x^3$.

**39** $y = x\{\log x + A\cos(\log x) + B\sin(\log x)\}$.

**40** $y = Ax + Bx^2 + \frac{1}{10}x^3\{\cos(\log x) + 3\sin(\log x)\}$.   **41** $y = A + B/x$.

**42** $y = \frac{1}{24}(\log x)^4 + A(\log x)^2 + B\log x + C$.   **43** $y = A\,e^{\sin x} + B\,e^{-\sin x}$.

**44** $y = (A + Bx)(1+x^2)^{-\frac{1}{2}} - \frac{1}{3}x(1+x^2)^{-1}$.

**45** $x = 2A\,e^t + B\,e^{-4t},\ y = 3A\,e^t - B\,e^{-4t}$.

**46** $x = 3A\cos 2t + 3B\sin 2t,\ y = A\sin 2t - B\cos 2t$.

**47** $x = 2A\,e^{-4t} - B\,e^{-7t} + 7e^t + e^{2t},\ y = A\,e^{-4t} + B\,e^{-7t} + e^t + \frac{7}{2}e^{2t}$.

**48** $x = \log t - At + Bt^{-1},\ y = A\,e^t + Bt^{-1} - 1$.

**49** $x = A\,e^t + B\,e^{-\frac{1}{2}t}\cos(\frac{1}{2}\sqrt3\,t + \alpha),\ y = A\,e^t + B\,e^{-\frac{1}{2}t}\cos(\frac{1}{2}\sqrt3\,t + \alpha + \frac{2}{3}\pi)$,
$z = A\,e^t + B\,e^{-\frac{1}{2}t}\cos(\frac{1}{2}\sqrt3\,t + \alpha + \frac{4}{3}\pi)$.

**50** $x = 2A\,e^{-2t} + \frac{3}{5}\{(2B+C)\cos t - (B-2C)\sin t\} + D\,e^{2t}$,
$y = A\,e^{-2t} + B\cos t + C\sin t$.

**51** $y = e^x(ax^2 + b)$.   **52** $y(1+x) = ax^3 + b$.

**53** $z'' + z = 0;\ y = x^{-\frac{1}{2}}(A\cos x + B\sin x)$.

**54** $n = -2,\ z'' + 2z' + 2z = e^{-x}\cos x;\ y = x^{-2}e^{-x}\{A\cos x + (B + \frac{1}{2}x)\sin x\}$.

**55** For given $(x,y)$, equation is a quadratic in $p$ whose roots have product $-1$.
$$\{y + \sqrt{(x^2+y^2)} - cx^2\}\{y + \sqrt{(x^2+y^2)} - c\} = 0.$$

**56** $y'' - (k+l)y' + kly = A\,e^{mx}$.   **57** $f(x) = e^x/(a - e^x)$.

**58** $y'\cos nx + ny\sin nx = \displaystyle\int_0^x f(t)\cos nt\,dt + nB$,

$y'\sin nx - ny\cos nx = \displaystyle\int_0^x f(t)\sin nt\,dt - nA$.

**59** $y = A\cos x + B\sin x + x\sin x + \cos x\,\log(\cos x)$.

## Exercise 6(a), p. 229

**1** $1/(n+1)$.   **2** $\frac{3}{4}\pi,\ \frac{7}{4}\pi$.   **3** $\frac{1}{3}(2a+b)$.

**5** $\sin^{-1}(2/\pi)$.   **6** $(1/n)^{1/(n-1)}$.   **7** $\frac{1}{2}$.

**8** $h^{-1}\{\sqrt{(a^2 + ah + \frac{1}{3}h^2)} - a\}$.   **9** $h^{-1}\log\{(e^h - 1)/h\}$.

**10** $0 < \dfrac{1}{h}\log\dfrac{e^h - 1}{h} < 1$.

**11** $f(x)$ is discontinuous in $a \leqslant x \leqslant a+h$, at $x = 0$.

**16** The points $x = a, b, c$ of the curve $y = f(x)$ are collinear; the curve generally has an inflexion somewhere between $a, b$.

## Exercise 6(b), p. 240

**1** $2^{n-1}\cos(2x + \frac{1}{2}n\pi)$.   **2** $\frac{3}{4}\sin(x + \frac{1}{2}n\pi) - (3^n/4)\sin(3x + \frac{1}{2}n\pi)$.

**3** $\frac{1}{2}e^x + \frac{1}{2}(-1)^n e^{-x}$.   **4** $\frac{1}{3}(-1)^n n!\{(x-1)^{-n-1} + 2(x+2)^{-n-1}\}$.

5  $e^x\{x^4 + 4nx^3 + 6n(n-1)x^2 + 4n(n-1)(n-2)x + n(n-1)(n-2)(n-3)\}$.

6  $(-1)^n e^{-x}\{x^3 - 3nx^2 + 3n(n-1)x - n(n-1)(n-2)\}$.

7  $(-1)^{n-1} n! \, x^{4-n}\left\{\dfrac{1}{n} - \dfrac{4}{n-1} + \dfrac{6}{n-2} - \dfrac{4}{n-3} + \dfrac{1}{n-4}\right\}$.

8  $2^{n-4}[2x\{3n(n-1) - 4x^2\}\cos(2x + \tfrac{1}{2}n\pi)$
$\qquad\qquad\qquad + n\{12x^2 - (n-1)(n-2)\}\sin(2x + \tfrac{1}{2}n\pi)]$.

9  $2^{n-2}\{\sin(2x + \tfrac{1}{2}n\pi) + 2^n \sin(4x + \tfrac{1}{2}n\pi) - 3^n \sin(6x + \tfrac{1}{2}n\pi)\}$.

15  $x - \tfrac{1}{3}x^3 + \tfrac{1}{5}x^5 - \tfrac{1}{7}x^7 + \dots$.

16  $a_{2n} = 2^{2n-1}\{(n-1)!\}^2, \ a_{2n-1} = 0; \ \dfrac{2}{2!}x^2 + \dfrac{2^3}{4!}x^4 + \dfrac{2^3 \cdot 4^2}{6!}x^6 + \dots$.

23  $\sin x + h\cos x - \dots + (-1)^n \dfrac{h^{2n}}{(2n)!}\sin x + (-1)^n \dfrac{h^{2n+1}}{(2n+1)!}\cos(x + \theta h)$.

24  $\cos x - h\sin x - \dots + (-1)^n \dfrac{h^{2n}}{(2n)!}\cos x + (-1)^{n+1}\dfrac{h^{2n+1}}{(2n+1)!}\sin(x + \theta h)$.

25  $a^x = 1 + x\log a + \dfrac{x^2}{2!}(\log a)^2 + \dots + \dfrac{x^{n-1}}{(n-1)!}(\log a)^{n-1} + \dfrac{x^n}{n!}(\log a)^n a^{\theta x}$.

## Exercise 6(c), p. 247

1 max.      2 inflex.      3 min.      4 inflex.

5 max.      6 inflex.

10 First approximation $x = \tfrac{3}{5}\pi \doteqdot 1\cdot8849$; second $\doteqdot 1\cdot8955 \doteqdot 1\cdot895$.

## Exercise 6(d), p. 252

1 $\xi = \tfrac{14}{9}, \ \xi_1 = \sqrt{\tfrac{7}{3}}, \ \xi_2 = \tfrac{3}{2}$.      2 $12\tfrac{1}{2}$.      3 $-2$.

4 1.      5 $\log 10 - 1$.      6 $2a/b$.      7 $-\pi^2/2e$.

8 2.      9 1.      10 1.      11 1.

12 1.      13 $e^2$.      14 0.      15 $-\tfrac{1}{3}$.

## Miscellaneous Exercise 6(e), p. 252

6 $\tfrac{1}{2}$.      7 $\tfrac{1}{8}$.      8 $-\tfrac{8}{3}$.      9 $\tfrac{1}{6}\pi$.

10 $e^{-\frac{1}{2}}$.      13 (i) $g(x) = x, \ h(x) = 1$;   (ii) $h(x) = 1$.

## Exercise 7(b), p. 269

1 (i) (a) $0\cdot7750$, (b) $0\cdot7828$;   (ii) (a) $0\cdot7833$, (b) $0\cdot7854$.

3 (i) Difference $= \displaystyle\int_{10}^{\infty} \dfrac{dx}{1 + x^4} < \int_{10}^{\infty} \dfrac{dx}{x^4}$;   (ii) $1\cdot07$;   (iii) $0\cdot04$;   (iv) $1\cdot11$.

## Exercise 7(c), p. 275

1 $\tfrac{4}{3}a^2$.      2 $3\pi a^2$.      3 $\tfrac{1}{15}$.      5 $\tfrac{9}{2}\pi a^2$.

6 $a^2$.      7 $\tfrac{1}{2}\pi a^2$.      8 $\tfrac{1}{12}\pi a^2$.      9 $2 - \tfrac{1}{2}\pi$.

10  $\frac{4}{3}a^2$.        11  $\frac{1}{4}a^2\log 3$.        12  (i) $\dfrac{1}{4a}(e^{\pi a}-1)$;  (ii) $\dfrac{1}{4a}(e^{5\pi a}-e^{4\pi a})$.

13  $(\frac{7}{12}\pi-\sqrt 3)\,a^2$.                    15  $\frac{1}{2}abu$.                16  $\frac{1}{4}\pi ab$.

17  $\frac{3}{8}\pi a^2$.        18  $\frac{1}{6}$.        19  $\frac{16}{35}$.        20  $\frac{1}{36}\pi a^3$.

21  $\frac{16}{105}\pi a^3$.        22  $5\pi^2 a^3$.        23  $\frac{1}{4}\pi l^3\tan^4\frac{1}{2}\alpha$.

24  $\frac{3}{10}\pi a^3$.        25  $\frac{1}{12}\pi$.

### Exercise 7(d), p. 279

1  $c\,\mathrm{sh}\,(x/c)$.        2  $6a$.        4  $\frac{8}{27}a\{(1+9c/4a)^{\frac{3}{2}}-1\}$.

5  $a\,e^{-s/a}$.        6  $\tan^{-1}\sqrt 2-\frac{1}{4}\pi+\sqrt 2$.        7  $\frac{4}{3}\sqrt 3$.

8  $8a$.        10  $(1/2a)\{r_1\sqrt{(a^2+r_1^2)}+a^2\,\mathrm{sh}^{-1}(r_1/a)\}$.        13  $2\pi a$.

14  $(\theta+c)^2=1+2a/r$.        15  $a^2 y^2-2\log y=4a(x+b)$.

### Exercise 7(e), p. 286

1  $4\pi a^2$.        2  $\frac{1}{3}\pi a^2$.        3  $\frac{6}{5}\pi a^2$.        4  $\frac{64}{3}\pi a^2$.

5  $\frac{2}{3}\pi l^2(\sec^3\frac{1}{2}\alpha-1)$.        10  $\dfrac{4a}{3\pi},\ \dfrac{4b}{3\pi}$.

11  (i) $\dfrac{a\sin\alpha}{\alpha}$,  (ii) $\dfrac{2}{3}a\,\dfrac{\sin\alpha}{\alpha}$ from the centre along the radius of symmetry.

12  $4\pi^2 a^2 b$.        13  $(\frac{5}{6}a,0)$, $\frac{7}{2}\pi^2 a^3$.        14  $(\pi a,\frac{4}{3}a)$, $\frac{32}{3}\pi a^2$.

15  $\frac{32}{105}\pi a^3$.        16  $\frac{20}{3}\pi$.

### Exercise 7(f), p. 292

1  $Ma^2$.        2  $M(\frac{1}{3}a^2+c^2)\sin^2\theta$.        3  $\frac{1}{2}M(a^2+b^2)$.

4  $\frac{1}{2}Ma^2$.        5  $\frac{1}{2}Mh^2$.        6  $\frac{3}{10}Mr^2$.

7  (i) $\frac{4}{5}Mab$;  (ii) $\frac{8}{35}Mb^2$.        8  $\frac{4}{3}Mab$.        9  $\frac{1}{2}Mr^2$.

10  $\frac{2}{3}Ma^2$.        11  (i) $\frac{1}{2}Ma^2$;  (ii) $2Ma^2$.        12  $\frac{4}{3}M(a^2+b^2)$.

13  $\frac{1}{4}Ma^2$.        14  $\frac{3}{2}Ma^2$.        15  $M(b^2+\frac{3}{4}a^2)$.        16  $\frac{1}{18}Mh^2$.

18  (i) $M(\frac{1}{4}a^2+\frac{1}{3}h^2)$;  (ii) $M(\frac{1}{4}a^2+\frac{1}{12}h^2)$;  (iii) $M(\frac{5}{4}a^2+\frac{1}{3}h^2)$.

19  (i) $\frac{2}{5}Ma^2$;  (ii) $\frac{13}{20}Ma^2$.        20  $\frac{12}{5}Ma^2$.

21  (i) $\frac{3}{5}Ma^2$;  (ii) $\frac{3}{10}Ma^2$;  (iii) $\frac{13}{10}Ma^2$.        22  $\frac{1}{8}\pi Ma^2$.

### Miscellaneous Exercise 7(g), p. 293

3  $\frac{1}{32}\pi a^2$.        5  $a(\alpha-\mathrm{th}\,\frac{1}{2}\alpha)$, $a^2(\frac{1}{2}\alpha-\mathrm{th}\,\frac{1}{2}\alpha)$.        6  $c\log\sin\psi$.

8  $\frac{1}{2}\pi c^2\{b-a+\frac{1}{2}c\,\mathrm{sh}\,(2b/c)-\frac{1}{2}c\,\mathrm{sh}\,(2a/c)\}$.

9  $\left(\dfrac{4a}{3\pi},\dfrac{4a}{3\pi}\right)$; $\left(\dfrac{2a}{\pi},\dfrac{8a}{3\pi}-\dfrac{a}{2}\right)$.

10  $(\frac{1}{8}\pi a\sqrt 2,0)$.        12  $2\pi a(2a+2b+\pi b)$, $\frac{1}{3}\pi a^2(4a+3\pi b)$.        13  $\frac{3}{32}\pi a^2$, $\frac{2}{21}\pi a^3$.

**14** $\pi a^3 \lambda(\pi + 2\sin^{-1}\lambda) + \frac{2}{3}\pi a^3(2+\lambda^2)\sqrt{(1-\lambda^2)}$.

**15** $\frac{3}{4}\pi a^2, \frac{1}{4}\pi^2 a^3$; $(\frac{5}{6}a, 0)$.        **17** $(\frac{1}{2}\pi, \frac{1}{8}\pi a)$, $\frac{2}{9}Ma^2$.

**18** $\frac{1}{3}Ma^2$.      **19** $\frac{3}{20}M(r^2+4h^2)$.    **20** $\frac{2}{3}Ma^2b^2/(a^2+b^2)$.

**21** $\frac{10}{21}Ma^2$.

## Exercise 8(a), p. 300

**1** $s = 8a\sin\frac{1}{3}(\psi - \frac{1}{2}\pi)$, if $s = 0$ when $\theta = 0$.

**2** $x = a(\theta + \sin\theta)$, $y = a(1-\cos\theta)$, where $\theta = 2\psi$.      **4** $y = \log\sec x$.

**5** $(x-a)^2 + (y-b)^2 = c^2$, where $a$, $b$ are arbitrary.

**6** $x = 2a(\cos\psi + \psi\sin\psi)$, $y = 2a(\sin\psi - \psi\cos\psi)$.

## Exercise 8(b), p. 303

**1** $n\theta + \frac{1}{2}\pi$.     **3** $\frac{1}{3}\pi$.       **4** $r^n = a^n\sin n\theta$.    **9** $r = a\theta + c$.

**10** $r(\theta + c) = a$.    **11** $r = ce^{\pm\theta}$.      **12** $r = a\sin(\theta + c)$.

**13** $r = k(1-\cos\theta)$.         **14** $r^2 = k^2\sin 2\theta$.    **15** $r^3 = k^3\sin 3\theta$.

**16** $r = k\sin\frac{1}{2}\theta\sqrt{\sin\theta}$.       **17** $x^2 + y^2 = k(x+y)$.

**18** $x^2 + y^2 = ky$.           **19** $y(3x^2+y^2) = k(x^2+y^2)^3$.

## Exercise 8(c), p. 306

**1** $1/p^2 - 1/r^2 = 1/a^2$.    **2** $pr^{n-1} = a^n$.      **3** $2l/r = 1-e^2+l^2/p^2$.

**4** $pr = 4a^2$.          **5** $a^2b^2/p^2 = a^2+b^2-r^2$.    **6** $r^2+3p^2 = a^2$.

**7** $r = a\sec^2(\frac{1}{2}\theta + c)$.     **8** $r = a\{1+\sin(\theta+c)\}$.    **9** $r = a\cos(\theta+c)$.

## Exercise 8(d), p. 312

**1** $c$.          **2** $a\tan\psi$.       **3** $s = c\sin\psi$.     **4** $(1/c)\text{sech}^2(x/c)$.

**5** $4a\cos\frac{1}{2}\theta$.     **6** $\frac{1}{6}at(4+9t^2)^{\frac{3}{2}}$.     **7** $x = c\psi$, $y = c\log\sec\psi$.

**10** $2r\sqrt{(r/a)}$.    **11** $r\cosec\alpha$.    **12** $a^2/3r$.       **15** $\frac{5}{24}\sqrt{5}$.

**16** $-\frac{3}{2}\sqrt{2}$.      **17** $\frac{2}{3}\sqrt{2}, -2\sqrt{2}$.    **18** $-\frac{1}{2}, \frac{5}{16}\sqrt{5}$.

**20** Circles of radius $c$.         **21** $\frac{1}{2}a$.

**22** $ay^2 = (ax+b)^2 + k$.

## Exercise 8(e), p. 318

**1** $(\frac{46}{3}, \frac{57}{8})$.     **2** $\left(\dfrac{a^2-b^2}{a}\cos^3\phi, -\dfrac{a^2-b^2}{b}\sin^3\phi\right)$.     **5** $1$.

**6** $\dfrac{a^2}{b}$; $\rho$ at $(0, -b)$.         **7** $\frac{3}{2}a$.        **8** $\frac{1}{4}$.

## Exercise 8(f), p. 326

**1** $4x^2y + 1 = 0$.      **2** $4x^3 + 9y^2 = 0$.

**3** $y = \frac{1}{2}v^2/g - \frac{1}{2}gx^2/v^2$; bounding parabola for projectiles with initial speed $v$.

**4** $x^{\frac{2}{3}} + y^{\frac{2}{3}} = a^{\frac{2}{3}}$.      **5** $x \pm y = \pm c$.      **6** $x^{\frac{1}{2}} + y^{\frac{1}{2}} = a^{\frac{1}{2}}$.

**7** $(x^2 + y^2 - c^2)^2 = 4a^2\{x^2 + (y-c)^2\}$.      **8** $r = a(1 + \cos\theta)$.

**11** $y = 0$ (which is also a cusp locus).

**12** $x = 0$, $y = 0$ (which are also cusp loci).

**14** $x = -\frac{1}{2}at^2(9t^2 + 2)$, $y = \frac{4}{3}at(3t^2 + 1)$.

**15** $x = \frac{1}{2}c(3t^4 + 1)/t^3$, $y = \frac{1}{2}c(t^4 + 3)/t$; $(x+y)^{\frac{2}{3}} - (x-y)^{\frac{2}{3}} = 2a^{\frac{2}{3}}$.

**16** $x = \{(a^2 + b^2)/a\}\operatorname{ch}^3 t$, $y = -\{(a^2 + b^2)/b\}\operatorname{sh}^3 t$; $(ax)^{\frac{2}{3}} - (by)^{\frac{2}{3}} = (a^2 + b^2)^{\frac{2}{3}}$.

**17** $x = at$, $y = a\operatorname{ch} t$; $y = a\operatorname{ch}(x/a)$.

**18** $x = a\cos t(1 + 2\sin^2 t)$, $y = a\sin t(1 + 2\cos^2 t)$; $(x+y)^{\frac{2}{3}} + (x-y)^{\frac{2}{3}} = 2a^{\frac{2}{3}}$.

## Miscellaneous Exercise 8(g), p. 326

**1** $\theta$, $\theta + \frac{2}{3}\pi$, $\theta + \frac{4}{3}\pi$.      **2** $0$, $\tan^{-1}\sqrt{2}$, $\pi + \tan^{-1}\sqrt{2}$.      **4** $r = ae^{\frac{1}{2}\theta^2}$.

**6** $dx/d\psi = f'(\psi)\cos\psi$, $dy/d\psi = f'(\psi)\sin\psi$;
$x = \frac{4}{5}a(5 - \tan^2\frac{1}{2}\psi)\sqrt{(\tan\frac{1}{2}\psi)}$, $y = \frac{8}{3}a(\tan\frac{1}{2}\psi)^{\frac{3}{2}}$.

**7** $\frac{16}{5}a\sin\frac{3}{2}\theta$; $32a$.      **8** $4r^2 = 3p^2 + 16a^2$; $12\pi a^2$.

**10** $x = a(\sin 2t + 2\sin t)$, $y = a(\cos 2t - 2\cos t)$.      **12** $\frac{1}{2}a$.

**14** $\{(1-n^2)r^2 + n^2 a^2\}^{\frac{3}{2}}/\{(1-n^2)r^2 + 2n^2 a^2\}$.      **16** $a^n p = r^{n+1}$.

**17** $x^2 + y^2 + 2ax - \frac{19}{3}ay + 10a^2 = 0$.      **18** $x^2 + y^2 = a^2$.

**19** $x(1 + \cos\theta) - y\sin\theta - a\theta(1 + \cos\theta) = 0$; $x = a(\theta - \sin\theta)$, $y = -a(1 + \cos\theta)$.

**21** $(x+a)^2 + y^2 = b^2$; $y = a\operatorname{ch}(b + x/a)$.

**22** Cycloids $x = b + \frac{1}{2}a(\theta - \sin\theta)$, $y = \frac{1}{2}a(1 - \cos\theta)$;
parabolas $(x - b)^2 = 4a(y - a)$.

**23** $a^2 y^2 = 1 + be^{ax}$.      **24** $-3\sqrt{2}$, $-\frac{15}{4}\sqrt{5}$.      **25** $(x^2 + y^2)^2 = 16c^2 xy$.

## Exercise 9(a), p. 337

**1** $6x - 2y$, $-2x + 10y$.      **2** $1/y$, $-x/y^2$.      **3** $2xy^3$, $3x^2 y^2$.

**4** $-y/x\sqrt{(x^2 - y^2)}$, $1/\sqrt{(x^2 - y^2)}$.      **5** $2x/(x^2 + y^2)$, $2y/(x^2 + y^2)$.

**6** $6x + 6y$, $6x$, $6y$.      **7** $-y\sin x$, $\cos y + \cos x$, $-x\sin y$.

**8** $y^2 e^{xy}$, $(1 + xy)e^{xy}$, $x^2 e^{xy}$.      **9** $\operatorname{ch} x\operatorname{ch} y$, $\operatorname{sh} x\operatorname{sh} y$, $\operatorname{ch} x\operatorname{ch} y$.

**15** $f''(z)\dfrac{\partial z}{\partial x}\dfrac{\partial z}{\partial y} + f'(z)\dfrac{\partial^2 z}{\partial x\,\partial y}$.      **16** $2u$.

**18** $\dfrac{y}{x}(x+y)^3 f'''(x+y)$.      **20** $z = x\dfrac{\partial z}{\partial x} + y\dfrac{\partial z}{\partial y} + \dfrac{\partial z}{\partial x}\dfrac{\partial z}{\partial y}$.      **21** $\dfrac{\partial^2 y}{\partial x^2} = \dfrac{\partial y}{\partial t}$.

**22** $\dfrac{\partial^2 u}{\partial x^2} = \dfrac{1}{c}\dfrac{\partial u}{\partial t}$.      **24** $\dfrac{\partial u}{\partial x}\dfrac{\partial v}{\partial y} = \dfrac{\partial u}{\partial y}\dfrac{\partial v}{\partial x}$.      **25** $Ae^{-c^2 p^2 y}$.

**26** $Ae^{py} + Be^{-py}$; $u = \dfrac{1}{1 - e^4}(e^{1-y} - e^{3+y})\sin x$.

**27** $A\cos ct + B\sin ct$.      **28** $A + B\log r$.

**29** (ii) $0$, $0$;   (iii) $+1$, $-1$;   (iv) the mixed derivatives are not equal when $x = 0$ and $y = 0$.

## Exercise 9(b), p. 341

1 $R\left(\dfrac{\delta T}{p}-\dfrac{T}{p^2}\delta p\right).$     3 $\dfrac{\pi}{\sqrt{(gl)}}\left(\delta l-\dfrac{l}{g}\delta g\right).$     5 $+0{\cdot}005\alpha.$

6 $\{(b-c\cos A)\,\delta b+(c-b\cos A)\,\delta c\}/\sqrt{(b^2+c^2-2bc\cos A)};\;\; C=\tfrac{1}{2}\pi.$

## Exercise 9(c), p. 346

1 $36t^3+24t^5.$     2 $\dfrac{y-(x+y)^2}{(x+y)^2-x}.$     3 $\dfrac{x(2ay^2-x^3)}{y(y^3-2ax^2)}.$

4 $\dfrac{\cos y-y\cos x}{\sin x+x\sin y}.$     5 $-y/x.$     9 $\xi\dfrac{\partial u}{\partial x}+\eta\dfrac{\partial u}{\partial y},\;\eta\dfrac{\partial u}{\partial x}+\xi\dfrac{\partial u}{\partial y}.$

10 $x\dfrac{\partial u}{\partial y}-y\dfrac{\partial u}{\partial x}.$     11 $\dfrac{\partial u}{\partial x}=y\,e^{xy}\dfrac{\partial u}{\partial\xi}+2x\dfrac{\partial u}{\partial\eta},\;\dfrac{\partial u}{\partial y}=x\,e^{xy}\dfrac{\partial u}{\partial\xi}+2y\dfrac{\partial u}{\partial\eta}.$

16 $-y/r^2,\;-r\sin\theta.$     17 $r/\sqrt{(r^2-y^2)},\;x\sec^2\theta.$

## Exercise 9(d), p. 351

1 (i) 0;   (ii) $-\tfrac{3}{2}u$;   (iii) 0.     2 Use Euler's theorem of second order.

4 $\pm1.$     5 $-3.$     9 $z=(1+\cos x)\cos y.$

10 $z=f(x+y)+g(x-y)-\tfrac{1}{4}\sin(x-y)\cos(x+y).$

## Exercise 9(e), p. 357

2 $-\dfrac{zf_x-xf_z}{zf_y-yf_z}.$     4 $\dfrac{f_x+f_zg_x}{1-f_zg_y}.$     8 $\left(\dfrac{\partial u}{\partial x}\right)_y=y,\;\left(\dfrac{\partial u}{\partial x}\right)_z=y-\dfrac{ax}{b}.$

9 $\left(\dfrac{\partial u}{\partial x}\right)_{yz}=2x,\;\left(\dfrac{\partial u}{\partial x}\right)_{yt}=2x+2yzt,\;\left(\dfrac{\partial u}{\partial x}\right)_{zt}=2x-\dfrac{2y^2}{x}.$

10 $3x(x-z).$     11 2, 1.

## Miscellaneous Exercise 9(f), p. 358

3 $xy(x+y)\,e^{x-y}.$     4 $2xy+c.$     10 $-\tfrac{3}{2}.$

12 $-\tfrac{1}{4}xt^{-\frac{3}{2}}f'(z),\;\tfrac{1}{4}t^{-1}f''(z).$     13 $(2\theta-\tan\theta)/\{(\pi-2\theta)\tan\theta-1\}.$

15 (i) $(x-x_1)f_x(x_1,y_1)+(y-y_1)f_y(x_1,y_1)=0;$

$$(x-x_1)f_y(x_1,y_1)=(y-y_1)f_x(x_1,y_1);$$

(ii) $-yf_y/f_x;\;-yf_x/f_y.$

18 $\left(\dfrac{\partial u}{\partial r}\right)^2+\dfrac{1}{r^2}\left(\dfrac{\partial u}{\partial\theta}\right)^2.$     21 $-\dfrac{1}{u^2}\cos\theta,\;-u^2\cos\theta.$

22 $\dfrac{\partial^2 u}{\partial\theta^2}-\dfrac{\partial u}{\partial r}.$     23 $\dfrac{\partial H}{\partial v}=0,\;\dfrac{\partial F}{\partial u}=0.$

24 Derive $\phi=0$ partially wo $x$ and wo $y$, then eliminate $\phi_u:\phi_v.$

25 (i) Use 11.43, Theorem I.

# INDEX TO VOLUME I

*Numbers refer to pages.*

\* *means 'Also see this entry in the Index of Vol. II'.*